Saline Lakes

Developments in Hydrobiology 59

Series editor
H. J. Dumont

Saline Lakes

Proceedings of the Fourth International Symposium
on Athalassic (inland) Saline Lakes, held at Banyoles, Spain, May 1988

Edited by
F. A. Comín and T. G. Northcote

Reprinted from Hydrobiologia, vol. 197 (1990)

Kluwer Academic Publishers
Dordrecht / Boston / London

Library of Congress Cataloging-in-Publication Data

```
International Symposium on Athalassic (Inland) Saline Lakes (4th :
   1988 : Banyoles, Spain)
     Saline lakes : proceedings of the Fourth International Symposium
   on Athalassic (Inland) Saline Lakes, held at Banyoles, Spain, May
   1988 / edited by F.A. Comin and T.G. Northcote.
         p.   cm. -- (Developments and hydrobiology ; 59)

     1. Salt lakes--Congresses.  2. Salt pans (Geology)--Congresses.
   I. Comin, F. A. (Francisco A.)  II. Northcote, T. G.  III. Title.
   IV. Series.
   QH95.9.I57  1988
   574.5'2636--dc20                                          90-4665
```

ISBN-13: 978-94-010-6759-1   e-ISBN-13: 978-94-009-0603-7
DOI: 10.1007/978-94-009-0603-7

Published by Kluwer Academic Publishers,
P.O. Box 17, 3300 AA Dordrecht, The Netherlands.

Kluwer Academic Publishers incorporates
the publishing programmes of
D. Reidel, Martinus Nijhoff, Dr W. Junk and MTP Press.

Sold and distributed in the U.S.A. and Canada
by Kluwer Academic Publishers,
101 Philip Drive, Norwell, MA 02061, U.S.A.

In all other countries, sold and distributed
by Kluwer Academic Publishers Group,
P.O. Bx 322, 3200 AH Dordrecht, The Neterhlands

Printed on acid-free paper

Contents

Hydrobiologia **197**: vii–viii, 1990.
F. A. Comín and T. G. Northcote (eds), Saline Lakes.

vii

Editors' preface

The IV International Symposium on Athalassic (Inland Saline Lakes took place in Banyoles (NE Spain) during 2–8 May 1988. Participants from Australia, Japan, India, Canada, USA, Mexico, Zambia, Israel, Austria, West Germany, France and Spain enjoyed the friendly atmosphere of Banyoles during the formal proceedings as well as the landscape and beauty throughout Monegros, Lake Gallocanta and La Mancha, during the excursion to visit Spanish saline lakes. The Symposium was funded by Ministry of Education and Science, Diputación General de Aragón, Generalitat de Catalynya and Diputació, and was honored by the City Hall of Banyoles.

The studies presented by the participants included an interesting range of subjects on different scientific aspects of saline lakes, now presented in this volume. Geochemical and related hydrological subjects are the focus of the first group of papers (1–9), which include both general overviews and detailed studies of particular interest. An example of the remarkable microstratifications and temporal changes which can occur in meromictic lakes is also presented (10–12). Two different approaches, review of historic data and satelite imagery, clearly establish the necessity and value of long-term studies of saline lakes (13, 14). Several papers take different types of organisms as the primary data source to focus on autoecology, population and community ecology (15–22), including experimental and descriptive, field and laboratory works. Paleoecological studies (23–25) complete the set of papers, which are representative of the saline lake research developed around the world during the three years between symposia. In addition to the formal presentations, informal discussions focussed on the past, present and future perspectives of saline lake research.

With a view towards future symposia on saline lakes we would urge workers in this field, especially from eastern and middle Asia, from Africa, from Central and South America and from eastern and southeastern Europe – all areas from which there have been very few participants at previous symposia – to contribute their studies. Furthermore international, national and local agencies from the private as well as the public sector should be the first to be encouraged to facilitate participation of researchers from these geographic regions in such symposia. Their support would do much to promote international communication which ultimately will be reflected not only in broadening of human knowledge but also in improvement of cooperation and economic conditions. An excellent opportunity lies ahead in 1991, with the V Saline Lakes Symposium being held in Bolivia (Convenor: Prof. Stuart Hurlbert, Department of Biology, San Diego State University, California 92182, USA).

Unfortunately, during the editing of these proceedings Peter Kilham died working on African lakes. We deeply regret this loss. The Kilham & Cloke paper in this volume is an outstanding and valuable scientific work and his vigorous contributions to discussions during presentations will be missed in future symposia.

Many people contributed to the organization of the IV Saline Lakes Symposium. Convenors of previous symposia – W. D. Williams, U. T. Hammer, J. M. Melack – gave useful advice. The Department of Ecology of the University of Barcelona provided the basic facilities and assistance. Special thanks are given to the postgraduate students Pilar Comín, Mayte Martin and Olga Delgado, who took care of the hundreds of details to be resolved before and during the Symposium, in addition to their management of technical, treasurership and travel viii

arrangements. Regarding the field excursions to Spanish saline lakes, we are particularly grateful for the contribution and guidance of J. J. Pueyo, R. Julia and J. A. De la Peña.

All the papers were critically refereed by two independent and well qualified peers, revised by the authors and edited before being accepted for publication. In this regard, on behalf of the authors and ourselves, we gratefully acknowledge the help of L. Albright, J. Armengol, W. Barnes, J. Bischoff, L. J. Borowitzka, E. R. Byron, P. Carbonel, E. Carmack, J. Cloern, P. DeDeckker, D. L. Galat, H. L. Golterman, K. J. Hall, S. Hurlbert, J. Imberger, R. Julia, P. Kilham, S. Kilham, J. Landwehr, W. Last, P. Lenz, B. Lyons, R. Margalef, S. McIntyre, J. M. Melack, T. Murphy, A. Nishri, A. Nissembaum, A. Parsons, T. F. Pedersen, N. Prat, R. Renaut, B. Timms, X. Tomás, D. Schindler, L. Sigg, G. G. E. Scudder, I. Valiela, C. C. van der Boch, R. F. Weiss and W. D. Williams.

The editing of the papers was greatly enhanced by the enthusiastic and meticulous assistance of P. Comín during the typing required for publication. The keen help of A. M. Domingo in preparing drawings is also appreciated.

Finally, we strongly encourage researchers on saline lakes to attend and participate in future symposia.

FRANCISCO A. COMÍN & THOMAS G. NORTHCOTE

Hydrobiologia **197**: ix–xii, 1990.
F. A. Comín and T. G. Northcote (eds), Saline Lakes.

List of participants

Adachi, Wakako
 Dpt. of Earth Sciences
 Nihon University
 3-25-40. Sakurajosui
 Setagayaku
 Tokyo 156
 Japan

Alcocer, Javier (46)
 Calle 15
 No. 64. San Pedro de los Pinos
 03800 México
 D.F. México

Alonso, Miguel (47)
 LIMNOS, S.A.
 Bruc 168 entlo. 2ª
 08037 Barcelona
 Spain

Amat, Francisco (40)
 Inst. Acuicultura Torre de la Sal.
  CSIC
 Ribera de Cabanes
 Castellón
 Spain

Anadon, Pere
 Inst. Geol. Jaume Almera. CSIC
 Marttí i Franquès s/n
 08028 Barcelona
 Spain

Arakel, Aro V. (42)
 Dpt. Applied Geology
 Queensland Institute of Technology
 GPO Box 2434
 Brisbane Q 4001
 Australia

Baltanas, Angel
 Museo Nacional de Ciencias Natu-
  rales
 Jose Guutierrez Abascal 2
 28006 Madrid
 Spain

Burk, Vincie (11)

Busson, Georges
 Laboratoire de Géologie du Muséum
  d'Historie Naturelle
 43 Rue de Buffon
 75005 Paris
 France

Cabrera, Lluis
 Dpt. Geologia Dinámica
 Fac. Geologia
 Univ. Barcelona
 08028 Barcelona
 Spain

Clavero, Vicente (18)
 Dpt. Ecología
 Fac. Ciencias
 Universidadde Málaga
 Campus Teatinos
 29071 Málaga
 Spain

Comín, Francisco A (49)
 Dpt. Ecologia
 Fac. Biologia
 Univ. Barcelona
 Diagonal 645
 08028 Barcelona
 Spain

Comín, Pilar (48)
 Dpt. Ecologia
 Fac. Biologia
 Univ. Barcelona
 Diagonal 645
 08028 Barcelona
 Spain

Dana, Gayle (14)
 Sierra Nevada Aquatic Research
  Lab
 Univ. of California
 Rt 1. Box 198
 Mammoth Lakes, CA 93546
 USA

Delgado, Olga (50)
 Dpt. Ecologia
 Fac. Biologia
 Univ. Barcelona
 Diagonal 645
 08028 Barcelona
 Spain

Escobar, Elba (38)
 Calle 15 No. 64
 San Pedro de los Pinos
 03800 México
 D.F. México

Everson, Robert Garth (37)
 Queenland Institute of Technology
 GPO Box 2434
 Brisbane
 Queensland Q 4001
 Australia

Everson, Floranne J. (17)

Fernandez, Jose A. (26)
 Dpt. Fisiología Vegetal
 Fac. Ciencias
 Univ. Málaga
 Campus Teatinos
 29071 Málaga
 Spain

Galat, David (21)
 Dpt. Zoology
 Arizona State University
 Tempe, AZ 85287–1501
 USA
 New Address:
 Missouri Cooperative Fish & Wild-
  life Research Unit
 112 Stephens Hall
 Univ. of Missouri
 Columbia, MO 65211
 USA

Numbers in parentheses refer to identities in the group photograph.

Galvez, Jose Angel (10)
Dpt. Ecología
Fac. Ciencias
Univ. Málaga
Campus Teatinos
29071 Málaga
Spain

Garcia, Carlos M. (8)
Dpt. Ecología
Fac. Ciencias
Univ. Málaga
Campus Teatinos
29071 Málaga
Spain

Hall, Ken J. (19)
Westwater Research Center
Univ. of British Columbia
Vancouver, B.C.
Canada V6T IW5

Hall, Shannon (20)

Hammer, U. Theodore (35)
Biology Dpt.
Univ. of Saskatchewan
Saskatoon, Sask
Canada SIN 0W0

Herbst, David B. (7)
Sierra Nevada Aquatic Research
Lab.
Univ. of California
Rt 1. Box 198
Mammoth Lakes, CA 93546
USA

Hontoria, Francisco (41)
Inst. Acuicultura Torre de la Sal
CSIC
Ribera de Cabanes
Castellón
Spain

Hurlbert, Stuart (2)
Dpt. Biology
San Diego State University
San Diego, CA 92182
USA

Jakher, Ganga R. (32)
Dpt. of Zoology
Univ. of Jodhpur
Jodhpur 342 003
India

Jehl, Joseph R. Jr. (3)
Sea World Research Institute
Hubbs Marine Research Center
1700 South Shores Rd.
San Diego, CA 92109
USA

Jimenez, Carlos (39)
Dpt. Ecología
Fac. Ciencias
Univ. Málaga
Campus Teatinos
29071 Málaga
Spain

Julia, Ramón
Inst. Geol.
Jaume Almera CSIC
Martí i Franquès s/n
08028 Barcelona
Spain

Kilham, Peter (12)

Kilham, Susan (53)
Dpt. Biology
Univ. of Michigan
Ann Arbor, MI 48109
USA

Konig, Dietrich (31)
Sandköppel 39
D-2300 Kronshagen b. Kiel
FGR

Lyons, W. Barry (36)
EOS University of New Hampshire
Durham, NH 03824
USA

Maest, Ann S. (13)
U.S. Geological Survey
MS 420
Menlo Park, CA 94025
USA

Margalef, Ramón (16)
Dpt. Ecologia
Fac. Biologia
Univ. Barcelona
Diagonal 645
08028 Barcelona
Spain

Martin, Mayte (51)
Dpt. Ecologia
Fac. Biologia
Univ. Barcelona
Diagonal 645
08028 Barcelona
Spain

Martinez, Asunción (27)
Dept. Zoología
Fac. Biologia
Univ. Sevilla
21080 Sevilla
Spain

Martino, Paloma
Dpt. Ecología
Fac. Ciencias
Ciudad Universiataria Canto Blanco
Univ. Autónoma de Madrid
28049 Madrid
Spain

Miracle, Rosa (28)
Dpt. Ecología
Univ. Valencia
Dr. Moliner 50
46100 Burjasot
Valencia
Spain

Montes, Carlos
Dpt. Ecologia
Fac. Ciencias
Ciudad Universitaria Canto Blanco
Univ. Autónoma de Madrid
28049 Madrid
Spain

Niell, F. Xavier (15)
Dpt. Ecología
Fac. Ciencias
Universidadde Málaga
Campus Teatinos
29071 Málaga
Spain

Nishri, Aminadav (54)
IOLR
The Kinneret Limnological Laboratory
POB 345
Tiberias 14102
Israel

Nissenbaum, Ari (24)
Weittzmann Institute of Science
76100 Rehovot
Israel

Northcote, Thoma G. (5)
Dpt. Zoology
Univ. of British Columbia
Vancouver, B.C.
Canada VGT 2A9

Northcote, Heather (1)

Papoulias, Diana (33)

Plana, Felicià
Inst. Geol. Jaume Almera CSIC
Martí i Franquès s/n
08028 Barcelona
Spain

Post, Fred (30)
Dpt. of Biology UMC 55
Utah State University
Logan, UT 84322
USA

Post, Jane

Pueyo, Juan José (4)
Dpt. Petrologia
Fac. Geologia
Univ. Barcelona
Martí i Franquès s/n
Spain

Real, Montserrat (45)
Dpt. Ecologia
Fac. Biologia
Univ. Barcelona
Diagonal 645
08028 Barcelona
Spain

Renaut, Robin W. (22)
Dpt. of Geological Sciences
Univ. of Saskkatchewan
Saskatoon
Saskkatchewan
Canada S7N 0W0

Ruttner-Kolisko, Agnes (29)
Biol. Station Lunz a Seee
A-3293
Austria

Sanz, M. Angeles
Instituto Pirenaico de ecología CSIC
Apto. 64 Jaca
22700 Huesca
Spain

Schmid, Reinhard M. (23)
UNZA School of Mines
Box 32379
Lusaka
Zambia

Servant-Vildary, Simone (9)
Laboratoire de Geòlogie
Museum National d'Historie Naturelle
43 Rue Buffon
75005 Paris
France

Stephens, Doyle (25)
U.S. Geological Survey
1745 West 1700 South
Salt Lake City, UT 84104
USA

Stiller, Mariana (52)
5 Haem Str.
Rishon Lezion 75240
Israel

Tominaga, Hiroyuki (43)
Water Research Institute
Nagoya University
Furocho
Chikusa-ku
Nagota City 464
Japan

Waisel, Y. (34)
Dep. of Boutany
Tel Aviv University
Tel Aviv 69978
Israel

Williams, William D. (44)
Dpt. Zoology
Univ. of Adelaide
GPO Box 498
Adelaide
South Australia

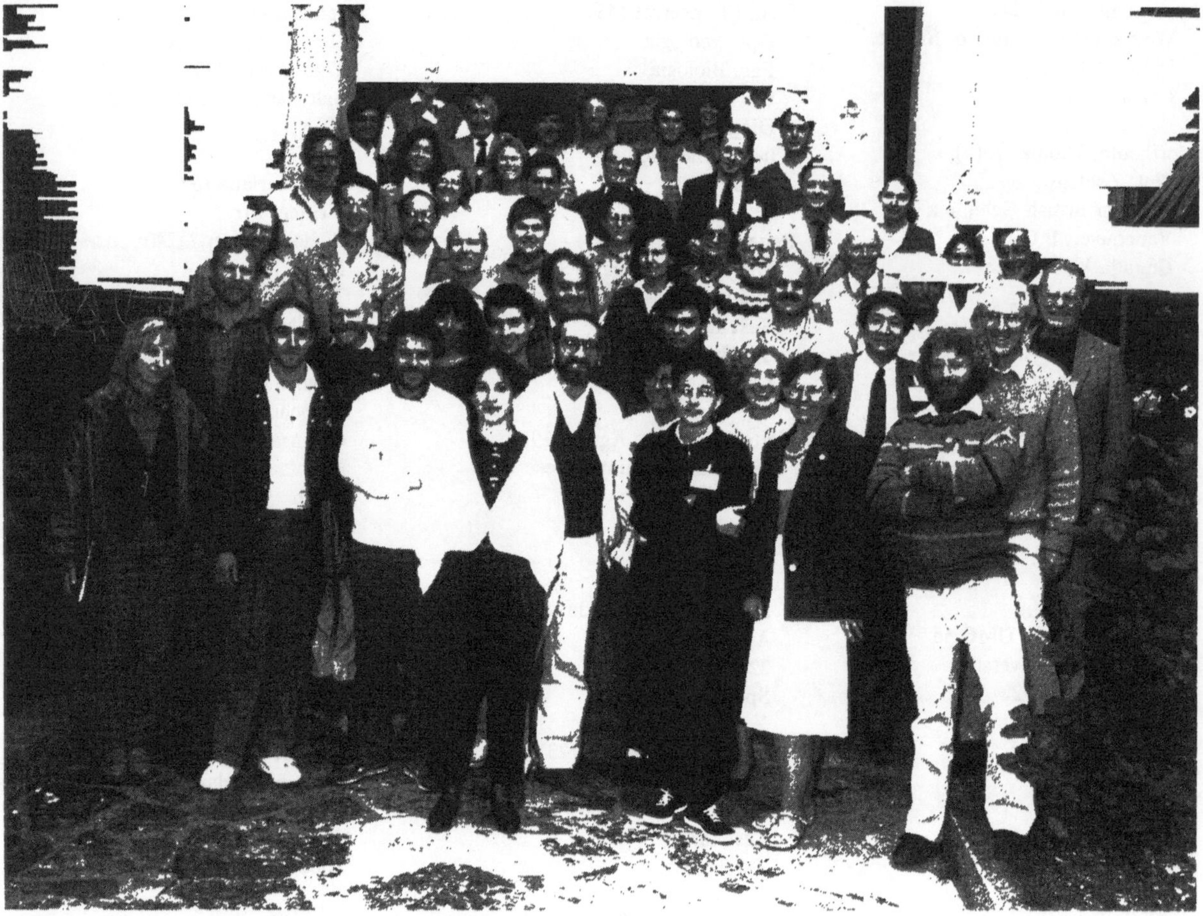

1
2
3
4
5
6
7
8
9
10
11
12
13
14
15
16
17
18
19
20
21
22
23
24
25
26
27
28
29
30
31
32
33
34
35
36
37
38
39
40
41
42
43
44
45
46
47
48
49
50
51
52
53
54

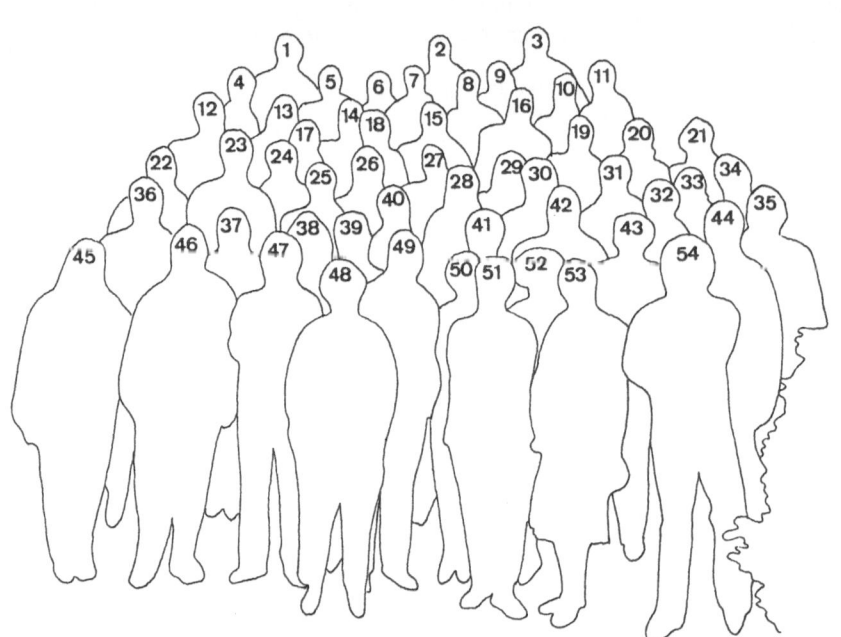

Hydrobiologia **197**: 1–12, 1990.
F. A. Comin and T. G. Northcote (eds), Saline Lakes.
© 1990 *Kluwer Academic Publishers.*

1

Sediment-water interaction as a control on geochemical evolution of playa lake systems in the Australian arid interior

A.V. Arakel[1], G. Jacobson[2] & W.B. Lyons[3]
[1]*Centre for Sedimentary & Environmental Geology, Department of Applied Geology, Queensland University of Technology. G.P.O. Box 2434*; [2]*Division of Continental geology. Bureau of Mineral Resources, Geology & Geophysics. G.P.O. Box 378. Canberra, ACT 2691, Australia*; [3]*Institute for the study of Earth, Oceans & Space. University of New Hampshire. Durham, N.H. 03824 USA*

Key words: playa lakes, geochemical evolution, arid-zone Australia

Abstract

Saline playa lakes represent major geomorphic and hydrologic components of internal drainage basins in the arid to semiarid interior of Australia. These lakes mark the outcrop areas of regional shallow groundwater; thus, they are effective hydro-chemical sinks for elemental concentration and authigenic formation of carbonate, evaporite, and silica/silicate minerals.

Field observations and petrochemical characterization of playa sediments from drainage basins in Western and Central Australia indicate that localized discharge of groundwater, from shallow aquifers in calcrete deposits, plays a fundamental role in geochemical evolution of playa-lake marginal facies. The available data indicates also that although evaporative concentration and salt recycling are major controls on geochemistry of the playas, yet a simple evaporative concentration model does not provide a complete explanation for brine evolution and particularly the geochemical process-product relationships observed in the individual playa lakes. The distribution of the chemical facies in the playas, in relation to geomorphic setting of the internal drainage basins, reflects a significant impact of variation in groundwater discharge pattern on the geochemical evolution of the playa lakes. Accordingly, the development of chemical facies in individual playas have progressed through repeated episodes of evaporative concentration, groundwater-level fluctuations and ion-exchange processes.

Introduction

Saline playa lakes are widespread in the arid and semiarid interior of Western Australia, South Australia, western and northwestern Victoria and the southwestern part of the northern Territory. These playa lakes form major geomorphic and hydrologic components of modern internal drainage basins; their irregular outline is characteristic, as is the occurrence of highly variable salinity playa waters from which an assemblage of chemical sediments are precipitated. The playa lakes are developed commonly in regional groundwater discharge zones, and are characterized by oxidizing, neutral to midly alkaline, saline to hyper-saline waters. Some of these lakes have been, in the past, considered unique as their major ionic compositions are similar to that of evaporated seawater. Airbone sea salt and/or recycled salt have been the most widely postulated sources of salt input to the inland lake systems (eg. Johns & Ludbrook, 1963; Williams, 1967; Williams & Buckney, 1976); although debate on the validity of the above or other sources remains continuous (cf. Lyons & Arakel, 1989).

The majority of the internal drainage basins, incorporating the playa lakes, date back to Tertiary times, when the climate was more humid and the basins incorporated active rivers and lake systems (Bowler, 1976). The existing literature indicates that the saline playa lakes have developed at the zone of regional groundwater discharge; they are sites of important hydrochemical and landform processes, and the past climatic regime has played an important role in concentration of the shallow groundwater solutions (Macumber, 1979; Teller et al., 1982; Bowler, 1986; Jankowski & Jacobson, 1988; Arakel, 1988). However, most of the debate on geochemical evolution of the Australian inland playa lake systems has been concentrated on the 'sources' of ionic input to the playa lakes, rather than on the processes involved.

This paper describes and compares the hydrogeology and physico-chemical features of sedimentary lithofacies in two major drainage areas namely Lake Way and Lake Miranda in Western Australia, and Lake Amadeus in Northern Territory (Fig. 1). Field and laboratory observations on chemical lithofacies and their distribution patterns in different hydrologic zones are utilized to assess the role of sediment-water interactions in geochemical evolution of the playa lake systems. Knowledge of the processes that control the evolution of these waters is required for the understanding of geochemistry of shallow water in major groundwater discharge zones, the nature and distribution of chemical precipitates in lacustrine systems, and Quaternary landscape evolution in the arid interior of Australia.

Description of sites studied

The drainage basins incorporating lakes Way and Miranda (Fig. 1), occur in the eastern part of an Archaean shield area ('Yilgarn Block', Williams, 1975), where the intrusive granites and gneisses cover approximately 70% of the terrain, with the remaining area occupied mainly by older north-northwest trending greenstone belts. The Yilgarn Block forms part of the Great Plateau of Western Australia (Jutson, 1950; Bettenay et al., 1979), which since early Tertiary has undergone significant landform modification, due to lateritization and arid-zone drainage basin development. These internal drainage basins are best exemplified by elongated paleo-drainage channels, which truncate the lateritic profiles and terminate in the present-day playa lake systems (Williams, 1975; van der Graaff et al., 1977). The alluvial/colluvial fill material in these drainage channels has been mostly replaced by Quaternary calcrete, along the groundwater flow paths. The groundwater calcrete deposits, in the trunk valleys of the paleo-drainage channels, contain lenses of groundwater, which are recharged by occasional flooding in the catchment areas and by direct infiltration of rainfall through their karstified upper horizons (Arakel & McConchie, 1982). The climate in the study areas is arid: annual rainfall is less than 200 mm, with the potential annual evaporation at about 4,000 mm.

The chain of playas examined in the Curtin Springs area of the Amadeus Basin (Fig. 1) form part of a vast paleo-drainage system, which during the Tertiary discharged to Lake Eyre via the ancestral Finke River (Wakelin-King & Arakel, 1988). The valley incorporating these playa lakes forms part of a Central Australian groundwater discharge zone that extends for more than 500 km, from Lake Hopkins, in Western Australia, to lakes Neale and Amadeus, in the Northern Territory (Fig. 1). Cainozoic lacustrine, alluvial/colluvial and aeolian sediments overlie intensely folded and fractured Proterozoic and Paleozoic sedimentary rocks of the Amadeus Basin sequence. The depression containing the paleo-drainages is a major linear feature on the topographic map of Australia, and may have formed from solution of Proterozoic evaporites, which formerly outcropped along an anticlinorial axis (Wells, 1980).

3

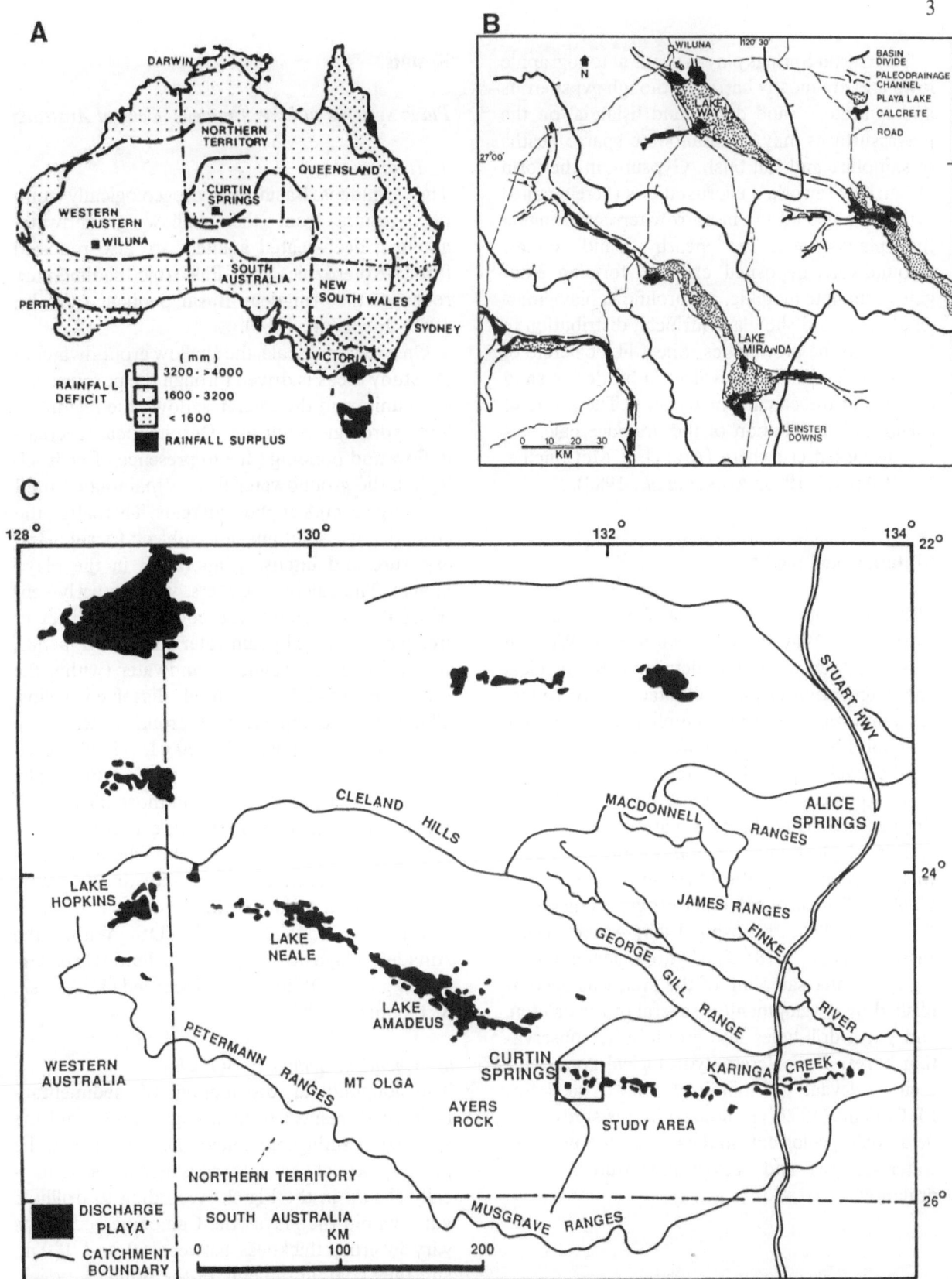

Fig. 1. Maps showing the locations of the study areas (A), geology and drainage features in the Lakes Way and Miranda (B), and in the Amadeus Basin (C).

The Curtin Springs playas occur at topographic lows and are mostly barren, although gypsiferous and quartzose sand dunes and 'islands' on the playa surfaces may accommodate sparse heaths of samphire and salt bush. Gypsum, in the form of partially reworked encrustations (herein called gypcrete), also occurs in narrow terraces fringing the playas and the nearby sand dunes. Groundwater-deposited calcrete, forming elongate carbonate mounds, are prolific in playa marginal areas and show a centripetal distribution in relation to the playa lakes. Sheet-like calcrete of vadose origin is also prevalent in bordering sand dunes and the catchment terrains. The mode of origin and distribution of the drainage calcretes are discussed elsewhere (Arakel & McConchie, 1982; Arakel, 1986; Arakel et al., 1988).

Materials and methods

Drill- and vibro-cores available from exploratory work in lake Way and Lake Miranda, in Western Australia, were utilized for detailed petrochemical characterization of playa lithofacies. Hydrological monitoring and water sampling of the existing drill holes and stock bores were undertaken during the 1985-86 period. The material for the study of the Amadeus playas was collected in the Curtin Springs area (Fig. 2), as a part of an ongoing hydrogeological investigation of the Amadeus Basin, instigated by the Division of Continental Geology, Bureau of Mineral Resources (Jacobson & Warne, 1986; Jacobson, 1988; Arakel, 1987; Jacobson et al., 1988). The hydrological monitoring and water sampling of the area was accomplished by development of several of the calcrete and playa drill holes into groundwater observation bores. The calcrete hydrological zones and their relevant terminology in Arakel & McConchie (1982) are utilized in this study. The field hydrogeological studies were followed by water analysis, and petrographic studies of the drillcores.

Results

Playa systems at lake Amadeus, Central Australia

(i) Hydrogeology
The Amadeus Basin is hydrogeologically complex: apart from the shallow groundwater aquifers, deep-seated aquifers are also reported from Proterozoic and Paleozoic sedimentary rocks of the Amadeus Basin Sequence (Wells, 1980; Jacobson, 1988).

On a regional scale, the shallow groundwater in the study areas is driven through the porous clay-sand units and the calcretes, down the regionally low hydrologic gradients. Despite local reversals in flow and ponding (due to presence of bedrock highs), the groundwater flow is maintained until discharge occurs at playa margins. Thereafter, the groundwater solutions are subject to subaerial exposure and intense evaporation in the playa chains. The calcrete aquifers, with a nearly centripedal flow pattern, are commonly subject to frequent recharge by rainwater infiltration; hence, they possess less saline groundwater (within the range of $0.5–1.5\ g\,l^{-1}$ total dissolved solids, TDS), when compared with groundwater in the clay-sand sand aquifers ($30–50\ g\,l^{-1}$ TDS) developed upstream from the calcrete aquifers. The highest salinities are attained in those playas that are downgradient from the direct influence of fresh groundwater discharge, and where a combination of evaporative concentration and admixing with the playa residual brines gives rise to salinities in excess of $200\ g\,l^{-1}$ TDS. Within the Amadeus playa beds porewater salinities in excess of $360\ g\,l^{-1}$ TDS have been recorded (Jankowski & Jacobson, 1988).

(ii) Chemical Sedimentary Facies
The longitudinal distribution of sedimentary facies with respect to hydrologic zones, and the associated authigenic mineral assemblages of the playa system at Curtin Springs area are schematically shown in the Fig. 2. According to drillhole data, within the playas the Cainozoic sediments vary in gross thickness between 50 and 120 m, and they contain the bulk of the brine. A typical

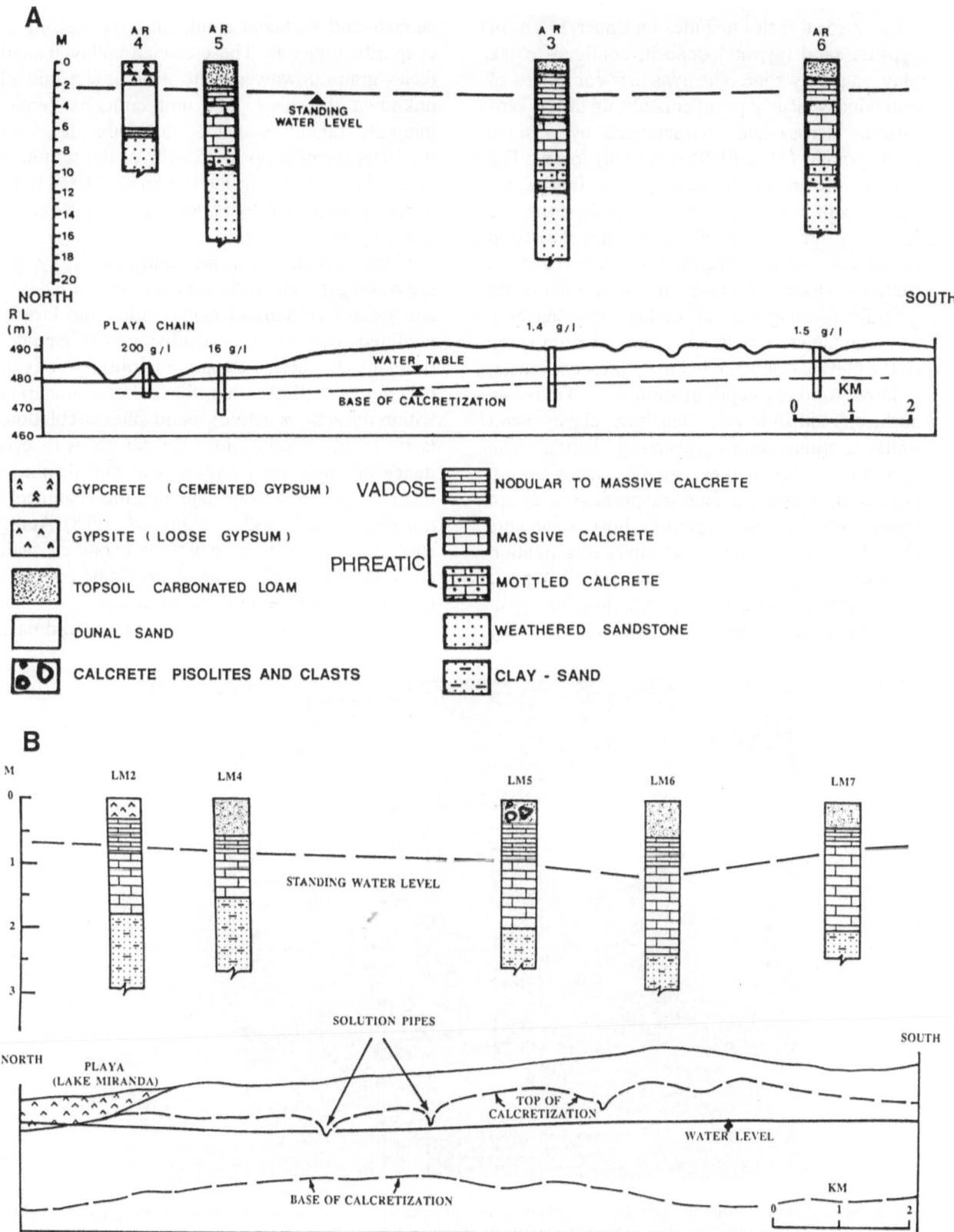

Fig. 2. Longitudinal section showing the lithofacies distribution, the hydrologic zones, and the authigenic mineral assemblages in different hydrologic zones of the Curtin Springs playa lake system (A) and the Lake Miranda area (B).

playa vertical facies includes an upper 1–2 m of gypsum sand (gypsite) deposit, confined to the playa capillary zone. The gypsite is comprised of individually clear gypsum crystals, up to few centimeters across and characterized by positive hemipyramid (111), (010) and (110) forms. The overall intergrown- to -interlocking fabrics together with the presence of partially corroded larger gypsum crystals, reflect episodes of gypsum dissolution and reprecipitation, due to fluctuations in local pore-water chemistry. Glauberite crystals, forming discrete nodules and lenses a few centimeters across, are recorded from some of the playas, below the halite efflorescent crusts.

Downward to a depth of around 8–10 m below surface, reddish-brown lacustrine clayey sand units alternate with gypsiferous bands. The gypsum crystals in these gypsiferous units are commonly confined to fenestral partings and pore spaces of the non-evaporitic host sediments (Fig. 3), a feature attesting to direct precipitation from concentrated pore solutions and subsequent crystal growth by displacement. Below the gypsiferous units the playa sediments are dominated by aeolian and colluvial sand, and are devoid of evaporitic minerals. These aeolian/colluvial sand facies grade downward into a dolomitic unit of unknown thickness. The unit contains zeolite minerals (mainly analcime and chabazite), with the latter forming up to 35 wt% of the sediment mode. These dolomitic facies are considered to be unrelated to the modern hydrologic and sedimentation cycle.

Playa marginal facies are characterized by calcrete and gypcrete duricrusts. The calcrete deposits are several tens of meters thick and largely confined to the zone of groundwater flow in paleo-drainage channels. Petrographic observations of drill cores indicate that the calcrete material, within the zone of active groundwater circulation, is invariably micritic, low-Mg calcite, with evidence of progressive replacement and displacement of the host clay-sand channel sediments (Jacobson et al., 1988; Arakel et al., 1988). At the surface, the calcrete is commonly karstified, with solution voids and sink holes providing conduits for effective recharge of the local aquifers.

Silica replacement of the calcrete groundmass

Fig. 3. SEM photograph showing an assemblage of gypsum crystals, nucleating within the pore spaces of non-gypsiferous playa sediments. Width of view is 0.4 mm long.

is prevalent throughout the calcrete drillcore profiles, and is most diverse within the zone of groundwater-level fluctuations (Fig. 4). Here, a variety of pore filling and pore lining opaline and chalcedonic silica occurs in conjunction with coarse sparry calcite. Petrographic evidences indicate that the authigenic silica precipitation proceeds concurrently with the dissolution of detrital quartz and feldspar grains (Arakel *et al.*, 1988). Neogenic silica also appears to replace other minor authigenic minerals (such as smectite, palygorskite and sepiolite) in the lower sections of the calcrete profiles.

The gypcrete deposits, comprised of an assemblage of corroded and abraded gypsum crystallites of different genetic varieties, from a distinct duricrust facies at the playa margins. The zone of gypcrete crust development is commonly a few tens of meters wide, confined to the playa-marginal terraces, but also extending upslope onto the flanks of partially vegetated quartzose and gypsiferous sand duns. The surficial crusts are commonly a few decimeters to a meter thick, and exhibit several diagenetic overprints related to dissolution, dehydratation and recrystallization (Jacobson *et al.*, 1988). Size-graded bedding features are apparent only in the thicker profiles, where the bedded gypcrete is interlayered with clayey and quartzose detrital material. Field observations and level surveys in the Curtin Springs area indicate that these crusts are largely above the direct influence of present groundwater level; their powdery appearance and surface shrinkage features also attest to the role of meteoric water in their modification.

Playa systems at lakes Way and Miranda, Western Australia

(i) Hydrogeology
Cainozoic sediments of the internal drainage basin at the lakes Way- Miranda area are the main host for regional shallow groundwater. This groundwater often 'outcrops' at the playa margins. The sediments, host to shallow aquifers, are comprised of lacustrine, aeolian and alluvial/colluvial clay-sand deposits, commonly incorporating groundwater-deposited calcrete and gypsite facies. Deeper aquifers in older calcretes (underlying the playa lake facies), appear to be inter-connected and truncate into Tertiary lateritized terrain. The shallow groundwater circulation in this area is principally a function of the physiographic setting and the distribution of the hydrologic units, with the latter defined by a net

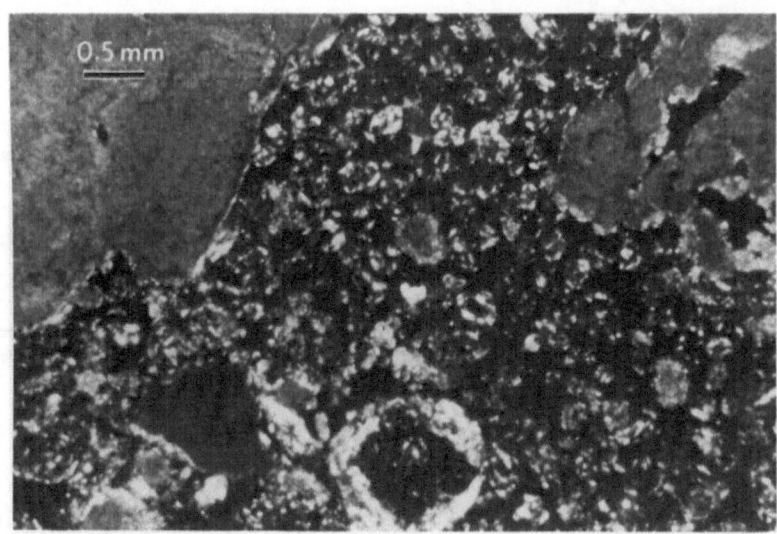

Fig. 4. Photomicrograph showing replacement of calcrete by silica within the shallow groundwater zone. The white patches and rims are opaline silica species, which replace the micritic calcite (grey colour particles) by dissolution and precipitation.

balance between total recharge and evaporative discharge. Recharge is by direct precipitation and infiltration; discharge is mainly by capillary pumping (on playa margins) and near-surface evaporation.

(ii) Playa Chemical Facies

In the lakes Way and Miranda area, the playa gypsiferous clay-sand facies laterally grade onto groundwater calcrete deposits (Fig. 2). Dolomite is a major component of the groundwater calcrete deposits, forming 21 to 28 wt% of the carbonate mode. Petrographic observations suggest that dolomite forms within a zone of alternate solution and reprecipitation, corresponding with the local groundwater-level fluctuation zone. Euhedral dolomite, as a pore-filling and fracture-lining cement, is best developed in lower sections of the drainage lines, where calcrete is shallow and spreads around the playa margins to form delta-shaped subdued carbonate mounds. As in the Curtin Springs area, neogenic silica appears to replace calcrete and other authigenic minerals, including smectite, palygorskite and sepiolite. Silica replacement is most pronounced in the lower sections of the calcrete profiles.

The distribution of playa marginal gypcrete facies is rather patchy, with the best development

recorded in areas away from the calcrete mounds, where some evidence of near-surface groundwater discharge is present.

Discussion

Hydro-geochemical processes and products

Generally, the distribution of playa chemical facies in the study areas corresponds closely with changes in hydrologic gradients and variation in chemistry of local groundwater (Fig. 5). Apart from the role of porosity and permeability of the playa beds in the subsurface fluid-flow pattern (and hence localized recharge/discharge), the other two major controls on geochemical evolution of playa lakes are the regional groundwater concentration trends and the nature of sediment-water reactions in the individual playas.

Down the regional hydrologic gradients, as the groundwater salinity increases, the removal of major ions from groundwater solutions follows the order of precipitation of chemical phases, namely the carbonate, silica/silicate, and evaporite minerals, in the near-surface sediments. The hydrochemical evolution of groundwaters in the study areas is best seen by the relationship

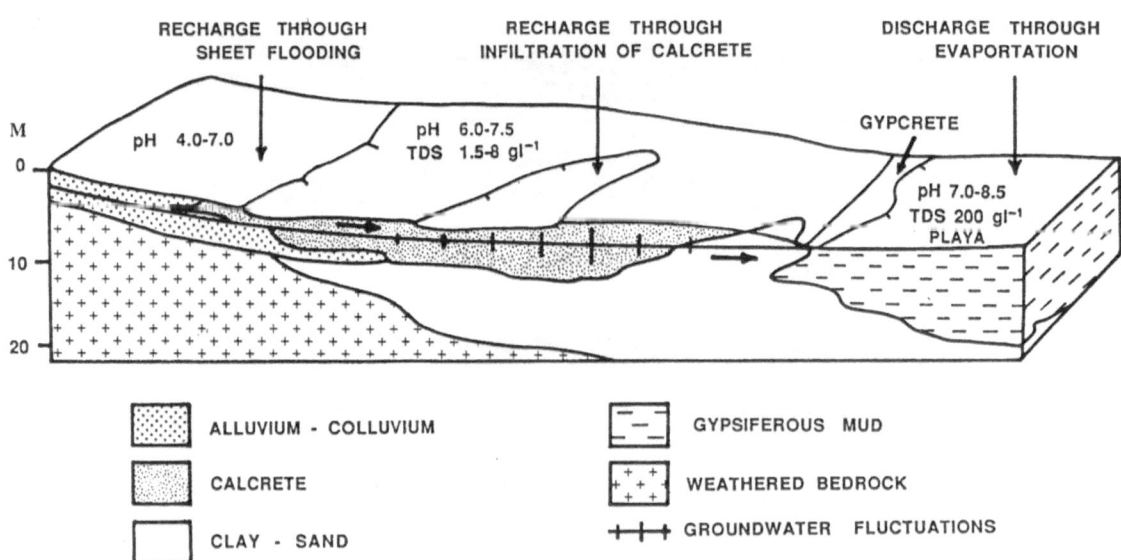

Fig. 5. Block diagram showing the overall lateral distribution of hydrologic zones in the internal drainage basins studied.

between major ions and total dissolved solids (Fig. 6). Accordingly, in the calcrete aquifers the fresher groundwater is typically a bicarbonate-chloride type of water and mildly alkaline, with sodium and calcium representing major cations. Precipitation of micritic low Mg calcite, initially within the pore spaces and thereafter by displacement and replacement of the aquifer material, results in progressive removal of calcium and bicarbonate from these groundwater solutions.

The ionic removal due to evaporative concentration and mineral precipitation occurs during lateral flow of groundwater through different lithofacies; thus promoting interactions between the pore water and the host sediments. In the Curtin Springs area, silica concentration (25–100 mg l^{-1}) in the calcrete aquifers remains constantly

high, close to saturation with respect to amorphous silica, despite a significant removal of the calcium and bicarbonate species. Petrographic examination of the calcrete drillcores indicate that the dissolution of the host detrital grains and reprecipitation of silica, as amorphous and/or chalcedonic coating and pore-lining cement, proceed concurrently in the same hydrologic zone (Arakel et al., 1988). Silica precipitation is most pronounced in the transitional zone between phreatic and vadose calcretes, where the co-precipitation of coarse calcite spar and silica species indicate significant fluctuations in local hydrochemical conditions.

The co-genesis of silica and calcite appears to be favoured by the presence of mildly alkaline pore solutions which engender elevated Mg:Ca

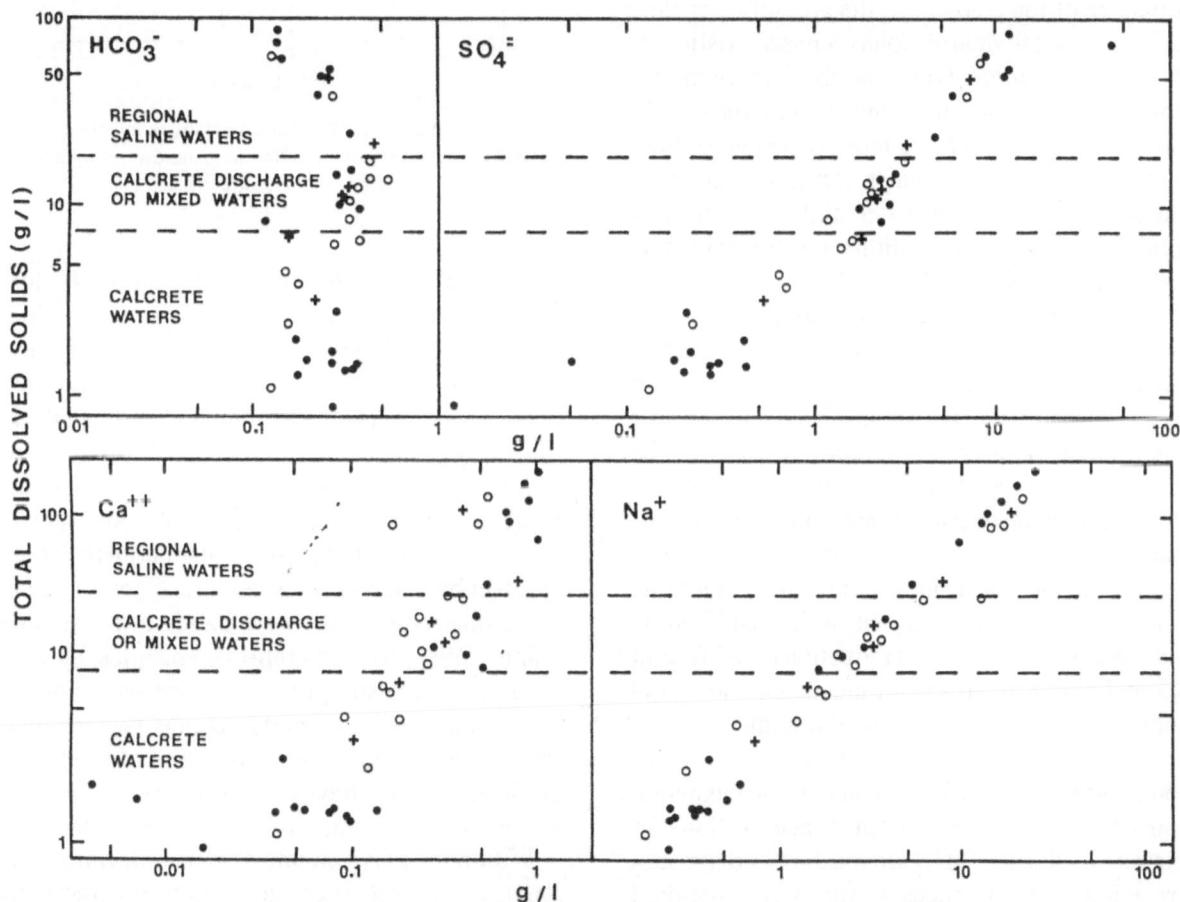

Fig. 6. Plots of major elemental concentrations against salinity (expressed in total dissolved solids) in different groundwater hydrologic zones of the drainages studied. For the symboles refer to Fig. 7.

ratios. Values of isotopic analyses of carbon-13 and oxygen-18 obtained for the calcrete profile are consistent with the reprecipitation of dissolved carbonate in groundwater, and this probably reflects loss of carbon dioxide. Significant fluctuations in pCO_2, particularly during intense dry-season evapotranspiration, evidently results in pH variation, and therefore, extension of the carbonate-silica precipitation front into the soil moisture zone (cf. Carlisle et al., 1978). In the margins of lakes Way and Miranda, the shallow calcrete aquifer broadens, forming 'chemical deltas' within which a variety of authigenic minerals, including dolomite, sepiolite, smectite, palygorskite and mixed-layer clays are precipitated (Mann & Horwitz, 1979; Arakel & McConchie, 1982). The patchy distribution of these authigenic mineral facies indicate the influence of the gradient of the groundwater flow and the porosity/permeability characteristics of the aquifer material. However, the vertical gradational fabrics of the authigenic precipitates in individual drillcore profiles is more reflective of fluctuations in the local water table, the solution chemistry of the pore water and the nature of interactions between solutions and the enclosing sediments (Arakel et al., 1988).

In the western Australia lakes studied, the pore solutions in the playa marginal sediments attain concentration ranges between $1.5\,g\,l^{-1}$ and $15\,g\,l^{-1}$ total dissolved solids (TDS) and become chloride-sulphate dominant with sodium as the major cation (cf. Mann & Deutscher, 1978). Precipitation of micritic calcite and amorphous silica is continuous; however, despite a high concentration of silica in groundwater solutions (Fig. 7), precipitation of silica is significantly reduced, probably due to a reduction in pH (about 6.3 at Lake Way area) and increasing elemental concentration, particularly magnesium.

At salinities greater than $15\,g\,l^{-1}$ TDS, sodium-chloride waters prevail and the elemental concentration increase, until concentrations in excess of $100\,g\,l^{-1}$ TDS are reached and gypsum precipitation commences at the playa marginal facies. The removal of calcium and sulphate from solutions is indicated by dominance of sodium,

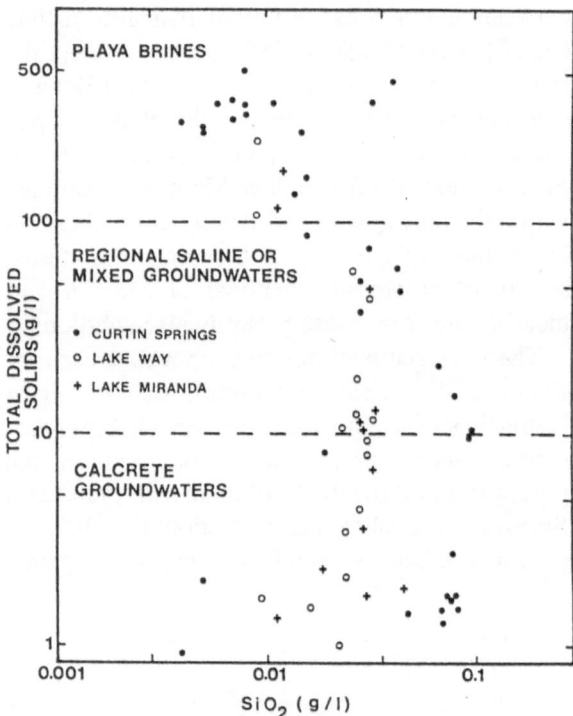

Fig. 7. Plot of silica concentration versus salinity (expressed in total dissolved solids) in different groundwater hydrologic zones of the study areas.

potassium and magnesium in the residual chloride-rich solutions. In the Lake Way area, minor amounts of primary dolomite and nodular aragonite co- precipitate with gypsum, at groundwater concentrations less than $200\,g\,l^{-1}$ TDS. In the Curtin Springs area, on the other hand, the playa- marginal gypsum deposits incorporate glauberite (Ca-Na sulphate) nodules and lenses. In the latter case, the gypsum/glauberite facies are overlain by a thin efflorescent crust, comprised of a centimetre-thick alternate bands of sylvite and halite. This probably reflects episodes of fluctuation in chemistry of near-surface waters due to geochemical reactions at the sediment-water interferance; and thence, the alternation of the halite-sylvite bands may have been promoted by removal of the sodium ion due to glauberite crystallization.

Within the playa chains of Curtin Springs area, intense evaporative concentration and recycling of the shallow regional groundwater results in brine pools, with concentrations in excess of

360 g l⁻¹ TDS. Here, apart from gypsite, glauberite and halite in the playa beds, the surficial sediments include also polyhalite, thenardite, hexahydrite and epsomite. Coring and brine monitoring indicate the influence of bedrock and calcrete on localized ponding and excessive saturation of these brines (Lyons & Arakel, 1989).

Conclusions

Field observations together with petrographic evidence and groundwater chemical trends from the study areas indicate that, despite a common geochemical evolutionary path down the regional hydrologic gradients (Jacobson *et al.*, 1988), a simple evaporative concentration models does not provide a complete explanation for brine evolution and geochemical processes-product relationships in some of the Australian playa lake systems. Apart from evaporative concentration and seasonal recycling of the surficial salts, other geochemical processes, such as sediment-water reactions have been previously considered as controlling parameters on playa water composition (Hardie & Eugster, 1970; Surdam & Sheppard, 1978; Mann & Deutscher, 1983; Mann, 1983). Hydrochemical observations of the drainage basins indicate that the hydrologic setting and intense evaporation play a major role in concentration of near-surface solutions. However, a major ion contribution due to weathering of the soil and/or bedrock in a low salinity groundwater medium and their subsequent cation segregation due to precipitation of authigenic minerals also contribute significantly to complexity of playa-brine characteristics (Mann, 1983; Lyons & Arakel, 1989). In the Curtin Springs area, feldspar and quartz particles within the fresh to brackish groundwater zone undergo significant dissolution in the upper parts of groundwater profile (Arakel *et al.*, 1989), causing abnormally high concentration of silica, calcium and potassium. The high silica and potassium values recorded from the lake Way drainage basin have been also attributed to rapid ongoing weathering of granite in the catchment areas of the playa lakes (Mann &

Deutscher, 1978). Thus, the available geochemical data indicate that although evaporative concentration and salt recycling remain major controls in the playa setting, chemical enrichment in the individual playa lakes may also be accounted for the influence of other geochemical processes, within both the vadose and phreatic environments of groundwater circulation. Accordingly, at playa marginal areas, because of the sluggish nature of the flow, the groundwater discharge front is significantly broadened and chemical sedimentation processes proceed through a combination of evaporative concentration, groundwater-level fluctuations and selective ion-exchange reactions (Mann & Deutscher, 1978). Repeated fluctuations in the groundwater level, and hence variation in pH and salinity gradients, evidently result in development of new porosity and permeability zones, and probably, significant variation in local groundwater head.

Acknowledgements

G. Jacobson publishes by permission of the Director of the Bureau of Mineral Resources, Australia.

References

Arakel, A. V. & D. McConchie, 1982. Classification and genesis of calcrete and gypsite lithofacies in paleodrainage systems of inland Australia and their relationship to carnotite mineralization. J. Sedim. Petrology 52: 1149–1170.

Arakel, A. V., 1986. Evolution of calcrete in paleodrainages of the Lake Napperby area, Central Australia. Palaeogeography, Palaeoclimatology, Palaeoecology 54: 283–303.

Arakel, A. V., 1987. Sedimentary petrology of BMR drill cores and shallow vibrocores from playas in the southern Amadeus Basin, Northern Territory. Bureau of Mineral Resources, Australia, Record 1987/61.

Arakel, A. V., 1988. Carnotite mineralization in inland drainage areas of Australia. Ore Geology Reviews 3: 289–311.

Arakel, A. V., G. Jacobson, M. Salehi & C. M. Hill, 1989. Silicification of calcrete in inland drainage basins of the Australian arid zones. Aust. J. Earth Sci. 36: 73–89.

Bettenay, E., R. E. Smith & C. R. M. Butt, 1979. Physical

12

features of the Yilgarn Block. In: 25th International Geological Congress, Excursion Guide No. 41C: 5–10.

Bowler, J. M., 1976. Aridity in Australia: Age, origins and expression in aeolian landforms and sediments. Earth-Sci. Rev. 12: 279–310.

Bowler, J. M., 1986. Spatial variability and hydrologic evolution of Australian lake basins: analogue for Pleistocene hydrological change and evaporative formation. Palaeogeography, Palaeoclimatology, Palaeoecology 54: 21–41.

Carlsle, D., P. Merifield, A. Orme & O. Kolker, 1978. The distribution of calcretes and gypcretes in southwestern United States. Based on a study of deposits in W. Australia and S.W. Africa (Namibia). Grand Junction, Colorado, Dept. Energy Report GJBX-29 (78), 274 pp.

Hardie, L. A. & H. P. Eugster, 1970. The evolution of closed-basin brines. Mineral. Soc. Am. Spec. Pap. 3: 273–290.

Jacobson, G., 1988. The hydrology of Lake Amadeus, a groundwater-discharge playa in Central Australia. BMR J. Aust. Geol. and Geophys. 10 (4): 301–308.

Jacobson, G., A. V. Arakel & Y. Chen, 1988. The Central Australian groundwaters discharge zones: evolution of associated calcrete and gypcrete deposits. Aust. J. Earth Sci. 35: 549–565.

Jacobson, G. & K. R. Warne, 1986. BMR drilling in the southern Amadeus Basin, Curtin Springs, Northern Territory, 1985. Bureau of Mineral Resources, Australia-record 1986/18.

Jankowski, J. & G. Jacobson, (manuscript). Hydrochemical evolution of regional groundwaters to playa brines in Central Australia. J. Hydrology (submitted).

Johns, R. K. & N. Ludbrook, 1963. Investigation of Lake Eyre. Geol. Surv. South Aust. – Repts. of Investigations, 31.

Jutson, J. T., 1950. The physiography of western Australia. West. Australian Geol. Survey Bull. 53: 830–840.

Lyons, W. B. & A. V. Arakel, (manuscript). The geochemical evolution of salt lake waters in inland Australia: A synthesis. Aust. J. Earth Sci. (submitted).

Macumber, P. G., 1979. The influence of groundwater discharge on the Mallee landscape. In R. R. Storrier and M. E. Stannard (Eds.), Aeolian landscapes in the semiarid zone of southeastern Australia. Aust. Soc. Soil Sci., Riverina Branch, Wagga Wagga: 76–85.

Mann, A. W., 1983. Hydrogeochemistry and weathering on the Yilgarn Block, Western Australia – ferrolysis and heavy metals in continental brines. Geochim. Cosmochim. Acta. 47: 569–577.

Mann, A. W. & R. L. Deutscher, 1978. Hydrogeochemistry of a calcrete- containing aquifer near lake Way, Western Australia. J. Geol. Soc. Aust. 26: 293–303.

Surdam, R. C. & R. A. Sheppard, 1978. Zeolites in saline, alkaline-lake deposits. In L. B. Sand & F. A. Mumpton (eds.), Natural Zeolites. Pergamon Press, New York: 145–174.

Teller, J. T., J. M. Bowler & P. G. Macumber, 1982. Modern sedimentation in Lake Tyrrel, Victoria, Australia. J. Geol. Soc. Aust. 29: 159-175.

Van de Graaff, W. J. E., R. W. A. Crowe, J. A. Bunting & M. J. Jackson, 1977. Relict Early Cainozoic drainages in arid Western Australia. Z. Geomorph. N.F. 48: 169–208.

Wakelin-King, G. & A. Arakel, (in press). Palaeoenvironments and chemical sedimentation in the Curtin Springs – Erldunda playa lake chain, south-east of Lake Amadeus, Northern Territory. Proceedings of SLEADS 1988 Conference, Lake Eyre, Arkaroola, Lake Frome, South Australia, August 1988.

Wells, A. T., 1980. Evaporites in Australia. Bureau of Mineral Resources, Bull. 198, 104 pp.

Williams, W. D., 1967. The chemical characteristics of lentic surface waters in Australia: a review. In A. H. Weatherley (ed.), Australian Inland Waters and their Fauna. ANU Press, Camberra: 18–77.

Williams, W. D. & R. T. Buckney, 1976. Chemical composition of some inland surface waters in South, Western and Northern Australia. Aust. J. mar. Freshwat. Res. 27: 379–397.

Williams, I. R., 1975. Eastern goldfields Providence. In Geology of Western Australia. West. Aust. Geol. Surv. Mem. 2: 33–54.

Hydrobiologia **197**: 13–22, 1990.
F. A. Comin and T. G. Northcote (eds), Saline Lakes.
© 1990 *Kluwer Academic Publishers.*

Metal concentrations in surficial sediments from hypersaline lakes, Australia

W.B. Lyons[1], A.R. Chivas[2], R.M. Lent[1], S. Welch[1], E. Kiss[2], P.A. Mayewski[1], D.T. Long[3] & A.E. Carey[4]

[1]*Institute for the Study of Earth, Oceans and Space, University of New Hampshire, Durham, New Hampshire 03824, USA*; [2]*Environmental Geochemistry Group, Australian National University, Canberra, A.C.T. 2601, Australia*; [3]*Department of Geological Sciences, Michigan State University, East Lansing, MI 48824, USA*; [4]*Gradient Corporation, Cambridge, MA 02138, USA*

Key words: trace metal, hypersaline, lakes, sediments

Abstract

We have conducted a preliminary survey of analyzed surface sediment samples from approximately twenty lakes in Australia for their Fe, Cu, Pb and Zn concentrations. In addition lake sediments from the gold mining areas of Western Australia (WA) were analyzed for Ag, and samples from NW Victoria were analyzed for Mn, Ni, Co, Cr and V. These lakes are discharge zones for regional and/or local saline groundwaters. The groundwaters entering many of these lakes have very low pH's (pH 3–4) and contain extremely high dissolved Fe concentrations. These low pH waters may also contain high concentrations of trace metals and radionuclides from the ^{238}U decay series.

The WA data yield the following information: (1) with the possible exceptions of Ag, these sediments show no trace metal enrichments above average shale/sandstone values; (2) some lakes draining Archean gold mining terrain have high Ag concentrations; (3) the highest Pb concentrations are in a lake where acid groundwater input occurs; and (4) the acid mineral alunite does not appear to be a metal 'sink' in the acid groundwater lakes.

In the two NW Victorian lakes, the sulfidic zone below the cyanobacterial mat is enriched above by about $100 \times$ for Cu, $30 \times$ for Zn and $78 \times$ for Pb compared to the sediments outside the discharge zone. The Victorian data set indicates a correlation between Cu and Zn and also between Cr and Ni concentrations. Three generations of 'ironstone-like' sediments were also analyzed. These iron-oxide rich sediments gained Cu but lost Co with age and maturation. The Mn concentration appears to be important in controlling the Zn and Cu in these 'ironstones'. It appears from our data that neither the oxide-rich nor the acid-rich authigenic sediment phases are major sinks for groundwater metal in these systems.

Introduction

Internally drained salt lakes occur widely in Australia especially in western and northwestern Victoria, in South Australia east of 135° E and in the south and central portion of Western Australia (Williams, 1967). The majority of the waters entering these lakes are Na$^+$ and Cl$^-$ rich brines (Williams, 1967).

Many of the 'salt lakes' in Australia are topographic lows where groundwaters are discharged. These lakes may contain standing water for only a few months of the year. DeDeckker (1988) has recently characterized Australian lakes into four basic categories: 1) large playa lakes, 2) small closed lakes or pans, 3) crater lakes and 4) coastal lakes. The data presented in this paper come from the first and second types of Australian lakes. Large playa lakes are defined by DeDeckker (1988) as being longer than 10 km, most often dry and occurring in old paleodrainage features of the landscape. On the other hand, the small closed lakes or pans are also ephemeral but are smaller and younger than the playas.

Due to the periodic desiccation of these lakes both detritral and authigenic minerals are deposited on the lake floor. The detrital material includes input from both fluvial and aeolian sources. There is no doubt that evaporitic minerals such as gypsum and halite account for the major portion of the authigenic minerals observed in these playas and pans, however, oxides/hydroxides of Fe^{3+}, acid sulfate minerals such as alunite, biogenically produced sulfides and probably a variety of aluminosilicate minerals are also formed authigenically in these lake sediments (Thornber et al., 1987).

Sediments of lakes from arid regions and the analysis of their chemical composition are the most promising of all types of lake sediments for the evaluation of differences in the geochemical backgrounds within and among drainage areas (Förstner, 1977). The bedrock composition of the drainage area maybe the major, if not the only, influence on the geochemistry of lake sediments in arid environments.

Some of the lakes in this study are regional and/or local discharge zones for very low pH (2.8–4.0) groundwaters (Mann, 1983; Macumber, 1983). Mann (1983) has proposed that the low pH of these waters is produced by ferrolysis of Fe^{2+}. Mann (1983) argued that these low pH, Cl^- rich groundwaters are capable of transporting high concentrations of economically important base and transition metals. Because these low pH groundwaters are neutralized within the lake sediment pile, the sediments in these lakes may act as 'sinks' for these metals. In other words, as the pH increases, the metals carried in these waters should be precipitated with authigenic phases within the lake sediments. This should be true particularly if the low pH groundwaters are also enriched in Fe because as neutralization of the brine occurs FeOOH or other Fe^{3+} minerals should precipitate. These minerals are known scavengers of transition and base metals (Tessier et al., 1985). In addition to elucidating the importance of base and transition metal removal, the nature of Fe^{3+} precipitation may provide an added insight into the mechanism of ancient banded iron formations (BIF's).

Description of sites

The samples discussed in this work were collected in two different locations in southern Australia. The first set of samples comes from NW Victoria in the Mallee region. All but three of these samples are from Lake Tyrrell (Fig. 1). The other three are from an area termed by Macumber (1983) as the Raak Boinka (Fig. 1). The hydrology, geomorphology and the groundwater geochemistry of this region have been studied in great detail (Macumber, 1983).

Lake Tyrrell and the playas and pans within the Raak Boinka are groundwater discharge loci. Much of the groundwater in this region is saline to hypersaline, of very low pH (2.8–4.0) and enriched in dissolved iron (Macumber, 1983). The waters on the western and southwestern portion of Lake Tyrrell are acidic while the waters discharging on the eastern side of the lake are approximately neutral. Sediment profiles were obtained on both the acid and the neutral portions of the lake. In addition, a number of surface samples from Lake Tyrrell and the Raak Boinka were also collected and analyzed.

In addition to the Victorian samples, surficial samples were collected from 18 lakes in the southern 'wheat belt' of Western Australia. These samples included sediments from small pans as well as from large playas such as Lake Moore, Lake Cowan and Lake Dundas. The hydrology,

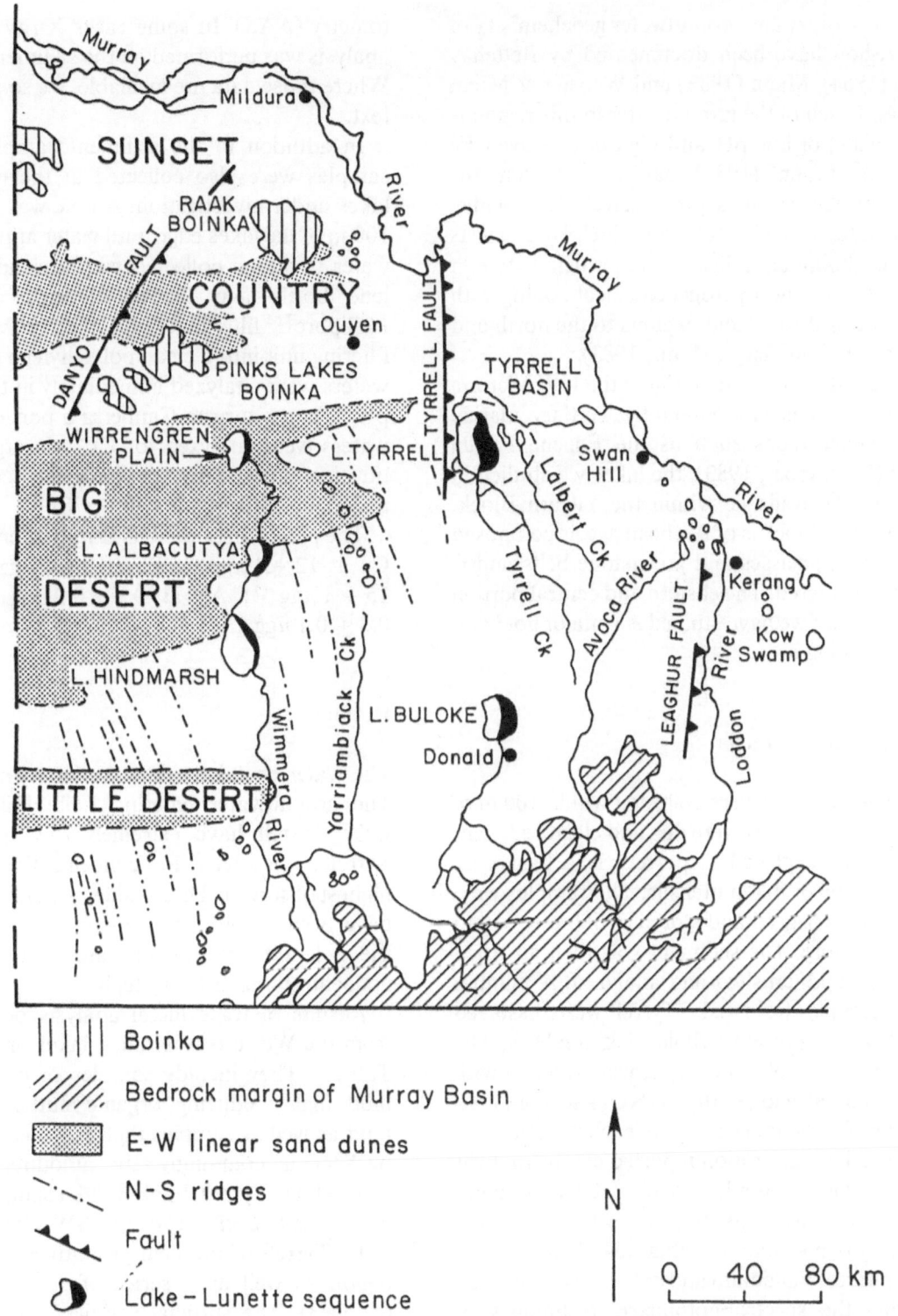

Fig. 1. Location map, sampling locations in NW Victoria (taken from Macumber, 1983).

geomorphology and groundwater geochemistry of this region have been documented by Bettenay *et al.* (1964), Mann (1983) and Webster & Mann (1984). Much of the groundwater in this region is also saline, of low pH and high in dissolved Fe and Al. Mann (1983) has argued that the groundwater becomes progressively less acidic, north of the so-called 'Menzies line'. He suggests that this boundary reflects a major delineation in soil moisture and hydrogeochemical zoning with the more arid, less 'acid' regions to the north and east of the boundary (Mann, 1983).

Unlike the aquifer material in the NW Victoria region which is predominantly Tertiary marine sedimentary rocks such as the Eocene Parilla Sand (Macumber, 1983), the lakes we studied in Western Australia lie within the Yilgarn Block. The Yilgarn Block is of Archean age, is composed of granites, gneisses and greenstone belts and is highly weathered. The eastern and central portion of this region we have studied is a major gold and nickel mining area.

Material and methods

Sediment samples were collected within 100 m of the lake edge. Holes were dug and clean sediment samples were placed in polyethylene bags with clean plastic or nylon utensils. The samples were then double bagged and stored until our return to the laboratory. The sediment was then air dried under a hood, crushed and analyzed. A number of the samples from Lake Tyrrell were dissolved in HF and represent 'whole-rock' analysis. The majority of samples, however, were leached with dilute mineral acid ($\sim 10\%$ HNO_3) at room temperature for 24 hours and the leachate analyzed. The metal concentration reported can be thought of as 'environmentally-active' metal and represents any metal associated with authigenic phases, organic matter or that adsorbed onto detrital phases. Vanadium and cobalt were analyzed utilizing the spectrophotometric techniques of Kiss (1973, 1975a, 1975b), Mn was determined using ICP whereas the rest of the metals were analyzed by flame atomic absorption spectropho-

tometry (AAS). In some cases X-ray diffraction analysis was performed on the sediment samples. Where those data are available, we so state in the text.

In addition to the sediment sampling, water samples were also collected at the majority of lakes under investigation. A hole was dug within 100 m of the lakes edge until water appeared. The water was then collected with a clean polyethylene bottle and filtered through a 0.45 μm Millipore™ filter in a polycarbonate Millipore™ filtering unit into a clean polyethylene bottle. The waters were analyzed immediately in the field for pH. Upon return to Canberra a portion of these waters were analyzed for Cl^- via argentimetric titration and Fe by the Ferrozine colorimetric technique (Murray & Gill, 1978).

The precision of the UNH measurements is for Cu at $12 \pm 1 \mu gg^{-1}$; Pb at $9 \pm 1 \mu gg^{-1}$; Zn at $15 \pm 1 \mu gg^{-1}$; Ag at $0.9 \pm 0.2 \mu gg^{-1}$; Cd at $0.3 \pm 0.1 \mu gg^{-1}$.

Results

The water chemistry data are shown in Table 1. The groundwater entering Lake Gilmore and Lake Tyrrell have extremely low pH's. Lake Tyrrell's dissolved Fe concentrations are the highest observed. These values are similar to what have been reported by Macumber (1983). With the exception of Baladjie and Tyrrell, the Cl^- concentrations are very high.

Results of trace metal analysis of sediments from the Western Australian lakes are shown in Table 2. They include samples of oxidizing red mud, highly reducing, organic matter rich, black mud as well as samples that have been identified by XRD as containing large amounts of alunite.

The Lake Tyrrell/Raak Boinka samples include a sediment profile from the SW portion of the Lake Tyrrell in the acid groundwater discharge region as well as a series of surficial samples (Table 3). The surficial samples are the 'whole rock' analyses whereas the profile represents acid leachable metal. Several of the whole rock samples contained material that was insoluble in

Table 1. Water Chemistry

Lake name	Map number	H^+ $\times 10^{-6}$ M/L	Cl^- M/L	Fe μM/L
Goongarrie	1	0.08	2.85	4.1
Koolkooldine	2	0.20	4.18	32.0
Green Lake Near Cunderlin	3	4.99	4.42	–
Baladjie	4	0.30	1.66	2.5
Wallambin	5	0.30	3.64	16.2
Moore (south end)	6	0.10	3.42	3.8
Moore (north end)	7	0.30	3.27	6.5
Austrin	8	0.10	2.99	4.4
Yarra Yarra	9	0.30	3.86	7.4
Grace	10	0.30	3.85	2.7
King	11	0.60	3.67	2.5
Gilmore	12	631	3.00	2.7
Cowan	13	0.30	3.47	13.5
Dundas	14	0.40	3.60	3.3
Lake Tyrell (west side)	–	–	1.28	107.3
Deborah	15	–	–	–
Cowcowing	16	–	–	–
Annean	17	–	–	–
Buchan	18	–	–	–
King	19	–	–	–
Hann	20	–	–	–

Table 2. Sediment Data From Western Australian (in μgg^{-1} dry wt)

Lake name	Sediment type*	Pb	Cu	Zn	Cd	Ag
Goongarrie	O	2	9	17	0.1	1.3
Deborah	O	6	10	21	<0.1	0.2
Koolkooldine	O	5	10	15	<0.1	<0.1
	R	5	6	12	<0.1	0.3
Baladjie	R	4	6	42	0.1	<0.1
Green	O	<1	<1	6	<0.1	<0.1
Wallambin	R	8	2	24	0.2	0.9
	O	21	5	15	0.1	<0.1
Moore-South	R	6	2	14	0.2	0.8
	O	4	5	19	<0.1	0.2
Cowcowing	A	4	<1	8	0.2	<0.1
Moore-North	O	1	6	15	<0.1	0.1
Annean	R	<1	2	19	<0.1	3.1
Austin	O	<1	6	26	<0.1	0.4
Yarra Yarra	R	2	6	10	<0.1	1.2
Grace	R	16	14	18	0.2	0.8
Buchan	R	10	11	17	0.2	0.5
King	R	<1	5	9	<0.1	0.8
Hann	A	<1	<1	9	0.1	<0.1
Gilmore	O	<1	<1	8	<0.1	<0.1
Cowan	O	1	11	16	0.1	0.3
Dundas	O	<1	11	14	0.1	<0.1
Gilmore	A	<1	1	8	<0.1	<0.1
Gilmore	A	<1	<1	9	<0.1	<0.1

* O = oxidized, R = reduced, A = alunite

Table 3. 'Whole Rock' Sediment Data Lake Tyrrell/Raak-Boinka in μgg^{-1} dry wt. Surficial sediments (0–2 cm) when not indicated.

Sample #	Location	Cu	Zn	Mn	Ni	Co	Cr	V
1	NE Lake Tyrrell	24	55	445	29	19	56	75
2	Neutral water discharge zone (0–2 cm) NE-LT	12	24	87	15	11	27	39
3	Neutral water discharge zone (204 cm) LT	11	24	70	16	11	28	38
4	Offshore black mud-SW-LT	17	50	96	24	14	49	75
5	*Acid discharge zone W-Lake Tyrrell (~15 cm)	10	38	68	22	6	67	84
6	*Acid discharge zone W-Lake (3–4 cm)	2	5	20	9	2	12	20
7	*Acid discharge zone W-Lake (508 cm)	2	9	28	12	9	19	26
8	Acid discharge zone W-Lake algal mat	24	54	181	43	29	33	52
9	°Acid discharge zone W-Lake redoxide	3	8	28	12	5	31	21
10	Raak Playa sulfide (2–5 cm)	6	12	218	11	19	8	16
11	Raak Playa ocide (~15 cm)	4	10	58	13	14	16	19
12	Ironstone LT	8	26	214	25	3	94	91
13	Older ironstone RP	15	10	51	14	4	11	1
	Average sandstone	16	–	–	2	0.3	35	20
	Average shale	95	45	850	68	19	90	130

* oxidized sediments, red in color
° 'new' ironstone

HF. These materials were tentatively identified as tourmaline and zircon and the highest amount (~5%) was found in sample 8. The results have not been corrected for the insoluble material.

Discussion

Western Australian Sediments. – The concentrations of acid leachable trace metals from the Western Australian salt lakes are similar to values obtained by the UNH group for other hypersaline systems (Table 4). The calculated means for the three gross sediment types (O, R, A) are compared to these other sediment regimes as well as the average sandstone and shale values.

In all cases, the metal concentration decreases in the order $Zn > Cu > Pb$. The WA data are comparable to those from Solar Lake and Laguna Ojo Liebre samples. These concentrations are equal to or lower than the average sandstone values and much lower than the average shale values. The WA data support the contention of Long *et al.* (1985) from the study of modern marine hypersaline lakes that, in general, hypersaline lakes do not have elevated trace metal con-

centrations and hence are not active loci of trace metal enrichment.

This appears to be the case even in lake systems where acid groundwater discharges occur. Lakes Cowcowing and Gilmore are active sites of acid water discharge and yet have low trace metal concentrations in their sediments. Mann (1983) has shown that acid groundwaters throughout the

Table 4. Comparison of Sediment and Rock Data from different Hypersaline Lakes for Pb, Cu and Zn (in μgg^{-1}).

		Pb	Cu	Zn
Western Australia	O	<4	<11	16
	R	<6	6	18
	A	<2	<1	8
Solar Lake (Gaudette & Lyons, 1984)		4	9	27
Goto Meer Bonaire (Gaudette & Lyons, 1984)		<5	29	44
Laguna OjoLiebre (Lyons, unpublished) Baja, Mexico		<1	1	7
Average sandstone (Mason, 1966)		7	–	16
Average shale (Mason, 1966)		20	45	95

'Wheat Belt' region of central-southern WA can have very high dissolved Pb concentrations. However, as shown in Table 2, the Pb concentrations in sediments in lakes from this region are very low. The alunite found in the acid-lakes has some of the lowest values of Pb measured. It is possible that dissolved Pb is removed from the groundwaters in the aquifer material and not in the lake sediments as hypothesized by Mann (1983).

Our values for Cu and Pb are lower than the values reported by Förstner (1977) for southern Western Australian lakes (his mean values, Pb = 28 μg/g and Cu = 33 μg/g). It must be emphasized, however, that these data represent 'whole-rock' analyses (via HF dissolution) of a less than 2 μm in sediment fraction, whereas our data are those metals that are leached from the bulk sediment. It is not surprising that our data are lower than his.

In three lakes, both oxidized and reduced sediments were obtained (Table 2). There are no discernable differences between the Pb, Cu, Zn concentrations in these two sediment types in Lakes Koolkooldine and Moore but the Pb concentration in the oxidized sediment and the Zn concentration in the reduced sediment of Lake Wallambin are considerably higher relative to the other sediment type. Whether this is truly a diagenetic effect or just an inhomogeneity in the trace metal distribution within the lake is unknown.

Probably the most interesting of these data are the relatively high Ag concentrations in a number of lakes. The Yilgarn Block is an important gold mining district of Australia. The petrological observations of the gold nuggets from this region indicate that gold is very pure, containing little to no Ag (Mann, 1984). This is particularly true in locations where the gold is associated with highly lateritic terrain (Mann, 1984). Mann (1984) has argued that the acid groundwaters in the region first dissolve gold and silver and then transport them as Cl^- complexes. The gold is then precipitated via reduction of the gold-chloride complex with Fe^{2+}. Reduction of the Ag-chloride does not occur, however, due to a difference in the redox potentials. Thus, the gold is separated and 'purified' by the removal of silver. This, of course, suggests that the Ag is then transported via the groundwaters. If Mann's interpretation is correct, a portion of this dissolved Ag could be transported to the ultimate 'sink' of these waters, the salt lake floors.

There are three lakes with Ag concentrations greater than 1.0 μgg^{-1} in their sediments. They are: Lakes Goongarrie, Annean and Yarra Yarra. Four other lakes, Wallambin, Grace, King and Moore, have values between 0.8–1.0 μgg^{-1}. Because the average shale value for Ag is only 0.07 μgg^{-1} (Mason, 1966), these sediments can be thought of as enriched in silver. The highest value of 3.1 μgg^{-1} was observed in Lake Annean. Lake Annean lies in a region where gold mining and prospecting are taking place currently. It is possible that Ag concentrations in salt lake sediments in WA could be a useful gold prospecting tool. Our data also support the ideas of Mann (1984) concerning the relative mobility of Ag.

Lake Tyrrell-Raak Boinka Sediments.
The trace metal abundances in whole rock sediments from Lake Tyrrell fall between the average sandstone and shale values (Table 4). With the exception of samples 5 and 8, the values from the acid discharge zone of the lake are very low. This is similar to observations in the WA acid lakes. Sample 5 was a very fine grained yellowish material that was jarosite. Sample 8 was associated with algal mat/sulfide rich material. Apparently the algal mat and the authigenic sulfides produced via sulfate reduction of the algal material can concentrate metals above background sediment values.

This is more readily observed in a Pb, Cu, Zn profile from the acid discharge zone in the SW portion of Lake Tyrrell (Fig. 2). The highest concentration of Zn is observed in the algal mat layer between 0–1 cm depth while the highest Pb values are found in the sulfide rich layer below the algal mat (1–3 cm). Interestingly, the highest Cu concentration observed was at the reducing/oxidizing boundary of the sediment at 3–4 cm depth (Fig. 2).

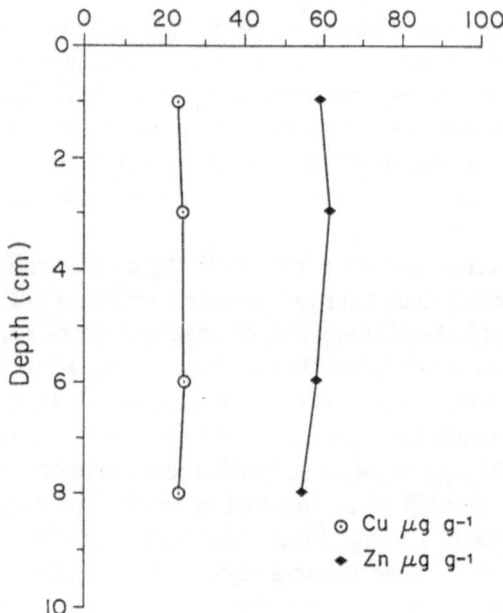

Fig. 2. Trace metal profiles from SW edge, Lake Tyrrell, Victoria.

This profile can be compared to a whole rock Cu and Zn profile from the eastern portion of the lake (Fig. 3) where there is no surface enrichment of either metal. In this location there is no indica-

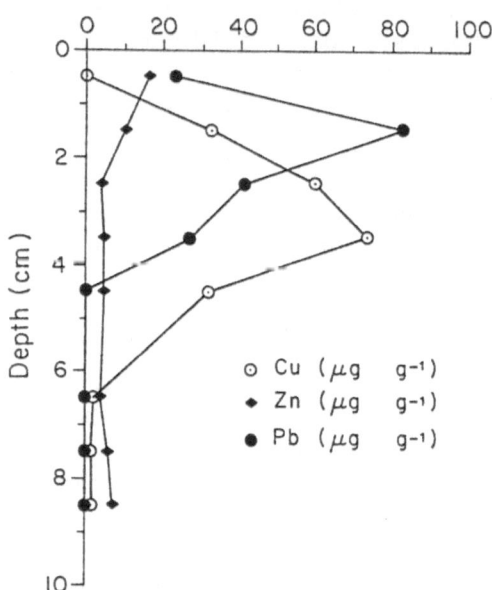

Fig. 3. Trace metal profiles, Pup Lagoon, NE portion, Lake Tyrrell, Victoria.

tion of groundwater discharge or of an algal mat presence. Algal mat/sulfide rich sediments on the eastern shore of the lake, where neutral pH water is discharged, yielded much lower Cu and Zn values (samples 2 and 3) than those observed in Fig. 3. These are the only data suggesting that this low pH, high Fe water discharged in the western shore of Lake Tyrrell is also transporting Cu, Pb and Zn. In turn, the Cu, Pb and Zn are removed from solution in these sediments. The removal appears to be by incorporation into organic matter and/or biogenic sulfides. Bowler and Teller (1986) have reported reduced sulfur concentrations as high as 5% in the top few cms of Lake Tyrrell muds. However, by our visual observations on three different visits to Lake Tyrrell, the extent of algal mats and sulfide rich environments in the acid discharge zones was determined to be small.

The mean surface values for Pb, Cu, Zn from surficial sediments (0–1 cm) taken in the same area as the core presented in Fig. 2, but where in the absence of algal mats were <10, 2 and 5 $\mu g g^{-1}$ respectively. Therefore, whether or not large amounts of metal are really accumulating in these acid discharge areas is still unknown. More recent sampling by one of us (DTL) may possibly answer this important question.

The mean values for our Lake Tyrrell whole rock data with the exception of Co, are lower than the means reported for Australian salt lakes by Förstner (1977). This, in part, is due to the fact that Förstner's data represent only the <2 μm sediment fraction (Table 5). The Lake Tyrrell mean values for Mn, Cu and Ni are 3.3, 3.4 and 5.0 times lower, respectively, than Förstner's values. These differences may be due to a true difference in the sources of these metals to Lake Tyrrell compared to lakes in other parts of the

Table 5. Comparison of Lake Tyrrell whole rock data and that of Förstner (1977) a number of Australian sale lakes (values in $\mu g\ g^{-1}$ dry wt).

	Cu	Zn	Mn	Ni	Co	Cr	V
LT	10	26	118	19	13	31	42
Förstner	33	90	400	95	11	60	–

continent or simply due to the differences in our analytical procedures.

Many hypotheses are available regarding the formation of banded iron formations (BIF's) (e.g. Drever, 1974; Kimberley, 1978; Walker, 1984; Francois, 1986). One of the most intriguing aspects of the low pH, hypersaline systems of NW Victoria is the development of ironstone-like sediments and rocks at Lake Tyrrell and the Raak Boinka. Although the outstanding work of Macumber (1983) has acknowledged the existence of these authigenic 'ironstones', there is little detailed understanding of how they form. As stated above, the groundwaters discharging onto the floors of these playas are rich in dissolved Fe. There are also high concentrations of dissolved Si in these waters (Macumber, 1983). Although many types of environments have been described as possible locations of BIF development, environments such as Lake Tyrrell have not been discussed. These low pH, high Cl^- and Fe rich waters may offer the missing link in BIF formation.

It is worthwhile to compare the geochemical differences of samples 9, 12 and 13. These samples are similar in appearance yet different in cohesiveness. We have termed all three samples 'ironstones'. All are coarse grained and deeply ferric iron stained. Sample 9 is an unconsolidated sediment; sample 12 is lithified. Sample 13 is obviously a much older and more lithified 'rock' than even sample 12, based on its degree of consolidation. If we can interpret these three samples as examples of an aging process with sample 13 being the end product of ironstone development in an acid, hypersaline environment, the trace metal geochemical evolution of this process can be estimated. The metals behave in very different ways. The Co concentration does not change throughout the aging process, while the Cu concentration increases with age (Table 3). The highest concentrations of the other metals (i.e., Zn, Mn, Ni, Cr and V) are found in the 'intermediate' sample. The 'oldest' sample contains extremely low Cr and V concentrations whereas the Zn and Ni concentrations are similar to those of the unconsolidated end member (sample 9). The 'oldest' sample has Mn concentration between

the 'youngest' and the 'intermediate' aged sample.

If the Zn, Ni, Cr and V concentrations are normalized to the Mn concentration, another pattern emerges. The Zn : Mn and Ni : Mn ratios change very little (0.2 ± 0.1 and ∼0.25 ± 0.12) while the Cr : Mn and V : Mn decrease dramatically with increasing age. These data suggest that Cr and V are lost from the ironstone with age while the Zn and Ni concentrations may be controlled, at least in part, by the absolute Mn concentrations. It has long been known that the concentrations of Zn and Ni are associated with Mn concentration in oceanic Mn-nodules (Cronan, 1976) so this result is not too surprising.

Concluding remarks

Although this work is a preliminary survey, the following conclusions can be drawn from this study:

1. with the exception of Ag, the trace metal concentrations in the Australian salt lakes under investigation are not enriched above background values;

2. high concentrations of sedimentary Ag are, in general, associated with Au mining areas;

3. in areas where acid groundwaters are thought to transport high concentrations of Pb, Cu and Zn (Mann, 1983), there is little to no metal enrichment in the sediments of the lakes;

4. metal removal appears to be associated with algal mat-authigenic sulfide rich sediments;

5. ironstone-like formations are currently being formed in the 'acid' lakes of NW Victoria.

Acknowledgments

We are deeply indebted to Dr. P.G. Macumber, who, not only introduced us to Lake Tyrrell, but shared many of this thoughts and much of his time with us. Much of the work on this manuscript was stimulated by him. We are very grateful to Dr. J.M. McArthur for conducting the pH measurements on the WA water samples and his help in collecting the WA sediment samples. We thank

J.M.G. Shelley for performing the Mn analyses. We also thank Dr. A.L. Herczeg for helping to collect the Lake Tyrrell and Raak Boinka samples. One of us (WBL) also acknowledges many stimulating conversations regarding the low pH waters of Australia with Dr. Herczeg. We appreciate the detailed reviews of the original manuscript by two unknown reviewers. This work was supported in part by funds from ANU to the senior author (WBL) while on leave from UNH. A very special thanks goes to Ms. Lyn Preble who typed the manuscript.

References

Bettenay, E., A. V. Blackmore & F. J. Hingston, 1964. Aspects of the hydrologic cycle and related salinity in the Belka Valley, Western Australia. Aust. J. Soil Res. 2: 187–210.

Bowler, J. M. & J. T. Teller, 1986. Quaternary evaporites and hydrological changes, Lake Tyrrell, north-west Victoria. Aust. J. Earth Sci., 33: 43–63.

Cronan, D. S., 1976. Manganese nodules and other ferromanganese oxide deposits. In Chemical Oceanography, Vol. 5. Academic Press, New York, p. 217–265.

DeDeckker, P., 1988. Biological and sedimentary facies of Australian salt lakes. Palaeogeo. Palaeoclim. Palaeoecol. 62: 237–270.

Drever, J. I., 1974. Geochemical model for the origin of Precambrian banded iron formations. Geol. Soc. Am. Bull. 85: 1099–1106.

Förstner, U., 1977. Mineralogy and geochemistry of sediments in arid lakes of Australia. Geol. Rundsch. 66: 146–156.

Francois, L. M., 1986. Extensive deposition of banded iron formations was possible without photosynthesis. Nature 320: 352–354.

Gaudette, H. E. & W. B. Lyons, 1984. Trace metal concentrations in modern marine sabkha sediments. In Microbial Mats: Stromatolites, A.R. Liss, Inc., New York, p. 425–434.

Kimberley, M. M., 1978. Paleoenvironmental classification of iron formations. Econ. Geol. 73: 215–229.

Kiss, E., 1973. Spectrophotometric determination of cobalt in silicates and meterorites. Analyt. Chim. Acta. 66: 385–396.

Kiss, E., 1975a. Selective spectrophotometric determination of vanadium in silicates with a new pyridylazophenol in the presence of hydrogen peroxide. 7: 205–221.

Kiss, E., 1975b. Selective spectrophotometric determination of cobalt in silicates and meteorites. Analyt. Chim Acta. 77: 320–323.

Long, D. T., W. B. Lyons & H. E. Gaudette, 1985. Trace metal concentrations in modern marine sabkhas. Chem. Geol. 53: 185–189.

Macumber, P. B., 1983. Interactions between groundwater and surface systems in northern Victoria. Ph.D. dissertation. University of Melbourne, Australia. 506 pp.

Mann, A. W., 1983. Hydrogeochemistry and weathering on the Yilgarn Block, Western Australia – ferrolysis and heavy metals in continental brines. Geochim. Cosmochim. Acta 47: 181–190.

Mann, A. W., 1984. Mobility of gold and silver in lateritic weathering profiles: Some observations from Western Australia. Econ. Geol. 79: 38–49.

Mason, B., 1966. Principles of Geochemistry, 3rd edition. J. Wiley & Sons. New York, 329 pp.

Murray, J. W. & G. Gill, 1978. The geochemistry of iron in Puget Sound. Geochim. Cosmochim. Acta 42: 9–20.

Tessier, A., F. Rapin & R. Carignan, 1985. Trace metals in oxic lake sediments: possible adsorption onto iron oxyhydroxides. Geochim. Cosmochim. Acta 49: 183–194.

Thornber, M. R., E. Bettenay & G. R. Russell, 1987. A mechanism of aluminosilicate cementation to form a hardpan. Geochim. Cosmochim. Acta 51: 2303–2310.

Walker, J. C. G., 1984. Suboxic diagenesis in banded iron formations. Nature 309: 340–342.

Webster, J. G. & A. W. Mann, 1984. The influence of climate, geomorphology and primary geology on the supergene migration of gold and silver. J. Geochem. Explor. 22: 21–42.

Williams, W. D., 1967. The chemical characteristics of lentic surface waters in Australia. In: Australian Inland Waters and their Fauna. ANU Press. Canberra, p. 18–77.

Hydrobiologia **197**: 23–33, 1990.
F. A. Comin and T. G. Northcote (eds), Saline Lakes.
© 1990 *Kluwer Academic Publishers.*

Heavy metals in the Dead Sea and their coprecipitation with halite

M. Stiller[1] & L. Sigg[2]
[1]*Weizmann Institute of Science, Rehovot 76100, Israel; Present address: 5 Haem str. Rishon LeZion 75240, Israel;* [2]*EAWAG, CH 8600 Dubendorf, Switzerland*

Key words: heavy metals, halite, coprecipitation, Dead Sea

Abstract

After a prolongued period of stratification (about 300 years) the Dead Sea overturned in 1979 and again in 1982. Its waters became saturated with respect to halite and the massive precipitation of halite which occurred in winter 1982/83 has been monitored. We followed the fate of the heavy metals during this period of physical and chemical changes.

The concentrations of Zn, Cd, Pb and Cu in the Dead Sea waters have been measured by anodic stripping voltammetry (ASV) which provided sensitive measurement of these elements after a minimal pretreatment of the samples (dilution 1:1 and acidification). In the meromictic lake (prior to 1979), the concentrations of all four elements were larger in the deep anoxic layers. With the onset of halite precipitation a decline in their concentrations was observed. Most dramatic was the decrease in Cd, which practically disappeared from the water column in 1985. The coprecipitation of heavy metals with halite – collected by sediment traps in 1983 – was examined, as well as that of older halite recovered from a sediment core. Although concentrations of heavy metals were somewhat larger in recent halite, all halite samples had the same coprecipitation pattern: the concentration of Pb was the largest, followed by Cd, and that of Cu was the smallest. The apparent distribution coefficient was larger for Cd than for Pb.

We estimated the amount of Cd which may have accompanied the deposition of halite during 1983–1985; it is compatible with its observed disappearance from the water column in 1985. The amounts of Pb and of Zn which are missing from the Dead Sea of 1985 are much larger than can be accounted for by coprecipitation with halite. A possible explanation is that the formation of halite crystals may have enhanced settling of particulates which in turn, may have scavenged Pb and Zn from the Dead Sea waters. Cu seems to be much less affected by the physical and chemical events which occurred in the Dead Sea during 1976–1985.

Introduction

The hypersaline Dead Sea overturned in 1979 (Steinhorn *et al.*, 1979; Steinhorn, 1985) after a prolongued period of about 300 years of meromictic stratification (Stiller & Chung, 1984). Then a short meromictic episode followed (Stiller *et al.*, 1984a) and the Dead Sea overturned again in 1982. Since then the Dead Sea has been monomictic (Anati *et al.*, 1987).

Until 1979 a relative dilution and oxic conditions prevailed in the upper water mass, while the deep waters were anoxic (Neev & Emery, 1967). The chemical composition of the major ions was very similar in both water masses: Cl > Br >> SO_4 > HCO_3 and Mg > Na > Ca > K. At the

1979 overturn the average salinity was 340 g l^{-1} and that of the most abundant ion, Cl 225 g l^{-1}.

During the overturn of 1979 precipitation of halite was observed at the Dead Sea surface (Steinhorn, 1983); in 1982 the brines became saturated with respect to halite (Anati *et al.*, 1987) and as documented by sediment traps (Stiller, Gat & Spencer, pers. commun.) the winter of 1982/1983 marked the onset of a heavy precipitation of halite. The published data on minor and trace elements in the Dead Sea are relatively scarce (Bentor, 1961; Schonfeld & Held, 1965; Brooks *et al.*, 1967; Nissenbaum, 1977) and sometimes controversial (Stiller *et al.*, 1984b).

The physical changes and the chemical evolution which took place recently in the Dead Sea have influenced the distribution of the heavy metals in its water column and may perhaps explain some of the controversy in the published data.

We have attempted to follow the fate of the heavy metals Zn, Cd, Pb and Cu in the brines of the Dead Sea during the years 1976–1985 and the extent of their coprecipitation with halite.

Description of study site

The Jordan River which combines the outflow of Lake Kinneret with that of the Yarmouk River supplied until the early 1960's two thirds of the Dead Sea's total inflow. Seasonal floods, small rivers, like the Arnon from the east, a few freshwater and saline springs and scarce rainfall – only 50–70 mm year^{-1} – accounted for the rest. In 1964 the outflow from Lake Kinneret was diverted; the diminished discharge of the Jordan River caused a steady decline in the level of the lake. In 1976 brines from the deep northern basin of the lake ceased to cross over the sill of the Lisan Straits into the very shallow southern basin (Fig. 1) which began to dry up.

At present, most of the southern basin has been transformed into a series of diked, solar evaporation ponds, into which brines from the deep northern basin are being pumped. Halite, and at a later stage carnallite, from which potash is then extracted, precipitate in the evaporation ponds. After the deposition of carnallite, the residual concentrated brines, called End Brines, are pumped back into the Dead Sea.

A chronology of the state of the water column in the northern basin during 1976–1985 is given below as a necessary background to interpret the heavy metals behavior. In January and in June 1976 the upper mixed layer was 80 m thick, the transition zone was between 80 and 110 m and the deep waters, 110 to 300 m were anoxic. In 1978 the meromictic pycnocline deepened to 175 m and the transition layer was between 175 to 200 m. The deeper waters, below 200 m were still anoxic as indicated by presence of divalent iron (Nishri & Stiller, 1984), but the odor of H_2S (there are no analytical data on its concentrations) was already very faint. In February 1979 as the Dead Sea overturned, the water column became homogeneous. The winter of 1979/80 brought many floods which initiated a short meromictic phase until December 1982. The upper 20 m were about 3% less saline in December 1980 than deeper waters. In September 1982 this upper layer became close to saturation with respect to halite and its salinity was comparable to that of the deep waters. At our last sampling in May 1985, after the turnovers of December 1983 and of December 1984, the water column was well mixed with the exception of a seasonal pycnocline at 15 m depth.

Materials and methods

Sampling and sample preparation

Dead Sea brines were sampled at about 31° 35' N, 35° 25' E in 1976–1980 and at the site of a permanently anchored buoy (Fig. 1) in 1982 and 1985. Hydrobios plastic samplers, equipped with protected and unprotected thermometers (manufactured by Gohla-Kiel) were used. The Dead Sea samples were collected in screw-capped glass bottles, precleaned with dilute hydrochloric acid and thoroughly rinsed with bidistilled water. The samples were stored in a dark, relatively cool (22–23 °C) place.

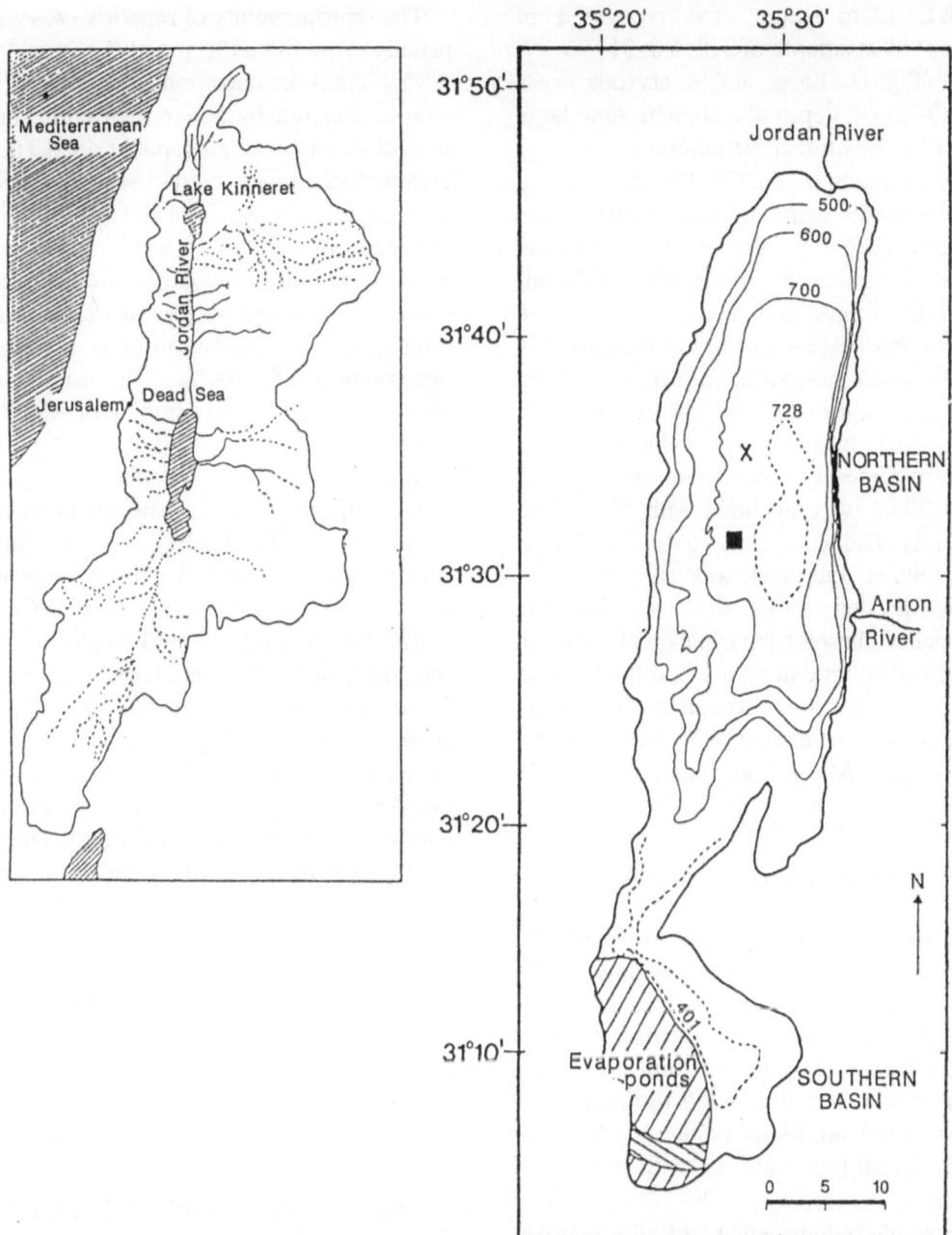

Fig. 1. Map of the Dead Sea and its catchment area (left). Shoreline corresponds to the − 398 m level; isodepths contours are meters below sea level; scale in km. Site of the anchored buoy is marked by a square. Sampling site of core B1 is marked by X.

Halite formed in the Dead Sea was collected by cylindrical sediment traps deployed at the site of the anchored buoy (Fig. 1) at 70 m depth of water, from February 8 to March 8, 1983. The halite crystals were separated from particulate material which settled at the bottom of the sediment trap,

by passing the slurry through a 200 μm nylon mesh and rinsing with about 200 ml Dead Sea brine from 70 m depth, immediately after recovery of the sediment traps. The halite crystals were then transferred from the nylon mesh into a pre-cleaned glass container.

Core B1, 1.8 m length, was recovered in December 1980 at a depth of 318 m at 31° 35′ N, 35° 27′ E (Fig. 1). Large halite crystals were found at 80–85 cm depth of sediment. One large crystal, 4.17 g, was taken for analysis.

The concentration of Zn, Cd, Pb and Cu were measured by anodic stripping voltammetry (ASV) with the Metrohm 646 VA Processor and by using the method of standard additions. The ASV analysis of all the brines and halite samples were performed at EAWAG – Dubendorf in May–July 1985, as well as the analysis of one sample of Lake Kinneret water and one sample of End Brine.

Prior to ASV measurement unfiltered brine samples were diluted 1 : 1 by volume with distilled water and 20 ml were acidified with 40 μl 5 M HCl (pH 1.1). Halite crystals, 5 to 25 g, were treated as follows: they were washed three times with triple distilled water, dried on Kleenex tissues, weighed, dissolved in triple distilled water and brought to volume in volumetric flasks. The exact weight of halite was checked by Cl analysis of these solutions. An aliquot of 20 ml was acidified with 40 μl 5 M HCl and taken for ASV analysis.

Anodic stripping voltammetry measurements

The solutions were purged initially with nitrogen for 10 minutes in the measuring vessel and for 60 seconds prior to each repetitive ASV run performed on the same aliquot. The measuring conditions were 90 seconds reduction time at a reduction potential of − 1.15 V or − 0.9 V. Fresh dilutions of standard additions (spikes) were prepared daily for all four elements with triple distilled water.

The measuring sequence was: 2 to 4 repetitive runs for the unspiked sample and 2 repetitive runs after each addition of a set of spikes (standard additions of all 4 elements). Two sets of standard additions were performed for each analysis. The final dilution due to the standard additions was 1%.

The analytical blank was triple distilled water acidified with 40 μl 5 M HCl. Cd, Pb and Cu were undetectable and Zn was about 0.1 μg l^{-1}.

The reproducibility of repetitive ASV runs was usually better than 3%, mostly between 0.3% and 1.8%. The concentrations of the heavy metals were calculated by linear regression, taking the data of all unspiked and spiked runs. The correlation coefficient was always better than 0.99. However, replicate analysis of different aliquots from the same brine sample were not as good (see Table 1) as one would expect from the analytical precision indicated above. Moreover, our methodology did not control sampling artifacts and the preservation of samples for the time lapse between sampling and the date of ASV measurement.

Typical ASV sensitivities, nA μg^{-1} l (nanoAmpere per microgram per liter), and peak potentials, mV (millivolts), *vs.* a Ag/AgCl reference electrode in the 1 : 1 diluted brines were: 2.1 and − 980 for Zn, 2.2 and − 640 for Cd, 1.1 and − 440 for Pb and 1.3 and − 250 for Cu. The sensitivity of the ASV analysis in the 1 : 1 diluted Dead Sea water was thus 50–70% (for different elements) of the sensitivity in a dilute ionic medium (0.01 M nitric acid). In the undiluted Dead Sea water, after acidification, the sensitivity was lowered by a factor of about 2. The reproducibility and the sensitivity of the data were best in 1 : 1 diluted, acidified Dead Sea water, with reduction at − 1.15 V. Some samples were measured with − 0.9 V reduction potential in order to avoid the reduction of Zn which was present at much higher concentrations than the other heavy metals.

Results

Complexes of heavy metals in the Dead Sea

Peak potentials obtained for the heavy metals in different media give some indications of their speciation. The most probable complexes in Dead Sea water are chloride complexes; in undiluted Dead Sea water, the concentration of chloride is approximately 6 M. The potential shift observed between the diluted and undiluted Dead Sea samples may be used to evaluate the existing complexes.

The potential shift with respect to the uncomplexed metal due to the complexation by a ligand X is given by the equation:

$$\Delta E = (RT/nF)\ln K - (RT/nF)\,p\,\ln[X] + (RT/nF)\ln(D_M)^{1/2}/(D_c)^{1/2}$$

where R = gas constant, T = temperature, F = Faraday constant, K = complex formation constant, n = number of electrons in reaction, p = number of ligands in complex MX_p and D_M, D_c are the diffusion coefficients of the free and of the complexed metal ion (Bard & Faulkner 1980).

If we assume that the potential shift between diluted and undiluted samples is mostly due to the difference in concentration of the ligand and that the same complex is predominant at both dilutions, information about the number of ligands p may be gained. The following average potential shifts ΔE between diluted and undiluted Dead Sea samples were found: i) for Zn, $\Delta E = 19$ mV, which corresponds to $p = 2$ and suggests the presence of $ZnCl_2^0$; ii) for Cd, $\Delta E = 39$ mV which corresponds to $CdCl_4^{2-}$ complexes (ΔE calc. = 36 mV). iii) for Pb, $\Delta E = 53$ mV, which is higher than calculated $\Delta E = 36$ mV for $PbCl_4^{2-}$ and may include some contributions from a difference in complexation constants; iv) the peak for Cu measured at the electrode corresponds to the oxidation of Cu(O) to Cu(I), as oxidation to Cu(II) cannot be observed in this medium. For Cu, $\Delta E = 46$ mV lies between the values calculated for $CuCl_2^{1-}$ (36 mV) and for $CuCl_3^{2-}$ (54 mV). Since Cu(II) is expected to be present in the original samples, these measurements do not allow conclusions about the actual speciation of Cu.

Heavy metals in the Dead Sea brines: 1976–1985

The concentrations of heavy metals measured in 8 samples of Dead Sea brines are summarized in Table 1.

In 1976 and in 1978 while the lake was still meromictic, the concentrations of the metals were generally higher in the deep anoxic waters, below 110 m and 200 m depth in 1976 and 1978 respectively, with the exception in 1976 for Cd, but the latter could be an artifact due to poor preservation (see Discussion). Also, the distribution of Pb in 1978 is unclear, either for the same reason or due to chemical interactions which accompanied the destratification. The samples of August 1978 were found to be contaminated with Cu by the sampling procedure: during this cruise Dead Sea brines were collected in Cu pipes for helium 3 analysis; prior to this sampling the pipes were flushed with the respective brine which was not discarded but stored for chemical analysis.

After 1979 the concentrations should and generally do (with the exception of Zn) indicate homogenation of the water column. Concentrations, estimated by mixing the content of the stratified water column in 1978, *vs.* measured concentrations in the post-overturn Dead Sea of 1980 ($\mu g\,l^{-1}$) were: Zn 259 *vs.* 298; Cd 0.73 *vs.* 0.75; Pb 16.7 *vs.* 17.8 and Cu 5.0 *vs.* 6.2 (the data of 1976 was taken for the estimate of Cu). In May 1985 the Dead Sea was well mixed and we consider the one sample which was analysed to be fairly representative for the state of the water column after two and a half years of continous halite precipitation. A very sharp decrease in concentration is observed for Cd and for Pb, a more moderate one for Zn, while Cu appears to be almost unaffected. It would have been useful, of course, to have some more post-halite data, e.g. for the water column in 1983 and 1984. Comparison with the scarce earlier data (summarized by Stiller *et al.*, 1984b) is quite difficult because most of them refer to samples taken at earlier dates than those of the present study. In the meantime, temporal and depth variations in the concentrations of the heavy metals might have taken place in response to the partial diversion of water from the Jordan River (the diversion caused continous lowering of the lake level and deepening of the meromictic pycnocline). However, the Zn content of the deep waters, namely 288 $\mu g\,l^{-1}$ (Brooks *et al.*, 1967), is comparable to that of 1976–1978 measured by us.

We consider the heavy metal data as representing fairly well the total concentrations, i.e. dis-

Table 1. Concentrations of heavy metals in the Dead Sea, measured by Anodic Stripping Voltammetry ($\mu g \, l^{-1}$).

Date	Depth	Zn		Cd		Pb		Cu	
		a	b	a	b	a	b	a	b
6/25/76	30 m	63		0.35		7.9		1.6	
		65		0.47		9.7		3.4	
1/20/76	130 m	334		0.19		45.3		6.7	
		314		0.33		41.4		6.7	
8/17/78	75 m	232	168[c]	0.90	0.57[c]	18.8	15.0[c]	d	
			132		0.60		6.3	d	
8/15/78	250 m	330	326[c]	1.04	1.06[c]	6.5	16.1[c]	d	
		313	293		0.76	16.1	12.0	d	
12/18/80	50 m	406	361[c]	0.76	0.89[c]	17.3	16.1[c]	14.8	1.3
			389	0.74	0.92	19.9		6.6	4.6
			299		0.86			7.7	
9/28/82	15 m	98		0.32		6.8		5.2	
9/28/82	200 m	730?		0.16		13.2		34?	
5/9/85	80 m	88		0.04		2.4		4.4	
		84		0.03		2.8		4.4	
7/25/83	End Brine	871		1.26		44.5		14.2	

[a] Measured after 1:1 dilution and acidification
[b] Measured after acidification in undiluted samples
[c] Measured without acidification
[d] Extremely large Cu concentrations; contamination from Cu pipes is suspected (see text)
? These large concentrations cannot be easily explained and could be due to contamination

solved and particulate. This assumption is based on the following facts: (i) the estimated concentrations of particulate ($\mu g \, l^{-1}$) Zn, Cd, Pb and Cu are 0.1, 0.024, 0.26 and 0.08 respectively; these constitute only a few percent of our measured values and are thus well within the experimental error between duplicate measurements (Table 1). The above estimates were obtained by multiplying the heavy metal content ($\mu g \, g^{-1}$) of the Dead Sea sediments (Nissenbaum, 1974): Zn 52, Cd 12, Pb 13 and Cu 43, by the concentration of particulates (excluding halite) larger than 0.4 μm which is usually about 2 mg l^{-1} (upper limit about 5 mg l^{-1}; Stiller, unpublished data). (ii) there was no systematic difference between the data of acidified and nonacidified samples suggesting that the concentration of heavy metals as particulates is indeed very small.

Concentrations of heavy metals in halite

The sediment traps collected two different types of halite: large crystals (about 1 cm in size) which grew on the exterior side of the trap wall and much smaller ones, which were collected at the bottom of the sediment trap together with settled particulate material. Both types of halite have been measured for heavy metal content, taking between 5 to 25 g for each analysis.

Halite which precipitated on the outside of the sediment trap is quite similar in its heavy metal content to that collected within the sediment trap (Table 2). The halite samples are extremely enriched with Pb and the sequence of coprecipitated metal concentrations is Pb > Cd > Zn > Cu.

The metal concentrations in the halite crystal taken from core B1 (see Fig. 1) are different from those of recently precipitated halite with less Pb,

Table 2. Heavy metals coprecipitated with halite from the Dead Sea (ng g^{-1}).

Sample description	Zn	Cd	Pb	Cu
Halite from Sediment	31.6	414	2393	4.8
Trap 'Wall', 70 m,	114.8	375	1694	5.2
Feb 8-Mar 8, 1983				
Halite from Sediment	53.8	368	1600	3.1
Trap, 70 m,				
Feb 8-Mar 8, 1983				
Halite crystal at	105.2	80	589	15.6
80-85 cm in core B1				
NaCl (analytical	38	0.12	30	26
grade)				

less Cd and more Cu. Core B1 has been dated at its bottom by ^{14}C and the inferred age at the depth of the halite crystals is about 265 years (Stiller & Chung, 1984). It seems that either different concentrations of heavy metals prevailed in the Dead Sea at that time of deposition or diagenetic processes that took place in the sediments have altered the heavy metal content of the buried halite.

For comparison with natural endogenic halite, concentrations of heavy metals in analytical grade NaCl are also given in Table 2.

Heavy metals in Lake Kinneret and in the End Brine

It is of interest to compare the concentrations of the heavy metals in Lake Kinneret, which until 1964 was a major source of water for the Jordan River and thus should represent typical concentrations for the inflow into the Dead Sea with those of the Dead Sea prior to the 1979 overturn. Surface water from Lake Kinneret sampled on April 8, 1985 contained $117.5 \pm 1.5 \mu g\,l^{-1}$ Zn, $0.040 \pm 0.012 \mu g\,l^{-1}$ Cd, $7.26 \pm 0.25 \mu g\,l^{-1}$ Pb and $3.6 \pm 0.5 \mu g\,l^{-1}$ Cu ($n = 2$). These data, compared with the oxic upper water mass of the Dead Sea in 1976, suggest that Cd had the longest residence time and Zn and Cu the shortest.

The concentrations of heavy metals in the sample of End Brine collected in July 1983 (Table 1) are larger than the average concentrations found in the Dead Sea of 1980 by a factor of 2.4 for Zn, 1.7 for Cd, 2.5 for Pb and 2.3 for Cu. Each liter of End Brine (total dissolved salts about $495\,g\,l^{-1}$) represents about 2.5 liters of original Dead Sea brine, from which about 250 g NaCl and 30 g KCl have been deposited in the solar evaporation ponds. This estimate of 2.5 liters is based on the concentrations of bromine and of calcium which behave as conservative elements during the evaporation process (only about 1 g of calcium sulphate is being deposited) and are found to be more concentrated in the End Brine by a factor of 2.6 and 2.3 respectively (Epstein, 1976). Our heavy metal data in the End Brine therefore seem reasonable and indicate that there is no loss of Zn, Pb and Cu during the evaporation process and only some of the Cd, which has the greatest affinity for coprecipitation with NaCl (see Discussion), is being lost.

However, it is unclear why the heavy metals are not coprecipitated with NaCl in the evaporation ponds. One possible answer could be that the coprecipitation process is pH dependent and in the evaporation ponds the pH is actually decreasing with increasing concentration of dissolved salts, from pH 6.2 in the Dead Sea to pH 4.5 in the End Brine (Stiller *et al.*, 1985). To confirm or refute our explanation, experiments of halite precipitation at different pH's and measurements of its heavy metal content as well as that of halite actually deposited in the evaporation ponds should be performed.

Discussion

The load of heavy metals in the Dead Sea: 1976–1985

It is evident that the physical and chemical changes which occurred in the Dead Sea during 1976–1985 have affected the vertical distribution and concentrations of the heavy metals. We have attempted to follow the temporal variations in the mass balances of the heavy metals during this period, although the number of samples which has been analysed is very limited; spatial variations, if any existed, are not represented.

Zinc

The concentrations of Zn in the deep waters in 1976 and in 1978 are similar, about 320 μg$\,l^{-1}$. The higher concentration of the upper layer in 1978, 150 or 176 μg$\,l^{-1}$, is due to the deepening of the pycnocline and is close to that predicted by mass balance, 167 μg$\,l^{-1}$. Consequently, both inventories of 1976 and of 1978 (Table 3) are about 32000 ton Zn/lake.

After the overturn of 1979 one would expect to find a Zn concentration of about 220 μg$\,l^{-1}$ throughout the mixed water column. But the Zn concentration at 50 m in December 1980 was 300 to 400 μg$\,l^{-1}$, which if representative for a mixed lake, would yield too large a load. The concentration of 730 μg$\,l^{-1}$ in the deep waters of 1982 is also very hard to explain and is attributed to contamination. We have attempted to estimate Zn load of the lake in 1982 by relying only on the concentration of the upper layer, 98 μg$\,l^{-1}$; it yields about 14000 ton/lake. This is in reasonable agreement with a concentration of 87 μg$\,l^{-1}$ at 80 m and a total load of 12500 ton/lake in May 1985.

The amount of halite estimated to have been formed within the Dead Sea during 1982–1985 is about 330 million tonnes (Stiller, unpublished data). The estimated coprecipitation of Zn with this amount of halite, being only about 20 tonnes, does not affect the Zn load of the lake. Other processes are responsible for the diminished mass of Zn; one of them could be scavenging by clay particles entrained to settle by the sedimentation of the halite crystals.

Cadmium

The measured concentration in the deep waters of 1976, 0.2–0.3 μg$\,l^{-1}$, is probably inaccurate as there is no apparent explanation for how it could have become larger by a factor of four in the deep isolated waters in 1978. Therefore we assume it to have been about the same as that measured in 1978, namely 1.05 μg$\,l^{-1}$. This assumption is also supported by a calculation of the expected Cd concentration in 1978 in the upper 175 m, as the pycnocline descended to that depth. By mixing the upper 80 m thick layer of 1976 containing 0.41 μg$\,l^{-1}$ with the 1976 transition zone (down to 110 m and having an intermediate content of 0.74 μg$\,l^{-1}$), and with the deep waters down to 175 m which supposedly contained 1.05 μg$\,l^{-1}$, an average concentration of 0.66 μg$\,l^{-1}$ is obtained. This is in reasonable agreement with the measured concentration in the upper 175 m in 1978: 0.57–0.61 μg Cd per liter. If this argument is acceptable then the load of Cd in 1976 becomes very similar to that of 1978, 116 and 106 tonnes respectively.

In 1980, the measured value of 0.75 μg$\,l^{-1}$ seems to represent well the homogenous water column and yields an inventory which is comparable with the two previous ones.

In September 1982, the Cd concentrations are unexpectedly smaller yielding a total mass of Cd which is well below that of earlier estimates. At that date, formation of halite had not yet started (it was first observed in November 1982) and no loss due to coprecipitation can be invoked. The Zn inventory of 1982 has the same unexplained feature; it could be that the samples of 1982 have been affected by mishandling.

Coprecipitation with 330 million tonnes of

Table 3. Estimates of heavy metals content in the Dead Sea 1976–1985 (tonnes/lake).

Year	Volume 3 km	Zn	Cd	Pb	Cu
1976	146.2	32130	116	4300	730
1978	146.2	32090	106	2440	?
1980	146.2	43550	109	2595	900
1982	145.0	14210	?	1800	755
1985	143.4	12330	4.3	373	635
Predicted coprecipitation with halite (approx.)		18–24	121–130	560–790	1

halite, from November 1982 to May 1985, can account for the loss of 125 tonnes of Cd from the water column (assuming 0.38 µg Cd per g NaCl, see Table 2). The inventory of Cd in 1985, of less than 5 tonnes seems therefore in agreement with expectation. It should be mentioned however that 125 tonnes could be an overestimate as the concentration of 0.38 µg Cd g^{-1} NaCl may have diminished as the water column became more and more depleted in Cd.

Lead

The Pb content of the lake in 1976, about 4300 tonnes is the largest of the 1976–1985 survey. Then a sharp decrease in the lake's load of Pb is observed in 1978. The progressive deepening of the pycnocline is reflected in the relatively larger Pb concentration of the upper layer in 1978, but its measured content is somewhat lower than that predicted by mass balance calculation (22.4 µg l^{-1}). It could be that this discrepancy, as well as the striking loss of Pb from the deep waters of 1978 (concentration decreased from 43 to 16 µg l^{-1}) are related to scavenging of Pb from the water column by settling iron hydroxide particles, freshly formed at the downward moving oxic/anoxic boundary. In freshwater Lake Zurich (Sigg et al., 1987) Cd, Cu and Zn are less affected by scavenging with iron hydroxides than Pb; the correlation coefficient between Fe and the heavy metals in settling material decreases from 0.78 for Pb, through Cu and Zn to 0.35 for Cd. It could be that a similar scavenging pattern takes place in hypersaline lakes as well. Concentrations and inventories of dissolved Pb 210 in August 1978 (Stiller & Kaufman, 1984) featured a similar behavior with that of stable Pb, namely a considerable loss from the water column.

In 1980 the Pb load is comparable with that of 1978, suggesting that the 1979 turnover did not affect it. In 1982, like the diminished loads of Zn and Cd, the decreasing load of Pb also remains unexplained.

The expected coprecipitation with halite during 1982–1985 is estimated to have possibly withdrawn between 500 to 800 tonnes Pb from the water column. But the inventory of 1985 indicates that about 1200 tonnes are missing. Unless the coprecipitated quantity is underestimated (it is based on a single sediment trap), and while variations in different areas of the lake and at different times cannot be ruled out, other scavenging processes such as that by clay particles must have been involved.

Copper

The loads of Cu are almost constant during 1976–1985, about 700–800 ton/lake and the amount expected to coprecipitate with halite in 1982–1985 is practically negligible, only about 1 ton. Thus it seems that the load of Cu was almost unaffected by changes in redox conditions and endogenic chemical precipitation which took place in the water column during the period of our survey.

However, there are slight fluctuations in the inventories of Cu which simultaneously resemble those of the other heavy metals: an increase is observed in 1980 and a decrease in 1982 which continues into 1985.

Coprecipitation of heavy metals with halite

The apparent distribution coefficient is estimated by dividing the ratio between the heavy metal and Na concentrations in halite to that in the Dead Sea brine. For the heavy metal content of the brines we will use average data from 1980 instead the data from 1982 when precipitation of NaCl had actually started, because we do not yet have a plausible explanation for the smaller concentrations measured at that later date. The Na concentration in the Dead Sea was 40.6 g l^{-1} (Steinhorn, 1985) and in halite it is 0.393 g g^{-1} halite.

Table 4 shows that although the Pb concentration in halite is the greatest, it is Cd which has the largest distribution coefficient. Two additional points should be noted: (1) the atomic radius of Cd^{2+} happens to be exactly the same as that of

Table 4. Estimated apparent distribution coefficients of the heavy metals for halite precipitating from the Dead Sea.

	Zn	Cd	Pb	Cu
Halite (ng g^{-1})[a]	66.7 ± 43.1	385.7 ± 24.8	1895 ± 433	4.4 ± 1.1
Dead Sea (μg l^{-1})[b]	363 ± 47	0.84 ± 0.08	17.8 ± 1.4	6.3 ± 1.6
K app	0.019	47.3	11.0	0.04

[a] Average of 3 measurements on halite collected by sediment traps in 1983 (see Table 2)
[b] average of heavy metals concentrations in the Dead Sea in 1980 (see Table 1)

Na^+, namely 0.97 Å, this may facilitate its incorporation into the halite crystal lattice; (2) the magnitudes of K app are almost in the same decreasing order $Cd > Pb > Cu > Zn$ as are the strengths of the chloro complexes of the heavy metals: $Cd > Pb > Zn > Cu$ (Long & Angino, 1977).

The decrease in the inventories of the heavy metals was generally larger than predicted by coprecipitation with halite (Table 3). The predictions assumed constant concentrations of heavy metals in the Dead Sea; this is an overestimate because in fact the coprecipitated amounts, especially for Cd and Pb, would even decrease with decreasing concentrations in the Dead Sea. One possible explanation for the discrepancy between predicted and observed loads of heavy metals in 1985 could be the following: the formation of halite particles, as it increases the number of particles per unit volume, causes a larger number of collisions between suspended particles, possibly weakening their colloidal stability and producing aggregates. The precipitation of halite may thus have enhanced the formation and settling of aggregates. If the heavy metals were somehow attached to otherwise stable colloidal particles, (bound to their surfaces; Sigg *et al.*, 1982) or were susceptible to being scavenged by these aggregates, this could be a possible way for their elimination from the water column. Supporting evidence for such a process is provided by sediment trap data (Stiller, Gat and Spencer, unpublished data). Prior to November 1982 the settling flux of particulates was usually about 0.06–0.11 mg cm^{-2} day^{-1} during summer and winter respectively. It has increased suddenly to about 0.9–1.1 mg cm^{-2} day^{-1} at the onset of halite precipitation in November 1982 and has persisted at this enhanced rate until April 1983. Then fluxes of particulates diminished back to the regular values mentioned above.

Acknowledgement

This study was performed while M. Stiller benefited from the hospitality of EAWAG on her sabbatical leave. Stimulating discussions with Prof. W. Stumm and the cooperative attitude of the colleagues at EAWAG are gratefully acknowledged.

References

Anati, D. A., M. Stiller, S. Shasha & J. R. Gat, 1987. Changes in the thermohaline structure of the Dead Sea: 1979–1984. Earth Planet. Sci. Lett; 84: 109–121.

Bard, H. J. & L. R. Faulkner, 1980. Electrochemical methods – Fundamentals and applications. J. Wiley, New York, 718 pp.

Bentor, Y. K., 1961. Some geochemical aspects of the Dead Sea and the question of its age. Geochim. Cosmochim. Acta 25: 239–260.

Brooks, R. R., B. J. Presley & I. R. Kaplan, 1967. APDC – MIBK extraction system for the determination of trace metals in saline waters by atomic absorption spectrophotometry. Talanta 14: 809–816.

Epstein, A. J., 1976. Utilization of Dead Sea minerals (a review) Hydrometallurgy 2: 1–10.

Long, D. T. & E. E. Angino, 1977. Chemical speciation of Cd, Cu, Pb and Zn in mixed freshwater, seawater and brine solutions. Geochim. Cosmochim. Acta 41: 1183–1191.

Neev, D. & K. O. Emery, 1967. The Dead Sea. Geol. Survey Isr. Bull. No. 41, 147 pp.

Nishri, A. & M. Stiller, 1984. Iron in the Dead Sea. Earth Planet. Sci. Lett. 71: 405–414.

Nissenbaum, A., 1974. Trace elements in Dead Sea sediments. Israel J. Earth Sci. 23: 111–116.

Nissenbaum, A., 1977. Minor and trace elements in Dead Sea water. Chem. Geol. 19: 99–111.

Schoenfeld, L. & S. Held, 1965. Spectrochemical methods for determining B, Ba, Li and Rb in Mediterranean and Dead Sea water. IAEC Report IA-1061, 17 pp.

Sigg, L., M. Sturm, J. Davis & W. Stumm, 1982. Metal transfer mechanisms in lakes. Thalassia Jugoslavica 18: 293–311.

Sigg, L., M. Sturm & D. Kistler, 1987. Vertical transport of heavy metals by settling particles in Lake Zurich. Limnol. Oceanogr. 32: 112–130.

Steinhorn, I., 1983. In situ salt precipitation at the Dead Sea. Limnol. Oceanogr. 28: 580–583.

Steinhorn, I., 1985. The disappearance of the long term meromictic stratification of the Dead Sea. Limnol. Oceanogr. 30: 451–472.

Steinhorn, I., G. Assaf, J. R. Gat, A. Nissenbaum, M. Stiller, M. Beyth, D. Neev, R. Garber, G. M. Friedman & W. Weiss, 1979. The Dead Sea: Deepening of the mixolimnion signifies the overture to overturn of the water column. Science 206: 55–57.

Stiller, M. & A. Kaufman, 1984. Pb-210 and Po-210 during the destruction of stratification in the Dead Sea. Earth Planet. Sci. Lett. 71: 390–404.

Stiller, M. & Y. C. Chung, 1984. Radium in the Dead Sea: a possible tracer for the duration of meromixis. Limnol. Oceanogr. 29: 574–586.

Stiller, M., J. R. Gat, N. Bauman & S. Shasha, 1984a. A short meromictic episode in the Dead Sea: 1979–1982. Verh. int. Ver. Limnol. 22: 132–135.

Stiller, M., M. Mantel & M. S. Rapaport, 1984b. The determination of trace elements (Co, Cu and Hg) in the Dead Sea by neutron activation followed by X-ray spectrometry and magnetic deflection of beta-ray interference. J. Radioanalyt. Nucl. Chem. 83: 345–352.

Stiller, M., J. S. Rounick & S. Shasha, 1985. Extreme carbon isotope enrichments in evaporating brines. Nature 316: 434–435.

Hydrobiologia **197**: 35–50, 1990.
F. A. Comín and T. G. Northcote (eds), Saline Lakes.
© 1990 *Kluwer Academic Publishers.*

The evolution of saline lake waters: gradual and rapid biogeochemical pathways in the Basotu Lake District, Tanzania*

Peter Kilham[1†] & Paul L. Cloke[2]
[1]*Department of Biology and the Center for Great Lakes and the Aquatic Sciences*; [2]*Department of Geological Sciences, The University of Michigan, Ann Arbor, Michigan 48109, USA; Present address: Science Applications International Corporation, Las Vegas, Nevada 89109, USA; (*address correspondence to S. Kilham, Department of Biology, University of Michigan, Ann Arbor, Michigan 48109–1048, USA)*

Key words: Saline lakes, biogeochemistry, geochemical evolution, sulfate reduction, CaCO3 precipitation, maars

Abstract

The biogeochemical evolution of solutes markedly alters the chemistry in the closed-basin maar lakes that comprise the Basotu Lake District (Tanzania, East Africa). Examination of 11 (out of 13) lakes in the Basotu Lake District identified two distinct evolutionary pathways: a gradual path and a rapid path. During the course of biogeochemical evolution these waters follow either the gradual path alone or a combination of the gradual and rapid paths. Solute evolution along the gradual path is determined by all of the biogeochemical processes that for these waters appear to be tightly coupled to evaporative concentration (e.g. mineral precipitation, sorption and ion exchange, CO_2 degassing, and sulfate reduction). Rapid evolution occurs when mixing events suddenly permit H_2S to be lost to the atmosphere. The chemistry of waters undergoing rapid evolution is changed abruptly because loss of every equivalent of sulfide produces an equivalent permanent alkalinity.

The Basotu Lake District in north central Tanzania is comprised of 13 maar lakes. They range in surface water conductivity from 592 to 24 000 μS cm^{-1} (at 20°). Within these lake basins only a few of the variety of geo- and biogeochemical processes known to occur in lakes of this type are actually responsible for the gain and/or loss of individual solutes. For example, potassium appears to be taken up in the formation of illite. Calcium is precipitated as calcite. Magnesium interacts with alumino-silicate precursors to form a variety of clay minerals that contain magnesium (e.g. stevensite). This process is also known as reverse weathering. Sulfate is reduced to sulfide and subsequently lost as H_2S and/or metal sulfides. Alkalinity is lost owing to calcite precipitation and as a consequence of reverse weathering. Alkalinity is gained in the form of extra permanent alkalinity when sulfide is lost from these waters (via metal sulfide precipitation or gaseous emission to the atmosphere). Rapid (punctuated) evolution can occur in any lake containing anoxic waters providing that mixing events take place which cause H_2S to be lost to the atmosphere.

[†]Peter Kilham died on March 20, 1989, in Kisumu, Kenya, while working as part of a research team on Lake Victoria.

Introduction

The Basotu Lake District (Tanzania, East Africa) consists of 13 volcanic crater lakes (Fig. 1). The same geochemical mechanisms thought to operate in closed-basin lakes everywhere control, in part, the geochemical evolution of solutes in these lakes. These geo- and biogeochemical processes are: evaporative concentration, cyclic wetting and drying, sorption and ion exchange, 'reverse weathering', mineral precipitation (predominantly carbonates and sulfides in moderately saline African waters), CO_2 loss by degassing, and sulfate reduction (Cerling, 1979; Drever, 1988; Eugster & Jones, 1979; Kilham, 1971a; 1984; Von Damm & Edmond, 1984).

Recognition of an additional process, however, is required to understand the biogeochemical evolution of waters in the Basotu Lake District, and that is the loss of sulfide from the system. The loss of sulfide from an aquatic system produces permanent alkalinity. Sulfide is lost for the most part to the atmosphere as H_2S (and other sulfur containing compounds) or to the sediments when metal sulfides precipitate. Second, physical mixing processes, which contribute to the formation, maintenance, and destruction of thermal and chemical stratification, strongly affect the chemistry of these lakes. Anaerobic bottom waters (hypo- and monimolimnia) develop following stratification. Subsequently, reduction of sulfate to sulfide initiates metal sulfide precipitation. Mixing events, caused by storms, can distribute H_2S throughout the mixed layers (epi- or mixolimnia) of a lake from which it is lost to the atmosphere (MacIntyre & Melack, 1982). When stratification eventually breaks down and turnover occurs, huge quantities of H_2S are suddenly lost to the atmosphere (P. Kilham, pers. observ.). Physical mixing processes strongly affect the chemistry of these lakes because one equivalent of permanent alkalinity is produced for every equivalent of sulfide lost.

The geochemical evolution of closed-basin lakes undergoing evaporative concentration is modeled as a gradual continuous process (Garrels & Mackenzie, 1967; Hardie & Eugster, 1970).

But, owing to the development and eventual destruction of stratification, the biogeochemical evolution of the Basotu Lakes is at times both rapid and discontinuous.

The chemistry of volcanic crater lakes in East Africa was first studied by Pappe & Richmond (1890). They analyzed a water sample collected on 17 June 1889 from Lake Katwe, Uganda, by H.M. Stanley's expedition. Stanley (1890) speculates on the origin of the high salt concentrations in Lake Katwe and concludes that evaporation was of primary importance. More recent investigations of the major ion chemistry of volcanic crater lakes in Africa are those of Prosser et al. (1968) in the Bishoftu crater lakes, Ethiopia, and Arad & Morton (1969) in the Katwe crater lakes, Uganda.

The chemistry of African Lakes is quite well known. Talling & Talling (1965), Kilham (1971b), and Wood & Talling (1988) have carried out extensive comparative investigations of the major ion chemistry of African lakes and rivers, while Eugster (1970), Jones et al. (1977), and Eugster & Maglione (1979) have intensively studied the geochemistry of a few selected lake basins (e.g. Lake Magadi, Kenya, and saline lakes northeast of Lake Chad, Chad).

General description of the Basotu Lake District

The Basotu volcanic field consists of ca. 30 explosion craters oriented mostly along two parallel NNE-SSW lines about 3 km apart. The crater lakes studied are shown in Fig. 1. The geology of the craters has been studied in detail by Downie & Wilkinson (1962). Eades & Reeve (1938) mapped the geology of the region west and south of Basotu. The craters are punched through the Precambrian Basement which is overlain by Tertiary and Recent deposits.

Figure 2 is a detailed geological section of the wall of Ghama Crater. It shows the stratigraphy of Tertiary and Recent deposits that overlie the Precambrian Basement. During the Miocene, the Basement Complex underwent peneplanation. A layer of quartz conglomerate, apparently depo-

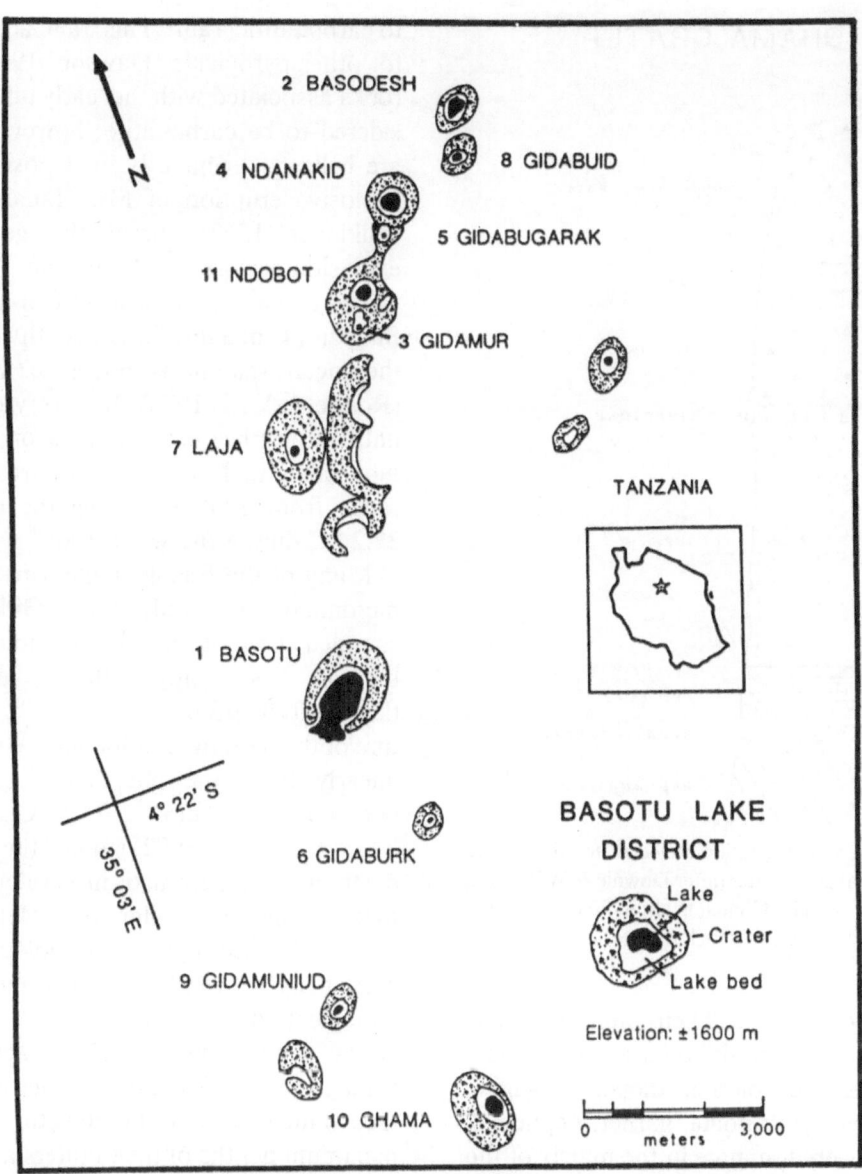

Fig. 1. Map locating explosion craters and maar lakes in the Basotu Volcanic Field. One additional crater, containing Lake Gawali, is ca 6 km south of Lake Ghama.

sited from fast flowing water, lies immediately above the peneplain. This conglomerate is overlain by 2.5 m of red clay (illitic), believed to be a flood-plain deposit (Downie & Wilkinson, 1962). The clay is in turn overlain by ca. 12 m of poorly bedded buff-colored tuff composed of numerous small crystals (up to 2 mm) of aegirine, biotite, quartz, and feldspars in a fine grained matrix of carbonate minerals, quartz, and plagioclase. The

'early tuff' is thought to have originated from Mt. Hanang, a large volcano 23 km east of Basotu. Capping the 'early tuff' is a layer (ca. 0.5 m) of compact fine-grained buff-colored limestone (ca. 75% calcite).

The cones of the Basotu explosion craters formed above the limestone that caps the 'early tuff' and consist of agglomerates and tuffs. The cone-forming pyroclasts range in size from

GHAMA CRATER

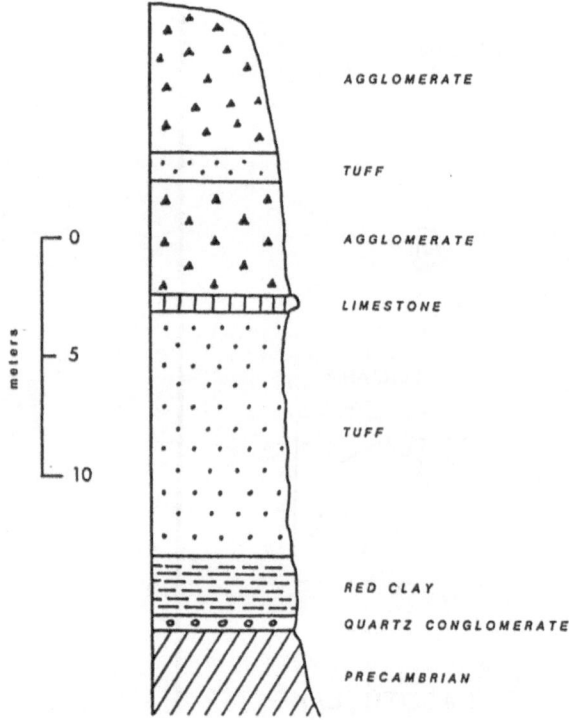

Fig. 2. Detailed geological section (selected for illustrative purposes only) of Ghama Crater (after Downie & Wilkinson, 1962). The surface of Lake Ghama is about 70 m below the layer of quartz conglomerate.

micrometers to meters. Their mineralogy is varied. Fragments contain quartz, microcline, plagioclase, perthite, biotite, diopside, augite, aegirine, magnetite, actinolite, garnet, nepheline, etc. Carbonates predominate in the matrix of the tuffs, but fine quartz and plagioclase also occur. Carbonates comprise from 15–45% of the cone-forming rock. Magnesian calcite and a calcite-dolomite mixture are the major carbonate minerals present in the matrix. The $MgCO_3$ content varies between 15–38%.

The carbonate rocks at Basotu are considered to have originated as carbonatite (Downie & Wilkinson, 1962; Dawson, 1964; Heinrich, 1966). Downie & Wilkinson (1962) propose that owing to local faulting the gas-charged magma exploded violently, causing eruptions, and the fragmentation of the carbonatitic magma gave rise

to carbonatitic 'rain'. This 'rain' acted as a matrix for other pyroclasts (Dawson, 1964). Carbonate rocks associated with the 'early tuff' are also considered to be carbonatite; however, these rocks are believed to have been deposited during the explosive eruption of Mt. Hanang. Downie & Wilkinson (1962) estimate the age of the Basotu explosion craters as Pleistocene.

The climate at Basotu is semi-arid steppe. At Singida (Tanzania; 70 km southwest of Basotu) the mean rainfall is 663 ± 167.7 SD mm yr^{-1} (Rodhe & Virji, 1976). It is very dry from June until November, and it rains on and off from November to May. Temperatures in the shade range from 12.8 °C during the dry season to 32.2 °C during the wet season.

Many of the Basotu Lakes are believed to be meromictic, but only Lake Gidamuniud was sampled at depth. In Lake Gidamuniud the mixolimnion is only three meters deep ($k_{20} = 10\,800\ \mu S\ cm^{-1}$; 20.3 °C). The temperature of the monimolimnion increases more or less linearly from the chemocline to the bottom ($k_{20} = 53\,600\ \mu S\ cm^{-1}$; 24.1 °C at 9 m). MacIntyre & Melack (1982) studied the formation and destruction of meromixis in a crater lake in Kenya that is generally similar to the lakes at Basotu. They identified basin morphology, the diurnal periodicity of winds and thermal stratification, biological decomposition, and seasonal and yearly changes in rainfall as factors that contribute to the maintenance of meromixis. The surface dimensions of the Basotu Lakes and the maximum depths of their craters are presented in Table 1.

Material and methods

Water samples for chemical analysis were collected in 500 ml polyethylene bottles that had been previously soaked in dilute HCl and repeatedly rinsed with deionized water. Field determinations of pH, calcium, magnesium, chloride, phosphate, silica, and hydrogen sulfide were made with a Portable Water Engineers Laboratory (model DR-EL, Hach Chemical Co.) using

Table 1. Surface dimensions of the Basotu lakes and the maximum depths of their craters from rim to lake surface (Downie & Wilkenson, 1962). Based on photographs, craters for which depth data are lacking are probably > 20 m deep.

Lake	Longest dimension of each lake (m)	Maximum depth of crater (m)
Basotu	700	12
Basodesh	440	18
Gidamur	100	–
Ndanakid	380	34
Gidabugarak	68	–
Gidaburk	100	–
Laja	180	–
Gidabuid	98	49
Gidamuniud	140	32
Ghama	290	96
Ndobot	280	50

colorimetric and titrimetric techniques. Water samples collected in East and Central Africa between May 1969 and January 1970 were shipped to Duke University for more detailed analysis. Even though agreement between field and laboratory data is good, only laboratory data are presented here unless otherwise noted. The unpreserved samples were analyzed within two years of collection.

Chemical techniques were selected for precision and availability. Sodium, potassium, calcium, magnesium, and silicon were determined by atomic absorption spectrophotometry (Perkin-Elmer 303: methods described in Perkin-Elmer, 1964). Sodium, potassium, calcium, and magnesium were analyzed in an air-acetylene flame; silicon, and sometimes calcium were done in a nitrous oxide-acetylene flame to preclude the formation of refractory compounds. Standards and aliquots for calcium and magnesium contained lanthanum. On occassion, calcium and magnesium were determined by EDTA titration (APHA, 1965). A Cotlove chloridometer (Lab. Glass and Instr. Co.) was used for concentrations of chloride > 2 meq l^{-1} and mercuric nitrate titration (APHA, 1965) was used for lower concentrations. Total alkalinity was determined by titration to the bromcresol green-methyl red endpoint

(APHA, 1965). Sulfate was determined turbidimetrically in the field (SulfaVer III, Hach Chemical Co.) and both turbidimetrically (using a precision spectrophotometer) and by ion exchange methods (Mackereth, 1963) in the laboratory. The different methods for calcium, magnesium, and sulfate agreed within 5% on samples analyzed by two methods. In the laboratory fluoride concentrations (see Kilham & Hecky, 1973) and pH were estimated electrometrically using appropriate electrodes and buffers. Conductivity was measured in the laboratory on a Philips meter (model PR9501, Philips Electronic Instr. Co.). All chemical analyses were carried out on unfiltered samples. The Basotu lakes were sampled in 1969, lakes Basodesh and Ndanakid were visited on 19 May and the remaining lakes were studied between 21–22 July.

Results and discussion

Patterns of solute behavior during the geochemical evolution of Basotu lake waters.

Chemical analyses for 11 of the Basotu lakes are given in Table 2. Hecky & Kilham (1973) and Kilham & Hecky (1973) published previously some of these analyses as part of their investigations concerning the diatom ecology and fluoride geochemistry in East African waters. This paper represents the first time the geochemistry of these lakes has been studied in detail.

The geochemical evolution of closed-basin lakes undergoing evaporative concentration has been studied in considerable detail. Garrels & Mackenzie (1967) modeled the isothermal evaporation of Sierra Nevada spring waters. Eugster, Jones, and others have carried out an extensive investigation of Lake Magadi, Kenya (Jones *et al.*, 1977; Eugster & Jones, 1979). In each of these studies the geochemical evolution of waters within one lake basin was studied. Cerling (1979), on the other hand, showed that chemically similar waters from numerous widely separated lakes in East Africa apparently evolve in a manner not unlike that originally predicted by Garrels &

Table 2. Surface water chemistry of the Basotu lakes. Conductivity (k_{20}, μS cm^{-1}). All ionic values are in meq^{-1} except silicon which is in μM. Alk = alkalinity (mainly HCO$_3^-$ + CO$_3^{2-}$).

Lake	No.	k_{20}	Na$^+$	K$^+$	Ca^{2+}	Mg^{2+}	SO$_4^{2-}$	Cl$^-$	Alk	Si	F$^-$	pH	S$^+$	S$^-$
Basotu	1	592	2.96	0.23	2.20	1.37	0.18	2.03	4.52	303	.054	6.6	6.76	6.73
Basodesh	2	927	3.70	0.14	3.19	1.54	1.50	2.81	4.16	393	.058	–	8.57	8.47
Gidamur	3	7,620	91.3	1.36	0.53	1.81	2.81	26.5	70.0	533	1.60	9.4	95.0	101
Ndanakid	4	7,000	82.6	1.38	0.60	3.62	11.0	45.1	30.2	463	.310	8.9	88.2	86.7
Gidabugarak	5	9,600	97.4	1.15	2.25	2.80	15.5	49.2	45.2	300	.600	8.9	104	111
Gidaburk	6	11,120	161	4.45	0.43	6.58	31.2	55.6	87.4	196	.490	9.0	172	175
Laja	7	11,800	157	0.34	0.46	3.37	14.6	57.8	85.2	178	2.80	9.4	161	160
Gidabuid	8	8,070	81.8	2.12	1.40	11.9	15.9	59.5	27.4	666	.260	8.5	97.2	103
Gidamuniud*	9a	10,800	122	4.83	0.70	8.89	13.8	81.0	42.8	340	.470	9.0	136	138
Gidamuniud†	9b	53,600	800	16.4	2.03	3.09	25.4	488	323	160	2.37	9.5	822	839
Ghama	10	17,800	254	1.77	0.28	1.32	12.1	93.1	163	566	2.25	10.5	258	271
Ndobot	11	24,400	324	3.35	0.17	4.86	10.7	188	144	340	1.38	9.5	332	344

* mixolimnion. † monimolimnion

Mackenzie (1967) for a hypothetical lake in the Sierra Nevada.

Because the Basotu lakes share a similar hydrology and climate, we believe we can treat each lake as representing different stages in the geochemical evolution of essentially a single parent water that is characteristic of the lake district, even though some variation in sodium and corresponding concentrations of sulfate and alkalinity seems to occur.

The behavior of major solutes during closed-basin brine evolution has been reviewed by Eugster & Jones (1979). They propose that individual solutes behave in a variety of ways during the course of evaporative concentration. The five basic patterns identified are shown in Fig. 3. Type I solutes are perfectly conserved species such as sodium and chloride that remain in solution throughout the concentration process. Types IIa and b represent solutes (an anion and cation) which combine in the precipitation of a mineral phase. Once saturation is reached, solute b is removed from solution and solute a continues to increase (at a rate less than for a conservative solute). Solutes following this pattern are calcium and carbonate. Type III behavior is shown by bicarbonate and carbonate combined. In this case the solute is gradually removed by a number of processes (e.g. mineral precipi-

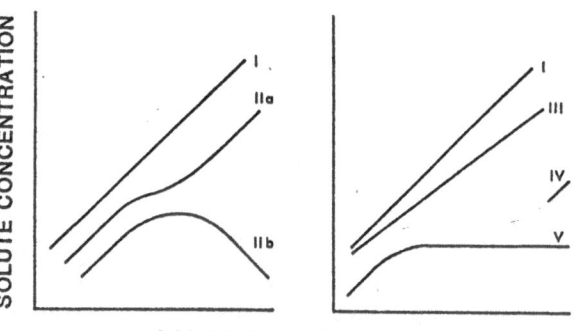

Fig. 3. Diagrammatic representation of the behavior of individual solutes in a solution undergoing evaporative concentration (redrawn from Eugster & Jones, 1979). Several mechanisms control the behavior of individual solutes. Identifiable patterns of solute behavior are labeled by Roman numerals and discussed in the text.

tation, sorption, degassing, etc.). The linear relationship which results has a slope less than that found for conservative solutes. Type IV solutes such as potassium and sulfate are removed from solution by mechanisms that Eugster & Jones (1979) believe are not strongly concentration dependent (e.g. ion exchange, sorption, and biogenic reduction). Because these removal processes act most effectively over the middle of the concentration range a sigmoid curve is produced (Fig. 3). These removal processes are generally thought to take place in the groundwater. Finally, Type V behavior is shown by solutes such as silicon which remain at relatively constant levels once saturation with respect to the corresponding mineral phase is reached.

We have plotted our data against chloride ion concentrations because chloride is an excellent conservative tracer of evaporative concentration in East African waters (Eugster & Jones, 1979). Chloride in these waters is thought to be largely derived from rainfall, but few data are available (see Hecky & Kilham, 1973). Chloride is not known to be removed from solution until halite saturation is reached. When plotted against one another on a log-log plot, conservative solutes concentrated by evaporation fall on a 45° line (slope = 1.0). The slope of the line is independent of the initial concentrations of the species plotted (Drever, 1988).

Sodium appears to be fairly conservative in the Basotu lakes (Fig. 4A, but see Yuretich & Cerling, 1983; Von Damm & Edmond, 1984). The slope of the line for sodium versus chloride is 1.05 and the correlation coefficient (r) is 0.98 (Table 3). The sodium to chloride molar ratio varies somewhat (mean = 2.05 ± 0.71 SD), but this still probably indicates that all of the lakes evolved from essentially a single parent water.

The behavior of potassium (Fig. 4B) is similar to that of alkalinity (Fig. 6B). The equations for potassium and alkalinity versus chloride (Table 3) have similar slopes and correlation coefficients. In contrast to the observations of Eugster & Jones (1979), potassium in the Basotu lakes is concentrated according to the Type III rather than the Type IV trend. This indicates that potassium is

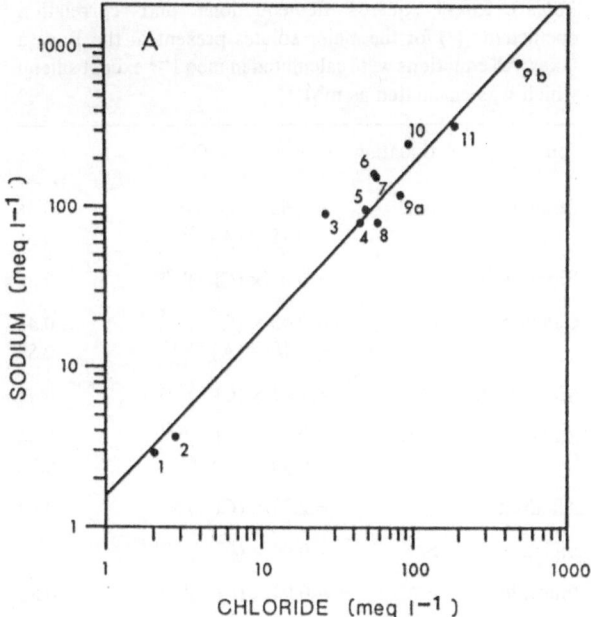

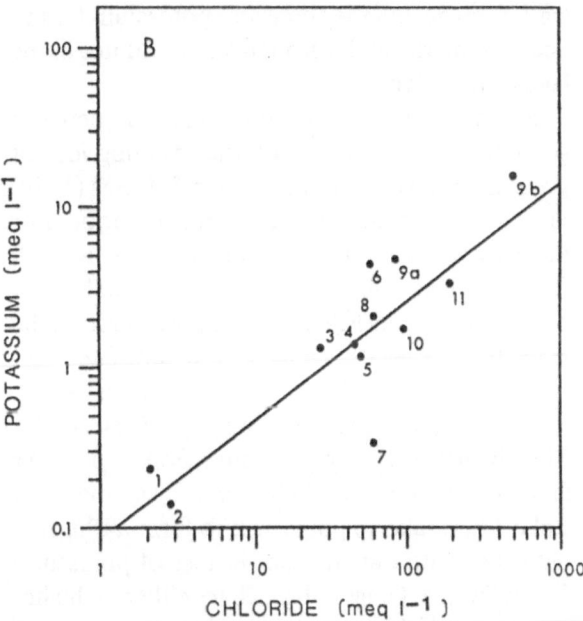

Fig. 4. Plots of sodium (A) and potassium (B) versus chloride for waters of the Basotu Lake District, Tanzania. Equations for regression lines shown are given in Table 3.

possibly being lost from solution within the individual lake basins. Eugster & Jones (1979) suggest that considerable amounts of potassium are differentially lost from solution over this concen-

Table 3. Least squares fit equations and correlation coefficients (*r*) for the major solutes present in the Basotu lakes. All equations were calculated in meq l^{-1} except silicon which was calculated as mM.

Ion		Equation	*r*
Sodium	Na$^+$	$= 1.63 \times (Cl^-)^{1.05}$	0.98
		$= 0.78 \times (A)^{1.22}$	0.97
Potassium	K$^+$	$= 0.08 \times (Cl^-)^{0.76}$	0.86
Calcium	Ca^{2+}	$= 2.43 \times (Cl^-)^{-0.29}$	-0.48
		$= 4.10 \times (A)^{-0.42}$	-0.59
Magnesium	Mg^{2+}	$= 1.41 \times (Cl^-)^{0.23}$	0.48
Sulfate	SO$_4^{2-}$	$= 0.36 \times (Cl^-)^{0.81}$	0.86
		$= 0.30 \times (A)^{0.84}$	0.76
Alkalinity	A	$= 2.35 \times (Cl^-)^{0.79}$	0.93
Silicon	Si	$= 0.44 \times (Cl^-)^{-0.07}$	-0.23
Fluoride	F$^-$	$= 0.04 \times (Cl^-)^{0.71}$	0.82

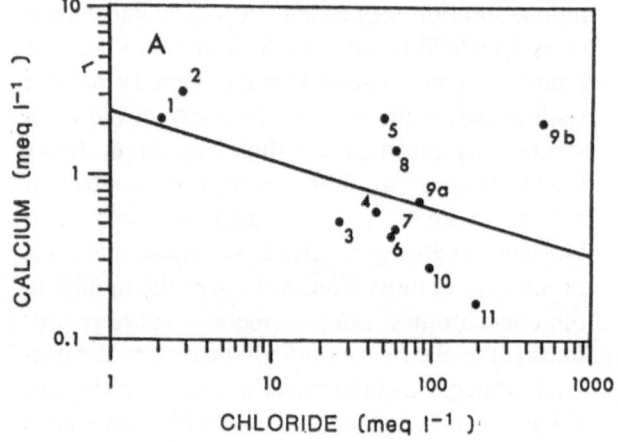

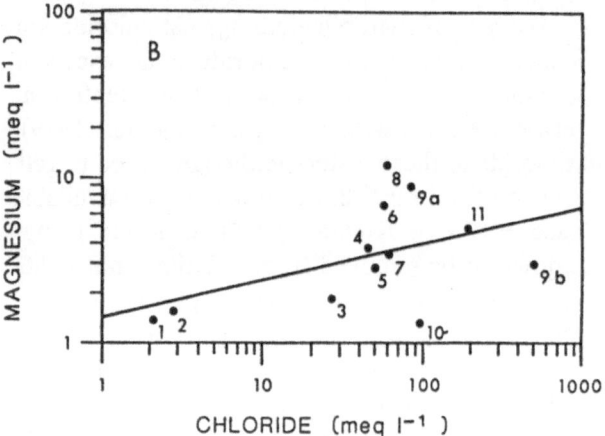

Fig. 5. Plots of calcium (A) and magnesium (B) versus chloride for waters of the Basotu Lake District, Tanzania.

tration range, but they consider potassium loss to occur primarily in the groundwater and not in the lakes themselves.

To demonstrate the fractionation between sodium and potassium, and the resulting loss of potassium from solution, Eugster & Jones (1979) plot the atomic ratio Na : K against chloride. For the Basotu Lakes this relationship is Na : K = 19 (Cl)$^{0.29}$. This slope, approximately half that observed for Lake Magadi, strongly resembles the slope (0.24) they determined for a combined plot for three lakes in Oregon (U.S.A., lakes Abert, Summer, and Goose; Eugster & Jones, 1979). This is not surprising because Jones & Weir (1983) have shown that clays in the sediments of Lake Abert take up potassium to form authigenic illite. They attribute the slower loss of potassium found in the Oregon basins to either a higher solution to solid surface ratio that decreases the effectiveness of the potassium removal mechanisms and/or to lower rates of chemical weathering, owing to the cooler climate, that may affect the fractionation process. Neither of these explanations can explain that slow loss of potassium in the Basotu Lake District because the climate is semi-arid and similar to that of the Magadi basin.

In the Basotu lakes fractionation between sodium and potassium is possibly rapid in the groundwaters because active surfaces for ion exchange and sorption (i.e. clays and various volcanic rocks) are common in the drainage basins of the lakes. However, considerable evaporative concentration probably occurs in the lakes themselves and clays are a substantial component of the sediments. Abundant illite and lesser quantities of kaolinite and chlorite are present (P. Stoffers, pers. comm.). Zeolites also appear. Therefore, abundant ion exchange sites occur in the sediments and these probably cause the gradual loss of potassium from solution. Many authors have suggested that potassium is lost from solution in African lakes containing illite-rich sediments (Singer & Stoffers, 1980; Von Damm & Edmond, 1984; Yuretich & Cerling, 1983). The potassium : chloride ratio of Lake Laja is exceptionally low (Fig. 4B, No. 7).

Calcium follows Type IIb behavior (Fig. 5A). Although the slope does not look precipitous, over 95% of the calcium in an initial water like those with the lowest chloride levels would be removed by evaporative concentration and precipitation during evolution to any of the other lake waters. Cerling (1979) also found that considerable amounts of calcium remain in solution in other East African lakes over this concentration range. Figure 5A could also be viewed as showing an initial drop in calcium followed by a more or less constant concentration.

Magnesium is at first lost from solution (Fig. 5B), but at chloride concentrations greater than ca. 40 meq l^{-1} magnesium begins to behave as a conservative constituent with a more or less constant concentration. According to Eugster & Jones (1979) the reasons why magnesium behaves as it does are not well understood. Jones & Weir (1983) found that the enrichment in magnesium of neoformed silicates in Lake Abert (Oregon) increased with increasing salinity. This is compatible with the observations for the Basotu lakes. Particularly high concentrations of magnesium are found in Lakes Gidabuid, Gidamuniud (mixolimnion), and Gidaburk.

The plot of alkalinity versus chloride shows the characteristic Type III pattern (Fig. 6B). Despite the apparent scatter the correlation coefficient is 0.93 (Table 3). Lake Gidamur has a high alkalinity to chloride ratio (2.64 by eq) while that of Lake Gidabuid (0.46) is very low.

Our plot of sulfate versus chloride shows the expected amount of scatter (Fig. 6A; Kilham, 1984). In the Basotu lakes sulfate follows the Type III rather than the Type IV pattern. This indicates that considerable amounts of sulfate (in the form of sulfides) are probably being lost in the individual lake basins (see below) rather than from the groundwater. Saturation with respect to gypsum is rarely reached in East African lakes because alkalinity values are high and calcium

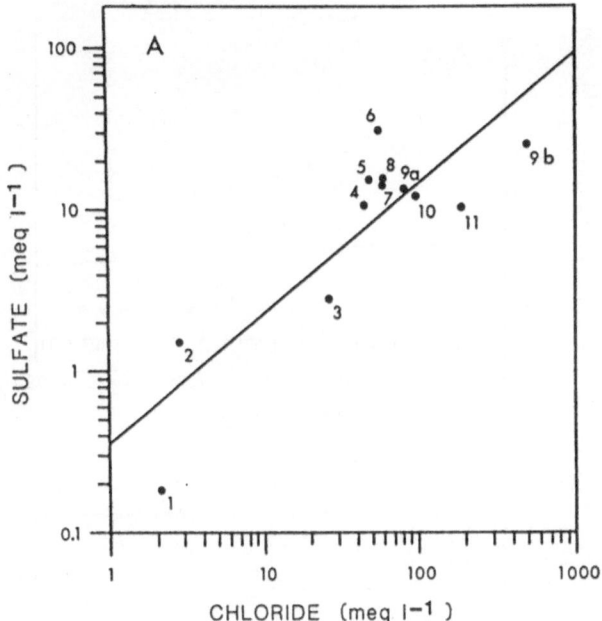

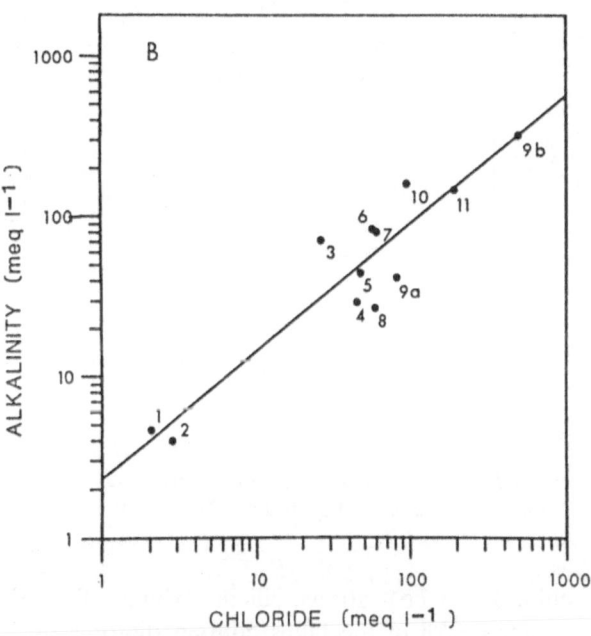

Fig. 6. Plots of sulfate (A) and alkalinity (B) versus chloride for waters of the Basotu Lake District, Tanzania.

values are generally low (Cerling, 1979; Wood & Talling, 1988).

Silicon concentrations remain relatively constant as chloride concentrations increase over two orders of magnitude (Fig. 7A). Over this range of

44

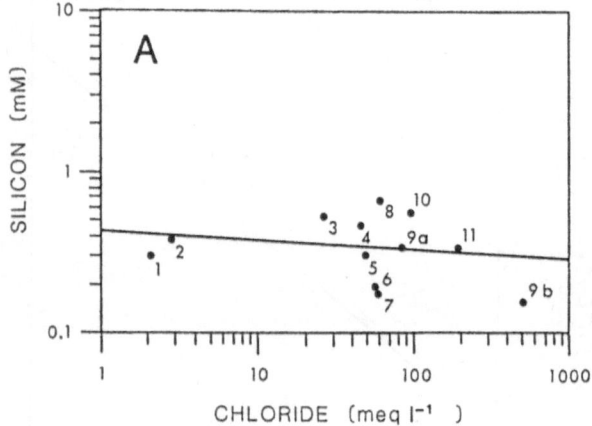

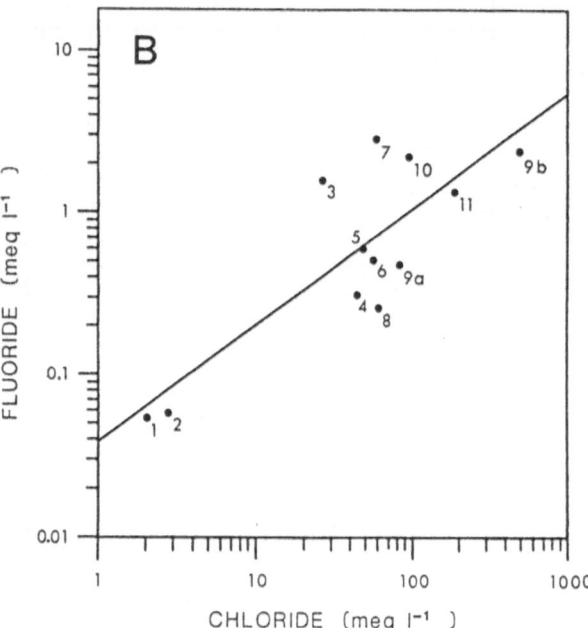

Fig. 7. Plots of silicon (A) and fluoride (B) versus chloride for the waters of the Basotu Lake District, Tanzania.

1973) fluoride behaves more conservatively than it does at Basotu. Some of the variability in our data may be the result of fluorite or fluorapatite precipitation. Saturation with respect to fluorite occurs in some of the Basotu lakes (calculated using WATEQ; Truesdell & Jones, 1974) and fluorite has been found in the sediments of other East African lakes (Cerling, 1979). Jones *et al.* (1977) are equivocal concerning processes removing fluoride in the more dilute waters at Lake Magadi, but they suggest that fluoride-rich waters may interact with minerals such as calcite and gaylussite to produce fluorite in the sediments of concentrated lakes.

Analyses of the Basotu lakes data with respect to the patterns of solute behavior thought to accompany the geochemical evolution of many closed-basin lakes indicate that most of the ions in these waters appear to follow the patterns predicted for a single parent water undergoing evaporative concentration. The major differences observed are for potassium and sulfate which follow the Type III rather than the expected Type IV pattern, presumably because these ions are being removed to a considerable extent in the lake basins themselves. However, there is a great deal of scatter in the plots of all of the ions (except sodium) versus chloride (Figs. 4–6). Much of the observed scatter can be explained geochemically if we examine the evolution of these waters in their individual lake basins.

Evolutionary processes

The ternary diagram (Fig. 8) shows the proportions of major anions. It also indicates the major pathways these waters may have followed as they evolved. All of these waters are thought to have had an initial anionic composition falling on or near the gradual evolution line, but some (stars in Fig. 8) have undergone rapid evolution and thus now fall below the line. Both evolutionary pathways are discussed below. Even though the Basotu lake waters appear to have originated essentially from a single parent water, the chemical compositions of several indicate that they

chloride concentrations, silicon exhibits Type V behavior. All of the lakes contain diatoms and they may be partly responsible for the removal of silicon from solution (Hecky & Kilham, 1973; Gasse *et al.*, 1983).

The plot of fluoride versus chloride shows considerable scatter (Fig. 7B). The slope of this Type III relationship is only 0.71. At Lake Magadi, Kenya (Jones *et al.*, 1977) and in the Momela Lakes, Tanzania (Kilham & Hecky,

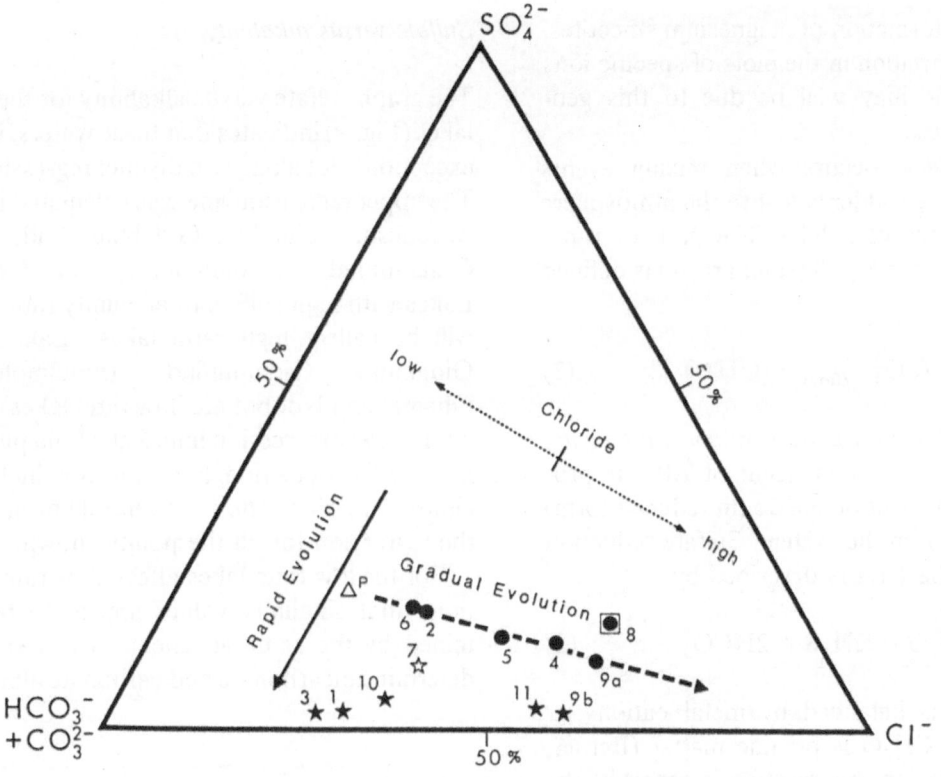

Fig. 8. Ternary diagram showing the proportions of equivalents of major anions for waters in the Basotu Lake District, Tanzania. The circles are high SO_4 : Alk ratio lakes and the solid stars are low SO_4 : Alk ratio lakes. Lake Laja (open star) is intermediate between the two lake types. Evolutionary trends are indicated by arrows. The open triangle gives a hypothetical anionic composition for the initial parent water (P). Lake Gidabuid (circle in square) is unusual. The lake is very steep-sided and well protected from the wind. It is therefore likely that few sediments are in contact with the mixolimnion. Under such conditions, sulfate reduction is probably reduced in the upper region of the lake.

evolved from waters with higher initial concentrations of sodium (and accompanying concentrations of sulfate and alkalinity) relative to chloride (these are referred to as low chloride waters, see Fig. 8). This indication can be seen in Fig. 4A, where the sodium values for lakes Gidamur, Gidaburk, Laja and Ghama lie appreciably above the regression line, which, as noted above, is essentially conservative. They also lie notably above the regression to a line (which fits the data well, but is not shown) constrained to be strictly conservative in sodium relative to chloride.

The term gradual evolution can be used to describe the geochemical pathway recognized by most previous workers to depict the gradual continuous evolution of alkaline, saline lakes. 'Reverse weathering' or neoformation of clay min-

erals is probably an important gradual process in many African lakes, but few quantitative data are available (Carmouze, 1983; Von Damm & Edmond, 1984; Wood & Talling, 1988). The following equation, for example, illustrates the formation of chlorite by the reaction of magnesium with kaolinite (presumably or detrital origin):

$$Al_2Si_2O_5(OH)_4 + 5\ Mg^{2+} + Si(OH)_4 +$$
$$10\ CO_3^{2-} + 5\ H_2O$$
$$\text{(Kaolinite)}$$
$$= Mg_5Al_2Si_3O_{10}(OH)_8 + 10\ HCO_3^- \qquad (1)$$
$$\text{(Chlorite)}$$

This equation indicates that 'reverse weathering' results in the loss of base cations, silicate, and alkalinity from solution. Similar reactions could be written for formation of magnesium smectites. Some of the variation in the plots of specific ions versus chloride may well be due to this geochemical process.

Rapid evolution occurs when mixing events cause H_2S to be suddenly lost to the atmosphere from the surface of a lake. The path of rapid evolution shown in Fig. 8 (solid arrow) is defined by (in eq)

$$(SO_4^{2-})_{initial} - (SO_4^{2-})_{final} = (OH^-)_{EPA} \qquad (2)$$

where EPA stands for *extra permanent alkalinity*. In other words, one equivalent of EPA is produced per equivalent of sulfate (in reduced form) stored or lost from the system. Sulfate reduction by anaerobic bacteria is described by

$$SO_4^{2-} + 2\,CH_2O \rightarrow \Sigma H_2S + 2HCO_3^- \qquad (3)$$

where SO_4^{2-} is balanced by metal cations or hydrogen and CH_2O is organic matter (Berner, 1971). Even though this reaction is generally the primary source of H_2S, some H_2S results from putrefaction (Dunnette *et al.*, 1985). EPA is formed when H_2S is lost to the atmosphere and when metal sulfides precipitate (see Goldhaber & Kaplan, 1974 for representative reaction; Kilham, 1984). Sulfate reduction, net-sulfate (or sulfide) loss, and the formation of EPA occur in all the Basotu lakes, but these processes are closely associated with evaporative concentration in the lakes falling on or above the gradual evolution line in Fig. 8. Mixing events resulting in sudden loss of H_2S (rapid evolution) rather than the evaporative concentration process *per se* have significantly influenced the geochemical evolution of waters falling below the gradual evolution line.

Even though rapid evolution is defined as occurring whenever H_2S is lost rapidly from a lake, the current chemical compositions of the Basotu lakes that have undergone rapid evolution are thought to be the result of numerous mixing events that have taken place over thousands of years.

Sulfate versus alkalinity

The graph sulfate versus alkalinity for the Basotu lakes (Fig. 9) indicates that these waters, with one exception, plot along two distinct regression lines. The upper regression line was calculated for lakes Basodesh, Ndanakid, Gidabuid, Gidabugarak, Gidamuniud (mixolimnion), and Gidaburk. Lakes with high sulfate to alkalinity ratios (in eq) will be called 'high ratio lakes'. Lake Basotu, Gidamur, Gidamuniud (monimolimnion), Ghama, and Ndobot are 'low ratio lakes' that fall on the lower regression line. Lake Laja plots in an intermediate position and was not included in either regression. The solid central line in Fig. 9 is the regression for all the points shown.

For the low ratio lakes affected by rapid evolution, final alkalinity values appear to be determined by the same geochemical processes that determine bicarbonate and carbonate alkalinity at

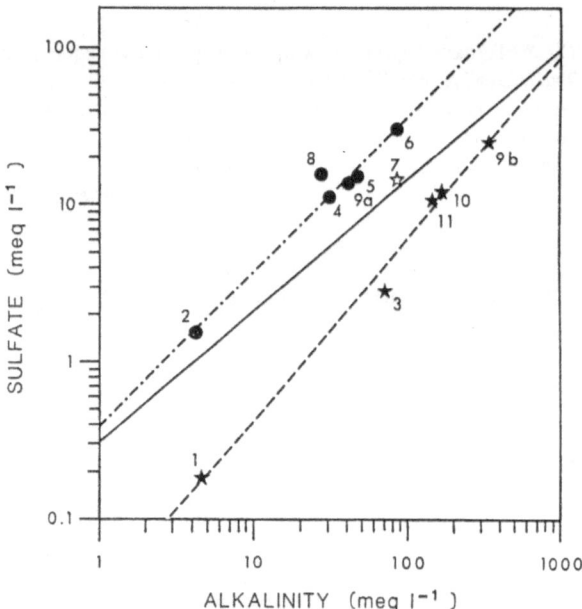

Fig. 9. Plot of sulfate versus alkalinity (Alk) for the Basotu lakes, Tanzania. The solid regression line includes all points (see Table 3). The upper regression line ($(SO_4) = 0.41\,(Alk)^{0.98}$, $r = 0.98$) encompasses the high ratio waters (circles). The lower regression line ($(SO_4) = 0.03\,(Alk)^{1.17}$, $r = 0.99$) encompasses the low ratio waters (solid stars). Lake Laja (open star) was not included in either regression (see text).

any given stage of evaporative concentration. Chloride enrichment (an increase in the molar ratio

$$(Cl^-)/((HCO_3^- + CO_3^{2-}) + (SO_4^{2-}) + (Cl^-))$$

with increasing concentration, as indicated by the chloride, (Eugster & Jones, 1979) is described by $0.12\ (Cl)^{0.29})$ $(r = 0.97)$ for the more saline low ratio lakes (Lake Basotu which contains freshwater was not included). Thus, chloride enrichment behaves as a regularly increasing function of chloride concentration. Chloride is enriched as carbonate and sulfate ions are lost from solution (Kilham, 1971a; Eugster & Jones, 1979). Carbonate precipitation (primarily calcite in the Basotu lakes, but a small amount of Mg-calcite (unspecified) was found in sediments of Lake Gidaburk, P. Stoffers, pers. comm.) decreases the total alkalinity.

The correlation between sulfate and alkalinity found for the low ratio lakes results primarily from the interplay between increases in the total alkalinity of a particular lake water (owing to EPA production), calcite precipitation, and magnesium loss. Because the Basotu lakes are supersaturated with respect to calcite (by a factor of ca. 10; WATEQ, Truesdell & Jones, 1974) most of the EPA produced is probably precipitated as calcite. The correlation between sulfate and alkalinity is not surprising, as Eq. 2 indicates. The final alkalinity is determined by the hydrolysis of alkaline silicates, calcite precipitation, magnesium loss, and the rate at which calcium and magnesium are supplied (from internal and external sources) to the lake.

Calcite precipitation

Calcite precipitation which is described by

$$Ca^{2+} + 2\,HCO_3^- \rightarrow CaCO_3 + CO_2 + H_2O$$

is particularly evident in the lakes strongly affected by sulfide loss. Calcium concentrations at every stage of evaporative concentration are lower in these lakes (Fig. 10). The relationship between calcium and chloride for the four more

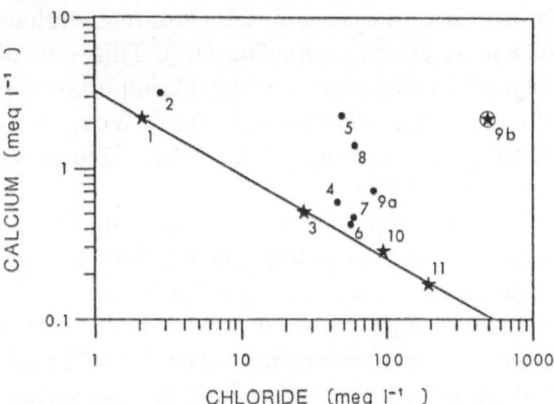

Fig. 10. Plot of calcium versus chloride for the Basotu lakes, Tanzania. The regression line is for four of the five low SO_4 : Alk waters (stars). Lake Gidamuniud (monimolimnion, circled star) has an anomalously high calcium concentration. The remaining waters (circles) include Lake Laja and the high SO_4 : Alk ratio lakes.

dilute lows ratio lakes is $3.3\ (Cl)^{-0.56}$ $(r = -1.0;$ in meq l^{-1}). Only Lake Gidamuniud (monimolimnion, 9b) does not fit this relationship (see below). Correlation coefficients for calcium and alkalinity $(r = -0.96)$ and calcium and sulfate (-0.98) are also very high. These data indicate that calcite precipitation induced by EPA production tends to smooth out much of the variability in the relationship between alkalinity and chloride or sulfate potentially caused by vicissitudes in the supply of calcium to the lakes, but other factors are probably also involved.

Magnesium, in addition to calcium, is being lost from solution somewhat more rapidly in the low ratio lakes than in the high ratio lakes. Magnesium concentrations for these waters plot on or below the regression line in Fig. 5B. Müller *et al.* (1972) have shown that the types of Ca-Mg carbonates that precipitate in lakes are largely controlled by the Mg : Ca ratio of the ambient water. Because the Mg : Ca ratios of the Basotu lakes range between 0.48 and 29.4 a variety of Ca-Mg carbonates can occur (i.e. low Mg-calcite, high Mg-calcite, aragonite, and sedimentary dolomite). Our sediment samples, which were admittedly limited in number, contained only calcite and Mg-calcite (unspecified; P. Stoffers, pers. comm.). Magnesium is also removed during the neoformation of magnesium-rich clays (e.g. various smectites, chlorite, sepiolite, etc.). This sink of magnesium may be of considerable importance in some alkaline, saline lakes (Jones & Weir, 1983; Yuretich & Cerling, 1983; Von Damm & Edmond, 1984).

We know that two of the water samples we obtained were originally anoxic, but became oxidized as a consequence of collection. Lake Gidamur had just turned over when it was visited and the surface water contained ca. 0.4 mM H_2S. The loss of hydrogen sulfide from the lake surface must have been enormous because the rotten-eggs smell was still detectable more than 5 km away. The surface of the lake was purple presumably due to a bloom of photosynthetic sulfur bacteria. The reduced monimolimnetic water of lake Gidamuniud also smelled strongly of H_2S.

Both of these waters are slightly anomalous. The monimolimnetic water of Lake Gidamuniud plots in the regression line of sulfate versus alkalinity for the low ratio lakes (Fig. 9). But, this water has a comparatively high calcium to chloride equivalent ratio (0.0042). The calcium concentration (2.03 meq 1^{-1}) found is not easy to explain given the data available, but it may be the result of plagioclase dissolution. The surface water of Lake Gidamur is greatly enriched in alkalinity (the Alk : Cl ratio is the highest of all the Basotu lakes: 2.64). This indicates that there probably had not been sufficient time since the lake turned over for calcium inputs to the lake to supply enough calcium to precipitate the EPA generated by the loss of sulfides from solution.

Owing to time constraints, the sample designated Lake Gidamuniud, monimolimnion (Table 2, No. 9b) was the only sample we obtained from an anoxic bottom water. The hydrogen sulfide concentration of this water was very high (> 9 mM). Owing to the insolubility of metal sulfides there is an inverse relationship in anoxic waters between the hydrogen sulfide concentration and the concentrations of metals that precipitate as metal sulfides (Fe^{2+}, Mn^{2+}, Zn^{2+}, etc.: Hutchinson, 1957; Cook, 1984; Nürnberg, 1984). This relationship indicates that ambient metal concentrations were probably extremely low in the monimolimnion of Lake Gidamuniud and that the flux of metal ions into this layer was not sufficient to check gradual increases in the concentration of hydrogen sulfide. Unlike metal sulfides, H_2S becomes increasingly soluble as alkalinity increases. Thus, very high concentrations of H_2S are sometimes observed in the anaerobic waters of meromictic alkaline, saline lakes (Hutchinson, 1957; Nürnberg, 1984).

Concluding remarks

Because rapid evolution will occur in any body of water that loses H_2S to the atmosphere, this phenomenon is undoubtedly widespread. Lakes in Indonesia, for example, mix only rarely, but when they do fish are sometime killed and the odor of H_2S is perceptible in the air (Ruttner, 1931; Hutchinson, 1957, p. 773). Similar events are known to occur frequently at Lake Bosumtwi, Ghana (Whyte in Beadle, 1981). MacIntyre & Melack (1982) also smelled H_2S during a mixing event in Lake Sonachi, Kenya. Mixing events in temperate freshwater lakes also cause H_2S to be released to the atmosphere (Folt et al., 1989). The process of rapid evolution, though it was not called that, was studied by King et al. (1974) in a series of laboratory microcosms which initially contained acidic water from a strip mine lake. They recognized the potential importance of rapid evolution in the alkalization of lakes affected by acid mine drainage.

The importance of sulfate reduction, sulfide loss, and the formation of permanent alkalinity in the biogeochemical evolution of lake waters have been recognized previously by Holland & Christie (1909; Lake Sambhar, India), Anderson (1945; Lake Corangamite, Australia), and Abd-el-Malek & Risk (1963, Wadi Natrum, Egypt). The biogeochemical sections of these papers are largely qualitative which may explain why they are not better known to limnologists.

Half of the Basotu lakes appear to have evolved largely according to the same slow biogeochemical processes thought to govern the gradual evolution of closed-basin lakes everywhere, but the other half have not. They have evolved rapidly at times. The phenomenon of rapid (punctuated) evolution probably occurs in many lakes worldwide and it may explain why plots of sulfate *versus* alkalinity or chloride often contain considerable scatter that has proven very difficult to interpret.

Hydrogen sulfide is a deadly gas, but we are not aware that the rapid degassing of lakes such as those at Basotu has ever killed anyone. However, until recently, one might have made the same observation for the carbon dioxide that escaped from lakes Nyos and Monoun in Cameroon, West Africa (Kling, 1987).

Acknowledgements

In Tanzania, this study was made possible by the National Parks System and the Mineral Resources Division. In East Africa valuable field assistance was supplied by the East African Fisheries Research Organization while it was under the direction of J. Okedi. We thank R.M. Baxter, B.F. Jones, S.S. Kilham, R.F. Stallard, and R.E. Stauffer for comments and suggestions. S.S. Kilham, C.M. Hecky, R.E. Hecky, T.J. Harvey, and D.A. Livingstone are thanked for their help and support in the field. This work was supported by NSF grants GB-8328X (to D.A. Livingstone) and OCE81-17377 (from the Oceanography Section to S.S. & P. Kilham). Final preparation of this paper was supported by the Max-Planck-Institut für Limnologie, Plön, West Germany and an Alexander Von Humboldt Fellowship (Senior U.S. Award) to the senior author.

References

Abd-el-Malek, Y. & S. G. Risk, 1963. Bacterial sulfate reduction and the development of alkalinity. J. appl. Bact. 26: 20–26.

Anderson, V. G., 1945. Some effects of atmospheric evaporation and transpiration on the composition of natural waters in Australia, 3. The waters of interior drainage catchments. J. Proc. Aust. chem. Inst. 12: 60–68.

American Public Health Association, 1965. Standard methods for the examination of water and wastewater, APHA, N.Y., 769 pp.

Arad, A. & W. H. Morton, 1969. Mineral springs and saline lakes of the western Rift Valley, Uganda. Geochim. cosmochim. Acta 33: 1169–1181.

Beadle, L. C., 1981. The inland waters of tropical Africa. 2nd ed., Longman, London. 475 pp.

Berner, R. A., 1971. Principles of chemical sedimentology. McGraw-Hill, New York.

Carmouze, J.-P., 1983. Hydrochemical regulation of the lake. In J.-P. Carmouze, J.-R. Durand & C. Lévêque (eds.), Lake Chad, Ecology and Productivity of Shallow Tropical Ecosystem. Dr. W. Junk, The Hague: 95–123.

Cerling, T. E., 1979. Paleochemistry of Plio-Pleistocene Lake Turkana, Kenya. Paleogeogr. Paleoclimatol. Paleoecol. 27: 247–285.

Cook, R. B., 1984. Distributions of ferrous iron and sulfide in an anoxic hypolimnion. Can. J. Fish. aquat. Sci. 41: 286–293.

Dawson, J. B., 1964. Carbonatitic volcanic ashes in Northern Tanganyika. Bull. volcan. 27: 81–91.

Downie, C. & P. Wilkinson, 1962. The explosion craters of Basotu, Tanganyika Territory. Bull. volcan. 14: 389–420.

Drever, J. I., 1988. The geochemistry of natural waters. 2nd ed. Prentice-Hall, Englewood Cliffs. 437 pp.

Dunnette, D. I., D. P. Chynoweth & K. H. Mancy, 1985. The sources of hydrogen sulfide in anoxic sediment. Wat. Res. 19: 875–884.

Eades, N. W. & W. H. Reeve, 1938. Explanation of the geology of degree sheet No. 29 (Singida). Bull. geol. Div. Tanganyika Dep. Lands Mines 11: 5–59.

Eugster, H. P., 1970. Chemistry and origin of brines of Lake Magadi, Kenya. Spec. Pap. mineralog. Soc. Am. 3: 215–235.

Eugster, H. P. & B. F. Jones, 1979. Behavior of major solutes during closed-basin brine evolution. Am. J. Sci. 279: 609–631.

Eugster, H. P. & G. Maglione, 1979. Brines and evaporites of the Lake Chad basin, Africa. Geochim. cosmochim. Acta 43: 973–981.

Folt, C. L., M. J. Wevers, M. P. Yoder-Williams & R. P. Howmiller, 1989. Field study comparing growth and viability of a population of phototrophic bacteria. Appl. envir. Microbiol. 55: 78–85.

Gasse, F., J. F. Talling & P. Kilham, 1983. Diatom assemblages in East Africa: classification, distribution, and ecology. Revue Hydrobiol. trop. 16: 3–34.

Garrels, R. M. & F. T. Mackenzie, 1967. Origin of the chemical composition of some springs and lakes. In Equilibrium Concepts in Natural Water Systems. Adv. Chem. Ser. 67: 222–242.

Goldhaber, M. B. & I. R. Kaplan, 1974. The sulfur cycle. In E. D. Goldberg (ed.), The Sea, 5. Interscience, N.Y.: 569–655.

Hardie, L. A. & H. P. Eugster, 1970. The evolution of closed-basin brines. Spec. Pap. mineralog. Soc. Am. 3: 273–290.

Hecky, R. E. & P. Kilham, 1973. Diatoms in alkaline, saline lakes: Ecology and geochemical implications. Limnol. Oceanogr. 18: 53–71.

Heinrich, E. W., 1966. The geology of carbonatites. Rand McNally & Co., Chicago.

Holland, T. H. & W. A. K. Christie, 1909. The origin of the salt deposits of Rajputana. Rec. geol. Surv. India 38: 154–186.

Hutchinson, G. E., 1957. A treatise on limnology, 1. J. Wiley & Sons, N.Y., 1015 pp.

Jones, B. F., H. P. Eugster & S. L. Rettig, 1977. Hydrogeochemistry of the Lake Magadi basin, Kenya. Geochim. cosmochim. Acta 41: 53–72.

Jones, B. F. & A. H. Weir, 1983. Clay minerals of Lake Abert, an alkaline, saline lake. Clays Clay Mineral. 31: 161–172.

Kilham, P., 1971a. The geochemical evolution of closed basin lakes. Abstrs. Progms. geol. Soc. Am. 3(7): 770–772.

Kilham, P., 1971b. Biogeochemistry of African lakes and rivers. Ph.D. thesis, Duke Univ., Durham (N.C.), 199 pp.

Kilham, P., 1984. Sulfate in African inland waters: sulfate to chloride ratios. Verh. int. Ver. Limnol. 22: 296–302.

Kilham, P. & R. E. Hecky, 1973. Fluoride: Geochemical and ecological significance in East African waters and sediments. Limnol. Oceanogr. 18: 932–945.

King, D. L., J. J. Simmler, C. S. Decker & C. W. Ogg, 1974. Acid strip mine lake recovery. J. Wat. Pollut. Cont. Fed. 46: 2301–2316.

Kling, G. W., 1987. Seasonal mixing and catastrophic degassing in tropical lakes, Cameroon, West Africa. Science 237: 1022–1024.

MacIntyre, S. & J. M. Melack, 1982. Meromixis in an equatorial African soda lake. Limnol. Oceanogr. 27: 595–609.

Mackereth, F. J. H., 1963. Some methods of water analysis for limnologists. Scient. Publ. Freshwat. biol. Ass. 21. 71 p.

Müller, G., G. Irion & U. Förstner, 1972. Formation and diagenesis of inorganic Ca-Mg carbonates in the lacustrine environment. Naturwissenschaften 59: 158–164.

Nürnberg, G., 1984. Iron and hydrogen sulfide interference in the analysis of soluble reactive phosphorus in anoxic waters. Wat. Res. 18: 369–377.

Pappe, A. & H. D. Richmond, 1890. A Central African salt lake. J. Soc. chem. Ind. Lond. 9: 734.

Perkin-Elmer, 1964. Analytical methods for atomic absorption spectrophotometry. Perkin-Elmer, Norwalk. Looseleaf, unpaginated.

Prosser, M. V., R. B. Wood & R. M. Baxter, 1968. the Bishoftu crater lakes: A bathymetric and chemical study. Arch. Hydrobiol. 65: 309–324.

Rodhe, H. & H. Virji, 1976. Trends and periodicities in East African rainfall data. Mon. Weath. Rev. U.S. Dep. Agric. 104: 307–315.

Ruttner, F., 1931. Hydrographische und hydrochemische Beobachtungen auf Java, Sumatra und Bali. Arch. Hydrobiol. Suppl. 8: 197–454.

Singer, A. & P. Stoffers, 1980. Clay mineral diagenesis in two East African lake sediments. Clay Mineral. 15: 291–307.

Stanley, H. M., 1890. In darkest Africa, 2. Charles Scribner's Sons, N.Y., 540 pp.

Talling, J. F. & I. B. Talling, 1965. The chemical composition of African lake waters. Int. Revue ges. Hydrobiol. 50: 421–463.

Truesdell, A. H. & B. F. Jones, 1974. WATEQ, a computer program for calculating chemical equilibria of natural waters. J. Res. U.S. geol. Serv. 2: 233–248.

Von Damm, K. L. & J. M. Edmond, 1984. Reverse weathering in the closed-basin lakes of the Ethiopian Rift and in Lake Turkana (Kenya). Am. J. Sci. 284: 835–862.

Wood, R. B. & J. F. Talling, 1988. Chemical and algal relationships in a salinity series of Ethiopian inland waters. Hydrobiologia 158: 29–67.

Yuretich, R. G. & T. E. Cerling, 1983. Hydrogeochemistry of Lake Turkana, Kenya: Mass balance and mineral reactions in an alkaline lake. Geochim. cosmochim. Acta 47: 1099–1109.

Hydrobiologia **197**: 51–66, 1990.
F. A. Comín and T. G. Northcote (eds), Saline Lakes.
© 1990 *Kluwer Academic Publishers.*

Hydrogeochemistry of Lake Gallocanta (Aragón, NE Spain)

F.A. Comín[1], R. Julià[2], M.P. Comín[1] & F. Plana
[1]*Department of Ecology. University of Barcelona. Diagonal 645. 08028 Barcelona, Spain;* [2]*Institute of Geology Jaume Almera. CSIC. Marti i Franqués s/n. 08028 Barcelona, Spain*

Key words: saline lake, hydrology, geochemistry

Abstract

Lake Gallocanta has undergone drastic changes during the last thirteen years. Water level changed from a high level (Z_{max} = 2 m) to total dryness in 1985. From 1986 to 1988 slow refilling occurred. The water volume fluctuations have been studied in relation to climatic variations recorded for that period. Variations in the major dissolved ions were related to water volume fluctuations from data at two different stages, one corresponding to the drying phase and another to the refilling phase. Mineralogical composition of the salts precipitated at different stages was examined by X-ray diffraction. Interstitial water and mineralogical composition of recent sediments were also studied along a transect through the lake.

The water column decrease from 1977 to 1985 is related to decreasing annual rainfall (500–250 mm respectively). The refilling in 1986-1988 is due to high annual rainfall (537 mm). In addition to these fluctuations, seasonal changes of the water level between 20 and 50 cm occurred every year.

Gallocanta is a Na-Mg-Cl-(SO_4) type lake. During the drying period a typical salt enrichment occurs with linear relationships between TDS, Cl, Na and K. Alkalinity is linearly correlated with Ca at relatively low salinities. As salinity increases a linear relationship between Ca and SO_4 is observed. Minerals formed from the brine are halite, bischofite, epsomite, hexahydrite, mirabilite, gypsum, aragonite, calcite and dolomite. The molar ratio Mg/Ca of the interstitial water changes from 1.5 along the shorelines, where calcite and aragonite precipitate, to 40 in the center of the lake. Sediment cores from the central part of the lake show aragonite in the top layers, magnesian calcite and low proportions of quartz and illite, while at 20 cm depth a high proportion of gypsum is present. In contrast, cores from the shore of the lake are mainly composed of low magnesium calcite in the top layers and low magnesium calcite together high magnesium calcite and dolomite between 30 and 70 cm depth. Gypsum deposits only occur in significant proportions at 80–100 cm depth.

The refilling process showed relationships between volume and salt concentration following the Langbein model. The salt mass in solution decreased about 50% from the drying to the refilling phase. However, Mg content decreased about 70% for the same period, suggesting a contribution of this element to the dolomite formation.

Introduction

Changes in saline lakes are strongly dependent on climatic factors. Seasonal and long term chemical changes in many saline lakes have been reported before (Bayly & Williams, 1966; Bayly 1970; Walker, 1973; Hammer, 1986). Geochemistry of saline waters has been described from many

52

endorheic regions. An extensive review of the geochemical processes in saline lakes with many examples from different parts of the world was provided by Eugster & Hardie (1978). However, most of the studies put together data collected from different lakes in a region or refer to a short period of time. The presentation of both hydrological, and geochemical changes in saline lakes has been the focus of several works (Langbein, 1961; Phillips & Van Denburgh, 1971; Kalk *et al.*, 1979).

This paper presents the hydrology and geochemistry of Gallocanta Lake for the last ten years. From 1977 on its water level decreased. The lake dried up completely in 1983-1986 and a slow refilling began in 1986. Rainfall, water level fluctuations and some chemical characteristics of

the water, together with sediment analysis permit an approach to the general geochemical changes occurring in the lake in relation to climatic factors. Previous information on Lake Gallocanta is given by Comín *et al.* (1983).

Material and methods

Rainfall was determined from the monthly average of five meteorological stations equipped with standard rain gauges and located around the lake all within 3 km of the lake shoreline. No significant differences were observed between stations. Monthly evaporation was estimated by the water volume decrease during months without rain. No estimates of groundwater inputs were done.

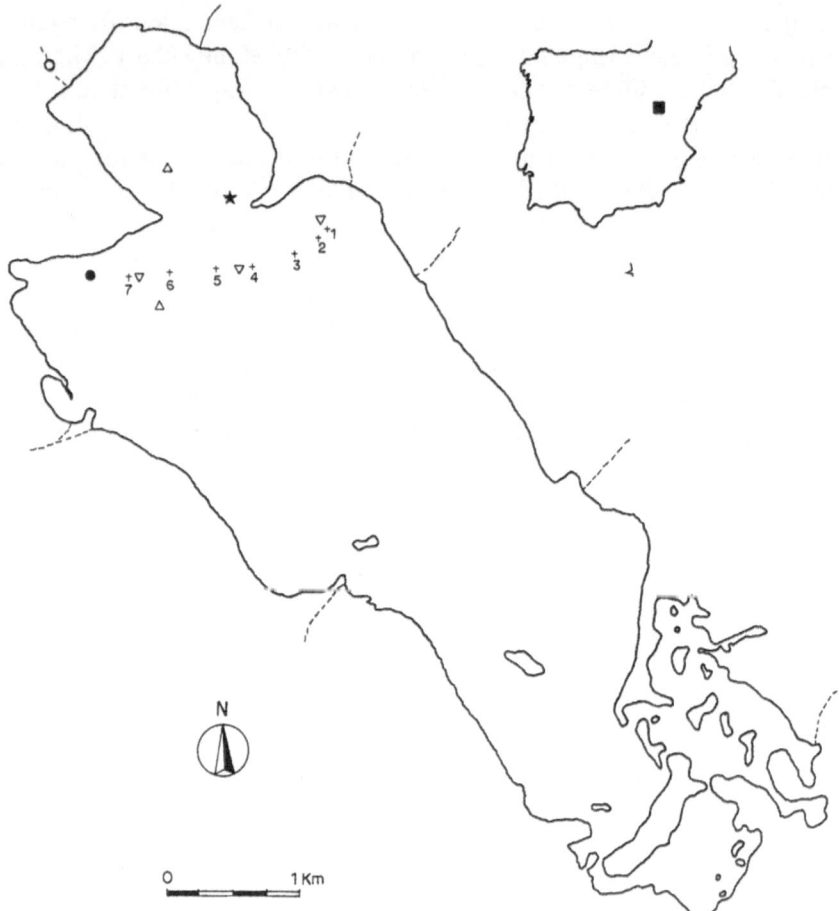

Fig. 1. Map of Lake Gallocanta showing the sampling points (△ surface water; ● pond; + pore water; ★ water level records; ▽ sediment cores). The inset shows the location of Lake Gallocanta in the Iberian Peninsule.

53

Water level is the height of the water as recorded visually from a graduated stick at the deepest part of the lake (Fig. 1). Volume of the water was deduced indirectly from aerial photographs at different stages of the lake (years 1957, 1977, 1983) and water level at the time of the photographs.

Water samples were collected directly from the lake at 20 cm depth monthly from September 1980 to October 1981 and from November 1986 to December 1987. Water samples were also collected from a stream and a pond located close to the main lake body at low water level (Fig. 1). Samples were filtered through glass fiber filters (Whatman® GF/C, preserved with 1 ml chloroform per liter of water (anions) or with HCl to pH 1 (cations). Cations were measured by atomic absorption (calcium, magnesium) or emission (sodium, potassium). Anions were determined volumetrically after titration with strong acid (alkalinity) (Golterman $et\,al.$, 1978), $AgNO_3$ (chloride) (APHA, 1980) or $Ba(ClO_4)_2$ (sulfate) (Fritz & Yamamura, 1955). Total dissolved solids were measured after water evaporation at 110 °C. Data are the average of sample analysis from two

localities (Fig. 2). No significant differences were observed between localities, variance being less than 2% of the average. Samples for pore water were collected by pumping with a peristaltic pump the seepage water that refilled holes excavated in the bottom mud (Fig. 1).

Mineralogical analyses were done from material retained in glass fiber filters (Whatman® GF/C) after filtering a volume of surface water and from sediment samples collected with a manual drilling core. The analyses were carried out by X-ray diffraction using copper radiation (Martinez & Plana, 1987).

Results

Recent lake level fluctuations

For the last eleven years Lake Gallocanta underwent a distinct level fluctuations related to drought. In 1977 the maximum depth was almost 2 m. Maximum annual depth decreased progressively from 1977 to 1983 (Fig. 2). During 1983–1986 the lake became completely dry

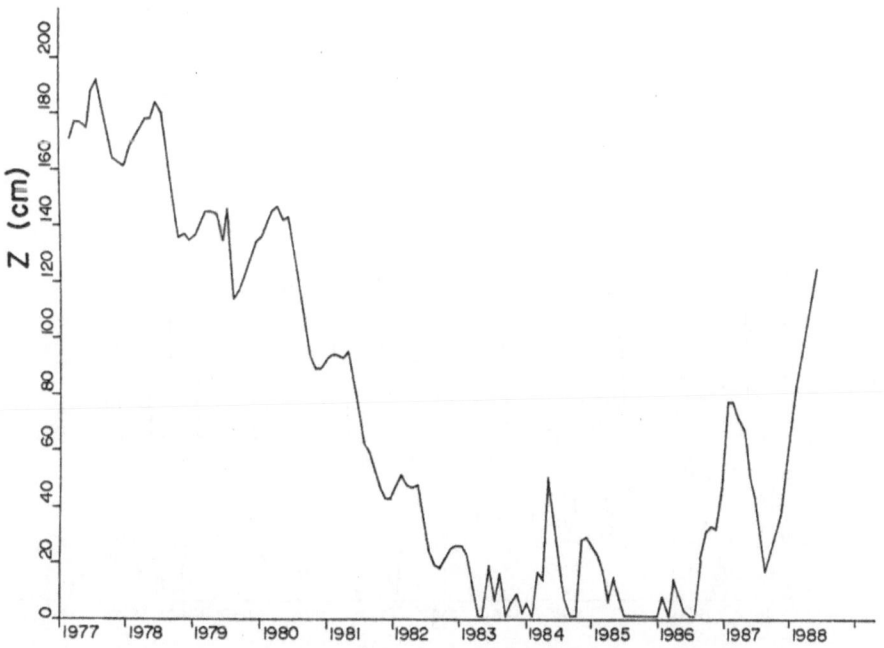

Fig. 2. Water level fluctuations in Lake Gallocanta, 1977–1988.

54

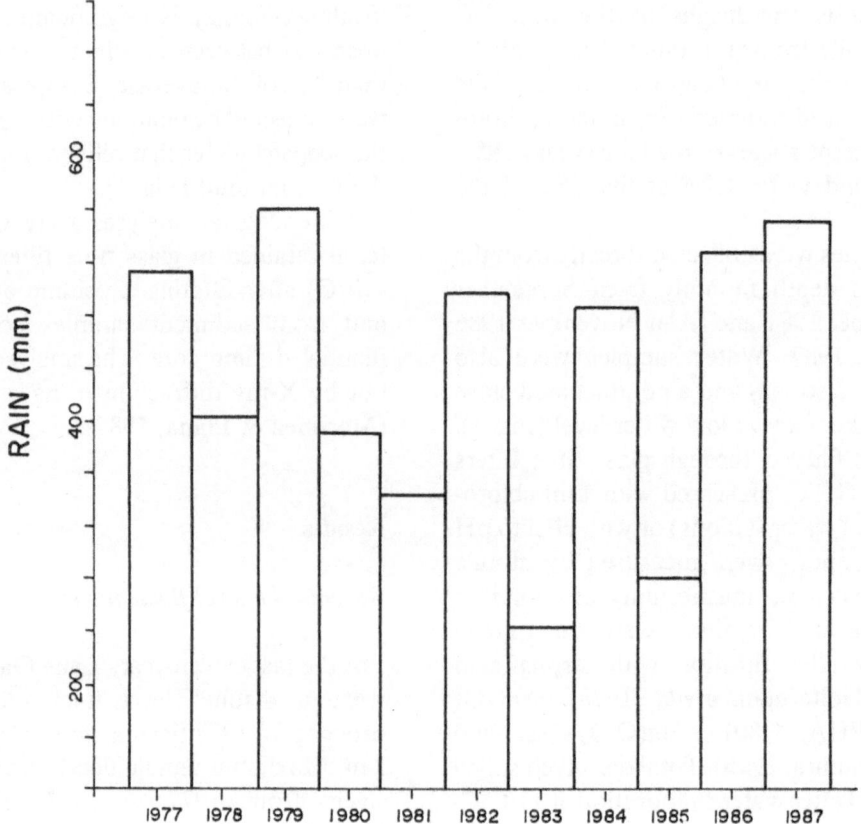

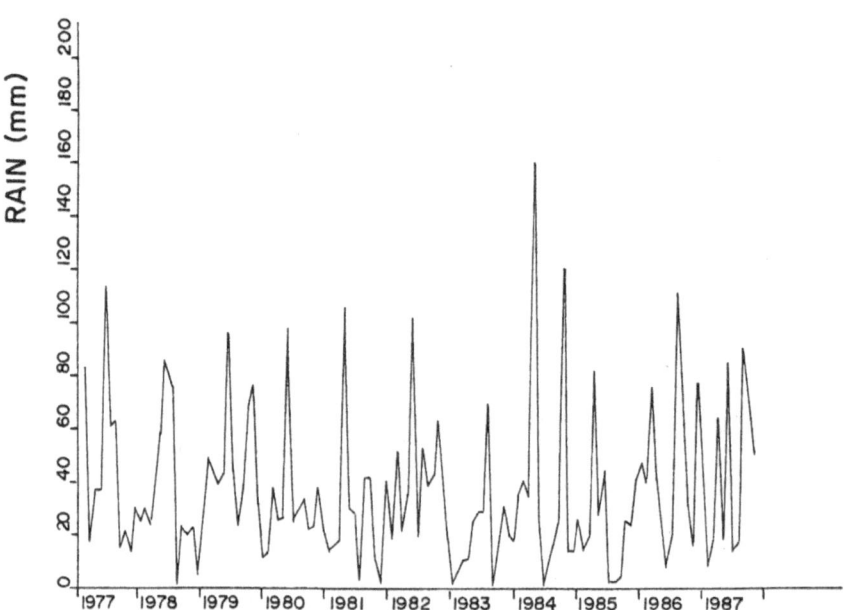

Fig. 3. Rainfall at Lake Gallocanta: a) annual, b) monthly.

several times. Occasionally a small area (less than 1000 m^2) was filled with brine about 20 cm deep. From September 1986 to June 1987 the water level increased 80 cm and by June 1988 the lake level was 120 cm (Fig. 2).

Seasonal fluctuations in level occurred every year following seasonal changes in the precipitation/evaporation budget. The maximum water level ocurred every year in early-mid spring (April–May). The minimum water level ocurred in early-mid autumn. The range of the fluctuation was between 20 and 60 cm, depending on the annual rainfall pattern.

The average annual rainfall during thirty-six years (1944–1980, data recorded at a locality 1 km south to the lake) was 488 mm. The maximum rainfall was 698 mm and the minimum 267. For the 1977–1987 period the average was 434 mm ($\pm$ 106 SD); the standard deviation was about 25% of the average annual rainfall. Year to year rainfalls are very irregularly distributed. Obviously (Fig. 3a) there is a large year-to-year variation in rainfall.

Maximum annual rainfall occurred in 1979 and 1987 (Fig. 3a) with minimum rainfalls in 1983 and 1985. Rainfalls in 1982 and 1984 were 495 and 485 mm respectively, about average for the longer period recorded. In spite of this, the lake dried up in April 1983 and July 1985. Rainfall in late spring-summer of 1982 and 1984 (Fig. 3b) accounted for 49 and 46% of the annual rainfall respectively. There was a large deficit in autumn and winter rains for the other years. However, the estimated evaporation from June to August was 0.5 m and rainfall for March–August was 250 mm, both in 1982 and 1984. The maximum monthly rainfall during spring was 159 mm (June 1984) and evaporation during the same month was 140 mm. These data indicate that rainfall during late spring-summer did not contribute to the water volume of the lake. In fact, the water level decreased dramatically after heavy rains in May 1982 and 1984 (Fig. 2 and 3b). Correlation analysis between monthly rainfall and water level show no significant correlation.

Wind is another important factor contributing to decreases in the lake water level. Strong winds blowing during a single day can decrease the water level about 3 cm. During low water level periods ($z = 20$–40 cm) strong and dry NW winds laterally displace the water mass onto dried up shore areas. This also contributes to decrease in water level by increasing evaporative surface, and promoting filtration through dry sediments.

The time response of Lake Gallocanta, calculated after Langbein (1961), for the differences in water volume between successive years of the decade 1977–1987 is 0.3–2.4 years. This range of values means that lake Gallocanta can fill or dry up in between 4 and 27 months depending on the previous stage. For example, four months is the time required for the lake at low level to dry up in years of irregularly distributed low rainfall as in 1983–1986. Twenty seven months would be the time for the lake at low level to refill to high water level if annual rainfall were high and, more importantly, distributed mainly during autumn and winter.

Lake water

Differences in ion concentrations between lake surface, pond, stream and rain water are very clear (Table 1). Rain water contained very low

Table 1. Range in total dissolved solids and major ion concentrations from different water sources in the Lake Gallocanta system, November 1986 to December 1987.

	Rain ($n = 2$)	Stream ($n = 13$)	Pond ($n = 13$)	Lake water ($n = 28$)
g l^{-1}				
TDS	.02–.10	.1– .6	0.30– 0.91	22– 150
meq l^{-1}				
Na$^+$	.73–.91	1.2–1.3	8.7 –17.2	317–1610
Mg^{2+}	.01–.05	1.1–1.9	2.7 – 5.7	115–860
K$^+$	.26–.41	.5–1.7	0.94– 2.1	6.4–23.2
Ca^{2+}	.15–.35	2.4–3.9	2.7 – 4.7	14– 43
Cl$^-$	.02–.19	0.7–2	2.3 – 8.2	293–1894
SO$_4^{2-}$	.09–.15	1.3–2.1	2.1 – 3.1	120– 570
Alk	.28	1.8–3.5	3.7 – 4.5	1.2–8.2

56

amounts of salts compared to Lake water. However, during some months of the year rain water contained a higher concentration of chloride and sodium than those of other ions. This occurred about the end of the summer, a period where the atmosphere probably contained more particles and aerosols from the sea. Stream water also contained relatively low amounts of solids and salts. The ionic proportions followed the freshwater type (Ca > Mg > Na > K; $CO_3 > SO_4 > Cl$). Pond water (water accumulated in a small spring close to the main lake body) (Fig. 1) showed higher concentrations of

dissolved solids and ions and a higher proportion of sodium and chloride than those of rain and stream water. The lake water had a relative by high ion abundance of the type Na > Mg > Ca > K; Cl > SO_4 > CO_3 and a TDS between one hundred and one thousand times higher than the water entering the lake. Only in October and November 1987, at the highest TDS, was the equivalent concentration of magnesium higher than those of other cations.

It is obvious from the lake sediment composition and salt precipitates analyzed during the dry years that most of the salts in the lake water come

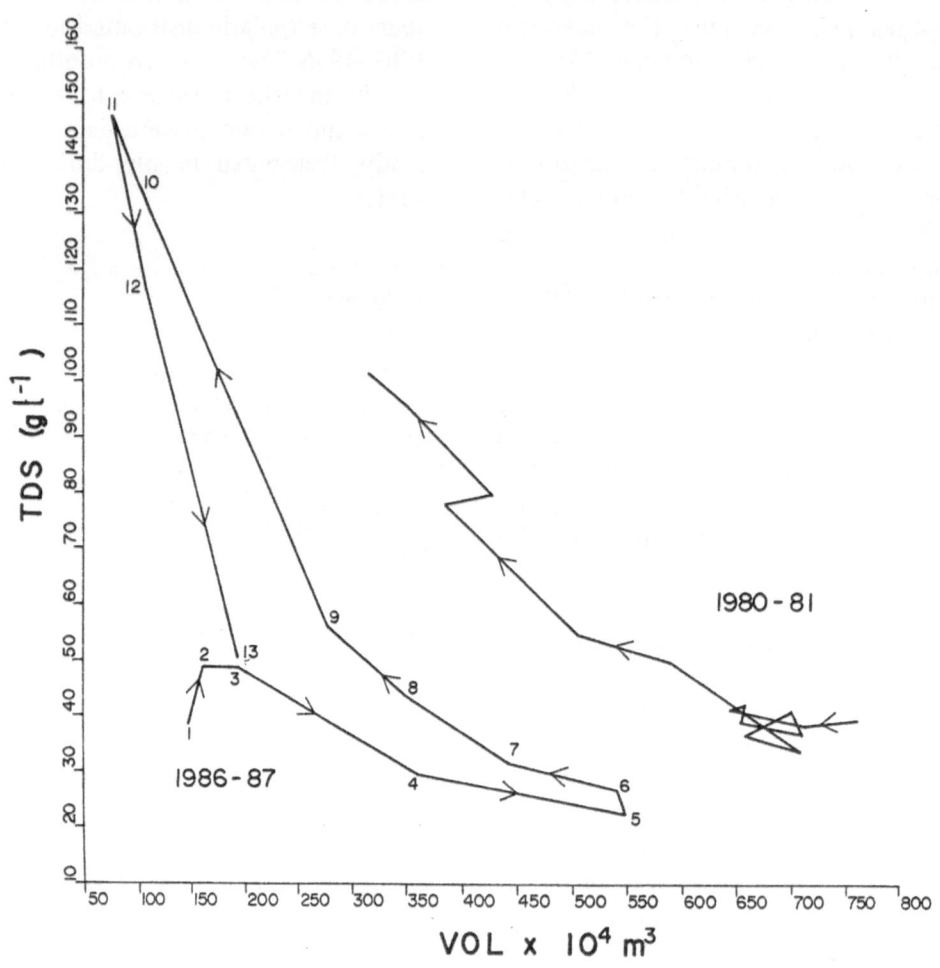

Fig. 4. Total dissolved solids versus water volume during the study period (October 1980–November 1981 and November 1986–December 1987) in Lake Gallocanta. The sampling dates during the second period of study are indicated: 1 = November, 2 = December, 3 = February, 4 = March, 5 = April, 6 = May, 7 = June, 8 = July, 9 = August, 10 = September, 11 = October, 12 = November, 13 = December.

from redissolution of salts previously precipitated. If one compares the amount of salts in the rain water with that in the lake for the period 1986–1987, only about 4% (probably an overestimate) of the TDS in the lake can be attributed to rainfall sources. The same type of calculation (based on differences of salts mass between 1981 and 1987) gives 60 000 years as the time necessary for the lake to increase the chloride concentration from 45 to 70 meq l^{-1}, if all environmental factors remained constant as in 1987.

The amount of total dissolved solids in the water of Lake Gallocanta changed in accordance with the general model of Langbein (1961) for saline lakes. Water volume and TDS fluctuated in opposite ways for most of the period studied (Fig. 4). This pattern is typical of gain-loss salt processes at relatively high water levels controlled mainly by dilution due to water entering the lake and concentration of salts by evaporation. The former occurred from 2 to 4 (Fig. 4) in 1986–1987 and the latter from 4 to 11 in 1986–1987 and during 1980–1981, with some obvious exceptions.

The increasing TDS concentration and decreasing water volume observed during 1981 continued as the lake became completely dry during 1983–1986. The highest TDS concentrations, measured closely to the driest phase, were between 500–600 g l^{-1} in August 1984.

The direct relation between both variables in November–December 1986 (1–2 in Fig. 4) is explained by dissolution of the salts precipitated during the dry phase (1983–1986). The 'salting level' (increasing salt concentration period) was reached after heavy rains in autumn, four months

being the time necessary for the water volume to reach the solubility capacity of the salt precipitates.

The dilution process (2 to 4 in Fig. 4) during the rainy period at low evaporation rates decreased the concentration of all the major ions (Fig. 5), together with TDS. Changes in major ion concentrations as water evaporated from the lake after April 1987 (5 in Fig. 5) differ according to their solubility capacity. Sodium, potassium, magnesium, chloride and sulfate changed (Figs. 5a, b, c, e, f) in a way similar to TDS (Fig. 4) and were highly correlated with it (Table 2). Calcium and alkalinity were not well correlated with TDS. It is clear that for the dilution and early concentration processes, calcium varied with TDS. However, after a concentration point (9 in Fig. 5d) calcium concentration decreased because of intensive precipitation. In fact, calcium and magnesium began to precipitate in May as carbonates (5 in Fig. 5d). This was corroborated by X-ray diffraction analysis of particulate suspended matter which showed changes in the relative amounts of precipitates in relation to the time sequence described here. The Mg/Ca ratio remained constant, about 7, from April to August (5 to 9 in Fig. 5h), making it evident that equivalent amounts of both ions were precipitating during these months. Later on, September (10 in Fig. 5), the corresponding analysis showed gypsum, epsomite, hexahydrite, bischofite and halite, but not carbonates.

The period from April–August (5 to 9 in Fig. 5) corresponded to the early precipitation phase and the period from August–October (9 to 11 in

Table 2. Correlation coefficients between the major ions concentrations and TDS in Lake Gallocanta for the two study periods.

	TDS	Cl	SO$_4$	Alk	Na	K	Mg	Ca
TDS	1	0.98	0.98	0.25	0.767	0.939	0.86	0.21
Cl		1	0.97	0.239	0.786	0.92	0.82	0.149
SO$_4$			1	0.279	0.745	0.93	0.89	0.271
Alk				1	−0.15	0.31	0.40	−0.15
Na					1	0.72	0.60	0.21
K						1	0.90	0.27
Mg							1	0.27
Ca								1

58

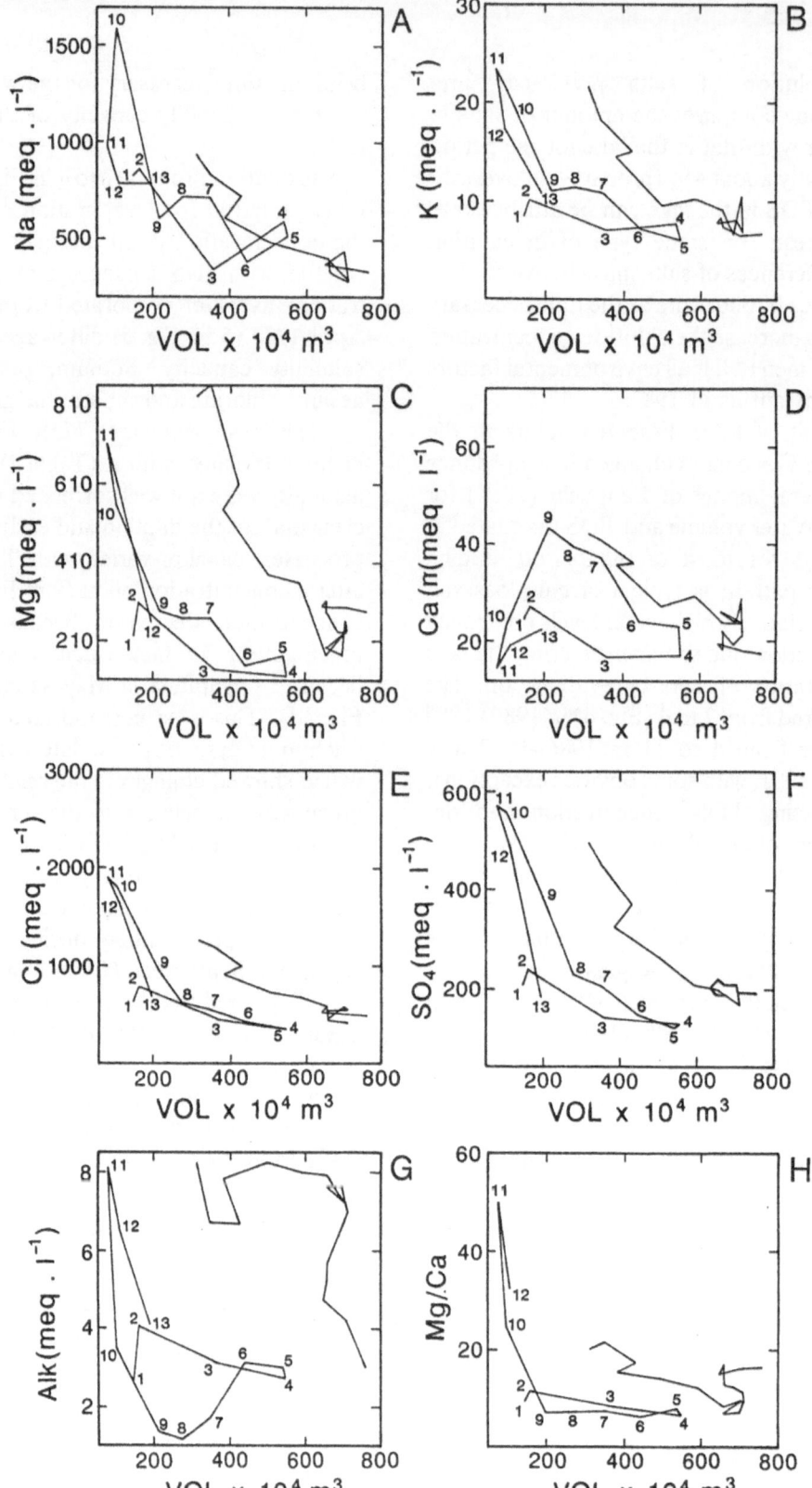

Fig. 5. Ion regimes in Lake Gallocanta in relation to water volume for the two periods of study (as in Fig. 4): A: sodium, B: potassium, C: magnesium, D: calcium, E: chloride, F: sulfate, G: alkalinity, H: magnesium/calcium ratio, versus volume.

Fig. 5) to the late phase of the Eugster & Hardy (1978) model. From August to October calcium precipitated in very large amounts and the Mg/Ca

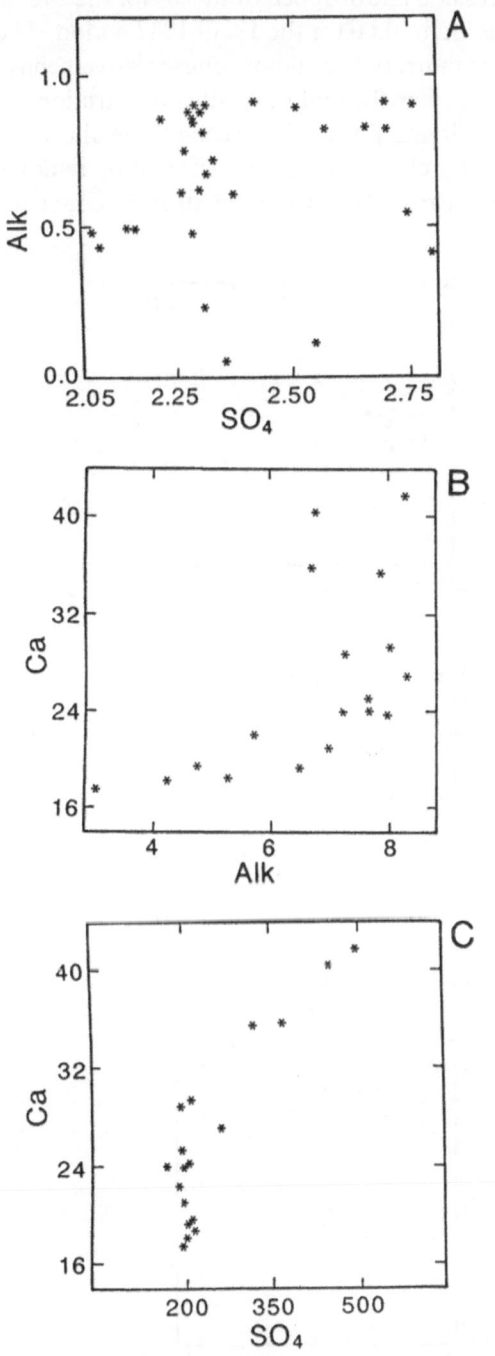

Fig. 6. Plots of alkalinity versus sulfate (A), calcium versus alkalinity (B) and calcium versus sulfate (C) for Lake Gallocanta. Top: scales are $\log_{10}$ of the concentrations. Middle and bottom: scales are meq l^{-1}.

ratio increased from 7 to 50. This was corroborated by the relationship between calcium and alkalinity and sulfate. Calcium and alkalinity were relatively well correlated at low concentrations (Fig. 6b). After the saturation point, and consequently carbonate precipitation, calcium was well correlated with a more soluble anion, sulfate (Fig. 6c).

A significant difference was observed for alkalinity between the two concentration periods 1980–1981 and 1986–1987 (Fig. 5 g). Alkalinity did not decrease as TDS increased during 1981. The two periods began at different initial conditions; the 1980–1981 period being one of dessication with high amounts of decomposing organic matter. The respiratory processes lead to increasing alkalinity, or the so-called 'extra permanent' alkalinity (Dunnette *et al.*, 198;, Kilham & Cloke, 1990), by itself or through sulfate reduction in anaerobic conditions.

Figure 7 shows the plot of the main ions versus chloride in Lake Gallocanta during the two periods of the study: September 1980–October 1981, drying phase, and November 1986–December 1987, refilling phase. Table 3 list the slopes of the regression lines and the correlation coefficients for the major ions in relation with chloride for both periods of time.

Sodium behavior is very conservative in relation with chloride during the drying phase (Fig. 7a). The slope of the sodium versus chloride regression line is 1.03 and the correlation

Table 3. Slopes of the regression lines and correlation coefficients of the major ions versus chloride concentrations for the two sets of data corresponding to the periods of the study October 1980–November 1981, drying phase, and November 1986–December 1987, refilling phase (see Fig. 7).

	Drying phase 1980–1981		Refilling phase 1986–1987	
Sodium	slope	r	slope	r
Potassium	1.03	0.94	0.49	0.74
Magnesium	1.04	0.93	0.66	0.94
Calcium	1.34	0.90	0.92	0.95
Sulfate	0.70	0.84	− 0.04	0.05
Alkalinity	0.93	0.96	0.87	0.85
	0.29	0.36	0.55	0.46

60

coefficient is 0.94. However, during the 1986-1987 period the relationship lost linearity ($r = 0.74$) and sodium did not keep its strong conservative character (slope = 0.49). The molar ratio of Na/Cl decreased at the highest concentrations indicating sodium carbonate and sulfate precipitation. The Na/Cl ratio varied widely (Fig. 7a), probably indicating a combination of different ionic sources or processes for the lake ionic composition.

Potassium and chloride were highly correlated both during the drying ($r = 0.93$) and refilling phases ($r = 0.94$). However the slope of the regression line dropped from 1.04 for the 1980–1981 period to 0.66 for the 1986–1987 period. The first one indicated an almost conservative behavior of potassium for that range of concentrations during the drying phase. The second one showed a behavior closer to the type affected by removal mechanisms which were not strongly concentration

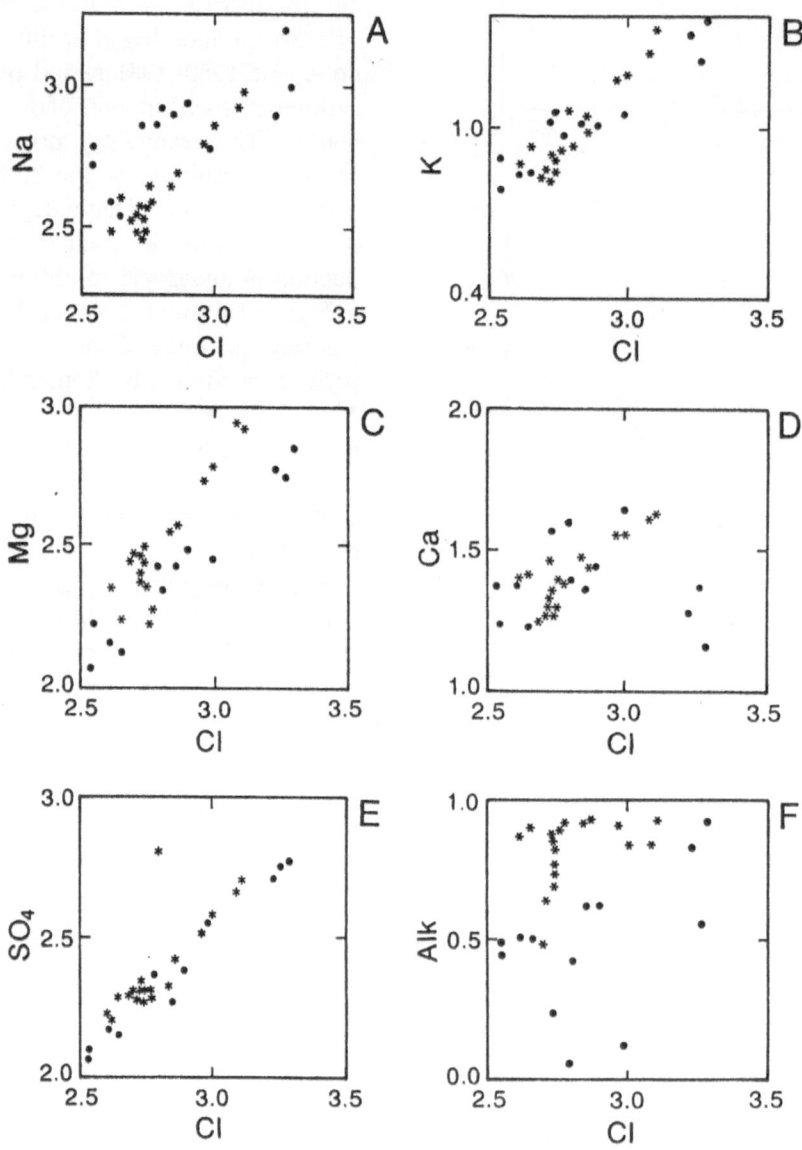

Fig. 7. Plots of sodium (A), potassium (B), magnesium (C), calcium (D), sulfate (E) and alkalinity (F) versus chloride for the waters of Lake Gallocanta during the 1980–1981 (o) and 1986–1987 (∗) periods of study. All scales are $\log_{10}$ of the concentrations.

dependent. However, the Na/k ratio hardly changed during the 1980–1981 period (45.4 ± 5.1) while chloride increased from 478 to 1250 meq l^{-1} (Na/K = 2.37 Cl$^{0.21}$; r = 0.04). For the 1986–1987 period the Na/k ratio varied over a wider range (67.8 ± 19.2), while chloride changeds from 623 to 1900 meq l^{-1}. The equation Na/K = 4.04 Cl$^{-0.66}$ (r = − 0.34) for this period indicated a weak increase of potassium in relation to sodium.

Magnesium was conservative for the range of concentrations observed during both periods in Lake Gallocanta (Fig. 7c, Table 3). As magnesium was one of the main components of the precipitates (Table 4, Fig. 10) it is not surprising that it should be lost from solution at lower concentration than the other ions noted. The correlation coefficient between Na and Cl was 0.77. The molar ratio Na/K remained within a narrow range (45.4 ± 5.1) during 1980–1981 drying period and 67.8 ± 19.2 during the 1986–1987 refilling period). Both observations indicated that sodium was lost from solution in relation to chloride.

The sulfate versus chloride plot showed an effective removal process which was strongly concentration dependent over the 436–575 meq sulfate l^{-1} range of the 1980–1981 period. It was lower for the 1986–1987 period, with some ion removal, but closer to a conservative behavior (Fig. 7e, Table 3).

Calcium and alkalinity versus chloride plots (Fig. 7d, f) reflected the combination of carbonate and calcium precipitation and depletion of calcium at the point of saturation. The wide scatter of the alkalinity concentrations indicated the importance of other processes in addition to relative amount of ion concentrations (photosynthesis, respiration, degassing) in the control of alkalinity variations.

Sediment

Groundwater samples collected along an E–W cross-section showed an increased in the amount of Mg at the central part of the lake while Ca did not showed major changes (Fig. 8). The Mg/Ca

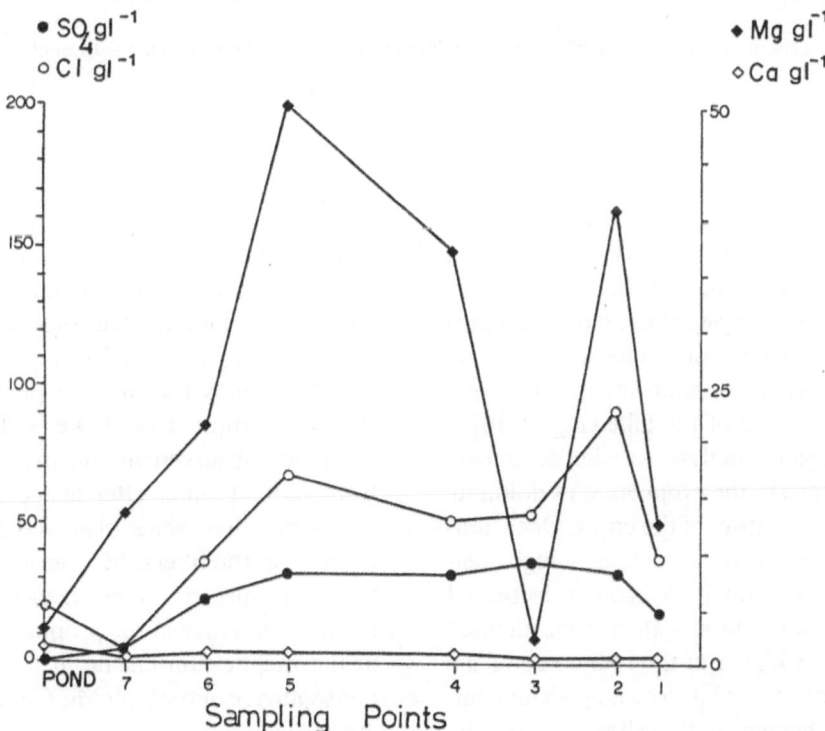

Fig. 8. Chloride, sulfate, magnesium and calcium concentrations in the pore water samples collected at different points in a E-W transect (see Fig. 1).

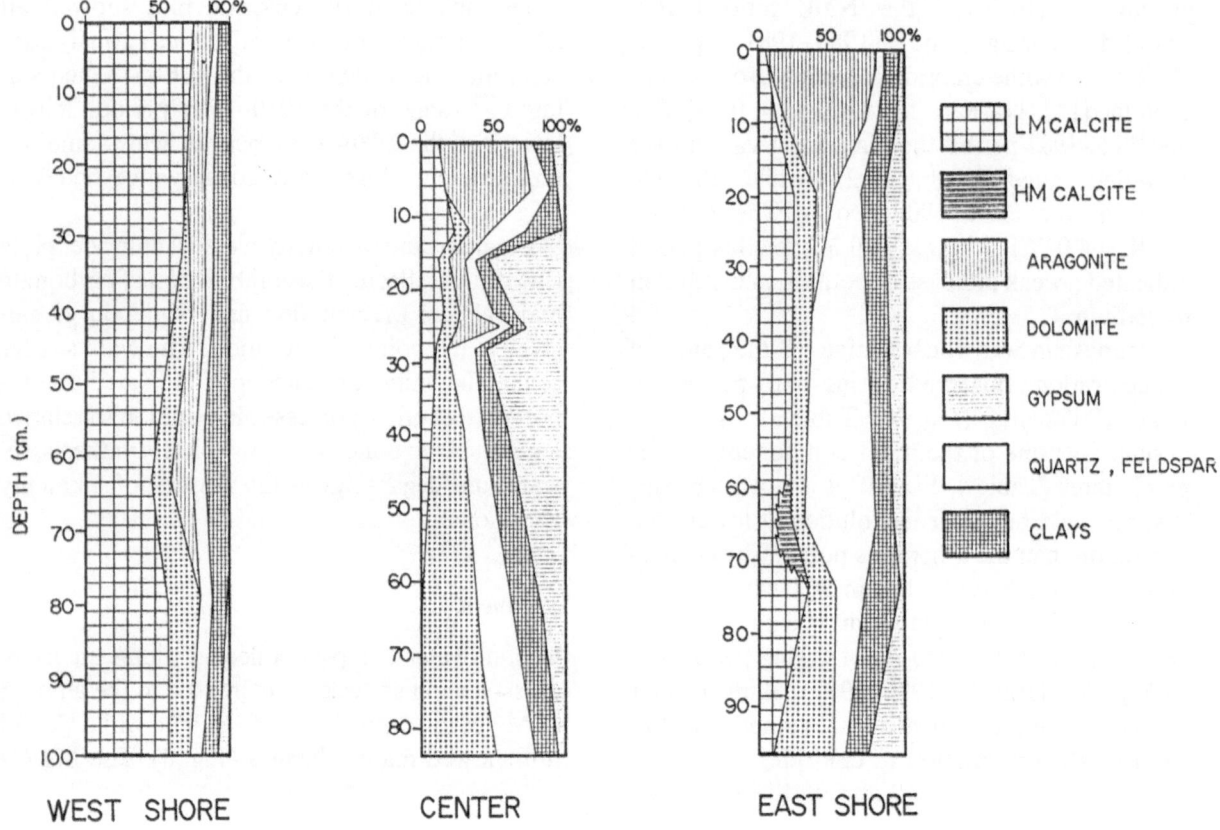

Fig. 9. Cumulative percentage of the major minerals in bottom lake sediments with depth in cores.

equivalent ratio in the groundwater changed from 1.5 at the shore of the lake to 40 in the central part of the lake. Chloride also increased from 2.8 to 146 g l^{-1} and sulfate from 9 to 230 g l^{-1}.

X-ray diffraction analysis of the sediment cores showed a high proportion of calcite and aragonite at the lake shore while dolomite and gypsum were dominant in the centre of the lake (Fig. 9, 10).

The minerological analysis of the sediments showed an increase in the proportion of dolomite from the top to the bottom of the cores. Occasionally carbonate precipitates containing a high magnesium content were found. Aragonite decreased vertically becoming absent at a well defined depth. Sulfate was higher in the sediments of the central part of the lake while low magnesium calcite was much higher in the shore sediments. Along the west shore, close to the major zone of

freshwater discharge, low magnesium calcite was dominant while in the east shore, where freshwater inputs were much less important, dolomite was present in only low amounts at the sediment surface and high magnesium calcite was evident to depths of 60–70 cm deep.

Table 4 lists the mineral phases identified in different samples from Lake Gallocanta. Silicates were not observed in the precipitates obtained from filtered water after evaporation at 100 °C. However, they were observed in the materials trapped in the filters. Magnesium calcite was not found in surface water. Geochemical reactions favor an increase in the proportion of magnesium in the magnesium calcite in the sediments, as a consequence of the high Mg/Ca ratios observed in groundwater.

Table 4. Mineral precipitates found in Lake Gallocanta from different samples collected between 1986 and 1988.

	Carbonates	Sulfates	Chlorides	Silicates
Minerals observed in materials trapped on filters after filtering lake water	$CaCO_3$, Calcita $(Ca,Mg)(CO_3)_2$, Dolomite	$CaSO_4 \cdot 2H_2O$, Gypsum $MgSO_4 \cdot 7H_2O$, Epsomite $MgSO_4 \cdot 6H_2O$, Hexahydrite $MgSO_4 \cdot 4H_2O$, Starkeyite $MgSO_4 \cdot H_2O$, Kieserite	$NaCl$, Halite $MgCl \cdot 6H_2O$, Bischofite	SiO_2, Quartz $KAl_2(Si_3Al)O_{10}(OH)_2$
Minerals observed in material precipitated after lake water evaporation at 110 °C	$CaCO_3$, Calcite $(Ca,Mg)(CO_3)_2$, Dolomite $CaCO_3$, Aragonite	$CaSO_4$, soluble Anhydrite $CaSO_4 \cdot 2H_2O$, Gypsum $MgSO_4 \cdot 7H_2O$, Epsomite $MgSO_4 \cdot 6H_2O$, Hexahydrite $MgSO_4 \cdot H_2O$, Kieserite $Na_{12}Mg_7(SO_4)_{13} \cdot 15H_2O$, Loeweite	$NaCl$, Halite $MgCl \cdot 6H_2O$, Bischofite	
Minerals observed in sediment cores	$CaCO_3$, Low Mg Calcite $CaCO_3$, High Mg Calcite $Ca_{65}Mg_{35}(CO_3)_2$, Dolomite $Ca_{52}Mg_{48}(CO_3)_2$, Dolomite $CaCO_3$, Aragonite	$CaSO_4 \cdot 2H_2O$, Gypsum $MgSO_4 \cdot 7H_2O$, Epsomite $MgSO_4 \cdot 6H_2O$, Hexahydrite $MgSO_4 \cdot 4H_2O$, Starkeyite $MgSO_4 \cdot H_2O$, Kieserite $Na_2Mg(SO_4)_2$, Bloedite $Na_2SO_4 \cdot 10H_2O$, Mirabilite Na_2SO_4, Thenardite	$CaCl$, Halite $MgCl \cdot 6H_2O$, Bischofite	SiO_2, Quartz $KAl_2(Si_3Al)O_{10}(OH)_2$, Illite $Al_4Si_4O_{10}(OH)_8$, Kaolinite $(Ca,Na)(Al,Si)Al2Si_2O_8$, Plagioclase

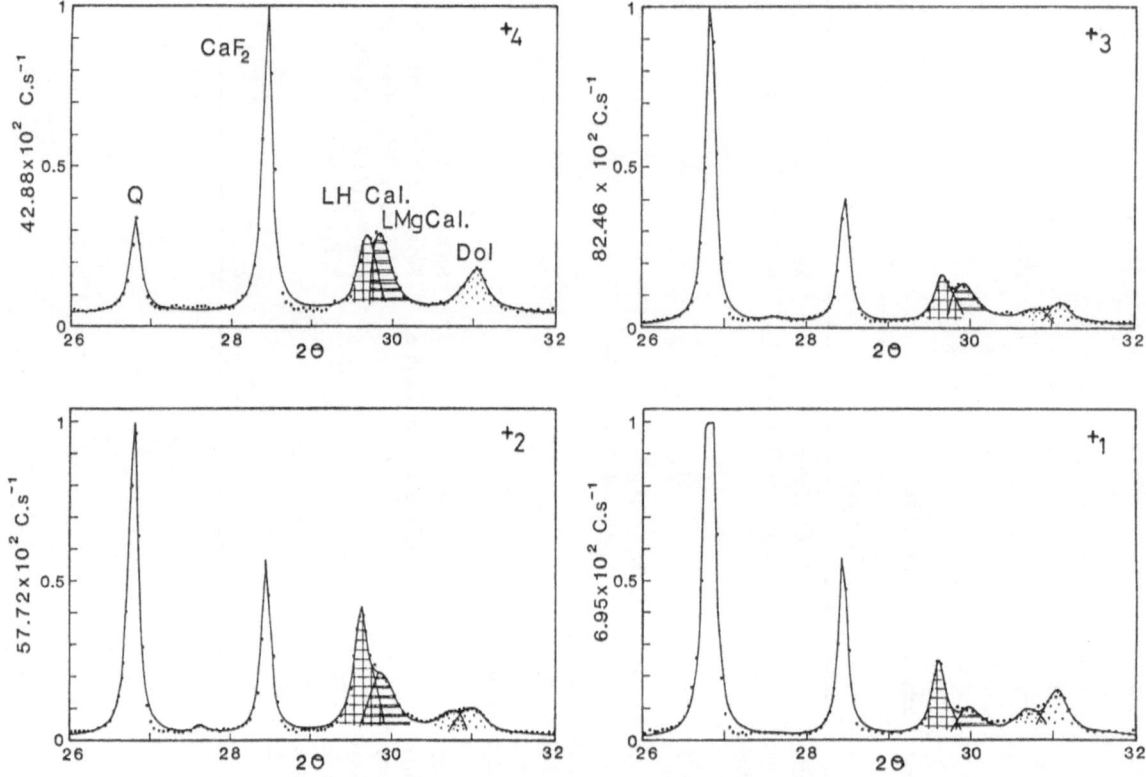

Fig. 10. Lorenzian fittings of the carbonate mineralogy in the sediment (see Fig. 1 for sample location).

Discussion

According to inhabitants of the villages in the Gallocanta basin the lake became completely dry four times during the present century. Thus, it is an ephemeral lake with water level fluctuating at periodicity intermediate between those which dry and refill every year, e.g. the Monegros lakes in the nearby Ebro River valley (Pueyo, 1978), and those which dry or refill much less frequently, e.g. Lake Eyre (Williams, 1988) or Lake George (Jacobson & Schuett, 1979).

Therefore, during this century the physical and chemical characteristics of the lake must have changed several times in a way similar to that described here. The biological communities must have changed in response to other environmental factors, likely in the way described previously (Comín *et al.*, 1983), with some additional changes after the agricultural expansion during the 1960s.

The geographical setting of the lake helps to explain the climatic fluctuations that affect it during this relatively short time. Lake Gallocanta is located at 1000 m.a.s.l. at the boundary between semiarid and semihumid climatic regions. This gives to the lake an unique character as, with the exception of only one permanent lake, all the other Spanish saline lakes dry every year because of the high difference between evaporation and precipitation in the Spanish semiarid interior (Comín & Alonso, 1988).

For longer periods of time little can be concluded without palaeoecological data. Field and aerial photographs show an ancient shoreline which corresponds approximately to an increase of 4 times the present lake area. For such a volume (after Bowler, 1981), the mean annual precipitation at that time must have been about 675 mm if mean annual evaporation was as today and the region remained tectonically stable.

Lake level changes such as those described in

this paper are a consequence of annual fluctuations in precipitation. However, the relationship between lake level and rainfall are not linear. Both total rainfall and frequency of rainfall during the year are important to define the water levels from year to year. For example, heavy rains during late spring do not increase the water level because of high evaporation at that time of the year. However, short and frequent rains during autumn and winter increase the water level significantly. In any case, the water level of Lake Gallocanta will not remain constant for more than two years if changes in the precipitation pattern alter the frequency and intensity of autumn and winter rains. The time response of Lake Gallocanta is very low indicating that major changes in the seasonal pattern of rainfall will cause marked lake level fluctuations in four to twentyseven months.

The difference in the absolute amount of salts dissolved in Lake Gallocanta between 1981 and 1987 indicates that very large amounts of salts are lost from solution during dry periods. A simple calculation comparing concentrations at equivalent volumes in 1981 and 1987 indicate losses in the order of 19 Tm of calcium and 64 Tm of sulfate.

The mechanisms removing salts during drying periods are mainly precipitation and aeolian flow (Langbein, 1961). Carbonates (calcite, aragonite, dolomite) and sulfates (gypsum, bloedite, thenardite, etc.) are the most important salts precipitating during the drying phases of the lake. They accumulate with spatial heterogeneity according to the salt saturation point as related to water volume. Sulfates are more abundant in the central part of the lake floor while carbonates dominate towards the shores. Furthermore, along shorelines where freshwater flows in, the proportion of magnesium in carbonate precipitates is lower than along shorelines where no important freshwater inputs occur. This is an example of a more generalized model of spatial salt accumulation proposed by several authors (Langbein, 1961, Eugster & Hardie, 1978).

Sulfate reduction must occur in the anoxic sediments of Lake Gallocanta during the drying phase as it has been noted in many other saline environments (Kilham, 1984; Jones et al., 1977). The loss of hydrogen sulfide through this reaction and by putrefaction (Dunnette et al., 1985) contributes to increase the alkalinity as in Gallocanta at the end of the 1980–1981 study period.

Lake Gallocanta shows a different salt concentration/water volume relationship at different lake stages because salts are lost from the water by precipitation and aeolian flow, among other processes. It is evident from the data presented that the same calcium concentration was observed for different volumes and vice versa.

Few geochemical differences can be observed from the relationships between ions at the different lake stages. The reason for these differences is that after drought periods, surface and groundwater inflows, redissolution, precipitation and other processes contribute to a wider range of types of water mixing in the lake, while during drying periods the geochemical evolution is mainly dependent on the ionic composition of the lake water mass. Thus, it is important to study the geochemical evolution of saline waters during different lake stages.

Acknowledgements

This work was supported by Diputación General de Aragón -Servicio Conservación Medio Natural and CAYCIT-CSIC (IO851). Thanks are given to Centro Meteorológico de Zaragoza for permitting the use of precipitation data.

References

APHA-AWWA-WPCF, 1980. Standard methods for the examination of water and wastewater. APHA, Washington. 1134 pp.

Bayly, I. A. E., 1970. Further studies on some saline lakes of south-east Australia. Aust. J. mar. Freshwat. Res. 21: 117–129.

Bayly, I. A. E. & W. D. Williams, 1966. Chemical and biological studies on some saline lakes of south-east Australia. Aust. J. mar. Freshwat. Res. 17: 177–228.

Bowler, J. M., 1981. Australian salt lakes. a paleohydrologic approach. Hydrobiologia 82: 431–444.

Comín, F. A., M. Alonso, P. López & M. Comelles, 1983.

66

Limnology of Gallocanta Lake, Aragón, Northeastern Spain. Hydrobiologia, 105: 207–221.

Comín, F. A. & M. Alonso, 1988. Spanish salt lakes: their chemistry and biota. Hydrobiologia 158: 237–245.

Dunnette, D. I., D. P. Chynoweth & K. H. Mancy, 1985. The sources of hydrogen sulfide in anoxic sediment. Wat. Res. 19: 875–884.

Eugster, H. P. & L. A. Hardie, 1978. Saline Lakes. In: A. Lerman (ed.). Lakes. Chemistry, geology, physics. Springer-Verlag. New York: 237–293.

Fritz, J. S. & S. S. Yamamura, 1955. Rapid microtitration of sulphates. Anal. Chem. 26: 1461–1464.

Golterman, H. L., R. S. Clymo & M. A. M. Ohnstad (eds.), 1978. Methods for physical and chemical analysis of freshwaters. Blackwell Sc. publ. Oxford. 214 pp.

Hammer, U., 1986. Saline lake ecosystems of the world. (Monographiae Biologicae, 59). Junk Publishers, 616 pp.

Jacobson, G. & A. W. Schuett, 1979. Water levels balance and chemistry of lake George, New South Wales. BMR J. Aust. Geol. Geophys. 4: 25–32.

Jones, B. F., H. P. Eugster & S. L. Rettig, 1977. Hydrochemistry of the lake Magadi Basin, Kenya. geochem. Cosmodrim. Acta 41: 53–72.

Kalk, M., A. J. McLachlan & C. Howard-Williams (eds.), 1979. Lake Chilwa. Studies of change in a tropical ecosystem. Dr. W. Junk Publishers. The Hague. 462 pp.

Kilham, P., 1984. Sulfate in African inland waters: sulfate to chloride ratios. Ver. int. Ver. Limnol. 22: 296–302.

Kilham, P. & P. L. Cloke, 1990. The evolution of saline lake waters: gradual and rapid biogeochemical pathways in the Basotu Lake District, Tanzania. Hydrobiologia 197: 35–50.

Langbein, W. B., 1961. Salinity and hydrology of closed lakes. U.S.G.S. Prof. Paper 412, US Gort. Print. Office, Washington, 220 pp.

Martínez, B. & F. Plana, 1987. Quantitative X Ray diffraction of carbonate sediments: mineralogical analysis through fitting of Lorentzian profiles to diffraction peaks. Sedimentology 34: 169–174.

Phillips, K. N. & A. S. Van Denburgh, 1971. Hydrology and geochemistry of Albert, summer and Goose Lakes and other closed basin lakes in South-Central Oregon. LL.S. G. S. Prof. Paper 502-B, 86 pp.

Pueyo, J. J., 1978. La precipitación evaporítica actual en las lagunas saladas del área: Bujuraloz, Sástago, Caspe, Alcañiz y Calanda (provincias de Zaragoza y Teruel), Rev. Inst. Inv. Geol. Dip. Prov. Barcelona 33: 5–56.

Walker, K. F., 1973. Studies on a saline lake ecosystem. Aust. J. mar. Freshwat. Res. 24: 21–71.

Williams, W. D., 1966. Conductivity and the concentration of total dissolved solids in Australia lakes. Aust. J. mar. Freshwat. Res. 17: 169–176.

Williams, W. D. & M. J. Kokkinn, 1988. The biogeographical affinities of the fauna in episodically fitted salt lakes: A study of Lake Eyre South, Australia. Hydrobiologia, 158: 227–236.

Hydrobiologia **197**: 67–81, 1990.
F. A. Comín and T. G. Northcote (eds), Saline Lakes.
© 1990 *Kluwer Academic Publishers.*

Recent carbonate sedimentation and brine evolution in the saline lake basins of the Cariboo Plateau, British Columbia, Canada

Robin W. Renaut
Department of Geological Sciences, University of Saskatchewan, Saskatoon, Saskatchewan, S7N 0W0, Canada

Key words: Saline lakes, brines, carbonates, chemical divides, mudflats

Abstract

There are more than 100 closed, saline lakes in the semiarid, intermontane plateaus of British Columbia. They range from shallow perennial lakes to ephemeral playas. Most are groundwater-fed and lie within glaciofluvial deposits and till. Some have permanent salts. Where underlain by basalts, sodium carbonate brines predominate. Magnesium sulphate brines occur where catchments lie within Paleozoic sedimentary rocks, metasediments and basic volcanics. A few sodium sulphate brines are also present.

A reconnaissance study of the sediments and mineralogy of 21 lake basins has shown that carbonates, including extensive magnesite and hydromagnesite deposits, and several occurrences of protodolomite, are widely precipitated in lake basins of each brine type. Analyses of stream, spring, ground and lake waters from the Cariboo Plateau region demonstrate that carbonate precipitation probably constitutes the major chemical divide responsible for producing the two dominant types of brine.

Introduction

In their model for the evolution of continental brines, Hardie & Eugster (1970) and Eugster & Hardie (1978) emphasized the importance of early precipitation of alkaline earth carbonates in providing a major chemical divide that controls the type of brine which ultimately is produced. In the saline lake basins of British Columbia, extensive recent carbonate sediments, including abundant magnesite and hydromagnesite, are associated with groundwaters and brines of widely varying composition. The lakes provide an unusual opportunity to examine the role of carbonate precipitation in brine evolution within a small geographical region.

The several hundred saline lakes of the semiarid, Interior Plateau of British Columbia range from small hypersaline lakes and playas, some with permanent evaporites, to large hyposaline and mesosaline lakes, some with stratified brines. These lakes probably have the greatest range of brine compositions of any small region in the world (Hammer, 1986). Although there are many sub-types, two brine types dominate: (i) highly alkaline brines, poor in Ca and Mg, with $Na-CO_3$ $-(SO_4)-Cl$ composition, and (ii) more neutral brines, poor in HCO_3 and CO_3, with $Mg-Na-SO_4$ composition. These belong to the two dominant brine types produced by the evaporative concentration of continental meteoric waters (Hardie, 1984).

68

Although aspects of their ecology have been studied (*e.g.* Scudder, 1969), with the exception of Nesbitt's (1974) work on the Basque Lakes, there have been only brief accounts of their sediments and mineralogy (Jenkins, 1918; Reinecke, 1920; Goudge, 1926a, 1926b; Cummings, 1940; Grant, 1987), and no attempts to explain the brine diversity. This paper presents preliminary results of an examination of the recent carbonate deposits on part of the Interior Plateau, and shows that their precipitation can account for the basic division into sulphate-rich and carbonate-rich brines.

Description of sites studied

The saline lakes of Interior British Columbia are clustered in five main groups: Osoyoos, Kamloops, Basque, Cariboo Plateau and Chilcotin Plateau (Fig. 1). The largest and most diverse group is that of the Cariboo Plateau.

The Cariboo Plateau, which lies between the Coastal Ranges and the Columbia-Rocky Mountains, is a gently undulating region at 1050 to 1250 m above sea level, characterized by disordered drainage and thousands of lakes, both

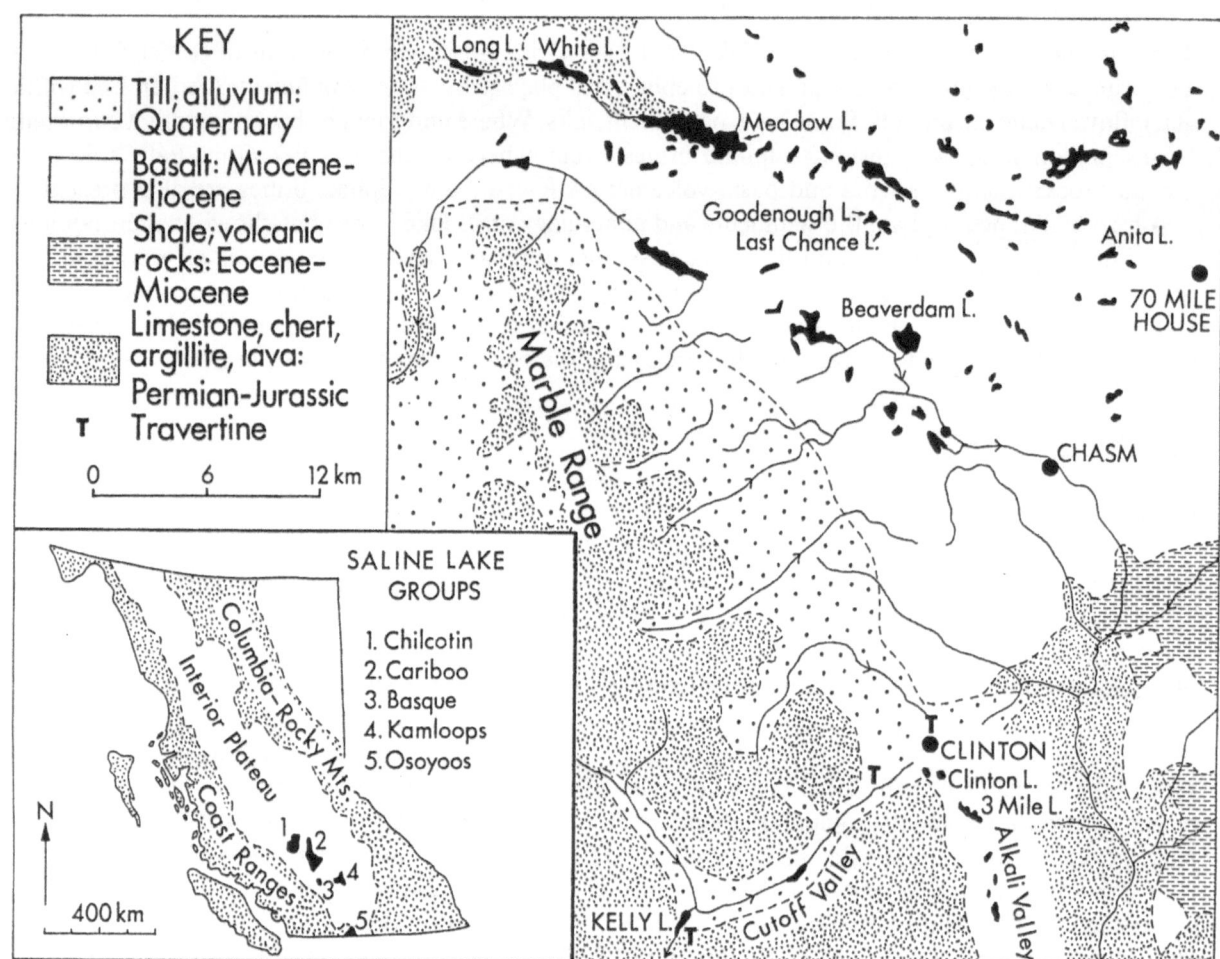

Fig. 1. Location and geological setting of the southern Cariboo Plateau lakes. Geology simplified after Campbell & Tipper (1971). Only the larger lakes are shown.

fresh and saline (Holland, 1964). Ice retreated from the region about 10 000 years ago (Clague, 1981; Fulton, 1984), leaving a thin (1–5 m) mantle of till and glaciofluvial sands and gravels. These overlie a thick series of Mio-Pliocene basalt flows. Limestones, marbles, cherts, argillites, basic lavas, greenstones and ultramafic rocks of the Cache Creek terrane (Permian-Jurassic) form the Marble Range which rises abruptly to 2 240 m elevation at the western edge of the Plateau (Fig. 1) (Campbell & Tipper, 1971; Monger, 1977; Cordey et al., 1987).

The climate is semi-arid to sub-humid with 300–400 mm yr^{-1} precipitation on the plateau (100–150 mm as snow), compared to > 1 000 mm yr^{-1} on the adjacent mountains. The annual moisture deficit is about 300–400 mm. Mean July temperatures range from 13 to 17 °C, compared with − 9 to − 11 °C in January. Only 60–90 days each year are frost-free (Valentine & Schori, 1980; Atmospheric Environment Service, 1982). Most of the saline lakes lie in clearings within the extensive coniferous forest that covers much of the plateau (Jones & Annas, 1978).

Saline lakes are most abundant in the region near Clinton and 70 Mile House (Fig. 1). Most are small (< 35 ha) and shallow (< 4 m), and lie within subcircular and elongate depressions within the glacial deposits, commonly aligned along axes of former glacial meltwater channels (Tipper, 1971). Most lakes have very small catchments and are predominantly groundwater-fed. Extensive areas of freshwater marsh demonstrate that high water tables occur across much of the plateau. The paucity of streams away from the Marble Range shows that the drift deposits are probably very permeable, acting as shallow, largely unconfined, aquifers. Depth of brine in the lakes varies seasonally and is usually greatest between May and July, and least during August and September. Most lakes are frozen between November and April. Many shrink in volume by > 50% annually, producing salt-encrusted mudflats surrounding a dense residual brine; others dry out periodically to produce extensive playa flats.

Representative analyses of Cariboo Plateau waters are given in Table 1. Further analyses are given by Topping & Scudder (1977), Northcote & Halsey (1969), Hudec & Sonnenfeld (1980) and Renaut & Long (1987). The general morphological and sedimentological characteristics of the lakes are described by Renaut & Long (1989). Only a few lakes, such as Last Chance Lake and Clinton Lake, have perennial evaporites (Figs. 1 and 4), but several precipitate subaqueous salts annually, by evaporative concentration, brine cooling and freeze-out. Near outcrops of Cache Creek rocks, including inliers on the plateau and at its margins near Clinton, Mg-Na-SO$_4$ and Na-Mg-SO$_4$-CO$_3$ brines predominate, from which epsomite (MgSO$_4 \cdot$7H$_2$O), bloedite (Na$_2$SO$_4 \cdot$ MgSO$_4 \cdot$4H$_2$O) and mirabilite (Na$_2$SO$_4 \cdot$ 10H$_2$O) form. Across much of the plateau, Na-CO$_3$-SO$_4$-Cl and Na-CO$_3$-Cl brines are more common, and natron (Na$_2$CO$_3 \cdot$10H$_2$O) is the dominant salt to precipitate (Renaut & Long, 1987).

Materials and methods

Water samples were collected in polyethylene bottles during August 1984, May–June 1985 and October 1986. All bottles were then sealed in plastic film to try to reduce vapour exchange. Formalin was added to duplicate samples to preserve organic matter and eliminate microbial activity after collection. More than 250 sediment samples and 24 short (< 1.5 m) cores were collected from 21 lakes during the same field seasons. All sediment samples were placed in self-sealing plastic freezer bags. Some lake waters and sediments were sampled during only one of the field seasons. Others, including Last Chance Lake and Goodenough Lake, were sampled during each field season. The pH was measured in the field using an Orion 201 portable pH meter. Salinities were recorded using a YSI Model 33 Salinity-Conductivity Meter. Water and sediment temperatures were measured in the field using thermistors.

More than 90 filtered water samples from streams, springs, groundwaters and lakes were

Table 1. Chemical analyses of Cariboo Plateau waters

	Na	K	Ca	Mg	HCO₃	CO₃	Cl	SO₄	SiO₂	pH	T°C	Date
Rivers												
Porcupine Creek	2.2	0.5	56	7.0	193	0	2.7	20	8.0	8.2	11	6/85
Cutoff Valley Ck.	5.6	1.9	71	23	294	0	5.4	33	5.1	8.1	10	6/85
57 Mile Creek	5.3	1.5	44	12	176	0	2.7	16	2.5	7.8	15	8/84
Stream 'BB'	16	2.1	30	32	280	0	2.5	4.9	3.1	7.9	12	8/84
Stream 'SLC'	32	3.3	23	18	232	0	1.8	5.7	2.2	8.0	14	8/84
Stream 'ML'	8.9	1.7	26	15	160	0	2.3	15	2.8	7.4	12	6/85
83 Mile Creek	42	7.2	30	80	655	0	2.4	8.7	2.3	8.1	15	8/84
Springs												
Goodenough No. 1	72	7.9	2.0	48	295	tr.	49	59	tr.	7.9	8	9/86
Clinton Creek	51	2.4	62	102	760	0	8.6	18	4.8	7.1	7	9/86
Clinton Lake	136	24	3.8	538	1090	255	7.9	1235	6.5	9.2	7	6/85
Lake marginal groundwaters												
Goodenough L.	380	45	12	85	900	140	85	200	tr.	8.6	10	8/84
Last Chance L.	250	15	11	76	800	0	34	180	5.0	8.5	12	8/84
3 Mile L.	5670	450	120	9845	270	0	1350	44500	tr.	8.1	11	6/85
Lakes												
Kelly	3.4	0.6	53	10	200	0	2.1	28	8.9	8.1	12	6/85
Leighwood	6.2	1.0	31	23	163	0	6.5	43	6.4	7.9	13	6/85
2½ Mile	61	10	29	251	634	33	83	910	3.0	8.7	13	6/85
Clinton Pond	136	24	3.8	538	1090	255	7.9	1235	0	9.2	15	6/85
Clinton Lake	21800	1750	45	61500	4500	0	3550	275000	10	8.1	18	9/86
3 Mile	3840	320	265	7310	0	0	395	36490	51	8.1	16	6/85
Anita marsh	155	41	16	96	880	45	55	tr.	tr.	8.3	11	6/85
Anita	45700	870	11	85	9880	48900	6950	4350	tr.	10.1	12	6/85
Beaverdam	17	4.3	10	70	412	0	2.4	5.9	tr.	8.1	13	8/84
Unnamed 'A'	110	13	7.5	130	870	22	3.5	38	12	8.2	16	8/84
White	1400	170	4.0	115	1960	850	280	550	30	8.7	16	8/84
Last Chance	21300	420	11	35	12350	15600	2890	7150	16	8.9	14	6/85
Last Chance	100400	885	tr.	86	23600	81900	17800	42600	72	10.4	18	8/84
Last Chance	45600	745	8.0	16	18000	40100	8670	10950	0	9.8	14	9/86
Goodenough	30050	975	4.1	60	15670	25400	5460	2860	tr.	10.0	15	8/84

analyzed by standard atomic absorption (cations, SiO_2), ion exchange (SO_4), titration (HCO_3, CO_3) and ion specific electrode (Cl) methods. Total dissolved solids (TDS) were determined by evaporation at 105 °C of water filtered through 0.4 micron membranes and weighing the salt residues to ± 0.1 mg. All analyses were made within 3 months of collection. A slight, but consistent, drop in pH with time was noted between collection and analysis for the most concentrated solutions. The theoretical state of saturation of the waters with respect to various mineral phases was calculated using the WATSPEC program (Wigley, 1977).

Mineralogy was determined by petrographic methods, X-ray diffraction (XRD), and scanning electron microscopy (SEM). Estimates of dolomite composition and Mg substitution in calcite were made using the method of Goldsmith & Graf (1958). Many of the salts (*e.g.* the 10-hydrate minerals) are sensitive to temperature and are unstable in the atmosphere. Although steps were taken to reduce changes between collection and analysis, such as storing samples in original brine and in sealed bags, some samples underwent change. The original mineralogy of some mineral suites was reconstructed by (i) examination of physical form of the samples in the field (including crystal habit and optical properties), (ii) X-ray diffraction analyses of moist and dry samples, and (iii) data indicating which phases are likely to be stable at the temperatures measured when the samples were collected.

Results

Modern carbonate sediments of the Cariboo Plateau region

Carbonate minerals and sediments are widespread and abundant on the Cariboo Plateau. Except for minor detrital carbonates eroded from Paleozoic limestones, almost all the carbonates are products of recent inorganic or biochemical precipitation. There are four main groups: soil carbonates, spring deposits, freshwater lake carbonates, and saline lake carbonates.

(i) Soils and near-surface sediments

Soils on glacial and glaciofluvial deposits on the Cariboo Plateau are commonly calcareous (Valentine & Dawson, 1978; Valentine & Schori, 1980). XRD analyses show the carbonate to be predominantly calcite, but with Mg-calcite (5–9 mol%, and 15–17 mol% $MgCO_3$), locally common. Calcite is most abundant in the upper metre of the sediment (up to 34 weight%), in the form of powdery micritic aggregates, micritic coatings on gravel clasts, rhizoliths, meniscus sparite (in sands), and crack-filling sparite in clay-rich till (Fig. 2). On slopes, some precipitated carbonate is probably derived from partial leaching of existing carbonate in tills. However, where shallow groundwaters are present in valley-bottom sites, much probably originates by capillary evaporation. Mg-calcite has only been found in the latter setting, commonly increasing both in amount and in Mg substitution upward in the profile, reaching a maximum concentration about 20 to 50 cm below the surface.

(ii) Travertines and spring deposits

Carbonate spring deposits are common at the margins of the plateau below the Marble Range (Fig. 1), but rare on the plateau itself. Minor travertine is forming today at Clinton Creek Spring, 1 km north of Clinton, from waters with a molar Mg/Ca ratio of 2.7, a temperature of 7 °C and a salinity of $1 \, \mathrm{g \, l^{-1}}$. Aragonite is the predominant polymorph, but is undergoing early neomorphism to calcite pseudosparite. Clinton Spring, 3 km west of Clinton, has waters with a Mg/Ca ratio of 0.5, and although there is a large calcite terrace-mound (Reinecke, 1920; Goudge, 1946; McCammon, 1958; Renaut & Long, 1986), there is no active precipitation. Both spring mounds are essentially fossil deposits. Although plant tufas are present, the mounds show structures, lithology and textures more commonly associated with hydrothermal springs.

On the plateau, springs are usually present as ephemeral seepages that discharge both onto and into the mudflats at the base of the hillslopes that enclose the basins. Well-defined spring orifices occur at Clinton, Goodenough and several other

Fig. 2. Section through cross-stratified glaciofluvial sediments underlying the Cariboo Plateau, showing the shallow zone of carbonate precipitation at 20 to 30 cm depth. Exposure lies in a gravel pit, approximately 5 km south of Beaverdam Lake.

lakes (Figs. 3 and 4). Spring and seepage waters on the plateau have Mg/Ca ratios from 8 to 42 and salinities from 0.5 to 3.5 g l^{-1} TDS. Although Goudge (1926a) reported tufa at Anita Lake (Fig. 1), most spring deposits take the form of efflorescent crusts precipitated on and within sediments adjacent to springs and their outflow channels (Renaut & Long, 1989). Minerals precipitated are mostly calcite, Mg-calcite and aragonite, but with local occurrences of proto-dolomite ($CaMg[CO_3]_2$), magnesite ($MgCO_3$), gypsum ($CaSO_4 \cdot 2H_2O$) and trona ($NaHCO_3 \cdot Na_2CO_3 \cdot 2H_2O$).

(iii) Freshwater lake carbonates

The dilute lakes of the Marble Range and several subsaline lakes on the Cariboo Plateau have extensive recent lacustrine carbonates. The best development is seen at Kelly Lake (Fig. 1), where shallow (< 1.5 m) carbonate platforms have developed on early post-glacial fan-deltas (Reinecke, 1920; Goudge, 1946). There is exten-sive carbonate precipitation associated with charophytes, planktonic green algae, agglutinated rhizopods and cyanobacteria. Laterally continu-ous cyanobacterial mats, locally > 50 m in extent, and abundant oncoids are present. All the sedi-ments are calcite, and are being precipitated from waters with a salinity of 0.3 g l^{-1} TDS and a Mg/Ca ratio of 0.3. Similar, but less extensive, carbonate platforms occur in several unnamed subsaline and hyposaline lakes in Alkali Valley (Fig. 1), and scattered on the plateau.

(iv) Saline lakes

Carbonate minerals are the dominant authigenic minerals in most of the saline lake basins, forming in the upper metre of the saline mudflats that surround the lakes, or forming the floor of playas, both as interstitial precipitates and surficial efflo-rescent crusts. Carbonates have also been precipi-tated subaqueously in several lakes.

Details of the depositional subenvironments of the lake basins are presented elsewhere (Renaut

Fig. 3. Goodenough Lake (Na-CO$_3$-Cl brine). Mudflats covered with efflorescent carbonates pass abruptly into forested slopes of glacial mounds. A small spring lined with microbial mats discharges in the centre foreground. Note the dense population of brine flies (*Ephydra* sp.) along shoreline.

Fig. 4. Clinton Lake basin (Mg-Na-SO$_4$ brine). Sediments surrounding the spring (foreground) are cemented by aragonite. Spring waters feed Clinton Pond (centre) where magnesite and hydromagnesite are forming. The carbonate-depleted waters then seep into Clinton Lake (background) where, epsomite, bloedite and mirabilite are precipitated.

& Long, 1989). Most basins consist of a vegetated hillslope subenvironment (commonly a linear glacial mound or an esker), a mudflat subenvironment (saline and dry mudflats of Hardie *et al.*, 1978), and a lake subenvironment (perennial or ephemeral) (Fig. 5).

The mudflats are the main zone of carbonate precipitation. The width of the mudflats varies from a few metres, where the lake is confined by hillslopes (Fig. 3) to several hundred metres (*e.g.* Meadow Lake, Fig. 1). By late summer, most mudflats are covered with efflorescent crusts. Groundwater normally fluctuates within a metre of the surface where crusts form, and most

precipitation can be attributed to capillary evaporation (Eugster, 1980), together with that produced by evaporation of receding lake waters during summer. Freeze-out precipitation of salts, including relatively insoluble carbonates, may occur as the soil zone freezes downward in winter (Renaut & Long, 1989). Field salinity measurements at Goodenough Lake (Fig. 1) have shown that the shallow marginal groundwaters are commonly subsaline (<3 g l^{-1} TDS) during spring, progressively increasing in salinity to >10 g l^{-1} TDS by late summer. High salinity gradients were recorded in traverses normal to the shoreline at Last Chance, Three Mile and other unnamed

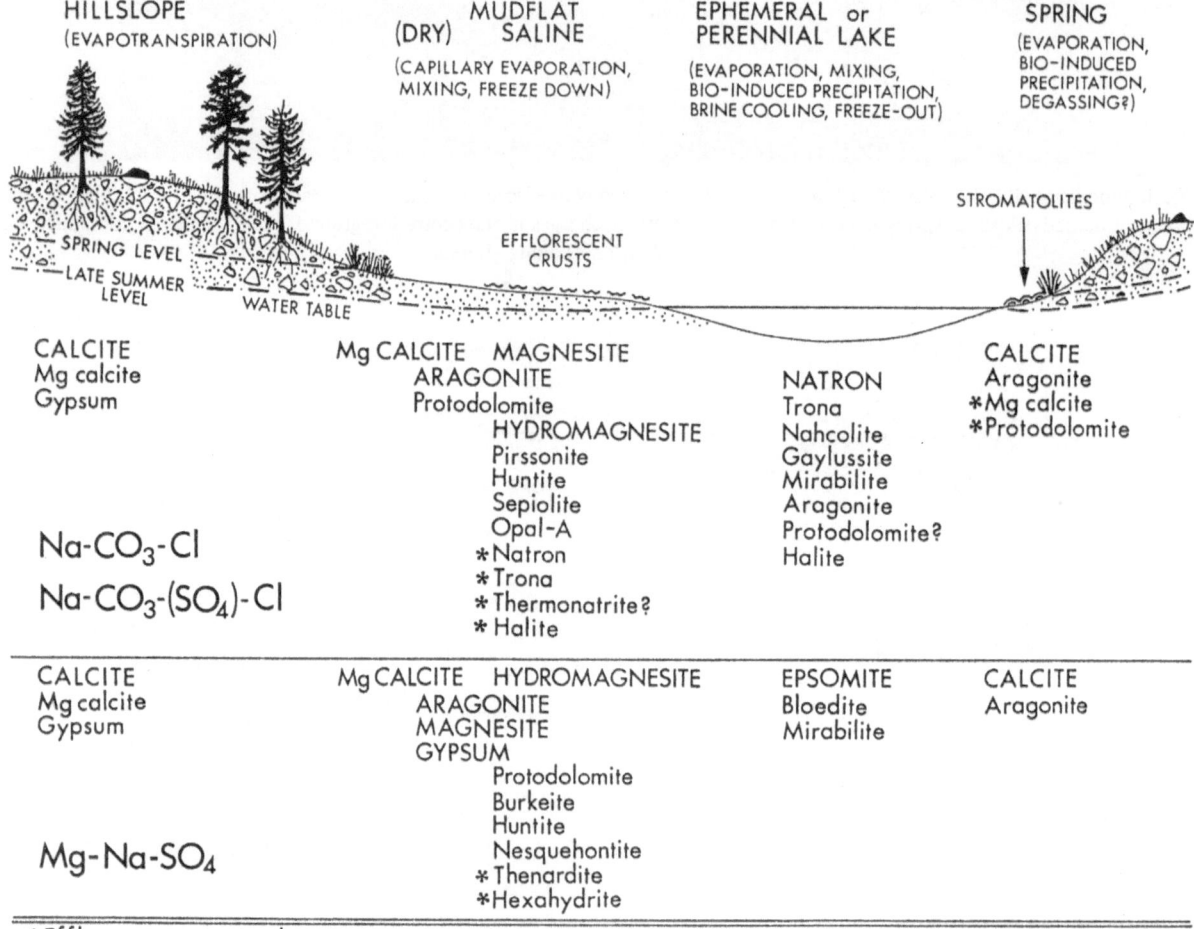

Fig. 5. Precipitated and diagenetic minerals associated with the different depositional subenvironments of the Cariboo Plateau lakes. Upper half shows typical association for Na-CO$_3$-Cl and Na-CO$_3$-SO$_4$-Cl lakes, the lower half for Mg-Na-SO$_4$ lakes. Dominant minerals for each subenvironment are capitalized. Not all minerals are necessarily present in each basin.

lakes suggesting that much of the recharge probably originates from shallow, subsurface waters seeping from the margins of the basins.

The parent muds of the supralittoral flats are mostly silty and sandy clays, composed of quartz, plagioclase, K-feldspar, volcanic glass and clay minerals. Preliminary XRD results indicate a predominance of smectite and illite on the plateau, with locally abundant chlorite, and generally minor kaolinite.

The minerals precipitated from groundwater vary from basin to basin. Lake marginal groundwaters in most closed basins have high molar Mg/Ca ratios, ranging from 3 to >350. With evaporative concentration, calcite, Mg-calcite, protodolomite, aragonite, magnesite and hydromagnesite $(Mg[OH]_2 \cdot 4MgCO_3 \cdot 4H_2O)$ are formed interstitially within the muds. Mg-calcite, magnesite and protodolomite (Ca_{45-52}) are present as aggregates of subhedral and anhedral crystals from 1 to 3 μm; the aragonite needles are from 1 to 5 μm long; hydromagnesite occurs as aggregates of subhedral platy crystals, 1 to 3 μm long. Details of their distribution, form and genesis will be presented elsewhere.

In basins with Na-CO_3-Cl and Na-CO_3-(SO_4)-Cl brines, the carbonates precipitated interstitially are mostly Mg-calcite, magnesite and, locally, protodolomite. Efflorescent crusts, in addition, include products of complete evaporation such as trona, natron, halite, aragonite, gypsum and thenardite (Na_2SO_4). Pirssonite $(Na_2CO_3 \cdot CaCO_3 \cdot 2H_2O)$ is locally abundant (e.g. at Goodenough Lake) and may have formed by interaction of interstitial brines with carbonate minerals or by the mixing of dilute waters percolating into the sediment with interstitial brine. A lateral zonation from Mg-poor to Mg-rich species is present in some basins (e.g. southern Goodenough Lake), but commonly one carbonate mineral dominates the mudflats because they are too narrow to permit much lateral differentiation by partial dissolution processes. Vertical zonations, such as hydromagnesite overlying magnesite, are common in many mudflats (e.g. south of Meadow Lake).

In Mg-Na-SO_4 lakes, such as Clinton Lake

(Fig. 1), Mg-calcite, protodolomite, aragonite, magnesite and hydromagnesite form interstitially, together with gypsum and burkeite (Fig. 4). Efflorescent crusts, in addition, may contain thenardite, huntite $(CaMg[CO_3]4)$, nesquehonite $(MgCO_3 \cdot 3H_2O)$ and hexahydrite $(MgSO_4 \cdot 6H_2O)$. Fig. 5 summarizes the distribution of authigenic minerals found around the Cariboo Plateau lakes.

The amount of carbonate precipitated in the supralittoral muds ranges from <5% to almost 100%. Small, hypersaline playa-lakes commonly lack extensive carbonate fringes. Where fractionation processes have already depleted the groundwaters of most of their alkaline earths or carbonate species, the amount of Ca-Mg carbonate able to form may be small. Thus, in the mudflats surrounding some strongly saline, alkaline lakes, interstitial, soluble sodium carbonates may form, but with little Ca-Mg carbonate (Renaut & Long, 1989). At the other extreme, very pure carbonates more than a metre thick have formed at Clinton Lake where discharging groundwaters have a $HCO_3 + CO_3/Mg + Ca$ ratio near unity.

Alkaline earth carbonates are also precipitated subaqueously within many saline lakes. Short cores from the margins of saline, alkaline lakes commonly contain calcite aragonite, protodolomite and gaylussite $(Na_2CO_3 \cdot CaCO_3 \cdot 5H_2O)$ within silty organic-rich muds. Some of the carbonates are finely laminated. Detailed studies have not yet been undertaken, but they appear to be more common where lakes are confined and lack extensive marginal mudflats. Northcote & Hall (1983) reported calcite precipitation in Mahoney Lake from the Osoyoos saline lake group.

Evolution of the Cariboo Plateau waters

The overall evolution of the waters on the Cariboo Plateau can be examined using the principle of chemical divides, established by Eugster & Hardie (1978). They emphasized that precipitation of a mineral phase, notably calcite or gypsum, has a profound effect on the subsequent evolution of the waters. For calcite, unless the molar ratio of Ca and CO_3 is one at the outset, waters must become enriched either in Ca or HCO_3 when calcite is precipitated. Furthermore, they showed that during precipitation of the alkaline earth carbonates, the Mg/Ca ratio of the waters increases, producing a concomitant increase in the Mg/Ca ratio of the solid phase.

Unlike studies of individual closed basins, the Cariboo Plateau lakes collectively provide an example where the chemical divergence to produce different brine compositions can be examined within a single geographical region. Fig. 6 shows a plot of Na versus Cl for 155 waters from the Cariboo Plateau and its western margin. The relationship is essentially linear, supporting mutual conservation of both ions and evaporative concentration as the principal method of increasing salinity. However, a considerable scatter is seen for dilute waters. Those with the lowest Na/Cl ratios are mostly streams draining Paleozoic limestones, cherts and argillites, whereas those plotting on, or above, the main evaporative trend are streams and dilute lakes on the plateau where the higher Na^+ concentrations may reflect its acquisition from feldspars and other sodic minerals in tills and basalts.

Most precipitation falling on the Marble Range infiltrates the thin soil rapidly, dissolving calcite and acquiring SO_4^{2-} by dissolution of local gypsum deposits (Cole, 1913) or by oxidation of sulphides (mostly pyrite) within the argillites and ultramafic rocks. Sampled runoff waters have low Mg/Ca ratios of 0.3 to 0.5, and calcite is precipitated in shallow valley-bottom sediments or dilute lakes. Where groundwaters emerge, carbonate spring deposits may form either as calcite or aragonite. According to WATSPEC, streams draining the Marble Range are slightly undersaturated with respect to calcite, but several dilute lakes, including Kelly Lake, are slightly supersaturated. Clinton Creek Spring waters, where both aragonite and calcite have formed recently, are saturated with respect to aragonite and dolomite. Although little Mg^{2+} is acquired on contact with the limestones, where groundwaters

encounter chloritic greenstones, basic lavas and ultramafic rocks of the Cache Creek terrane, higher Mg/Ca ratios may be obtained than those recorded for runoff.

On the plateau, runoff waters infiltrate permeable glacial deposits. They acquire Mg^{2+} from weathered mafic minerals in the till and thus have higher Mg/Ca ratios (0.7 to 4.3) than runoff on the limestones of the plateau margins. Early evaporation in shallow sediments, and evapotranspiration under the forest cover, produces calcite cements. Although most stream waters on the plateau are undersaturated with respect to carbonate minerals, some are saturated with respect to calcite and dolomite, as are several dilute lakes.

Removal of $CaCO_3$, both as cement or biochemically within dilute lakes, increases the Mg/Ca ratio in the groundwaters. Where they emerge, either in springs or lake marginal seeps, the ratio has increased to between 3 and about 40, varying both with location and time of year. East of Anita Lake (Fig. 1), a seep with an Mg/Ca ratio of 5.6 in May-June had increased to 12.3 by September, but with a substantially reduced discharge. Degassing of CO_2 and evaporation, probably aided by microbial activities, leads to a precipitation of a range of carbonates (calcite, Mg-calcite, aragonite, protodolomite) around the springs.

Evaporative concentration of the shallow groundwaters around the margins of lakes or within playa floors, leads to further precipitation of carbonates, increasing the Mg/Ca ratio to > 300, in some examples. This phase represents the major chemical divide in the development of the two dominant brines. Most waters toward the end of this stage are either relatively enriched in Mg^{2+} or carbonate species. In sulphate lake basins, gypsum precipitation in the mudflats removes most remaining calcium, further increasing the Mg/Ca ratio. At this stage, if $Mg \gg HCO_3 + CO_3$, then further evaporation eventually produces a Mg-Na-SO_4 brine (e.g. Clinton and Three-Mile lakes). If $HCO_3 + CO_3 \gg Mg$, then further evaporation or mixing with dilute waters depletes much of the remaining Mg^{2+}, producing a Na-CO_3-SO_4-Cl brine (e.g.

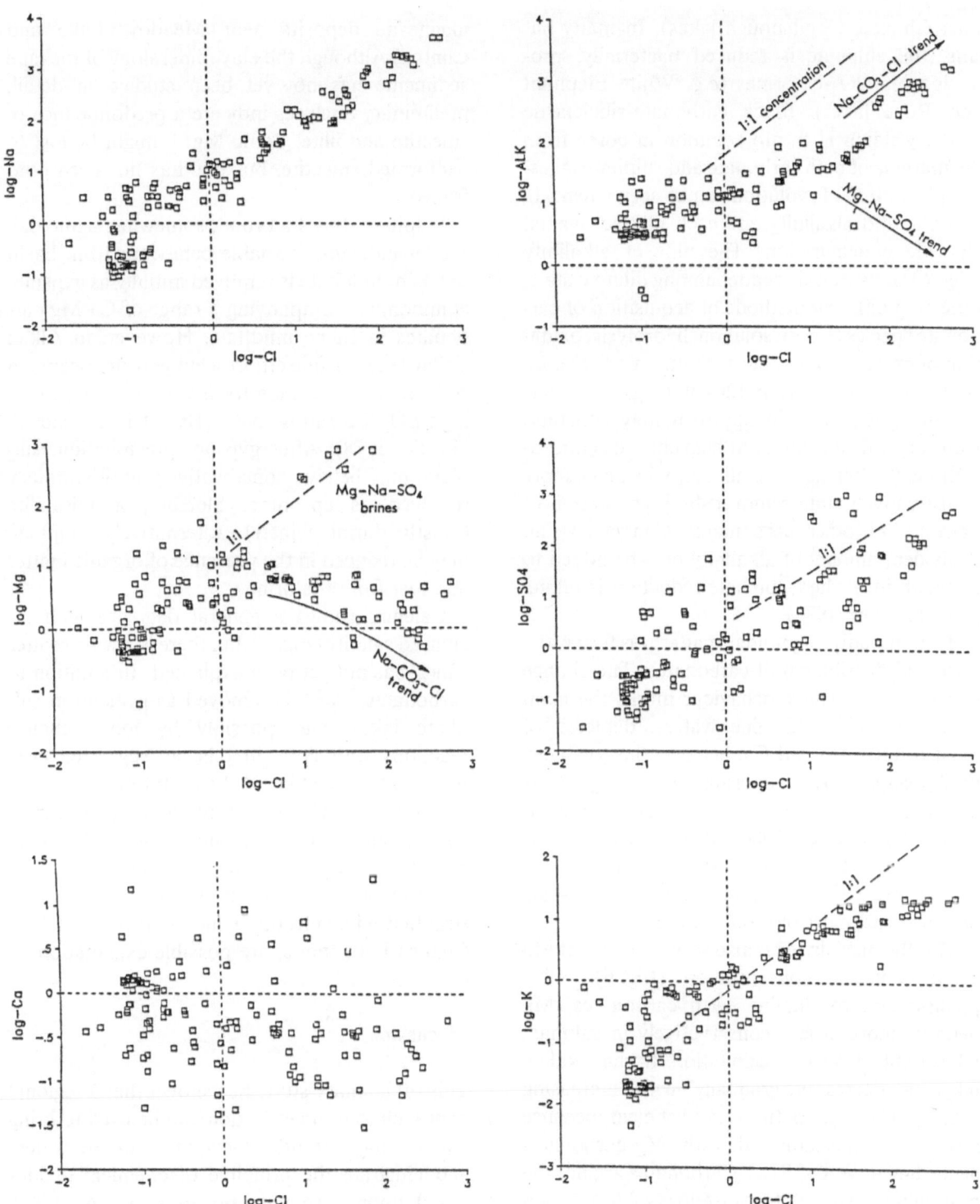

Fig. 6. Log-log plots (molar basis) of selected ions against chloride for 155 Cariboo Plateau region waters. Waters plotted include surface waters for the lakes within the study area listed in Topping & Scudder (1977), together with 93 original analyses. The dashed lines show the slope for mutual conservation of both ions plotted, for comparative purposes. ALK: alkalinity $(HCO_3 + CO_3)$.

Last Chance, Goodenough lakes). In many basins, the sulphate is reduced bacterially, producing Na-CO₃-Cl brines (e.g. White Elephant and Rose lakes). Black carbonate-siliciclastic muds, yielding H_2S, are common in cores from the margins of both carbonate and sulphate lakes.

This pattern of evolution can be seen when Mg and Ca, and alkalinity, are each plotted against chloride concentration. The plot of alkalinity (Fig. 6) shows a wide scatter among dilute waters, reflecting different methods of acquisition of carbonate species (e.g. dissolution, hydrolysis, oxidation of organic matter, etc.). At intermediate concentrations of chloride (20–300 mg l^{-1}), carbonate species remain approximately constant, while chloride increases. At high chloride concentrations (> 300 mg l^{-1}), alkalinity behaves more conservatively until removed by precipitation of natron and other carbonates. Conversely, an equivalent amount of alkalinity may be added to anoxic brines where sulphate reduction is taking place (Drever, 1988).

This broadly sigmoidal pattern reflects the observed distribution of carbonates. The change to more conservative behaviour marks the main chemical divide. Alkaline waters, depleted of most of their Mg and Ca, behave conservatively until precipitated as evaporites. Degassing of CO_2 may occur across much of the range of concentration (cf. Eugster, 1980). In contrast, the sulphate-rich waters are depleted in most of their HCO_3 and CO_3 after precipitation of Mg-carbonates, and most plot off the graph.

The chemical divide is also seen on a plot of Mg versus chloride concentration (Fig. 6). After precipitation of alkaline earth carbonates, Mg behaves more or less conservatively in sulphate lakes until epsomite precipitation. In the alkaline lakes it decreases gradually with increasing salinity. The cause of this is not yet clear, because most are supersaturated with Mg-carbonates according to WATSPEC. Carbonates are not the only mineral species to cause loss of Mg^{2+}. An earthy talc-like mineral, which from XRD data appears to be a disordered sepiolite ($Mg_2Si_3O_8 \cdot nH_2O$), occurs with opaline silica as veinlets within the upper part of some hydro-

magnesite deposits near Meadow Lake and Clinton. Although the clay mineralogy of the lake sediments has not yet been studied in detail, preliminary analyses indicate a predominance of smectite and illite. Some Mg^{2+} might be lost to neoformed smectite, but this has not been confirmed.

Sulphate shows a broad, somewhat sigmoidal, scatter indicating variable behaviour from basin to basin. Sulphate is removed initially as gypsum, commonly accompanying a range of Ca-Mg carbonates in saline mudflats. However, in Alkali Valley (Fig. 1), fine efflorescent gypsum occurs in soils, perhaps forming from waters with initially high SO_4/Ca ratios (path IIIC of in Eugster & Hardie, 1978). After gypsum precipitation, sulphate may behave conservatively until variously removed as epsomite, bloedite or mirabilite (mostly during winter). Alternatively, sulphate may be reduced in the presence of organic matter and lost from the brine.

Calcium shows a general decrease with increased salinity (Fig. 6), but there is a wide scatter which has not yet been explained. In addition to carbonates, Ca^{2+} is removed as gypsum in sulphate lakes, and possibly by ion exchange reactions with Na^+ in tills, as suggested elsewhere by DeDeckker & Last (1989).

Potassium behaves essentially as a conservative element until high salinities (Fig. 6). Removal by ion exchange or sorption (Eugster & Jones, 1979; Drever, 1988), or by silicates such as interstratified illite-smectite, which is present in Last Chance Lake muds, are possible explanations.

Discussion

This preliminary study has shown that a regional approach to brine evolution can be used to bring out the major trends in evolution of the waters and illustrate the principle of chemical divides based upon carbonate precipitation. As Fig. 5 shows, the minerals precipitated during the earliest stages of brine evolution (hillslope and mudflat) are common to both carbonate and sulphate lakes. Waters then appear to follow paths

IIIA or IIIB1b of the Eugster & Hardie (1978: 244, Fig. 5) model. Freeze-out precipitation can abruptly remove solutes from the waters, producing rapid fractionations of brines. However, with the exception of Last Chance Lake, Clinton Lake and a few minor saline ponds, most salts precipitated annually by freeze-out (natron, mirabilite and, possibly, hydrohalite) redissolve in the spring.

Although the general trends are evident, individual basins may follow somewhat different patterns, according to local variations in water composition, and processes and environments of sedimentation. For example, Clinton Lake, is fed both by evolved Mg-rich spring water and less evolved dilute groundwater seepage, leading to mixing and spatially anomalous salt distribution (Renaut, unpublished data).

Although demonstrating the existence of a major chemical divide, the analyses do not explain the reasons for the divergence. As emphasized by Eugster & Hardie (1978), the initial composition of runoff is the main control in brine evolution, even though carbonate precipitation may provide the means of division. The compositional differences between runoff waters originating on the plateau lavas and those on the Cache Creek rocks may, in part at least, explain the diversity of the brines.

Runoff waters draining the Paleozoic limestones, cherts and argillites have molar $HCO_3/Mg + Ca$ ratios from 1.7 to 2.0 and SO_4/Cl ratios of 1.5 to 3.5. In contrast, runoff waters on the plateau have ratios of 2.0 to 2.9 and 0.7 to 1.5, respectively. In other words, the dilute waters more commonly associated with sulphate lakes are initially slightly higher in SO_4 and relatively poorer in HCO_3 than those on the till-covered basalts. Secondly, because the ratio of $HCO_3/Mg + Ca$ is initially closer to unity (partly reflecting dissolution of the limestones and marbles), relatively greater carbonate precipitation might be expected associated with the sulphate lakes. This appears to be the case.

Runoff waters on the Plateau initially have higher Mg/Ca ratios, but the highest Mg concentrations are recorded from the sulphate lakes. Ca is removed by gypsum precipitation, as well as carbonate in the sulphate lake basins, perhaps increasing the Mg/Ca ratio more rapidly than in the alkaline lakes. This may account for a relatively greater abundance of magnesite and hydromagnesite with sulphate-rich waters. However, the high Mg/Ca ratios in sulphate basins may also reflect acquisition by circulating groundwaters of Mg from greenstones, basic lavas and ultramafic rocks of the Cache Creek terrane. No runoff waters from these rocks could be sampled for analysis.

Although Topping & Scudder (1977) were hesitant to follow Cummings (1940) in correlating brine composition with geology, it is notable that $Mg-Na-SO_4$ lakes are present near outcrops of Paleozoic sedimentary rocks, metasediments, and basic volcanics at Clinton, Basque (Nesbitt, 1974; Monger & McMillan, 1984) and in adjacent Washington (Bennett, 1962), whereas sodium carbonate lakes are predominant where underlain by till and volcanic rocks in British Columbia, Washington and Oregon (Allison & Mason, 1947). Long Lake, which has $Na-Mg-SO_4$ brines (Topping & Scudder, 1977), lies within a Paleozoic inlier in the plateau basalts (Fig. 1). The origins of these brines have not been investigated. One possible explanation is that because of the mixed bedrock within the catchment, the inflow waters may have a relatively higher HCO_3/SO_4 ratio than the $Mg-Na-SO_4$ lakes, so that more Mg can be removed as carbonate, leaving sulphate as the main anion and sodium the main cation. A few sodium sulphate lakes are also found at Basque (Goudge, 1926b) and Kamloops (Cummings, 1940).

Concluding remarks

The basic controls of brine diversity on the Cariboo Plateau appear to be firstly, bedrock composition, which controls the availability and initial ratio of most ions in the dilute runoff waters, and secondly, extensive carbonate precipitation which acts as the major chemical divide. Initial precipitation of alkaline earth carbonates takes place in soils, shallow sediments, dilute lakes and at springs, with an accompanying increase in the Mg/Ca ratio. Shallow groundwaters discharging into the lakes across and within the saline mudflats, precipitate extensive carbonate minerals, including abundant magnesite and hydromagnesite, leading to relative enrichment of the waters either in carbonate species or magnesium. Waters depleted in carbonate evolve to become Mg-Na-SO_4 brines, whereas those depleted in alkaline earths become Na-CO_3-(SO_4)-Cl brines.

Acknowledgements

This research was supported by a grant from the Natural Sciences and Engineering Research Council of Canada. I thank Peter Long and Tony Gonzales for help with field work and water analyses.

References

Allison, I. S. & R. S. Mason, 1947. Sodium salts of Lake County, Oregon. Short Pap. Oregon Dept. Geol. Miner. Ind. 17, 12 pp.

Atmospheric Environment Service, 1982. Canadian climate normals 1951–1980: Temperature and precipitation (British Columbia). Environment Canada, Vancouver, 268 pp.

Bennett, W. A. G., 1962. Saline lake deposits in Washington. Wash. Div. Geol. Bull. 44, 129 pp.

Campbell, R. B. & H. W. Tipper, 1971. Geology of the Bonaparte Lake map-area, British Columbia. Mem. Geol. Surv. Can. 363, 100 pp.

Clague, J. J., 1981. Late Quaternary geology and geochronology of British Columbia. Part 2: Summary and discussion of radiocarbon-dated Quaternary history. Pap. Geol. Surv. Can. 80–35, 41 pp.

Cole, L. H., 1913. Gypsum in Canada. Rep. Mines Branch Can. Dept. Mines 245.

Cordey, F., N. Mortimer, P. Dewever & J. W. H. Monger, 1987. Significance of Jurassic radiolarians from Cache Creek terrane, British Columbia. Geology 15: 1151–1154.

Cummings, J. M., 1940. Saline and hydromagnesite deposits of British Columbia. Bull. B.C. Dept. Mines 4, 160 pp.

DeDeckker, P. & W. M. Last, 1989. Modern non-marine dolomite in evaporitic playas of western Victoria, Australia. Sed. Geol. 64: 223–238

Drever, J. I., 1988. The geochemistry of natural waters, 2nd Edition. Prentice-Hall, Englewood Cliffs, N.J., 448 pp.

Eugster, H. P., 1980. Geochemistry of evaporitic lacustrine deposits. Ann. Rev. Earth Planet. Sci. 8: 35–63.

Eugster, H. P. & B. F. Jones, 1979. Behaviour of major solutes during closed-basin brine evolution. Am. J. Sci. 279: 609–631.

Eugster, H. P. & L. A. Hardie, 1978. Saline lakes. In: A. Lerman (ed), Lakes: chemistry, geology, physics. Springer, N.Y.: 237–293.

Fulton, R. J., 1984. Quaternary glaciation, Canadian Cordillera. In: R. J. Fulton (ed), Quaternary stratigraphy of Canada – A Canadian contribution to I.G.C.P. Project 24. Pap. Geol. Surv. Can. 84–10: 39–47.

Goldsmith, J. R. & D. L. Graf, 1958. Relation between lattice constants and composition of the Ca-Mg carbonates. Am. Mineral. 43: 84–101.

Goudge, M. F., 1926a. Sodium carbonate in British Columbia. Rep. Mines Branch Can. Dept. Mines 642: 81–102.

Goudge, M. F., 1926b. Magnesium sulphate in British Columbia. Rep. Mines Branch Can. Dept. Mines 642: 62–80.

Goudge, M. F., 1946. Limestones of Canada: their occurrence and characteristics, Part V: Western Canada. Rep. Mines Branch Can. Dept. Mines 811, 233 pp.

Grant, B., 1987. Magnesite, brucite and hydromagnesite occurrences in British Columbia. Open File Rep. B.C. Geol. Surv. Branch 1987–13, 68 pp.

Hammer, U. T., 1986. Saline lake ecosystems of the world. Junk, Dordrecht, 616 pp.

Hardie, L. A., 1984. Evaporites: marine or non-marine? Am. J. Sci. 284: 193–240.

Hardie, L. A. & H. P. Eugster, 1970. The evolution of closed-basin brines. Spec. Pap. Mineral. Soc. Am. 3: 273–290.

Hardie, L. A., J. P. Smoot & H. P. Eugster, 1978. Saline lakes and their deposits: a sedimentological approach. In: A. Matter & M. E. Tucker (eds), Modern and ancient lake sediments. Blackwell, Oxford: 7–41.

Holland, S. S., 1964. Landforms of British Columbia. Bull. B.C. Dept. Mines 48, 138 pp.

Hudec, P. & P. Sonnenfeld, 1980. Comparison of Caribbean solar ponds and inland solar lakes of British Columbia. In: A. Nissenbaum (ed), Hypersaline brines and evaporitic environments. Elsevier, Amsterdam: 101–114.

Jenkins, O. P., 1918. Spotted lakes of epsomite in Washington and British Columbia. Am. J. Sci. 46: 638–644.

Jones, R. K. & R. Annas, 1978. Vegetation. In: K. W. G. Valentine, P. N. Sprout, T. E. Baker & L. M. Lavkulich (eds), The soil landscapes of British Columbia. Resour. Anal. Branch, Ministry Environ. B.C., Victoria, B.C.: 35–45.

McCammon, J. W., 1958. Limestone deposits in the Ashcroft-Clinton area. Rep. Minister Mines Can. 1958: 90–92.

Monger, J. W. H., 1977. Upper Paleozoic rocks of the western Canadian Cordillera and their bearing on Cordilleran evolution. Can. J. Earth Sci. 14: 1832–1859.

Monger, J. W. H. & W. J. McMillan, 1984. Bedrock geology of Ashcroft (92 I) area. Open File Rep. Geol. Surv. Can., 980.

Nesbitt, H. W., 1974. The study of some mineral-aqueous solution interactions. Ph.D. Thesis, Johns Hopkins University, Baltimore, Maryland (unpublished).

Northcote, T. G. & K. J. Hall, 1983. Limnological contrasts and anomalies in two adjacent saline lakes. In: U. T. Hammer (ed), Proc. 2nd Int. Symp. Athalassic (Inland) Saline Lakes, Devel. Hydrobiol. 16, Junk, The Hague: 179–194.

Northcote, T. G. & T. G. Halsey, 1969. Seasonal changes in the limnology of some meromictic lakes in southern British Columbia. J. Fish. Res. Bd Can. 21: 1763–1787.

Reinecke, L., 1920. Mineral deposits between Lillooet and Prince George, British Columbia. Mem. Geol. Surv. Can., 118.

Renaut, R. W. & P. R. Long, 1986. Post-glacial travertine deposits of the Clinton area, Interior British Columbia. Prog. Abstr. Joint Annu. Meet. Geol. Assoc. Can., Miner. Assoc. Can. 11: 117.

Renaut, R. W. & P. R. Long, 1987. Freeze-out precipitation of salts in saline lakes – examples from Western Canada. In: G. L. Strathdee, M. O. Klein & L. A. Melis (eds), Crystallization and precipitation. Pergamon, Oxford: 33–42.

Renaut, R. W. & P. R. Long, 1989. Sedimentology of the saline lakes of the Cariboo Plateau, Interior British Columbia, Canada. Sed. Geol. 64: 239–264.

Scudder, G. G. E., 1969. The fauna of saline lakes on the Fraser Plateau in British Columbia. Verh. Int. Ver. Limnol. 17: 430–439.

Tipper, H. W., 1971. Glacial geomorphology and Pleistocene history of central British Columbia. Geol. Surv. Can. Bull. 196, 89 pp.

Topping, M. S. & G. C. E. Scudder, 1977. Some physical and chemical features of saline lakes in central British Columbia. Syesis 10: 145–166.

Valentine, K. W. G. & A. B. Dawson, 1978. The Interior Plateau. In: K. W. G. Valentine, P. N. Sprout, T. E. Baker & L. M. Lavkulich (eds), The soil landscapes of British Columbia. Resour. Anal. Branch Ministry Environ. B.C., Victoria, B.C.: 121–134.

Valentine, K. W. G. & A. Schori, 1980. Soils of the Lac La Hache – Clinton area, British Columbia. Rep. B.C. Soil Surv. 25, 118 pp.

Wigley, T. M., 1977. WATSPEC: a computer program for determining speciation of aqueous solutions. Tech. Bull. Brit. Geomorphol. Res. Group 20, 48 pp.

Hydrobiologia **197**: 83–89, 1990.
F. A. Comin and T. G. Northcote (eds), Saline Lakes.
© 1990 *Kluwer Academic Publishers.*

Nutrients in pore waters from Dead Sea sediments

Arie Nissenbaum[1], Mariana Stiller[1] & Aminadav Nishri[2]
[1]*Isotope Research Department, Weizmann Institute of Science. Rehovot 76100, Israel;* [2]*Allon Kinneret Laboratory, Israel Oceanographic and Limnological Research Organization, Tabkha, Israel*

Key words: Dead Sea, pore waters, early diagenesis, Ammonia, phosphate, carbonate, nutrients

Abstract

Pore waters were separated from 50 cm-long cores of Dead Sea sediments raised from waters depths of 25, 30 and 318 m. The salinity of the pore water is close to that of the overlying water at 225–230 g l^{-1} chloride. The titration alkalinity of the pore water is about 60% of the overlying water, and sulfate is also depleted. Ammonia and phosphate concentrations are higher than those of the water column with up to 50 mg l^{-1} N-NH$_3$ (ten times increase) and 350 μg l^{-1} P-PO$_4^{3-}$ (four to eight times increase). Early diagenetic reactions are a result of decomposition of organic matter and of water-sediment interactions, resulting in aragonite precipitation, phosphate removal to the sediments, probably by absorption on iron-oxyhydroxides followed by remobilization, reduction of sulfate and formation of iron sulfides and accumulation of ammonia. Mass balance calculations show that pore water contribute about 80% of the ammonia and 30% of the phosphate input into the Dead Sea water column. On the other hand, the sediments act as a sink for carbonate and sulfate.

Introduction

The Dead Sea occupies the deepest part of the Jordan–Dead Sea–Arava Rift Valley which is northernmost extension of the East African Rift Valley. With its surface at − 406 meters below sea level, it is the lowermost exposed surface on the face of the earth.

The Dead Sea is a prime example of a closed lake. It has no outlet for its water, except by evaporation, and its salt content is about 340 g l^{-1}. The lake is about 80 km long and 11 km wide. However, the dimensions of this water body vary considerably owing to changes in water level. About 20 years ago the lake was divided into two basins; a Northern basin 330 m depth and a Southern basin with a depth of less than 8 m. The

two basins were separately by a shallow sill. The recent drop in lake level by natural causes, but accelerated by the diversion of water from the Jordan River, its major supply of water, resulted in a 10 m decrease in level and as a result the Southern basin completely dried up to become a salt playa, except for an area into which Dead Sea water are pumped to form a solar evaporation pond for the production of potash.

The Dead Sea has been a point of interest for more than 2300 years (Nissenbaum, 1979). The most thorough investigation of the lake in recent times was made by Neev & Emery (1967) for the period 1959–1963. At that time the lake was stratified with the pycnocline at about 40 m depth. The pycnocline separated an oxygen containing (ca. 1 mg l^{-1}) upper water mass from a reducing, sulfide-containing lower water body. This stratification which existed for about 260 to 280 years (Stiller & Chung, 1984) was destroyed in 1979 when the lake overturned (Steinhorn *et al.*, 1979) which resulted in marked changes in the chemistry and the physical structure of the water column.

The average chemical composition of the water (in 1977) was reported by Beyth (1980) as (in g l^{-1}): Mg = 44; Ca = 17.2; Na = 40.1; K = 7.7; Cl = 224.9; Br = 5.3; SO_4 = 0.45; HCO_3 = 0.1; TDS = 339.6; density = 1.232.

The biota of the Dead Sea is extremely limited in variety and the dominant (and perhaps the only) forms of life are the green alga *Dunaliella* and the bacterium *Halobacterium* (Nissenbaum, 1975; Oren, 1981). The distribution of microorganisms is quite patchy in space and time. Bacterial counts may reach a value of 1×10^7 to 5.5×10^3 cells per ml and algal counts of between 0 and 8800 cells per ml have been found (Oren, 1981). The organisms are probably at the limits of survival due to total salt content and chemical composition of the brine (Oren, 1981) and lack of phosphate (Nissenbaum, 1975).

The sediments of the Dead Sea are laminated deposits of allochthounous material such as clay, calcite and dolomite and authochthonous aragonite, gypsum (in near-shore deposits) and halite.

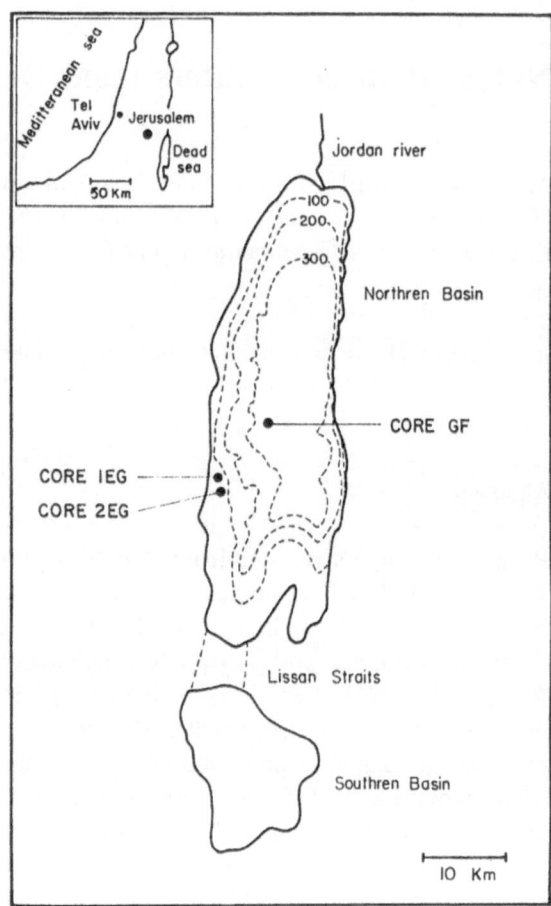

Fig. 1. Location map for the studied cores. Isobaths are in meters.

Materials and methods

Sediment samples for the present study were collected by gravity corers from three locations (Fig. 1) at the following depths: core 1-EG from 25 m; core 2-EG from 30 m depth in the shallow part of the lake and core GF from 318 m in the deep central part of the basin. Sediments at the shallower sites are composed of aragonite, detrital minerals and varying amounts of gypsum (usually a few percent but occasionally up to 20%). The sediments are reducing and two distinct black zones of iron sulfide were noted. The deep core, GF, is composed of 70% detritus and distinct white laminae of aragonite. In the top 20 cm of the cores gypsum concentration may reach 20% but below 20 cm very little gypsum is found.

Pore waters were separated from the sediments, after sectioning, by centrifugation. The whole operation, from sectioning to collection of pore water took a few minutes. The waters were filtered through 0.45 μm Millipore filter and analyzed for alkalinity by potentiometric titration, ammonia by colorimetry and phosphate by co-precipitation with aluminium hydroxide followed by dissolution and colorimetric determination as molybdophosphate. Sulfate was determined gravimetrically by precipitation as $BaSO_4$.

Results

Carbonate

A detailed description of the carbonate system in Dead Sea water and sediments is given by Nishri (1982) and the pore water data is summarized below. Titration alkalinities as measured in the pore water in both the shallow and deep sediments are on the average 2.5×10^{-3} meq 1^{-1}. The pH as measured by 'punching' an electrode into the wet sediment was always lower than 6.

The titration alkalinity of the overlying water is significantly higher at about 3.5×10^{-3} meq 1^{-1} and the pH is about 6.25. However, because the titration alkalinity refers to all protonated species, contributions from other sources (e.g. borate) must be considered. Direct determination of dissolved inorganic carbon (Stiller & Talma, unpublished data) gives values of around 1×10^{-3} meq 1^{-1}).

If, as a first approximation, the contribution to titration alkalinity from non-carbonate sources is the same in the pore water as in the water column, then the CO_3^{2-} concentration as calculated for the pore water is about one third that of the overlying water.

Phosphate

Data on phosphate concentrations in our cores are given in Tables 1 and 2. Phosphate is quite depleted in the water column of the lake with dissolved $P\text{-}PO_4^{3-}$ values around 30 μg 1^{-1} and particulate $P\text{-}PO_4^{3-}$ of 10 to 50 μg 1^{-1} (Garber, Stiller & Nissenbaum, unpublished data). The concentrations in the pore water exceed those of the water column by a factor of four to eight, reaching values of 200 to 350 μg 1^{-1}, with the deep core showing on the average the highest values. The distribution with depth is somewhat erratic, partly due to analytical difficulties. However, depletion of phosphate appears to occur already in the top few cm in all sites (Fig. 2).

Ammonia

Ammonia is highly enriched in Dead Sea water, with values of 6.1 to 7.7 mg 1^{-1}. All other species of nitrogen are minor (Garber, Stiller & Nissenbaum, unpublished data). The pore waters (Tables 1 and 2) show values higher by a factor of two to three in the shallow cores and nine to ten in the deep core. The distribution with depth is concave (Fig. 2) probably due to a diffusion controlled process.

Table 1. Ammonia and Phosphate data from two shallow cores of Dead Sea Sediments

Station	Depth in Core (cm)	N-NH$_3$ (mg/l)	P-PO$_4$ (mg/l)
Core 1-EG, 25 m	overlying water	6.0	n.d
	0 – 1.7	8.0	n.d
	1.7– 4.7	9.2	n.d
	4.7– 9.4	13.1	0.24
	9.2–10.7	16.6	n.d
	10.7–16.2	15.8	0.26
	16.2–20.9	17.0	0.26
	20.9–27.9	17.5	0.34
	27.9–30.5	18.5	0.35
	30.5–33.0	18.7	n.d
	33.0–35.8	18.0	0.22
	35.8–39.9	18.0	n.d
Core 2-EG, 30 m	overlying water	6.0	0.04
	0.0– 0.5	n.d	n.d
	0.5– 3.0	7.8	0.20
	3.0– 6.5	8.8	n.d
	6.5– 9.0	9.9	0.17
	9.0–11.2	10.1	0.27
	11.2–13.0	9.8	0.20
	13.0–15.5	10.0	0.18
	15.5–18.3	8.5	0.24
	18.3–22.2	9.5	0.17
	22.2–25.6	10.0	0.23
	25.6–28.6	8.5	0.24
	28.6–29.4	10.0	n.d
	29.4–31.6	8.4	0.10
	32.0–35.5	10.5	n.d

n.d – not determined

86

Table 2. Nutrient and other data on pore water from deep core of Dead Sea Sediments.

Station	Depth in core (cm)	pH	Cl (gr/l)	N-NH$_3$ (mg/l)	P-PO$_4$ (mg/l)
GF, 318 m depth	overlying water	6.1	230	21.6	0.04
	0.0– 7.0	5.8	232	24.5	0.13
	7.0– 8.0	5.7	240	26.5	n.d
	8.0–12.5	5.6	240	35.5	0.53
	12.5–14.8	5.5	236	39.0	n.d
	14.8–17.0	5.7	232	36.5	0.27
	17.0–20.4	n.d	234	40.0	0.22
	20.4–26.0	5.3	223	41.5	0.25
	26.0–30.0	n.d	225	45.0	n.d
	30.0–41.0	5.4	234	50.0	0.33
	41.0–49.2	5.4	221	58.5	0.27

n.d – not determined

Sulfate

Information on the biogeochemistry of sulfur in Dead Sea sediments was provided by several authors. It is added here in order to consolidate all the data on Dead Sea pore waters in one place.

Neev & Emery (1967) pointed out that even though gypsum crystallized occasionally from the upper water mass, and constituted a considerable portion of the deposits in sediment traps, it was almost completely absent from the deep, anaerobic, sediments. This was attributed to post-depositional reduction by sulfate-reducing bacteria. Lerman (1967) calculated the rate of sulfate reduction in sediment from the southern Basin and found it to be comparable to rates observed in sediments from the marine basins of Southern California.

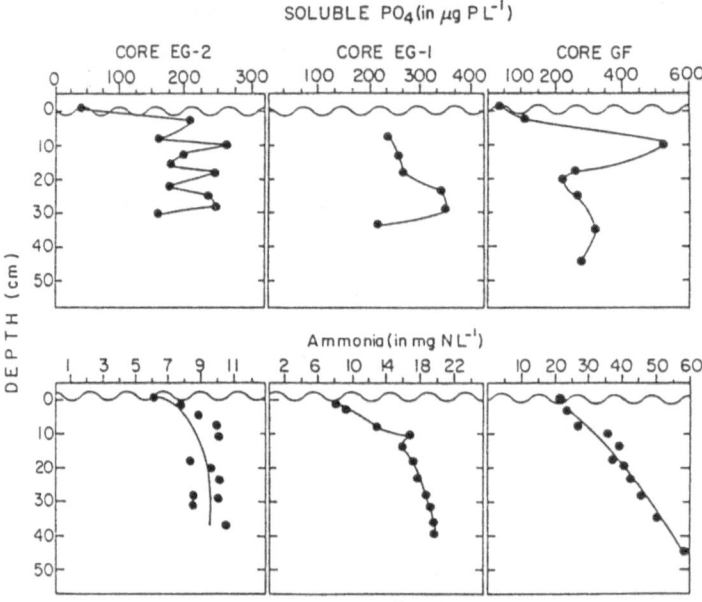

Fig. 2. Distribution of ammonia and phosphate in the pore water.

Table 3. $\delta^{34}S$ Values for sulfur species in Dead Sea Water (before overturn of 1979) and pore water ($\delta^{34}S$ values vs. CDM standard)

	Station	Depth	Sulfate	Dissolved H_2S	Acid volatile sulfide	Organic sulfur
Water Column	Ein Gedi	300 m	+ 14.2‰	− 21.7‰		
	Massada	120 m	+ 15.0‰	− 19.6‰		
Sediments	Ein Gedi	330 m	+ 20.6‰	− 16.3‰	− 16.3‰	− 19.6‰
	Massada	165 m	+ 10.0‰	− 12.7‰	− 15.9‰	− 19.6‰

After Nissenbaum & Kaplan (1976).

Nissenbaum & Kaplan (1976) reported the sulfate concentration of pore water from grab samples of deep sediments to be about 40 mg l⁻¹ or about 10% of the values in the overlying water. Levy (1980) reported values of 230 to 335 mg l⁻¹ from pore water of two cores. Although these values are higher than those reported by Nissenbaum & Kaplan (1976) they are still depleted by 30 to 50% as compared with the overlying waters.

Isotopic studies by Nissenbaum & Kaplan (1976) using $^{34}S/^{32}S$ ratios indicated the role of sulfate reduction by biological activity both in the then stratified, lower water column and in the sediments (Table 3).

The sulfide which is produced by the reduction of sulfate is fixed into the sediments as poorly crystalline iron sulfides. Pyrite and elemental sulfur occur in only minor amounts. The isotopic fractionation factors (α) between the sulfate and reduced sulfur in pore waters and in the sediments are 1.0029 to 1.0030. These values are similar to those found in the basins of Southern California (Kaplan *et al.*, 1963). The nature of the organism responsible for sulfate reduction is unknown at present.

Discussion

Early diagenesis

The data presented show that early diagenesis results in marked difference in the concentration of nutrients in the pore water as compared with the overlying water. Two types of processes are involved. They are: biological activity and water-sediment interactions.

Biological activity is expressed in the decrease of sulfate by sulfate-reducing bacteria and the formation of iron sulfides. Reduction of sulfate is accompanied by consumption of organic matter and Nishri (1982) showed a decrease in the organic carbon of the sediments from roughly 0.7% organic C in a layer deposited 600 years ago to 0.34% in a 2 100 year old sample to 0.23% in a 7 000 year old layer. This decrease (if we assume constant flux of organic carbon to the sediments and a uniform rate of deposition) can not be accounted solely by consumption of organic matter by sulfate reducers. The major source for the sulfate is probably not through diffusion of sulfate from the impoverished overlying water but through the dissolution of gypsum which is 'rained' into the sediment from the water column. The mineralization of organic matter is also responsible for the release of ammonia and phosphate into the pore water. The ratio of $N-NH_4^+$ to $P-PO_4^{3-}$ in the pore water is roughly ten times higher than expected from the Redfield ratio of 16. As a first approximation, the ammonia content can be considered to behave conservatively, as evidenced from ammonia accumulation in the Dead Sea water column. The depletion of phosphate seems to be controlled by solubility products of mineral phases or other types of sediment-water interactions. Nishri (1982) showed that pore water from the Dead Sea have phosphate concentration profiles which parallel those of Fe. Thus, some of the phosphate which is released by the mineralization of organic matter can be absorbed by iron oxyhydroxides. Reduction of the oxides will result in remobilization of iron in the sediment column accompanied by that of phosphate.

Using the apparent stability constants for the carbonate system in Dead Sea water (Sass & Ben Yaakov, 1977), and ignoring contribution to titration alkalinity from non- carbonate species, Nishri (1982) reported a calculated constant value of 4×10^{-4} M l^{-1} CO$_3^{2-}$ in pore water from core EG-1 and values three times higher for the overlying water at about 1.2×10^{-3} M l^{-1}.

The constancy of dissolved carbonate concentration in the pore water, and the fact that 50 to 60% of the sediments are composed of aragonite and calcite, suggests that the pore water may be in equilibrium with respect to aragonite. The concentration of calcium in the pore water is the same as in the overlying water. Nishri (1982) calculated that the overlying water is supersaturated with respect to aragonite by a factor of three. This conclusion is supported by the observation that addition of aragonite to Dead Sea water (in the laboratory) results in decrease of pH, as expected

from precipitation of aragonite from a supersaturated solution. It is therefore possible that carbonate diffuses from the water column into the sediments resulting in the diagenetic accumulation of aragonite derived from pore water below the sediment-water interface.

Inputs-Outputs

The pore water of the Dead Sea can act both as sources and as sinks for nutrients. The increased concentration of ammonia and phosphate in the pore water results in upward diffusion of those nutrients. The gradient of nutrients from the different cores have been calculated based on the linear gradient in core GF (Table 4).

Using D values of 1.98×10^{-5} cm^2 sec^{-1} for ammonia and 4×10^{-6} cm^2 sec^{-1} for phosphate (Berner, 1980) at 25 °C, which is very close to the actual temperature of the sediments, we obtain annual inputs from the pore water into the water column of 5 000 to 6 000 ton yr^{-1} ammonia and 44 ton yr^{-1} phosphate. As compared with other sources (floods, river water, springs), the pore water are, by far, the largest contributor of ammonia (estimated at 80–85% of the total input), but are a less important source for phosphate (33% of total input).

The sediments act as a sink for sulfate and carbonate, and very probably there is a downward diffusive flux of those species into the sediment. However, the involvement of other parameters such as particulate transport of gypsum and aragonite, does not allow the quantitative estimate of the contribution of those species from the overlying water into the pore water.

Finally, it is interesting to note that the Dead Sea behaves in many respects as a 'marine' system where the sediments are a very important source of nutrients. This differs from the situation in many lakes where various types of surface water input can be a major source of nutrient.

Table 4. Gradients of ammonia and phosphate in the three cores from Dead Sea.

Core	Water depth	Ammonia $\Delta C/\Delta z$ (in mg cm^{-4})	Phosphate $\Delta C/\Delta Z$ (in mg cm^{-4})
GF	318 m	0.81×10^{-3}	$2.5\text{-}5.0 \times 10^{-5}$
1-EG	25 m	0.93×10^{-3}	3×10^{-5}
2-EG	30 m	$1.0 \ \times 10^{-3}$	9×10^{-5}

Acknowledgements

We wish to warmly acknowledge Naomi Bauman for very skillful assistance in the laboratory. This research was supported by a grant from the Gesellschaft für Strahlen und Ummwitforschung, München, Germany and the National Council for Research and Development, Israel.

References

Berner, R. A., 1980. Early Diagenesis: a theoretical approach. Princeton Univ. Press. Princeton, N.J., 241 pp.

Beyth, M., 1980. Recent evolution and present stage of Dead Sea brines. In: Hypersaline Brines and Evaporitic Environments. In: A. Nissenbaum (ed.), Elsevier Pub. Co., Amsterdam, 155–166.

Kaplan, I. R., K. O. Emery & S. C. Rittenberg, 1963. The distribution and isotopic abundance of sulphur in recent marine sediments off Southern California. Geoch. Cosmoch. Acta, 27: 297–331.

Levy, Y., 1980. Chemistry of bottom sediments and interstitial water from the Dead Sea: Geol. Survey of Israel. report MG/8/80, 11 pp.

Lerman, A., 1967. Model of chemical evolution of a chloride lake – the Dead Sea. Geoch. Cosmoch. Acta 31: 2309–2330.

Neev, D. & K. O. Emery, 1967. The Dead Sea: depositional processes and environments of evaporites. Geol. Survey of Israel: report no. MG/41/67. 147 pp.

Nishri, A., 1982. The Geochemistry of manganese and iron in the Dead Sea. Unpublished Ph. D. thesis, Weizmann Institute of Science.

Nissenbaum, A., 1975. The microbiology and biogeochemistry of the Dead Sea. Microb. Ecol. 2: 139–161.

Nissenbaum, A., 1979. Life in a Dead Sea – Fables, Allegories and Scientific Search. Bioscience 24: 153–157.

Nissenbaum, A. & I. R. Kaplan, 1976. Sulfur and carbon isotopic evidence for biological processes in the Dead Sea ecosystem. In: Environmental Biogeochemistry, 1. (ed. J. O. Nriagu). Ann. Arbor Sci. Pub., Mich., 309–325.

Oren, A., 1981. Approaches to the microbial ecology of the Dead Sea. Kieler Meerforschung. Sonderheft. 5: 416–434.

Sass, E. & S. Ben Yaakov, 1977. The carbonate system in hypersaline solutions: Dead Sea brines. Mar. Chem. 5: 183–199.

Steinhorn, I., G. Assaf, J. R. Gat, A. Nishri, A. Nissenbaum, M. Stiller, M. Beyth, D. Neev, R. Garber, G. M. Friedman & W. Weiss, 1979. The Dead Sea: deepening of the mixolimnion signifies the overture to overturn of the water column. Science 206: 55–57.

Stiller, M. & Y. Chung, 1984. Ra-226 in the Dead Sea a possible tracer for the duration of meromixia. Limnol. Oceanogr. 29: 574–586.

Hydrobiologia **197**: 91–97, 1990.
F. A. Comín and T. G. Northcote (eds), Saline Lakes.
© 1990 *Kluwer Academic Publishers.*

Influence of salinity on the concentration and rate of interchange of dissolved phosphate between water and sediment in Fuente Piedra lagoon (S. Spain)

V. Clavero[1], J.A. Fernández[2] & F.X. Niell[1]
[1]*Laboratorio de Ecología. Facultad de Ciencias, Universidad de Málaga, Campus de Teatinos, 29071 Málaga, Spain;* [2]*Laboratorio de Fisiología Vegetal, Facultad de Ciencias, Universidad de Málaga, Campus de Teatinos. 29071 Málaga, Spain*

Key words: Sediment, water column, phosphate, interchange, salinity

Abstract

Short (60 minutes) and long term (3 hours) experiments were performed to measure the final equilibrium phosphate concentration in water and the net fluxes of phosphate interchange between water and sediment at different salinities. The rate of phosphate release from the sediment increases with the salinity increment, as well as the final equilibrium phosphate concentration. In both short and long term experiments, the net rate of dissolved phosphate removal follows a saturation kinetics except for long term experiments at 70 g 1^{-1} salinity. In this case, the relationship between net removal and dissolved phosphate concentration is linear. The experiments show that salinity stimulates phosphate release from sediment.

Introduction

Relationships between water and sediment have been recognized as important factors in the phosphorus cycle of fresh and seawater systems. Fluxes between water and sediment are influenced by physical, chemical and biological factors (Mortimer, 1941; Kamp- Nielsen, 1974).

The solubility of phosphate in interstitial water depends mainly on redox conditions of the sediment (Mortimer, 1941). In some cases phosphorus release and removal exhibit a fluctuating pattern that can be correlated with continuous changes, in Eh and pH of the sediment ranked in order of importance (Anderson & Gahnstrom, 1985). Nothing is known about the influence of the ionic strength of the interstitial water on the

magnitude of these fluxes. In certain sediments specially saline lakes and estuaries, periodic changes in the ionic concentration in the sediment interstitial water can be observed.

Some authors have found a strong correlation between salinity and the concentration of dissolved phosphate. For instance, a decrease of 0.3 μM in dissolved phosphate has been reported by Atkinson (1987) when the salinity increased from 35 to 65 g 1^{-1} in a hypersaline bay of Australia. In ten localities of York Peninsula, also in Australia Tominaga *et al.* (1987) showed an increase in the concentration of dissolved phosphate with increasing water salinity.

In the Fuente Piedra Lagoon of southern Spain water salinity is negatively correlated with the rainy season. Salt concentration ranges from 10 g

l^{-1} in winter to saturation level in summer (Herberg, 1986; Lucena *et al.*, in press). Among the ions investigated by these authors sodium and chloride exhibit a major change in concentration. The results of these authors show that salinity, and probably changes in NaCl amount, could exert a strong influence on the equilibrium phosphate concentration in the water.

The aim of this work is to investigate the influence of salinity in: a) the kinetics of phosphate interchange between water and sediment, and b) the final equilibrium phosphate concentration in water with different salinities, in systems where salinity is not dependent on NaCl, using sediment as source of phosphate.

Material and methods

The experiments were performed in aquaria of $20 \times 20 \times 20$ cm. The sediment for the experiments was collected in Fuente Piedra Lagoon, and athalassic lake near Málaga (southern Spain) by means of plastic cores 25 cm long and 15 cm internal diameter, and stored out of direct sunlight at temperatures below $15°$ C. Some characteristics of the sediment are shown in Table 1 (see also Vargas *et al.*, 1983).

Wet, homogeneous sediment was placed in the bottom of the aquaria and water containing different amounts of NaCl(10, 30, 50 and 70 g l^{-1})

Table 1. Some chemical characteristics of the surface sediment in the sampling area. Data for $CaCO_3$, total nitrogen, organic matter, potassium, total phosphorus, sodium and chloride are expressed in $mg\,g^{-1}$ and conductivity is expressed in μmhos.

pH	8,00
Co_3Ca	0,45
N-Total	0,001
Organic matter	0,015
K^+	1,66
P-Total	0,13
Na^+	5,04
Cl^-	1,8
Conductivity	4,5

was added. The proportion between sediment and water was $1:5$ (w/w).

The system was allowed to equilibrate for 24 hours at room temperature (20 °C) under white fluorescent lamps ($100\,\mu E\,m^{-2}\,s^{-1}$), after which a known amount of phosphate was added to water to produce a final phsophate concentration of 1, 3, 5 and 7 μM.

This moment was considered time zero in every experiment. After phosphate addition, two types of experiments were carried out: short term and long term. In short term experiments measurements were made every 5 minutes for one hour and in long term experiments every 30 minutes for three hours. The water was gently mixed before a 2 ml sample was taken to measure the phosphate concentration. The samples were filtered immediately using Whatman GF/C filters and dissolved inorganic phosphate (DIP) was measured by the malachite green method (Fernández *et al.*, 1985) in a Technicon AAII autoanalyzer.

Eh measurements were performed using two platinum electrodes, one placed in the water 2 cm above the sediment and the other 1 cm below the sediment-water interface. The electrodes were connected to a CRISON model 501 pH-Eh meter.

Phosphate concentration was plotted against time and the net fluxes were computed considering the decrease of phosphate concentration in water (insolubilization or/and adsorption) as a positive rate and the increase of phosphate concentration in water (solubilization/desorption) as a negative rate. These calculated rates were plotted against phosphate concentration, the origin of such flux.

DIP flow rate versus phosphate concentration were fitted to a saturation equation:

$$V = V_{max}(1 - EP/P) \qquad Eq1$$

Where V_{max} is the maximum flow rate, EP is the equilibrium concentration at which no net phosphate flow occurs, P is the phosphate concentration and V is the flow rate. Data were transformed in order to be fitted to the linearized form

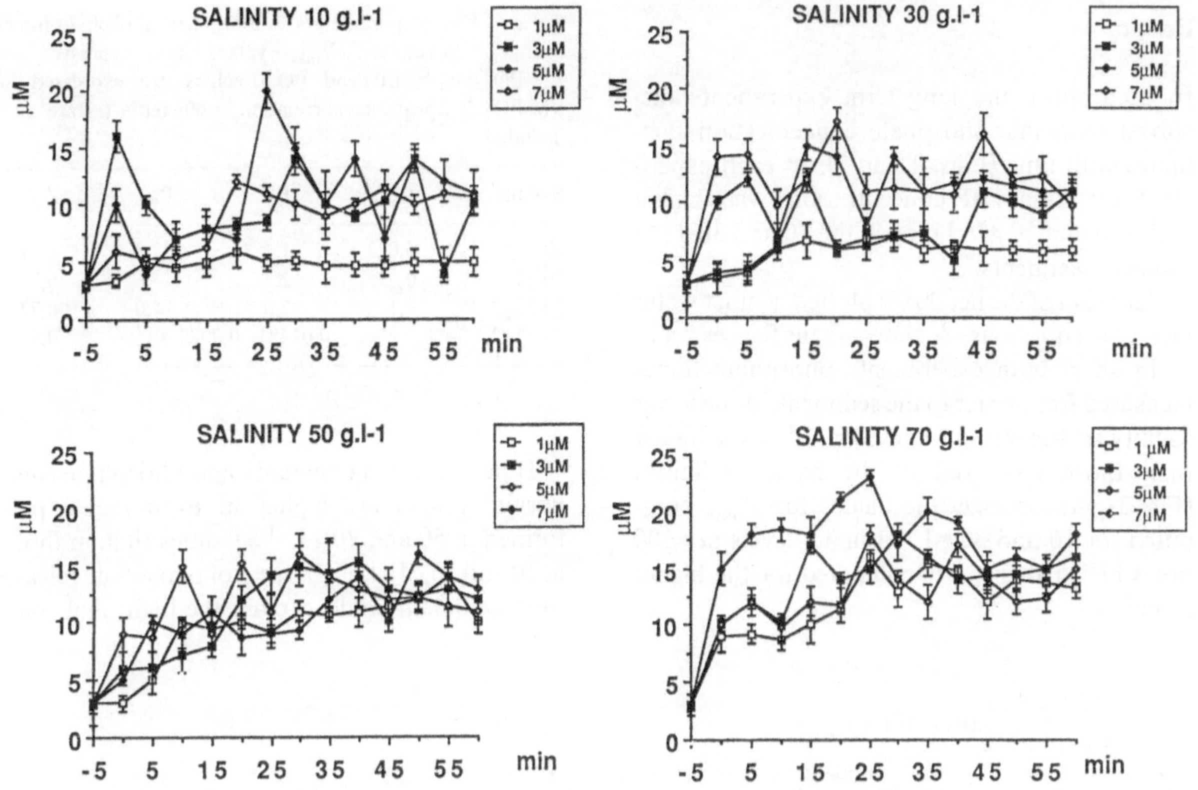

93

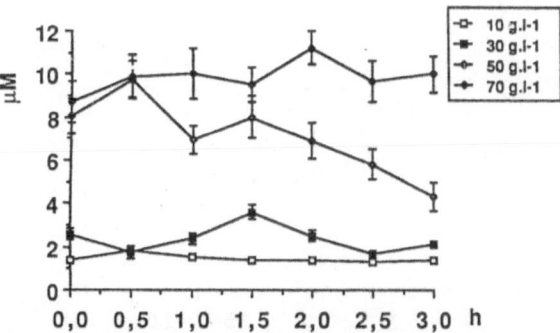

SALINITY 10 g.l-1 — 1μM, 3μM, 5μM, 7μM

SALINITY 30 g.l-1 — 1μM, 3μM, 5μM, 7μM

SALINITY 50 g.l-1 — 1μM, 3μM, 5μM, 7μM

SALINITY 70 g.l-1 — 1μM, 3μM, 5μM, 7μM

Fig. 1. Variation with time of dissolved phosphate concentration (DIP) in water with different salinities in short term experiments adding different amounts of phosphate: 1, 3, 5 and 7 μM. The initial point, marked with a negative number shows the equilibrium phosphate concentration before addition of a known amount of phosphate.

of equation 1:

$$V \cdot P = V_{max} \cdot P - V_{max} \cdot EP \qquad Eq2$$

V_{max} is the computed slope and the equilibrium point (EP) is the Y-intercept divided by the slope of equation 2.

Half saturation constants ($P_{0.5}$) were computed from equation 1, i.e. the value of DIP that yields a V equal to $V_{max}/2$.

Fig. 2. Variation with time of dissolved phosphate concentration (DIP) in water with different salinities in long term experiments without initial phosphate addition.

Table 2. Kinetic parameters expressed in $\mu M\ PO_4^{-3}\ h^{-1}$ (V_{max}) and $\mu M\ PO_4^{-3}$ (EP and $P_{0.5}$) along with Spearman correlation coefficients in short term experiments (less than 1 hour). The final figures of Eh in water (Ehw) and sediment (Ehs) are also included.

Salinity	V_{max}	EP	$P_{0.5}$	r	Ehw	Ehs
10 g l^{-1}	1,4	8,5	17,0	0,72	124	− 153
30 g l^{-1}	1,0	8,8	17,7	0,64	63	− 238
50 g l^{-1}	0,021	12,0	23,9	0,78	70	− 286
70 g l^{-1}	0,03	14,8	29,6	0,74	50	− 291

Results

In both short and long term experiments dissolved inorganic phosphate concentration fluctuates with time (Figs. 1 and 2). In each experiment, the initial DIP concentrations were higher in the 70 and 50 g l^{-1} than in the 10 and 30 g l^{-1} salinity treatments.

Variation of the net fluxes plotted against water DIP concentrations are shown in the figures 3 and 4. In short term experiments phosphate fluxes measured from water to the sediment, at the lower salinity treatments (10 and 30 g l^{-1}) were higher than those measured at the higher salinities (Fig. 3). Accordingly the values for V_{max} computed for 10 and 30 g l^{-1} salinity levels are 100 times higher than those computed for the higher salinities.

Table 3. Kinetic parameters in long term phosphate interchange experiments. V_{max} values are expressed in $\mu M\ PO_4^{-3}\ h^{-1}$; EP and PO.5 values are expressed in $\mu M\ PO_4^{-3}$; Spearman correlation coefficients (r) are also included.

Salinity	V_{max}	EP	$P_{0.5}$	r
10 g l^{-1}	0,022	1,4	2,8	0,81
30 g l^{-1}	0,025	2,3	4,6	0,73
50 g l^{-1}	0,1	7,0	14,0	0,77
70 g l^{-1}	$V = -0,0148 + 0,0024.$ P			0,97

Half saturation constants and phosphate equilibrium points are higher in experiments performed in 50 and 70 g l^{-1} salinities than in those at 10 and 30 g l^{-1}. High rates of phosphate release from sediment could explain the high final equi-

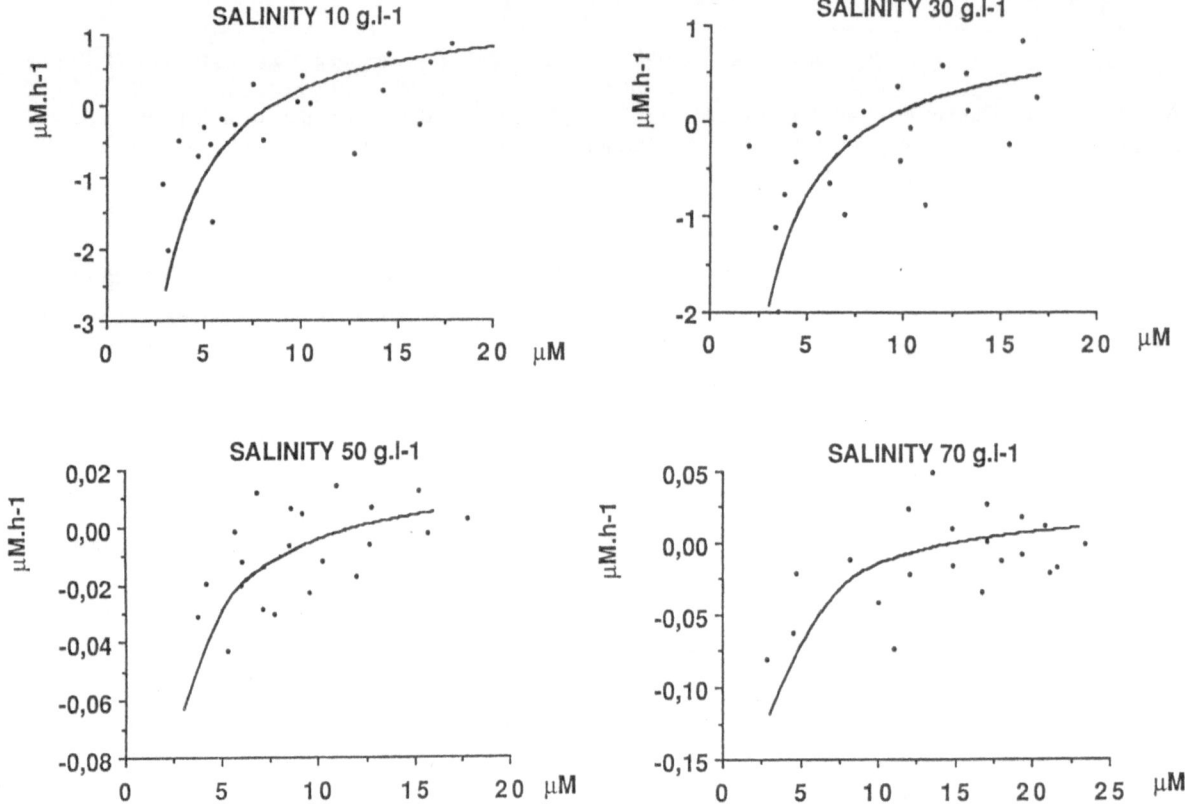

Fig. 3. Variation of the flow rate of dissolved phosphate with phosphate concentration in water with different salinity treatments in short term experiments with different phosphate amounts added to water. Positive values indicate phosphate removal from water and negative ones indicate phosphate release from sediment.

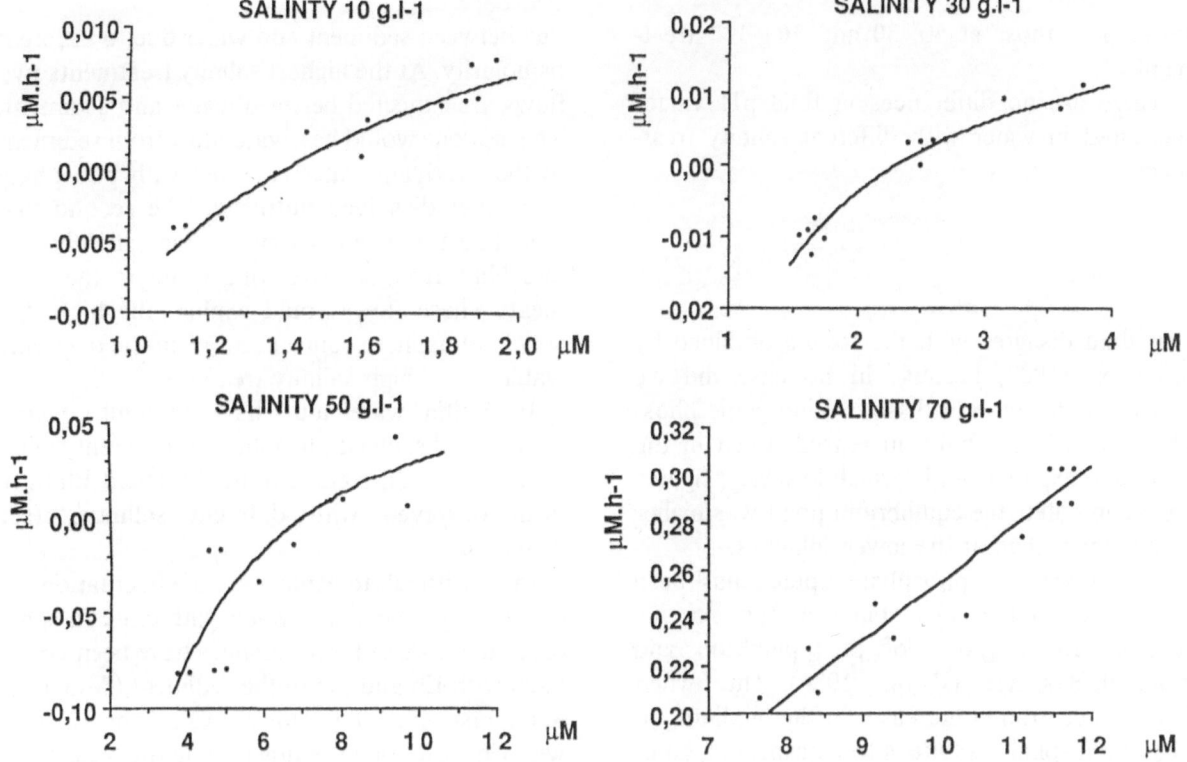

Fig. 4. Variation of the flow rate of dissolved phosphate with phosphate concentration in water with different salinity treatments in long term experiments. Positive values indicate phosphate removal from water and negative values indicate phosphate release from sediment.

librium phosphate concentration observed in the 50 and 70 g l^{-1} treatments.

The values for the equilibrium points and half saturation constants in the long term experiments follow the same pattern as in the short term experiment, i.e. highest values phosphate equilibrium points and half saturation constants occur in the highest salinity treatments. In contrast, V_{max} data show the opposite pattern, with lowest rates in the 10 and 30 g l^{-1} salinity treatments. Phosphate removal at 70 g l^{-1} of salt concentration in the long term experiments does not show saturation kinetics (Fig. 4). Instead it fits a linear model.

Eh variations in both water and sediment do not show any correlation with phosphate concentration variations in water and do not exhibit a fluctuating pattern (Fig. 5).

Eh values measured in the sediment at the end of experiments in 70 g l^{-1} salinity treatment are

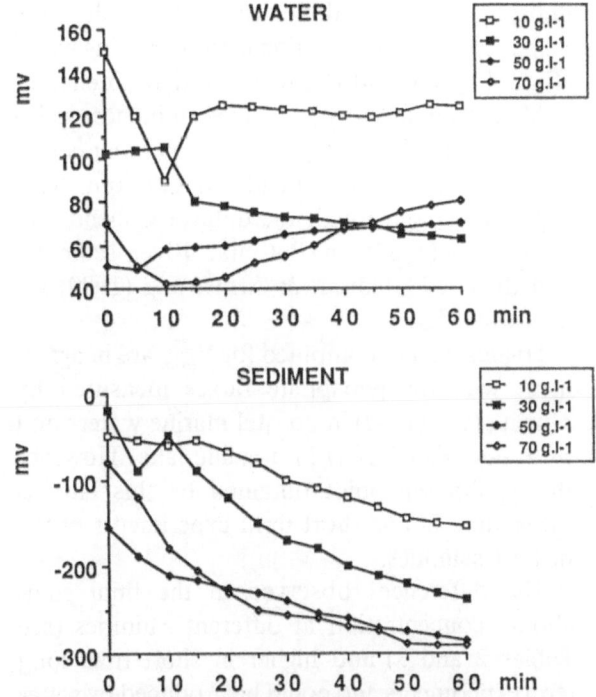

Fig. 5. Eh variation with time in water and sediment in short term experiments at different salinities.

lower than those at 50, 30 and 10 g l^{-1} treatments.

There are no differences in final pH values measured in water with different salinity treatment.

Discussion

Our data disagree with the results obtained by Atkinson (1987) because in no case did we observe a decrease in dissolved inorganic phosphate when the salinity increased. Even in the long term experiments in which low V_{max} figures were computed, the equilibrium point was higher at the upper than at the lower salinities.

The process of phosphate uptake has been related to sodium availability in the external medium for a group of phytoplankton and macroalgal species (Raven, 1984). Thus when Na^+ concentration increases in the medium, an higher phosphate uptake is to be expected, compared with that at low Na^+ concentrations.

Kelderman (1984) shows a linear relationship between phosphate accumulation and phosphate concentration in water. Our results point out that phosphate accumulation in the sediment depends on phosphate concentration in the overlying water (Figs. 4 and 5) and follow saturation kinetics.

Maximum fluxes (V_{max} values, at highest salinity treatments in our short term experiments and at lowest salinity treatments in our long term experiments agree with the data of Callender & Hammond (1982) for Potomac River sediment and those of Holdrem & Armstrong (1980) for freshwater lakes.

Highest values computed for V_{max} are in agreement with the phosphate fluxes measured by Nixon et al., (1980) in coastal marine waters and by Kelderman (1984) in a saline lake. However the equilibrium point obtained by this later is lower than in our short term experiments at the highest salinities.

The differences observed in the final equilibrium concentration at different salinities (see Tables 2 and 3) and higher in short than long term experiments and could be produced by water flux between sediment and water due to different osmolarity. At the highest salinity treatments two flows are expected between water and sediment. The first one would be a water flow from sediment to the overlying water enriched with phosphate and other dissolved nutrients. The second one would be a flow of conservative ions, mainly Cl^- and Na^+ from the overlying water to the sediment. These flows could explain the high dissolved phosphate concentrations in the overlying water at the high salinity treatments.

Interstitial water must have different concentration of dissolved phosphate at different salinities, because differences in the sediment Eh have been observed with different salinities (see Table 2).

More difficult to explain are the fluctuations of dissolved phosphate in water with time. In some cases these short terms changes have been correlated with Eh and pH of the sediment (Anderson & Gahnstrom, 1985). Our values do not correlate with Eh variation in sediment or in the water. The explanation for such a phenomenon, because it is dependent of the dissolved phosphate concentration in water, could be a short term mass balance related to physical process of adsorption and desorption by sediment particles as noted by Krom & Berner (1980). The saturation kinetics that we obtained agree with the pattern obtained by these latter author for adsorption and desorption processes.

Acknowledgements

This work has been supported by grant No. PB86-0677 of the Spanish Comission Interministery of Science and Technology (C.I.C.Y.T.)

References

Anderson, G. & G. Gahnstrom, 1985. Effects of pH on release and sorption of dissolved substances in sediment-water microcosms. Ecol. Bull. 37: 301–318.

Atkinson, M. J., 1987. Low phosphorus sediments in a Hypersaline Marine Bay. Estuar. coast. Shelf. Sci. 24: 335–347.

Bostrom, B., M. Janson & C. Forsberg, 1982. Phosphorus release from lake sediments. Arch. Hydrobiol. 18: 5–59.

Bostrom, B. & K. Petersson, 1982. Different patterns of phosphorus release from lake sediment in laboratory experiments. Hydrobiol. 92: 415–129.

Callender, E. & D. E. Hammond, 1982. Nutrient exchange across the sediment-water interface in the Potomac River estuary. Estuar. coast. Shelf. Sci. 15: 395–414.

Fernández, J. A., F. X. Niell & Lucena, 1985. A rapid and sensitive automated determination of phosphate in natural waters. Limnol. Oceanogr. 30 (1): 227–230.

Gallepp, G. W., 1979. Chironomid influence and phosphorus release in sediment-water microcosms. Ecol. 60: 547–556.

Herberg, O., 1986. Valoración del impacto provocado por el Arroyo Santillán en la Laguna de Fuenta Piedra (Málaga). Tesis de Licenciatura. Universidad de Málaga, 122 pp.

Hesslein, R. H., 1980. In situ measurements of pore water diffusion coefficients using tritiated water. Can. J. Fish. aquat. Sci. 41: 1609–1617.

Holdrem, G. C. & D. E. Armstrong, 1980. Factors affecting phosphorus release from intac lake sediments cores. Envir. Sci. Technol. 14: 79–87.

Kamp-Nielsen, L., 1974. Mud-water exchange of phosphate and other ions in undisturbed sediment cores and factor affecting the exchange rates. Arch. Hydrobiol. 73: 218–237.

Kelderman, P., 1984. Sediment-water interchange in Lake Grevelingen under different environmental conditions. Neth. J. Sea. Res. 18 (3/4): 286–311.

Krom, M. D. & R. A. Berner, 1980. Adsorption of phosphate in anoxic marine sediment. Limnol. Oceanogr. 25 (5): 797–806.

Lucena, J., F. X. Niell & O. Herberg (in press). Transformación del input orgánico en la laguna atalasohalina de Fuente Piedra. Limnética.

Mortimer, C. H., 1941. The exchange of dissolved substances between mud and water in lakes. Ecol. 29: 280–329.

Nixon, S. W., J. R. Kelly, B. N. Furnas & C. A. Oviatt, 1980. Phosphorus regeneration and the metabolism of coastal marine bottom communities. In K. R. Tenore & B. C. Coull (eds.). Marine Benthic dynamics. Univ. South Carolina Press, Columbia, 219–242.

Nurnberg, G. K., 1984. The prediction of internal phosphorus load in lakes with anoxic hypolimnion. Oceanogr. 29: 111–124.

Quigley, M. A. & J. Robbins, 1986. Phosphorus release processes in nearshore southern Lake Michigan. Can. J. Fish. Aquat. Sci. 43: 1201–1207.

Raven, J.A., 1984. Energetics and transport in Aquatic Plants. Alan R. Liss, N.Y., 587 pp + IX.

Tominaga H., N. Tominaga & N. D. Williams, 1987. Concentration of some Inorganic Plant Nutrients in Saline Lakes on the Yorke Peninsula, South Australia. Aust. J. Mar. Freshw. Res. 38: 301–305.

Twinch, A. J. & R. H. Peter, 1984. Phosphate exchange between littoral sediments and overlying water in a oligotrophic North- Temperate Lake. Can. J. Fish. aquat. Sci. 41: 1609–1617.

Vargas, J. M., M. Blasco & A. Antunez, 1983. Los vertebrados de la Laguna de Fuente de Piedra. I.C.O.N.A. Monografias, 28. 228 pp.

Hydrobiologia **197**: 99–104, 1990.
F. A. Comin and T. G. Northcote (eds), Saline Lakes.
© 1990 *Kluwer Academic Publishers.*

Solubility of oxygen in the Dead Sea brine

A. Nishri[1] & S. Ben-Yaakov[2]
[1]*Israel Oceanographic & Limnological Research, Kinneret Limnological Laboratory, P.O. Box 345, Tiberias, 14102, Israel;* [2]*Department of Electrical and Computer Engineering, Ben-Gurion University of the Negev, Beer-Sheva, 84105, Israel*

Key words: Oxygen, solubility, brines, temperature, methods

Abstract

The solubility of the O_2 in the Dead Sea brine was determined over the temperature range 5 °C–50 °C using a modified Winkler titration, volumetric analysis, and a polarographic sensor. The solubility at room temperature and 1 atmosphere pressure was *ca.* 1 mg l^{-1}, and the temperature coefficient 0.006 mg l^{-1} °C^{-1}. The data are nearly consistent with sea water solubility extrapolated to Dead Sea brine salinity.

Introduction

Dissolved oxygen (DO) concentration is associated with a multitude of chemical and biological processes. Hence DO analysis in saline lakes, such as the Dead Sea, can help to better characterize and understand the chemistry and the biology of the lakes. Furthermore, DO concentration can be used as a tracer to follow water masses and to better understand mixing and circulation (Craig, 1971; Ben-Yaakov, 1972).

Theoretical calculation of oxygen solubility in hypersaline solutions such as the Dead Sea waters (DSW) are presently difficult due to the uncertainty in exact formulation of all the interactions in mixed electrolytes of high ionic strength. Previous attempts to describe oxygen solubility in seawater (Green & Carrit, 1967; Weiss, 1970) applied experimental equations similar to the Setschenow exponential relationship (Harned & Owen, 1958) or a polynomial expansion (Weiss, 1970). It is therefore evident that

a critical examination of oxygen solubility in DSW must include a careful study to determine the solubility experimentally.

The objectives of the present research are to determine oxygen solubility in DSW and to evaluate the feasibility of using different analytical methods for determining the concentration of dissolved oxygen in this brine. The study used three analytical approaches; a modified Winkler titration (APHA, 1971); membrane covered dissolved oxygen electrode (Clark *et al.*, 1953; Ben-Yaakov, 1979, 1981); and gas stripping method.

Material and methods

The salinity of the Dead Sea water used during experiments was 280‰ (280 g kg^{-1} or 345 gr l^{-1}) and the chemical composition: Cl^- 190, Ca^{2+} 14, Mg^{2+} 32.5, Na^+ 33.3, K^+ 7, Br^{-1} 4 and SO_4^{2-} 0.35 gr kg^{-1}. Such DSW are about saturation with respect to gypsum.

Winkler titration

The Winkler or modified Winkler titration is widely used for determining dissolved oxygen concentration in natural and waste waters (APHA, 1971; Strickland & Parsons, 1972). The Winkler method is carried out in two steps: first, O_2 is reduced by oxidation of Mn^{2+}, and second the oxidized manganese is reduced as it oxydizes iodide to iodine. The iodine is then back titrated by thiosulfate using starch for end point detection. In order to ensure that the reactions be completed at a sufficient rate, the first reaction must, in DSW (Nishri, 1984), occur at relatively high pH whereas the second reaction must occur at low pH. These requirements are difficult to meet in DSW due to composition and high ionic strength. For example, attempts to raise the pH above *ca.* pH 8 by adding base is futile because of massive precipitation of Mg-hydroxides, which entrap dissolved oxygen. Another difficulty that might be encountered is precipitation of gypsum ($CaSO_4 \cdot 2H_2O$) which will be formed if manganous sulfate is used for the first Winkler reagent and when acidifying the solution by H_2SO_4 to oxidize the iodide. To reduce these problems a modified procedure was applied to DSW. It has precipitation of *ca.* $\pm 2.5\%$ at 1 mg O_2 l^{-1}.

The following procedure was used during Winkler titration of DSW:

1. 1 ml of reagent 1 ($MnCl_2$ 3 M) was added into a *ca.* 300 ml BOD bottle so that the initial Mn^{2+} was about 500 mg l^{-1}. This was followed by stirring for 2–3 sec.
2. 1 ml of reagent 2 (Strickland & Parsons, 1972) was introduced at the bottom of the bottle, the stopper replaced and the solution stirred for 5–10 min. If O_2 was present in a level more than 0.2 ml l^{-1}, the solution turns to a pale yellow color.
3. 1 ml of concentrated H_2SO_4 was added, the stopper replaced and the solution stirred for 2–3 min. The solid suspended is gypsum.
4. 10 ml were drawn from the top of the bottle and discarded. 2 ml of starch solution were then added and stirred for a few seconds.
5. Titration was carried out with thiosulfate (0.02 or 0.025 N) into the continuously stirred BOD bottle.

Membrane covered electrode

The Clark type (Clark *et al.*, 1953) membrane covered polarographic electrode is an electrochemical pO_2 sensor in which O_2 is reduced cathodically on the surface of a noble metal. The resulting electrical current is proportional to the rate at which O_2 is reaching the surface of the cathode. The current is thus a measure of the chemical potential (i.e. pO_2) driving the diffusion of O_2 across an inert membrane which separates the sample from the internal electrolyte of the sensor. Hence, this DO electrode can measure DO concentration only if the relationship between pO_2 and concentration (i.e. O_2 solubility) is known or if calibrated against an independent analytical method. Difficulties may arise when Clark electrodes are applied to DSW. Considerable diffusion of water vapor, due to large differences in the partial pressure of water, across the electrode membrane, may take place. Due to the high viscosity of DSW, adequate stirring could also be a problem.

The performance of a polarographic sensor was studied by comparing its response against the Winkler titration. The sensor used was similar to that described by Ben-Yaakov (1979); it has a gold cathode and a Ag/AgCl anode, a $1*10^{-3}$ m thick Teflon membrane and a 3 M KCl filling solution. A two membrane, flow-insensitive construction (Ben-Yaakov & Ruth, 1980), was also tested but later abandoned because it was less stable than the one membrane sensor.

The temperature sensitivity of the polarographic sensor was tested by monitoring the response of the electrode in DSW which was left to equilibrate with air at various temperatures.

Intercalibration between the Winkler titration and the membrane covered sensor was performed on air saturated samples through which nitrogen gas was bubbled for a few seconds to drive off some of the oxygen.

Gas stripping

A gas stripping method was employed using a special vacuum line on a DSW sample and transferring the released gas to a manometer via a toepler pump. Water vapor was trapped by freezing it with a mixture of dry ice and acetone.

Results and discussion

The solubility of O_2 in DSW was determined on temperature-controlled samples which were left, gently stirring, for 24 hours to equilibrate with air. The results obtained for the Winkler titration procedure are shown in Table 1, both in mg l^{-1} and ml l^{-1} units.

Cross calibration results between the Winkler titration and the gas stripping method are given in Table 2. It is apparent that the methods are consistent to within 10% when DSW solutions supersaturated with oxygen are used. The gas stripping method did not yield consistent results with an air saturated solution (in which nitrogen was present) for unexplained reasons. The fact that the precision of the Winkler titration was better than $\pm 2.5\%$ (Table 1) seems to indicate that the rather poor intercalibration results (Table 2) are mainly due to the lower precision of the gas stripping method.

The response of the DO electrode (Fig. 1) was found to correlate well ($r^2 = 0.97$) with that of the modified Winkler titration.

Table 2. Intercalibration of modified Winkler titration and gas stripping analysis, determined in DSW solutions in which pure oxygen gas was bubbled for different periods.

Sample No.	Stripping O_2 ml l^{-1}	Winkler O_2 ml l^{-1}	Deviation %
A	3.67	3.72	1.4
B	5.44	5.21	− 4.2
D	4.06	3.62	− 10.8
G	3.60	3.30	− 8.3

The temperature sensitivity of the DO electrode (Fig. 2) follows the behaviour found in more diluted solutions, which is attributed to the temperature dependence of the O_2 diffusivity through

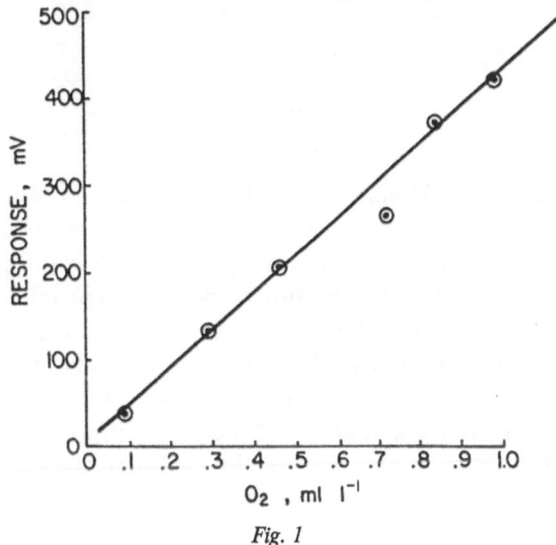

Fig. 1

Table 1. Dissolved oxygen concentrations in air saturated DSW as determined by the modified Winkler titration at an average barometric pressure of 0.99 atmospheres.

Temperature °C	Number of samples	Dissolved oxygen	
		ml l^{-1}*	mg l^{-1}
4	10	1.105(± 0.01)	1.65
25	10	0.981(± 0.006)	1.45
46	5	0.788(± 0.012)	1.15
47	6	0.798(± 0.0078)	1.15
51	5	0.778(± 0.01)	1.16

* Numbers in parenthesis are standard deviations.

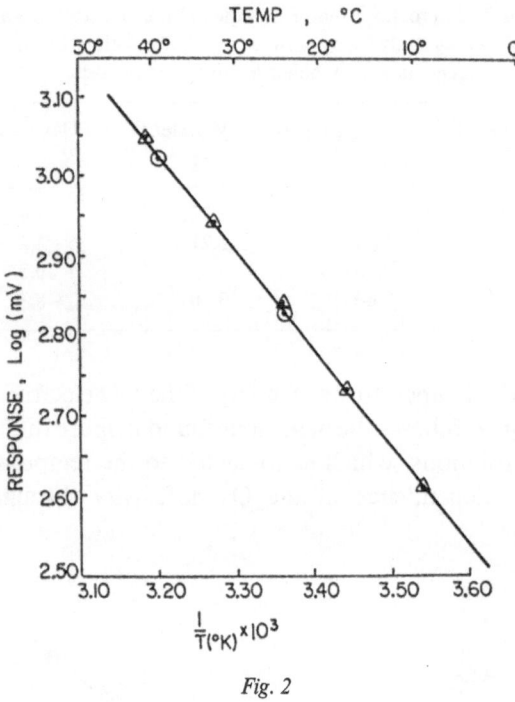

Fig. 2

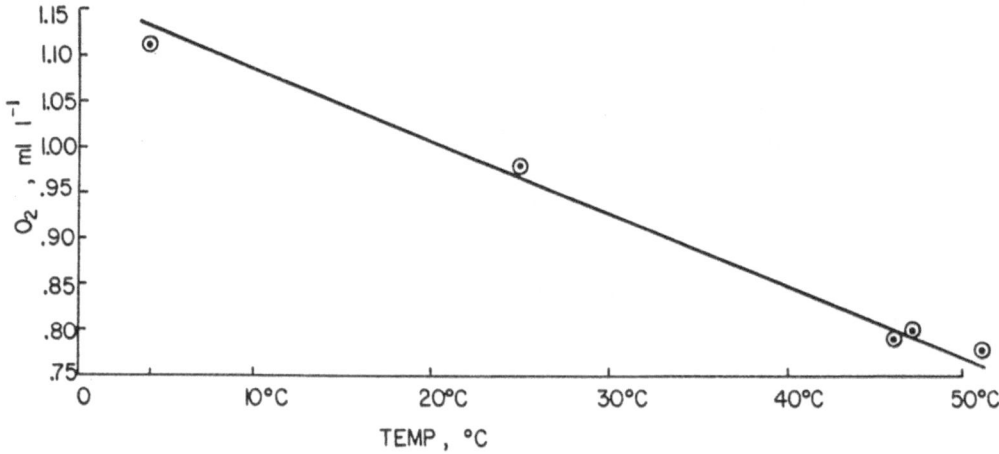

Fig. 3

plastic membranes (Ben-Yaakov, 1981). Although the response of the polarographic DO sensor appears normal, its application in routine analyses of DSW was difficult because of instability and apparent poisoning of the cathode surface. The useable lifetime of an electrode was only a few hours. Thereafter its response would drop dramatically and calibration would then be impossible. Rejuvenation was possible by polishing the sur-

face of the cathode and by replacing both the membrane and the inner electrolyte.

We suggest that two processes affect the response of the electrode. Water is probably lost from the thin electrolyte film which is normally present between cathode and the overlying membrane. This loss is driven by the difference in water vapor pressure between the inner 3 M KCl solution and the high ionic strength DSW. Diffusion through the membrane would deplete the water from the thin electrolyte layer to the point of KCl precipitation. Another interference is likely the poisoning of the cathode surface as DSW penetrates the membrane via pinholes or imperfect seals.

The concentration of dissolved O_2 in air saturated DSW is low and the temperature effect on the solubility is extremely small (Fig. 3), i.e. $ca.$ 0.006 ml l^{-1} $°C^{-1}$. In seawater the temperature effect is about 0.1 ml l^{-1} $°C^{-1}$. However this is not expected if one extrapolates O_2 solubilities to DSW concentration by the experimental equation derived by Weiss (1970) for seawater:

$$\ln C = A_1 + A_2 (100/T) + A_3 (\ln(T/100)) + A_4 (T/100) + S\permil\, B_1 + B_2 (t/100) + B_3 (T/100)^2$$

where C is the solubility either in ml (STP) l^{-1} kg^{-1} in air saturated seawater at one atmosphere, and the A's and the B's are constants, T is the

103

absolute temperature and S‰ is the salinity. Oxygen solubility in DSW is consistent with the extrapolated solubility as calculated from the above equation (Fig. 4).

Although the general behaviour of O_2 solubility at DSW salinities is predictable from the experimental equation, there is a large deviation between the extrapolated and measured values. This is not surprising considering the large difference in ionic composition between seawater and DSW.

Slightly higher (about 5%) oxygen saturation values were recently obtained in a series of experiments conducted by Shatkay & Gat (pers. commun.).

Weiss & Price (in litt.) have calculated a predicted value for solubility of oxygen in DSW by using an average of a Bunsen coefficient that was measured for argon and helium in diluted DSW. Their value of oxygen saturation (at $P = 1$ atm. $T = 25\,°C$) is about 6.8% higher than our measurement.

For a single temperature, the experimental equation of Weiss (1970) reduces to the form

$$Ln\,C = A + S\,B\,,$$

or:

$$C = exp\,A + S‰\,B = C_o\,exp\,S‰\,B$$

where A and B and C_o are temperature-independent constants. The function is nonlinear, imply-

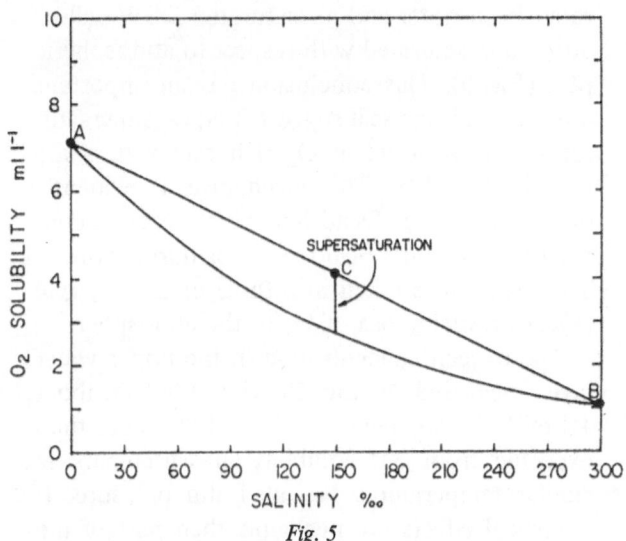

Fig. 5

ing that a mixture of two air-saturated brines will not be in equilibrium with atmospheric pO_2. This behaviour is depicted in Fig. 5, from which the degree of supersaturation of mixtures can be deduced. For example a 1 : 1 mixture of air-satu-

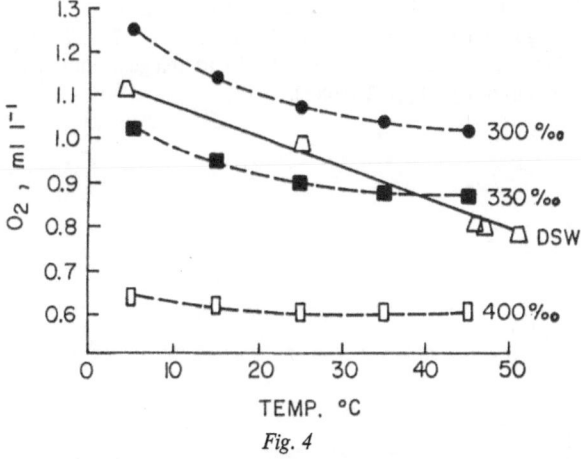

Fig. 4

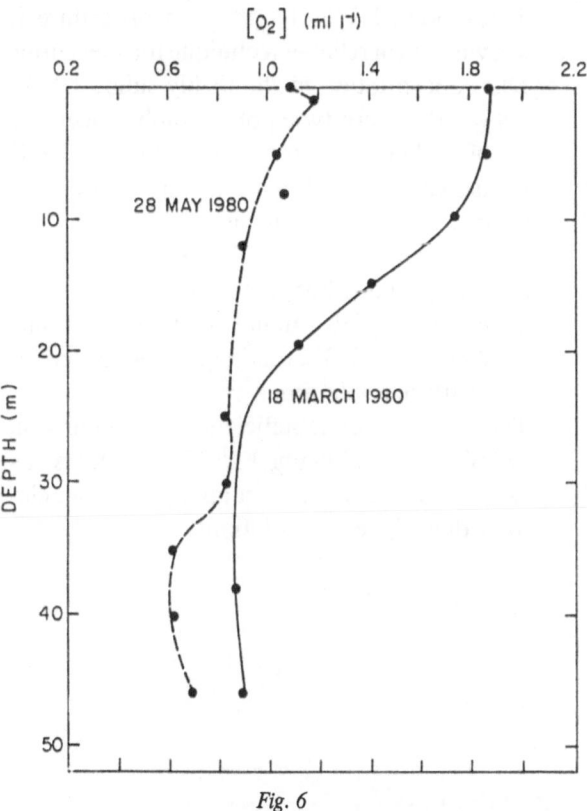

Fig. 6

rated fresh water and air-saturated DSW will be 30% supersaturated with respect to atmospheric pO_2 (Fig. 5). This conclusion has an important implication because it suggests that O_2 supersaturation will arise when O_2 rich rain waters are mixed into DSW. This mechanism is evidently operational in the Dead Sea which receives considerable amounts of flood water during winter. The O_2 profiles taken in a three month interval (Fig. 6) reveal a loss of O_2 to the atmosphere.

The oxygen concentration in the upper water layer measured on the 28 May 1980 is about $1.1 \, ml \, l^{-1}$. This figure is about 10% more than expected from our solubility measurements at similar temperatures but at 1 atm pressure. If biological effects are negligible then part of the difference may be explained by the mean surface barometric pressure prevailing at the Dead Sea surface, of about 1.048 atm.

Conclusions

1. The modified Winkler titration procedure is shown to be a reliable technique for measuring DO concentration in the highly saline DSW.
2. As yet, the Clark type polarographic electrode is of value only for short duration DO measurements in DSW; its application for routine analysis is difficult.
3. The solubility of O_2 in DSW (at 25 °C) is lower, by about 20%, than the extrapolated solubility of a 280 salinity solution, according to Weiss's (1970) experimental equation for much fresher solutions.
4. The temperature coefficient of solubility in DSW is *ca.* $0.006 \, mg \, l^{-1} \, °C^{-1}$, and is, as expected from Weiss's equation, much smaller than that of fresher solutions.

5. As O_2 solubility is non-linear dependent on salinity, mixtures of air saturated brines of different salinity may yield a brine which is supersaturated with respect to atmospheric O_2. This phenomenon evidently occurs in the Dead Sea system.

References

American Public Health Association, American Water Work Association & Water Pollution Control Federation, 1985. Standard methods for the examination of water and wastewater. 16th ed. APHA, AWWA, WPCF, Wash. D.C., 1268 pp.

Ben-Yaakov, S., 1979. A portable dissolved oxygen analyzer for the fish pond industry. Bamidgeh 31: 69–77.

Ben-Yaakov, S., 1981. Electrochemical instrumentation. In: Whitfield, D. & Janger, D. (eds.), Marine Electrochemistry. John Wiley, New York, 99–122.

Ben-Yaakov, S. & E. Ruth, 1980. A method for reducing the sensitivity of a polarographic dissolved oxygen sensor. Talanta 27: 391–395.

Clark, L. C. Jr., R. Wolf, D. Granger & Z. Taylor, 1953. Continuous recording of blood oxygen by polarography. J. Appl. Physiol. 6: 189–199.

Craigh, H., 1971. The deep metabolism: oxygen consumption in abyssal ocean waters. J. Geophys. Res. 76: 5978–086.

Green, E. J. & D. E. Carrit, 1967. Oxygen solubility in sea water: Thermodynamic influence of sea salt. Science 157: 191–193.

Harned, H. S. & B. B. Owen, 1958. The physical chemistry of electrolytic solution. Reinhold Brook Corp. New York.

Nishri, A., 1984. The geochemistry of manganese in the Dead Sea. Earth Planet. Sci. Lett. 718: 415–426.

Strickland, J. D. H. & T. R. Parsons, 1972. Practical Handbook of seawater analysis. Bull. Fish. Res. Bd Can. 167. 310 pp.

Weiss, R. F., 1970. The solubility of nitrogen, oxygen and argon in water and seawater. Deep Sea Res. 17: 712–735.

Weiss, R. F. & B. A. Price (in litt.). Dead Sea gas solubilities (Manuscript 9 pp, 2 tables).

Hydrobiologia **197**: 105–114, 1990.
F. A. Comin and T. G. Northcote (eds), Saline Lakes.
© 1990 *Kluwer Academic Publishers.*

Vernal microstratification patterns in a meromictic saline lake: their causes and biological significance

T. G. Northcote[1] & K. J. Hall[2]
[1]*Department of Zoology, The University of British Columbia, Vancouver, B.C., V6T 2A9, Canada;*
[2]*Westwater Research Centre, The University of British Columbia, B.C., V6T 1W5, Canada*

Key words: saline lakes, meromixis, microstratification, climatic regulation

Abstract

Periodic high spring runoff, in addition to lake surface snow and ice melt, is shown to be a major cause of sharp secondary chemocline formation in a small (20 ha) lake in arid south-central British Columbia. Initially detected in 1982 at about 1 m and enhanced by high inflow of low salinity meltwater in spring 1983, the secondary chemocline gradually deepened and broke down over four subsequent years. Associated microstratification layers (major changes within a few cm of depth), exhibited very high temperatures (> 30 °C), and very high dissolved oxygen ($> 200\%$ saturation) as well as very low (close to 0% saturation) levels. Oxygen supersaturation resulted from photosynthetic production at the microstratification boundaries. In the springs of 1982 and 1983, formation of an anoxic layer between regions of high oxygen concentration, separated the phytoplankton and zooplankton communities into two layers above the primary chemocline. The several year persistence of the secondary chemoclines and associated interface processes (concentration of particulate organic matter, bacterial decomposition, nutrient regeneration, phytoplanktonic production) attest to their functional importance in this meromictic lake.

Introduction

Long term changes in stratification patterns of meromictic lakes are well known (see for example Walker, 1974; Kimmel *et al.*, 1978; Hammer, 1986) and often are caused by man's alteration of watershed hydrology over a period of several decades. In continental temperate regions, which frequently experience marked year to year differences in extent of lake ice and snow cover as well as spring precipitation and runoff, short-term (year to year) changes in stratification might also be expected. Hammer *et al.* (1978) illustrate such effects on the conductivity and temperature pro-files of Waldsea Lake in central Canada between 1971 and 1977, but the differences were largely in depth and salinity of the mixolimnion and there was little evidence of microstratification changes occurring on a cm depth scale. Indeed in meromictic Big Soda Lake, the vertical gradients in water density, temperature and dissolved oxygen are said to be highly predictable, making it ideal for study of pelagic microbial processes (Zehr *et al.*, 1987).

In contrast we found that a secondary micros-tratification layer formed during spring in near-surface waters of Mahoney Lake was not only extremely sharp in the first few years of study, but also that it changed greatly from year to year. Our objectives in this paper are to demonstrate the sharpness of the layer and its temporal variability, to examine causes for its occurrence and change, and to comment on its biological significance. Planktonic production and decomposition processes associated with the layer are examined in a subsequent paper (Hall & Northcote, 1989).

Materials and methods

All water samples were collected from the middle of Mahoney Lake (see Northcote & Hall 1983 for location and morphometric characteristics of the lake) with an electric pump using a weighted 24 mm ID clear plastic hose with a horizontal four directional sucking head. When the flat-bottomed 4.3 m long work boat was well anchored and wind minimal, accurate sampling and *in situ* measurements could be made at 5–10 cm intervals in zones of microstratification.

Vertical temperature, conductivity, and salinity measurements were made with a regularly calibrated (standard 0.01 M KCl) YSI Model 33 meter, taking care to clean the probe in acidic isopropanol solution (10 parts dist. H_2O to 10 parts isopropanol to 1 part conc. HCl) after immersion in the H_2S rich monimolimnion of Mahoney Lake. Some of the later (1985 and 1986) temperature and conductivity readings were made with a microprocessor controlled temperature-conductivity-depth profiler (Applied Microsystems STD–12) which helped to verify the manual readings by duplicated profiles with intercalibrations, as well as provide greater detail on microstratification structure. Because of the marked vertical differences in chemical composition and concentration in the lake, it was necessary to determine separate temperature–conductivity adjustment curves for a series of depths in order to make appropriate corrections to express the conductivity at a standard temperature of 25 °C (Hall & Northcote, 1986).

Vertical dissolved oxygen readings were taken with a YSI Model 57 meter calibrated by air saturation for each series, correcting *in situ* readings for salinity. The nomogram of Mortimer (1981) was used to determine percent oxygen saturation at the altitude of Mahoney Lake (471 m), making an additional correction for salinity by means of a nomogram developed from salinity-oxygen solubility tables given in Standard Methods (A.P.H.A. *et al.*, 1985).

Meteorological data between late 1982 and 1987 were obtained from the closest recording station, McLean Creek, Okanagan Falls, situated at an elevation of 455 m about seven km northeast of Mahoney Lake. Data between December 1981 and March 1982 came from the Penticton Airport station (elevation 344 m, 19 km to the north), making adjustment by regression where possible (mean maximum monthly air temperature, accumulated day degrees below 0 °C) between stations. Mahoney Lake ice depth measurements were available for six different years but only two of these were in the 1982–87 study period. Ice depth was regressed against accumulated day degrees below 0 °C between December and the date of ice measurement ($y = 0.024x + 9.386$, y = ice depth, x = day degrees; $r^2 = 0.73$; $n = 6$) following approaches given in Williams (1963), Shulyakovskii (1966) and Michel (1971). In this way we were able to estimate maximum ice depth at the end of February for all years of the study period.

Water yield from the Mahoney Lake watershed over the spring period (1982–1987) was estimated from the WRENSS (Water Resources Evaluation of Non-point Silvicultural Sources; EPA, 1980) model adapted by Dr. D. L. Golding, Hydrologist, Faculty of Forestry, University of British Columbia. The Okanagan (Region 6) sub-model was chosen for use as it seemed to fit best the particular combination of forest type and climate in the Mahoney Lake area. The model inputs included seasonal precipitation (cm), silvicultural condition by area (%), seasonal evapotranspiration (cm), basal area of forest stand ($m^2\ ha^{-1}$), leaf area index of stand, and rooting depth modifier coefficient of stand; output gave water available for stream flow (cm over area).

Lake water level was measured from an iron pin fixed into a rock point at the southern shoreline of the lake. Discharge from meltwater channels was measured by diverting flow into a calibrated container and measuring time for filling. Phytoplankton and zooplankton sampling as well as counting followed methods outlined in Northcote & Hall (1983).

Microstratification patterns

Conductivity

The primary chemocline separating the mixolimnion from the monimolimnion of Mahoney Lake starts between 7–8 m and extends to about 10–11 m, with conductivities in the monimolimnion consistently within the 55–60 000 μS cm^{-1} range at 25 °C (Northcote & Halsey, 1969; Northcote & Hall, 1983; Hall & Northcote, 1986). The upper portion of this primary chemocline is clearly evident between 1982 and 1987 (Fig. 1). In addition, over this period there also has been a secondary (and on some years a less obvious tertiary) chemocline within the mixolimnion (Fig. 1). For example, between 1982 and 1984, the upper secondary chemocline was as sharp as the primary one, although the extent of its conductivity change was less, usually never more than about 15000 μS cm^{-1}.

Although other features of the mixolimnion conductivity profiles have changed greatly over recent years, there has been a consistently isohaline lower leg at about 25000 μS cm^{-1} with a depth no more than 2 m recently (1984–87), but about twice that depth in 1982 and '83 (Fig. 1).

The occurrence and depth of a secondary chemocline within the mixolimnion as well as conductivity of the uppermost isohaline limb, when present, have changed greatly over the six year period of our study. From April to May 1982 it was only isohaline within the upper half metre at conductivities between 12 500 and 15 000 μS cm^{-1} (Fig. 1). The following spring the uppermost 1.5 m was isohaline at the lowest surface conductivities ever recorded in the lake (less than 7 500 μS cm^{-1}). In the springs of 1984 and '85 the upper layer was about 2 m deep at conductivities

close to 9000 and 11 000 μS cm^{-1} respectively, although those under the ice in mid February were slightly higher, as might be expected. Depth of the upper layer of low conductivity was more difficult to discern in 1986 because of the multiple small chemoclines, but in 1987 it was evident at a depth of 4 m and at conductivity of over 17 000 μS cm^{-1} (Fig. 1). Details of the six year sequence of change in early spring conductivity structure for the upper 5 m of the lake (Fig. 2) clearly shows the initial (1982) shallow layer of intermediate salinity at the lake surface, followed by its deepening to over 1 m of very low salinity water (1983), and its further deepening as well as increase in salinity between 1984 and 1987.

Temperature

Not surprisingly, the temperature profiles in the mixolimnion of Mahoney Lake (Fig. 1) reflect the annual changes in conductivity structure noted above. In 1982 and 1983 when there was a shallow (< 2 m) and sharp secondary chemocline present, a sub-surface high temperature lens 1–2 m deep rapidly developed in spring after the disappearance of ice cover. By early summer, temperature maxima in this lens were above 26 °C in 1982 and slightly above 30 °C in 1983, with the sharpest thermocline lying at about the depth of the secondary chemocline (close to 1 m in 1982 and about 1.75 m in 1983). Temperature maxima recorded in the upper water column of Mahoney Lake were higher than those recorded for shallow heliothermal lakes in south central British Columbia by Hudec & Sonnenfeld (1980), but well below those at middepth in shallow Hot Lake in north central Washington (Anderson, 1958).

In the following two years when the secondary chemocline lay at about 2.5 m, there was also evidence of thermal increase at this depth (even in mid winter under the ice), but the sharpness of stratification was considerably reduced from that evident previously.

In 1986 a slight thermal increase eventually developed by May at a depth of about 3 m, coincident with a slight increase in conductivity at the

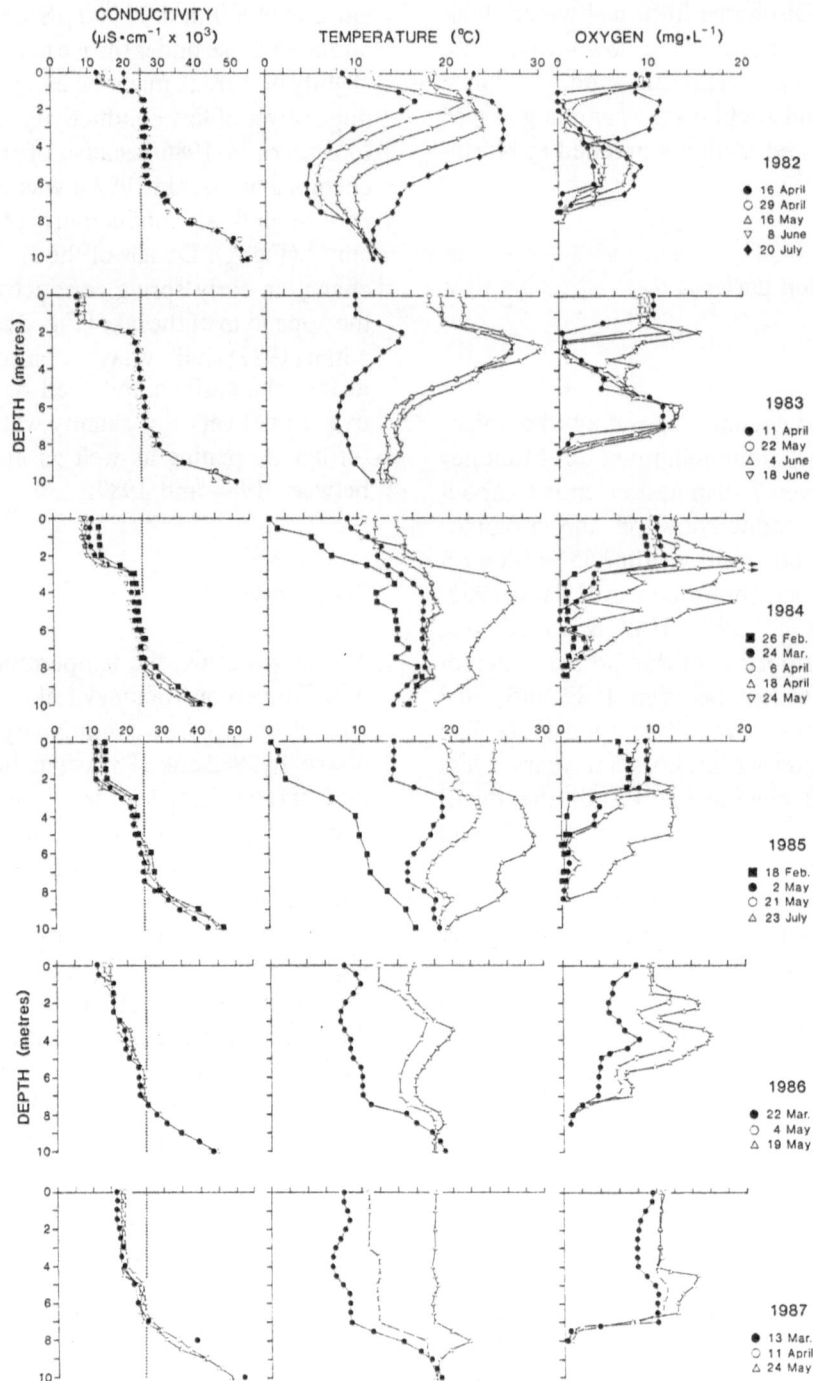

Fig. 1. Winter to summer conductivity, temperature, and dissolved oxygen profiles for the upper 10 m at Station 1, Mahoney Lake, 1982–1987. Maximum depth during study period approximately 16.5 m; arrows on 24 May 1984 oxygen concentrations indicate values off scale of meter (exceeding 20 mg l^{-1}).

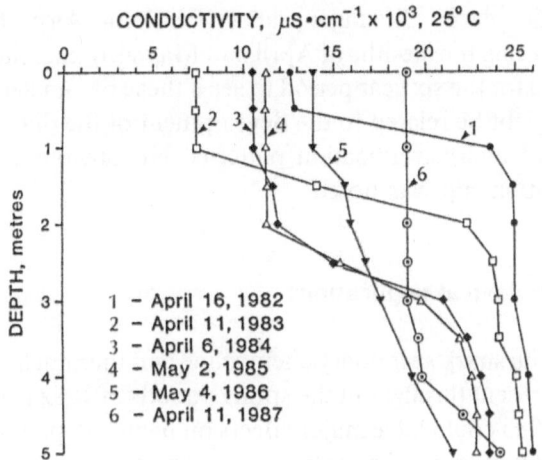

CONDUCTIVITY, $\mu S \cdot cm^{-1} \times 10^3$, 25°C

1 – April 16, 1982
2 – April 11, 1983
3 – April 6, 1984
4 – May 2, 1985
5 – May 4, 1986
6 – April 11, 1987

Fig. 2. Conductivity structure of the upper 5 m of Mahoney Lake, Station 1, in early spring, 1982–1987.

same depth (Fig. 1), and in the following year the deeper (ca 4.5 m) and even weaker chemocline may have been responsible for the very small thermal increase that developed at that depth by late May.

In many of the thermal profiles a slight temperature increase and decrease (Fig. 1), usually about 1 °C, occurred near the depth of the bacterial plate (ca 7.5, 8.2, 8.4, 8.3, 8.1, 8.1 m, springs of 1982–87 respectively).

Dissolved oxygen

The seasonal and annual changes in dissolved oxygen profiles of the lake were large, ranging from a complex three-layered system in 1982 and 1983 to a mainly unimodal one in more recent years (Fig. 1). Throughout the springs of 1982 and 1983 an upper shallow (0.5, 1.0 m respectively) well oxygenated layer was followed by a 1 m layer with very low oxygen concentration (0–1 mg l^{-1}), and then by another layer where oxygen increased again, eventually up to near surface levels or higher, before declining once more to zero at the approximate depth of the primary chemocline (near 8 m).

In the winter as well as spring of 1984 there was still evidence of weak bimodality in the dissolved oxygen profiles (Fig. 1), with highest concentra-

tions (> 20 mg l^{-1}) being reached just below 2 m. Although such concentrations are high for many North American lakes (Wetzel, 1983), they are still well below the maximum level (nearly 36 mg l^{-1}) recorded by Birge & Juday (1911) in some Wisconsin lakes. In 1984 a secondary deep dissolved oxygen peak developed between 6.5 and 7 m in Mahoney Lake. Although minor peaks developed at various depths in subsequent years, a sharp bimodality in oxygen concentration never reappeared.

Microstratification causes

The intrusion and gradual disappearance of a surficial layer of low salinity water would seem to be a primary cause for the marked annual changes in vernal microstratification of Mahoney Lake between 1982 and 1987. There are three main sources of such water, (1) early spring melting of ice and snow cover on the lake itself, (2) rainfall on the lake surface, and (3) spring inflow from watershed snowmelt and precipitation (no permanent streams enter the lake).

Total annual snowfall during the six year period has ranged from a maximum of over 72 cm in the 1981–82 winter to a minimum of 8 cm in the 1986–87 winter (Fig. 3). Estimated maximum ice depth over the same period has ranged from 28 cm (1985) to 15 cm (1983). No obvious relationship is evident between either of these sources of low salinity water and the occurrence of the low salinity layer in 1983. Precipitation and estimated inflow from the lake's watershed were much higher in 1983 than any other of the study years (Fig. 3). On that year runoff could have contributed 2.2 m of low salinity water to the lake surface. There were three surface inflow channels on 11 April 1983 carrying together a total of 900 l per minute of dilute (672–924 μS cm^{-1} at 25 °C) meltwater at temperatures between 7.8 and 12.0 °C onto the lake surface. The following spring (1984) estimated inflow was much reduced (Fig. 3); on 6 April metered discharge from the same channels was only 90 l per minute. In the spring of 1985 estimated inflow was further reduced and only one of the meltwater channels was carrying surface inflow, estimated at 2.5 l per minute in early May. Even this channel was dry after mid May. None of the channels showed any evidence of carrying surface inflow to the lake during the springs of 1986 and 1987.

Reduced precipitation and surface inflow since 1983 have been accompanied by a gradual drop in lake surface level of about 1 m (Fig. 3). The upper sharp density boundary (secondary mixolimnion interface) when first noted in mid-April 1982 was at 1 m or less (Figs. 1, 3), reached maximum sharpness and lowest salinity in spring 1983, and subsequently has become deeper, less sharp, and higher in salinity, almost disappearing by 1987.

Annual changes in mean maximum air temperature (McLean Creek) and average wind speed in the nearby Okanagan valley (Penticton Airport) during the months of April and May were examined for the six year period to see if these parameters might be related to the development of the different microstratification patterns. No obvious relationship was noted.

Biological implications

The sharp secondary chemocline and thermocline present throughout the spring periods of 1982 and 1983 could have major effects on bacterial, phytoplanktonic and zooplanktonic communities of the upper mixolimnion. Indeed the strong density barrier to mixing afforded by the secondary chemocline at about 1 and 2 metres respectively on those years resulted in a severe upper level

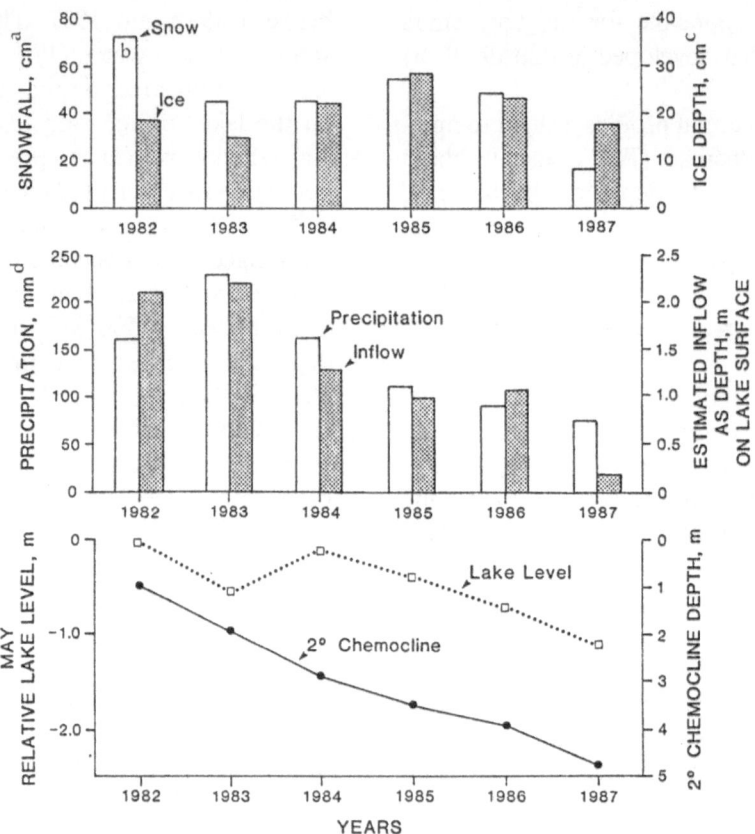

Fig. 3. Annual changes (1982–1987) in conditions related to development and depth of the secondary chemocline interface in Mahoney Lake. *a* total snowfall (Dec.–March) at McLean Creek; *b* 1982 data for Penticton Airport; *c* maximum estimated depth at end Feb.; *d* total rainfall Nov.–March.

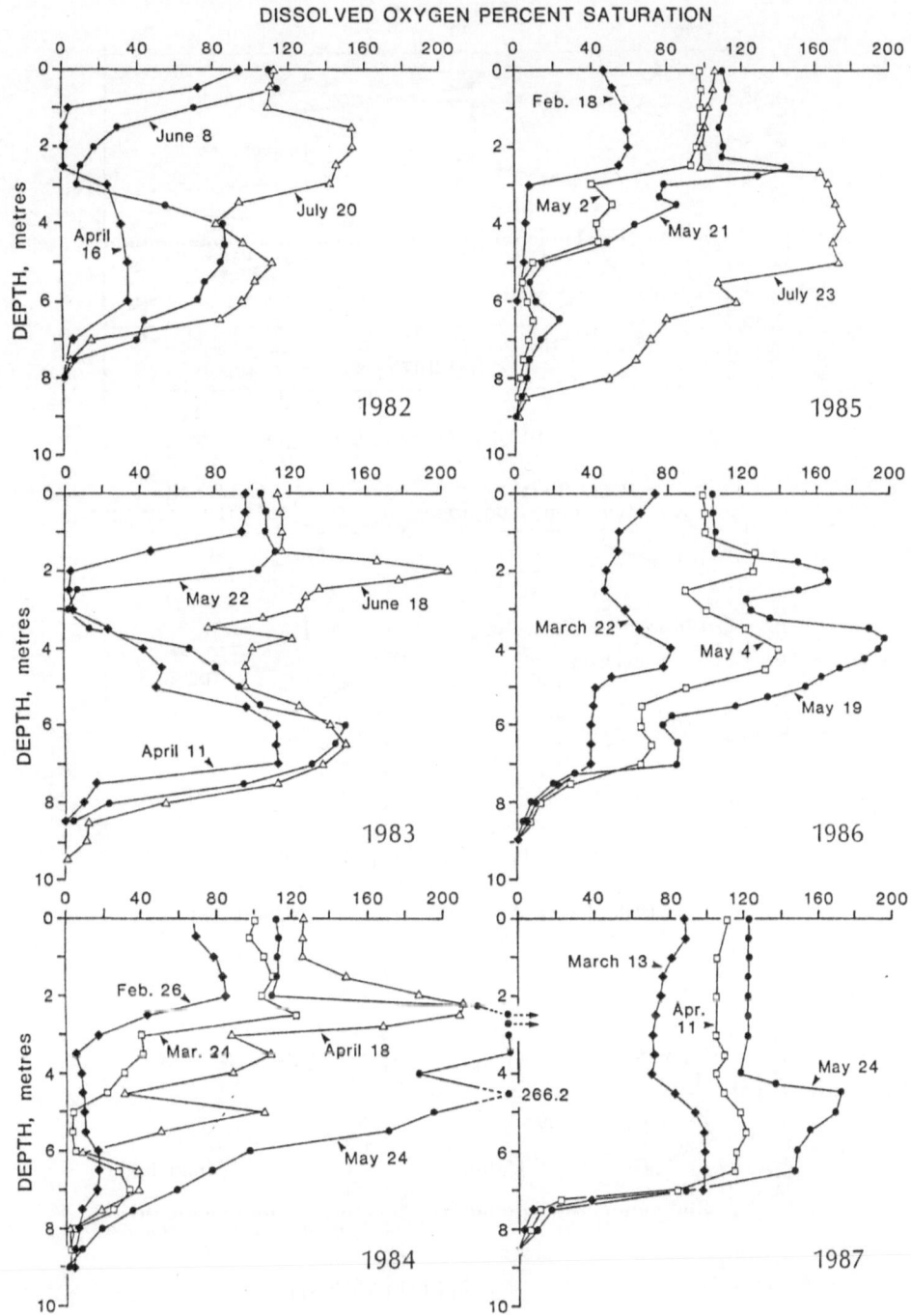

Fig. 4. Oxygen percent saturation profiles at Station 1, Mahoney lake, 1982–1987.

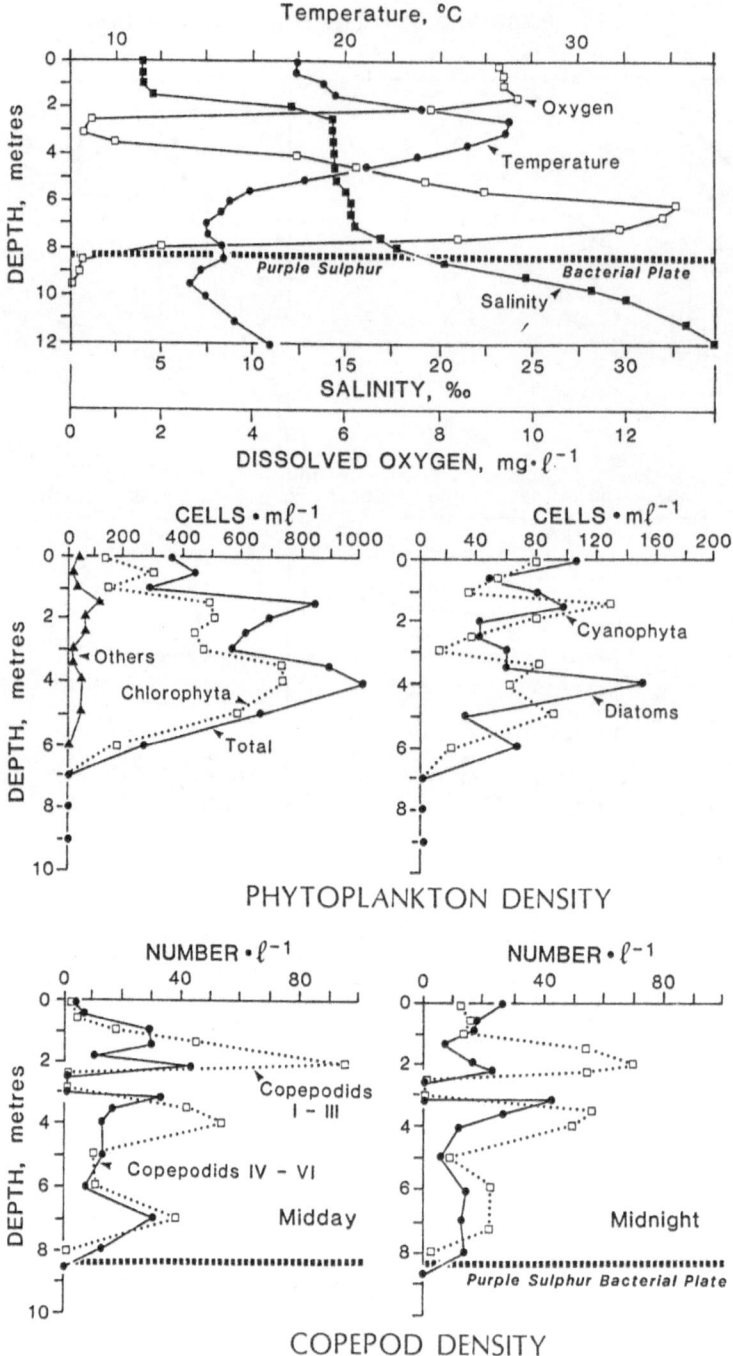

Fig. 5. Vertical microstratification profiles in the mixolimnion of Mahoney Lake, Station 1, 22 May 1983. Upper: physical and chemical features; middle: phytoplankton; lower: *Diaptomus connexus* copepodid stages I–III, IV–VI.

oxygen depletion, probably a result of rapid bacterial decomposition of organic material settling out temporarily on the density boundary (Fig. 1), where temperature also was high. During Secchi depth determination the level of the sharp density layer could be detected visually by the sudden 'shimmering' effect on the disc. At this same depth there were also noted many small fragments of partially decomposed *Chara* sp. and associated whisps of periphyton, presumably carried out into the lake along the density boundary from the littoral zone. This upper oxygen depletion layer must have resulted largely from the relatively high heterotrophic activity occurring there, combined with the low level of planktonic primary production in that zone (Hall & Northcote, 1989). Furthermore in the later years of the study when the secondary chemocline was reduced in sharpness and eventually disappeared, heterotrophic activity decreased, primary production increased, and there was no severe oxygen depletion layer.

High rates of bacterial decomposition at this boundary probably resulted in considerable nutrient mineralization. That zone subsequently became a region of high photosynthetic activity by phytoplankton, with oxygen supersaturation reaching over 150% in 1982 and over 200% in 1983 (Fig. 4). Even in more recent years when the secondary chemocline descended and became less sharp (Figs. 1, 2), increasing levels of oxygen supersaturation indicative of photosynthetic production were evident at this boundary (Fig. 4). Thus the formation of a secondary chemocline by heavy early spring runoff on one or two years apparently had a marked effect on thermal and chemical microstratification, on bacterial decomposition and nutrient mineralization, as well as on photosynthetic production, and extended for several years after initial formation. Indeed development of a sharp secondary chemocline within the mixolimnion may be an important means of increasing production within that layer, without which organic material could settle into the monimolimnion with little chance for recirculation and enhancement of primary production in the mixolimnion (see Hall & Northcote, 1989). However, Cloern *et al.*, (1987) show that in Big Soda Lake

only about 30% of the particulate carbon is retained by the permanent chemocline between 30 and 40 m where conductivity increases approximately $110\,000\ \mu S\ cm^{-1}$ over a 6 m depth interval. In Mahoney Lake the secondary chemocline at its sharpest development (1983) increased about $6\,500\ \mu S\ cm^{-1}$ over a 0.8 m depth interval. Unfortunately we have no sediment trap data to compare with the studies on Big Soda lake, but the high rates of bacterial productivity and heterotrophic uptake as well as bacterial cell densities at the secondary chemocline suggest that high rates of organic matter mineralization should be occurring there (Hall & Northcote, 1989).

The development of a severe oxygen depletion layer in the near-surface waters of the mixolimnion for about a two month period in the spring of 1982 and 1983 partitioned the phytoplankton and zooplankton communities into highly stratified systems (Fig. 5). Although total phytoplankton cell number was considerably reduced in the anoxic zone at about 3 m, as was the major constituent (Chlorophyta), Cyanophyta and diatoms also showed major reductions in cell density in that depth zone. The common calanoid copepod population in Mahoney Lake, *Diaptomus connexus*, including nauplii (not shown) as well as copepodid stages, was separated into two zones, one within the upper 2 m and the other between 3 and 8 m, with virtually none in the intervening low oxygen layer (2–3 m). The major features of this separation remained intact at night as well as during the day (Fig. 5). Bimodal (near-surface and near-chemocline) concentration zones of copepod life stages are not unusual in meromictic lakes (Swift & Hammer, 1979; Elgmork & Langeland, 1980) but rarely does one find the zones separated by anoxic water with virtually no copepods in the intervening layer.

Acknowledgements

We thank Dr. M. A. Chapman, Dr. J. D. Green, Dr. P. R. B. Ward, Mrs. S. Hall and Mrs. C. H Northcote for assistance in data collection. Field laboratory facilities and hospitality were kindly provided by G. B. Northcote. Dra. E. Cornejo made phytoplankton analyses as did Dr. M. A. Chapman for zooplankton. Dr. J. R. MacKay gave helpful advice in estimating ice depth from air temperature data for years when no direct measurements were available. Dr. D. L. Golding provided means for estimating water yield from the Mahoney watershed over the study period. Funding was obtained from NSERC grants 67–3454 (TGN) and 67–8935 (KJH).

References

American Public Health Association, American Water Work Association & Water Pollution Control Federation, 1985. Standard methods for the examination of water and wastewater, 16th ed. APHA, AWWA, WPCF, Wash. D. C., 1268 pp.

Anderson, G. C., 1958. Some limnological features of a shallow saline meromictic lake. Limnol. Oceanogr. 3: 259–270.

Birge, E. A. & C. Juday, 1911. The inland lakes of Wisconsin. The dissolved gases of the water and their biological significance. Bull. Wis. Geol. Nat. Hist. Survey 22, Sci. Ser. 7, 259 pp.

Cloern, J. E., B. E. Cole & S. M. Wienke, 1987. Big Soda Lake (Nevada). 4. Vertical fluxes of particulate matter: seasonality and variations across the chemocline. Limnol. Oceanogr. 32: 815–824.

Elgmork, K. & A. Langeland, 1980. *Cyclops scutifer* Sars – one and two-year life cycles with diapause in the meromictic lake. Blankvatn. Arch. Hydrobiol. 88: 178–201.

EPA, 1980. An approach to water resources evaluation of non-point silvicultural sources (a procedural handbook). U.S. Envir. Prot. Agency, Envir. Res. Lab., Off. Res. Dev., Athens, Georgia, EPA–600/8–80–012, Chap. 3, 173 pp.

Hall, K. J. & T. G. Northcote, 1986. Conductivity-temperature standardization and dissolved solids estimation in a meromictic saline lake. Can. J. Fish. aquat. Sci. 43: 2450–2454.

Hall, K. J. & T. G. Northcote, 1990. Production and decomposition processes in a saline meromictic lake. Hydrobiologia 197: 115–128.

Hammer, U. T., 1986. Saline lake ecosystems of the world. Kluwer Academic Publishers Group, Dordrecht, The Netherlands, 614 pp.

Hammer, U. T., R. C. Haynes, J. R. Lawrence & M. C. Swift, 1978. Meromixis in Waldsea Lake, Saskatchewan. Verh. int. Ver. Limnol. 20: 192–200.

Hudec, P. P. & P. Sonnenfeld, 1980. Comparison of Caribbean solar ponds with inland solar lakes of British Columbia. In: Brines and Evaporitic Environments. A. Nissenbaum [ed.] Elsevier, Amsterdam. pp. 101–114.

Kimmel, B. L., R. M. Gersberg, L. P. Paulson, R. P. Axler & C. R. Goldman, 1978. Recent changes in the meromictic status of Big Soda Lake, Nevada. Limnol. Oceanogr. 23: 1021–1025.

Michel, B., 1971. Winter regime of rivers and lakes. Cold Regions Science and Engineering Monograph 111–B1a, 131 pp.

Mortimer, C. H., 1981. The oxygen content of air-saturated fresh waters over ranges of temperature and atmospheric pressure of limnological interest. Mitt. int. Ver. Limnol. 22, 23 pp.

Northcote, T. G. & K. J. Hall, 1983. Limnological contrasts and anomalies in two adjacent saline lakes. Hydrobiologia 105: 179–194.

Northcote, T. G. & T. G. Halsey, 1969. Seasonal changes in the limnology of some meromictic lakes in southern British Columbia. J. Fish Res. Bd. Can. 26: 1763–1787.

Shulyakovskii, L. G. [ed.], 1966. Manual of forecasting ice-formation for rivers and inland lakes. Manual of Hydrological Forecasting No. 4, 245 pp.

Swift, M. C. & U. T. Hammer, 1979. Zooplankton population dynamics and *Diaptomus* production in Waldsea Lake, a saline meromictic lake in Saskatchewan. J. Fish. Res. Bd. Can. 36: 1431–1438.

Walker, K. F., 1974. The stability of meromictic lakes in central Washington. Limnol. Oceanogr. 19: 209–222.

Wetzel, R. G., 1983. Limnology. Saunders College Publishing, Philadelphia, USA, 859 pp.

Williams, G. P., 1963. Probability charts for predicting ice thickness. Engineering Journal, E.I.C. 46: 31–35.

Zehr, J. P., R. W. Harvey, R. S. Oremland, J. E. Cloern & L. H. George, 1987. Big Soda Lake (Nevada) 1. Pelagic bacterial heterotrophy and biomass. Limnol. Oceanogr. 32: 781–793.

Hydrobiologia **197**: 115–128, 1990.
F. A. Comín and T. G. Northcote (eds), Saline Lakes.
© 1990 *Kluwer Academic Publishers.*

Production and decomposition processes in a saline meromictic lake

K. J. Hall[1] & T. G. Northcote[2]
[1]*Westwater Research Centre, The University of British Columbia, Vancouver, B.C., V6T 1W5, Canada;*
[2]*Department of Zoology, The University of British Columbia, Vancouver, B.C., V6T 1W5, Canada*

Key words: primary production, heterotrophic activity, bacterial productivity, purple phototrophic sulphur bacteria, meromixis, saline lake

Abstract

Bacterial and phytoplankton cell number and productivity were measured in the mixolimnion and chemocline of saline meromictic Mahoney Lake during the spring (Apr.-May) and fall (Oct.) between 1982 and 1987. High levels of bacterial productivity (methyl ^{3}H-thymidine incorporation), cell numbers, and heterotrophic assimilation of ^{14}C-glucose and ^{14}C-acetate in the mixolimnion shifted from near surface (1.5 m), at a secondary chemocline, to deeper water (4–7 m) as this zone of microstratification gradually weakened during a several year drying trend in the watershed. In the mixolimnion, bacterial carbon (13–261 μgC l^{-1}) was often similar to phytoplankton carbon (44–300 μgC l^{-1}) and represented between 14–57% of the total microbial (phytoplankton + bacteria) carbon depending on the depth interval. Phototrophic purple sulphur bacteria were stratified at the permanent primary chemocline (7.5–8.3 m) in a dense layer (POC 250 mg l^{-1}, bacteriochlorophyll *a* 1500–7000 μg l^{-1}), where H$_2$S changed from 0.1 to 2.5 mM over a 0.2 m depth interval. This phototrophic bacterial layer contributed between 17–66% of the total primary production (115–476 mgC m^{-2} d^{-1}) in the vertical water column. Microorganisms in the phototrophic bacterial layer showed a higher uptake rate for acetate (0.5–3.7 μgC l^{-1} h^{-1}) than for glucose (0.3–1.4 μgC l^{-1} h^{-1}) and this heterotrophic activity as well as bacterial productivity were 1 to 2 orders of magnitude higher in the dense plate than in the mixolimnetic waters above. Primary phytoplanktonic production in the mixolimnion was limited by phosphorus while light penetration appeared to regulate phototrophic productivity of the purple sulphur bacteria.

Introduction

Mahoney Lake, located in south-central British Columbia, is a strongly meromictic saline lake which contains a dense layer of purple phototrophic sulphur bacteria at a permanent mid-depth (ca 8 m) chemocline (Northcote & Halsey, 1969). Studies in the early 1980's (Northcote & Hall, 1983) indicated that the mixolimnetic waters of Mahoney were developing heliothermal condi-

tions similar to those observed in some shallow saline lakes in interior British Columbia (Hudec & Sonnenfeld, 1980) and in Hot Lake, Washington (Anderson, 1958). Our initial observations indicated that the shallow density stratification, which initiated these elevated temperatures, served as a favorable environment to stimulate microorganism activity in the spring. It has been suggested that these sharply stratified boundaries may provide a mechanism to maintain nutrients in the euphotic zone that would otherwise be lost to the monimolimnion and sediments (Culver & Brunskill, 1969).

Over the past six years vernal changes in the microstratification of Mahoney Lake have been investigated during a drying trend in the watershed (Northcote & Hall, *MS*). Associated with these physical changes in the lake, we have investigated the dynamics of microbial production and decomposition processes in an attempt to understand the biological significance of these sharp salinity gradients. Primary production measurements and pigment analyses have been used to determine the activity and abundance of phototrophic microorganisms while the importance of heterotrophic microorganisms have been determined by labelled organic solute assimilation and direct enumeration.

Materials and methods

Vertical temperature, conductivity and oxygen measurements were made and the profiles graphed in the field to aid in selection of sampling depths for microbial studies (see Northcote & Hall, 1983 for instrumental details). Water samples were collected with an electric pump using a weighted 24 mm ID clear plastic hose with a horizontal four-directional sucking head.

Alkalinity was determined on 5 or 10 ml samples diluted to 100 ml with distilled water and titrated with 0.02 N H_2SO_4 to pH 8.3 and 4.5 using a pH meter (APHA *et al.*, 1985). Sulphide samples were collected in 60 ml glass-stoppered bottles, preserved with 1 N zinc acetate and 6 N NaOH, and quantified by the iodometric titration procedure (APHA *et al.*, 1985). It was necessary to correct the alkalinity titration for sulphide concentrations below the primary chemocline to accurately determine the bicarbonate concentrations.

Bacteria were enumerated during 1982 by direct counting of formalized samples using epifluorescence microscopy and the DAPI staining technique (Porter & Feig, 1980). Subsequent bacteria samples (1983–87) were enumerated by the acridine orange direct count (AODC) technique. Nuclepore filters (0.2 μm, 25 mm dia.) were stained with irgalan black solution (2 g l^{-1} in 2% acetic acid) which were rinsed with membrane filtered lake water prior to sample filtration. The bacteria were filtered onto the prestained filters, air dried for 8 h in petri dishes and transported to the laboratory for counting. The filters were stained with acridine orange and counted under oil immersion at 1565 × magnification on a Leitz Ortholux microscope fitted with epifluorescence accessories (Hobbie *et al.*, 1977).

Microbial activity and bacterial productivity were made on samples incubated *in situ* in 20 ml disposable plastic syringes. Net bacterial activity was measured at one solute concentration since studies have indicated a good relationship between V_{max} and uptake at one solute conc. when comparative measurements were made (Griffiths *et al.*, 1977). Uniformly labelled substrates, namely ^{14}C-glucose (170–346 mCi $mMol^{-1}$) and ^{14}C-acetate (45–60 mCi $mMol^{-1}$) representing between 4 and 8 μgC l^{-1} of carbon, were added to water samples from different depths for the microbial activity measurements. One ml of diluted isotope was added to 9 ml of lake water in the syringe. Two active and one blank sample, which was killed with gluteraldehyde at 2% final conc., were incubated at each sampling depth for 2–3 h. The samples were filtered through a 0.2 μm, 2.5 cm cellulose nitrate membrane filter placed in a Nuclepore filter assembly attached to the syringe. The filters were washed with 10 ml of lake water to remove any soluble label, immediately placed in scintillation vials and inactivated with PCS scintillation solution (Amersham). They were counted on a Nuclear Chicago Isocap scintillation counter using an external standard to correct for quenching.

Bacterial productivity was estimated by the incorporation of [methyl-^{3}H]thymidine (26–69.4 Ci $mMol^{-1}$). Ten μCi of the isotope (one ml) was added to 9 ml of sample in duplicate live and inactivated samples in 20 ml disposable plastic syringes and incubated for 2 h at the sampling depth. After incubation, the syringes were immersed in ice for one min and then 10 ml of ice cold 10% TCA was added to extract the pool of nucleic acids. After a 5 min extraction period on ice, the samples were filtered through 0.2 μm, 25 mm dia. membranes in a Nuclepore filtration assembly. The filters were rinsed twice with 2 ml of ice cold 5% TCA, dissolved in PCS scintillation solution, and counted on the scintillation counter (Fuhrman & Azam, 1982). Thymidine incorporation (pMol 1^{-1} h^{-1}) was converted to a production rate (μgC 1^{-1} h^{-1}) by multiplying by $2 \cdot 10^{18}$ cells per mole for thymidine incorporation and by 10^{-8} μgC cell^{-1} (Riemann et al., 1982; Lovell & Konopka, 1985; Fuhram et al., 1986).

Phytoplankton biomass was estimated by chlorophyll a measurements. Samples were filtered on 0.45 μm membrane filters, layered with 1% $MgCO_3$, and either extracted immediately or frozen in petri dishes wrapped in Al foil. Initial measurements of chlorophyll a were made on 90% acetone extracts with a Perkin Elmer Hitachi spectrophotometer (APHA et al., 1985). Later measurements were made on methanol : chloroform extracts (Wood, 1985) and quantitated fluorimetrically on a Turner Designs Fluorimeter calibrated with standard chlorophyll (Strickland & Parsons, 1972). The chlorophyll a concentrations were multiplied by 25 to convert to algal carbon biomass (Cloern et al., 1987). Phytoplankton enumeration followed the procedure outlined by Northcote & Hall (1983).

Stratified samples at 5 cm intervals were collected at the chemocline to characterize the purple sulphur bacteria plate. The particulate cellular material was estimated by loss on ignition at 550 °C in a muffle furnace by samples filtered through GFC filters placed in Gooch crucibles. The gelatinous mass of the sulphur bacteria was so dense that no cells passed through these filters. Samples for pigment analysis were filtered through GFC filters and extracted with acetone at room temperature in the dark for two days. The absorption spectrum was scanned from 320–800 nm on SP8-100 UV spectrophotometer (Pye Unicam). On a series of samples from the bacterial plate a comparison was made between absorption at 772 nm ($\lambda_{\max}$ for bacteriochlorophyll a) and fluorescence. This correlation (Fluorescence = 208.8 (Absorbance) + 1.20; $r^2 = 0.991$), allowed fluorescence measurements to be converted to absorbance values and finally to bacteriochlorophyll a (BChl a) concentrations using the equation developed by Takahashi & Ichimura (1970). Primary production was measured by the incorporation of ^{14}C-HCO_3^- (10 μCi per 60 ml glass stoppered bottle) in triplicate samples which were incubated in situ in a horizontal position on a plexiglass holder for a 6–8 h period. A dark formalized blank was used as a control. Samples were filtered on 0.45 μm membrane filters, treated with HCl to remove any precipitated $CaCO_3$, dissolved in PCS scintillation solution, and counted similarly to the bacterial activity samples. A series of samples was integrated over the depth profile to determine the production in mgC m^{-2} h^{-1} and then the incubation period was scaled over the total daily insolation from a Belfort pyroheliometer record to determine net photosynthetic daylight production (Wetzel & Likens, 1979). Vertical light transmission was measured with a Licor Radiometer (LI 185A).

Nutrient stimulation studies were conducted by preincubation of water samples with nutrients in 18 l clear plastic containers at the sampling depth for three days prior to the primary production measurements. Stimulation with phosphorus (100 μg PO_4-P 1^{-1}) and with nitrogen + phosphorus (500 μg NH_4-N 1^{-1} + 100 μg PO_4-P 1^{-1}) were compared to controls (no nutrients added) and to the open water column.

Poly-β-hydroxybutyrate (PHB) was extracted from a centrifuged, freeze dried pellet of purple sulphur bacteria and analyzed by a method of Braunegg et al., (1978) which consisted of depolymerization of PHB with sulphuric acid and conversion of hydroxybutyrate monomers to volatile hydroxybutyric acid methyl esters with acidified methanol. The PHB was analyzed on a gas chromatograph (Hewlett Packard 5880A with a flame ionization detector) fitted with a silinized glass column 1.83 m × 2 mm ID, and packed with Chromosorb W-AW DMCS 80-100 mesh coated with 5% Carbowax M20 TPA (Chromatographic Specialties).

Results

Microbial phototrophic organisms

Vertical profiles of primary productivity and pigment distribution were made during several spring (Apr.–May) and fall (Oct.) periods between 1982–87 (Fig. 1). The vertical distribution of phytoplankton cell numbers were presented for 16 October 1983 and 19 April 1984 since no chlorophyll a data were available. For April 1982 and 1984 primary productivity tended to be higher in the upper 4 m of water. In the spring of 1985 there was a shift in peak primary production into deeper water of mixolimnion (4–7 m). This trend seemed fairly consistent with the vertical distribution of phytoplankton biomass (chlorophyll a). In 1986-87, although some minor peaks in primary productivity were present, a more uniform vertical distribution of primary productivity and chlorophyll a were observed. The more detailed phytoplankton enumeration showed microstratification that was not apparent in the wider spaced chlorophyll a measurements.

During the initial study period (April 1982) water transparency was low (Secchi depth 2.8 m);

however, in subsequent years the Secchi disc transparency was much greater (5.6–7.5 m). Vertical light transmission measurements over the study period indicated that between 9–33% of the incident radiation was recorded at 4 m and between 2–13% reached 7 m providing light for photosynthetic activity down to the permanent chemocline on most occasions in the spring and fall.

Primary production in the mixolimnion varied from 0.3 to 8 μgC l^{-1} h^{-1} over the study period with the highest values occurring at 4–6 m during October. Estimates of primary production associated with the dense purple sulphur bacteria plate varied from 14–168 μgC l^{-1} h^{-1} with a median value of 35 ($n = 7$). These values were measured by incubating the samples at the upper surface of the dense plate to prevent any light limitation. Therefore this procedure could overestimate the contribution of the bacterial plate to primary production in the water column. Vertical integration of primary production gave a range of 115.1–476.1 mgC m^{-2} d^{-1} with an average value of 214.7 (Table 1). Primary production by the phototrophic sulphur bacteria (PSB) contributed an average of 40 percent (range 17–66%) of carbon fixed in the lake.

The purple PSB were found in a dense plate 20–25 cm thick at the boundary were sulphide concentrations increased rapidly (Table 2). Over

Table 1. Primary production in Mahoney Lake.

Date	Primary Production (mgC m^{-2}d^{-1})			Percent PSB[1] Contribution
	Phytoplankton	PSB[1]	Total	
16 Oct. '83	232.7	46.7	279.4	16.7
19 Apr. '84	58.3	–	–	–
3 May '85	54.8	60.3	115.1	52.3
23 May '85	226.8	121.5	348.3	34.8
19 Oct. '85	161.8	314.4	476.2	66.0
20 May '86	58.8	48.9	107.7	45.4
25 May '87	89.3	29.3	118.6	24.7
3 Oct. '87	124.2	89.7	213.9	41.9
Average			214.7	40.3

[1] **PSB** = phototrophic sulphur bacteria. In depth profile integration assumed that bacterial plate had a thickness of 0.25 m (see Table 2).

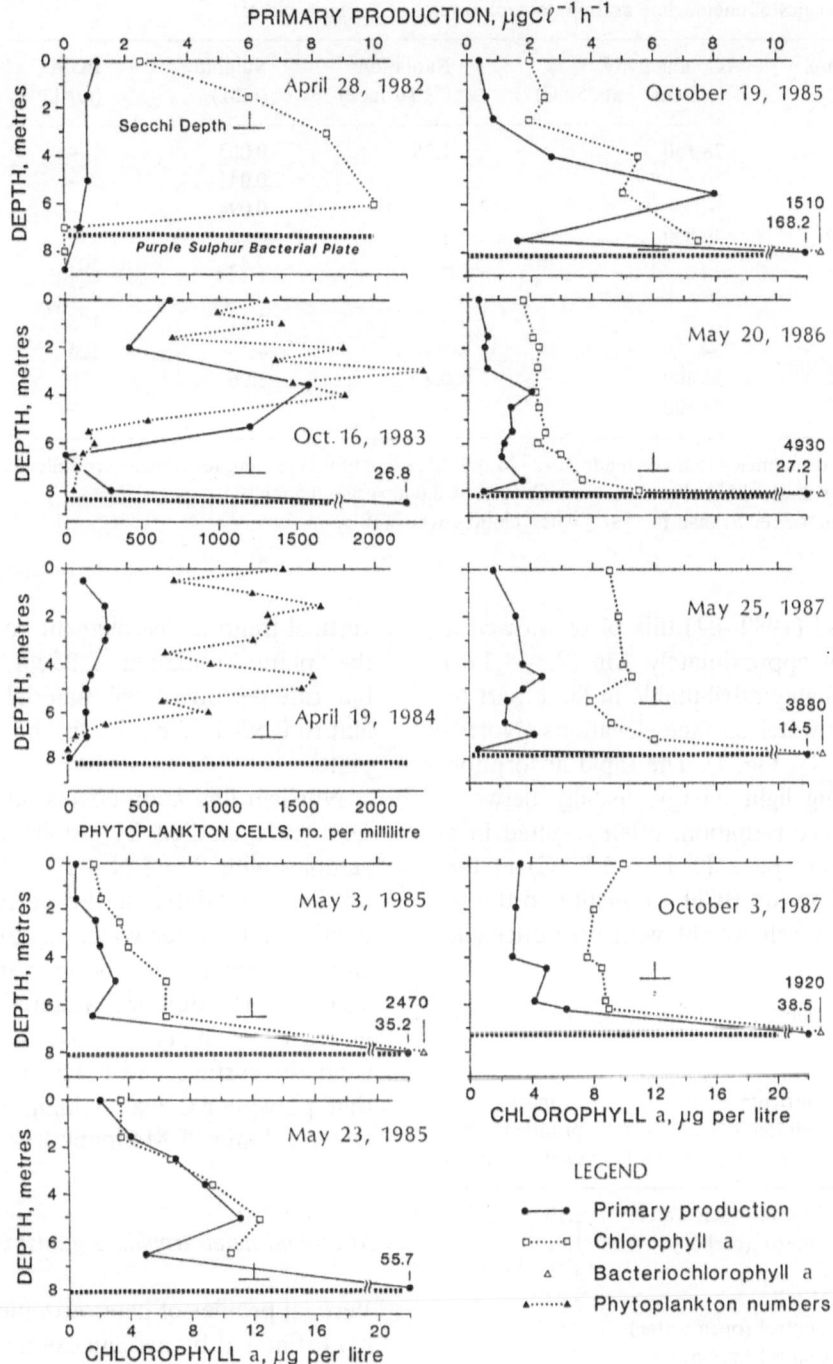

PRIMARY PRODUCTION, $\mu gC\ell^{-1}h^{-1}$

Fig. 1. Vertical distribution of primary productivity and photosynthetic pigments (primary production values represent average for $n = 3$ at each depth. Pigment concentrations are average values for duplicate samples at each depth).

Table 2. Characteristics of chemocline and phototrophic sulphur bacteria plate[1].

Depth (m)	Temp. (°C)	Conductivity ($\mu S\ cm^{-1}$ at 25 °C)	Radiation[2] (% surface)	Sulphide (mM)	POM (mg l^{-1})	BChl *a* (mg l^{-1})
8.0	18.2	28,150	1.25	0.003	–	–
8.1	–	–	–	0.041	–	–
8.2	–	–	–	0.094	–	–
8.25	19.2	29,800	1.1	–	7	0.47
8.3	–	–	–	2.43	307	2.84
8.35	–	–	–	–	475	7.09
8.4	–	–	–	3.17	175	3.07
8.45	–	–	–	–	125	1.42
8.5	20.0	31,400	0.2	3.96	70	0.71
9.0	19.0	39,100	–	–	–	–

[1] Samples collected and measurements made 21–23 May 1985, except for H_2S samples which were collected 20 May, 1986 when plate maximum still at 8.35 m. No data = —. Oxygen at 8.0 m = 0.2–0.5 mg l^{-1}.

[2] Surface radiation = 3600 μEinst. $m^{-2} s^{-1}$ when cloud cover 4/10's.

the study period (1982–87) this plate showed a vertical shift of approximately 1 m (7.3–8.3 m) which was probably attributable in large part to the variation in water surface elevations (Northcote & Hall, *MS* – Fig. 4). The rapid absorption of the remaining light energy, usually between 2–10% of surface radiation, often resulted in a small temperature peak (+1 to 1.5 °C) at the plate. In the plate the BChl *a* constituted 0.9 to 1.7% of the dry cell weight with no noticeable

Table 3. Effect of nutrient enrichment on phytoplankton productivity[1].

Water depth (m)	Nutrient[2] addition	Primary productivity[3] ($\mu gC\ l^{-1} h^{-1}$)
0	Control (open water)	1.10
	Control (carboy)	1.01
	P	4.28
	N + P	3.97
6.5	Control (open water)	2.51
	Control (carboy)	2.45
	P	8.32
	N + P	6.19

[1] Samples collected 21 May, 1985 and incubated for 3 days *in situ* with nutrients prior to production measurements.

[2] Incubation with P at 100 μg l^{-1} PO_4-P and N at 500 μg l^{-1} NH_4-N.

[3] Average values presented, $n = 3$.

vertical trend in cell pigment content. Cells from the plate contained 2.25 μg poly-β-hydroxybutyrate per mg of cell material which is equivalent to 0.69–1.06 mg l^{-1} in the dense area of the plate.

Nutrient limitation effects on primary production were studied by pre-incubation of water samples with P and N + P (Table 3). There was no significant difference between the primary production in the open water column and the plastic carboys after a 3 day confinement period. Both P and N + P additions resulted in a 3–4 fold increase in primary productivity in water samples from the surface and 6.5 m in spring indicating that phosphorus was a limiting nutrient in the water column of Mahoney Lake.

Microbial heterotrophic organisms

Vertical profiles of heterotrophic microbial activity, estimated by net glucose and acetate assimilation, were measured between 1982 and 1987 (Fig. 2). During the period of secondary microstratification in the mixolimnion (see Northcote & Hall, *MS*) heterotrophic activity was high and concentrated near the surface (1.5 m). As this microstratification gradually deteriorated (1984–85) there was a shift in maximum micro-

HETEROTROPHIC ACTIVITY, $\mu gC\ell^{-1}h^{-1}10^{-2}$

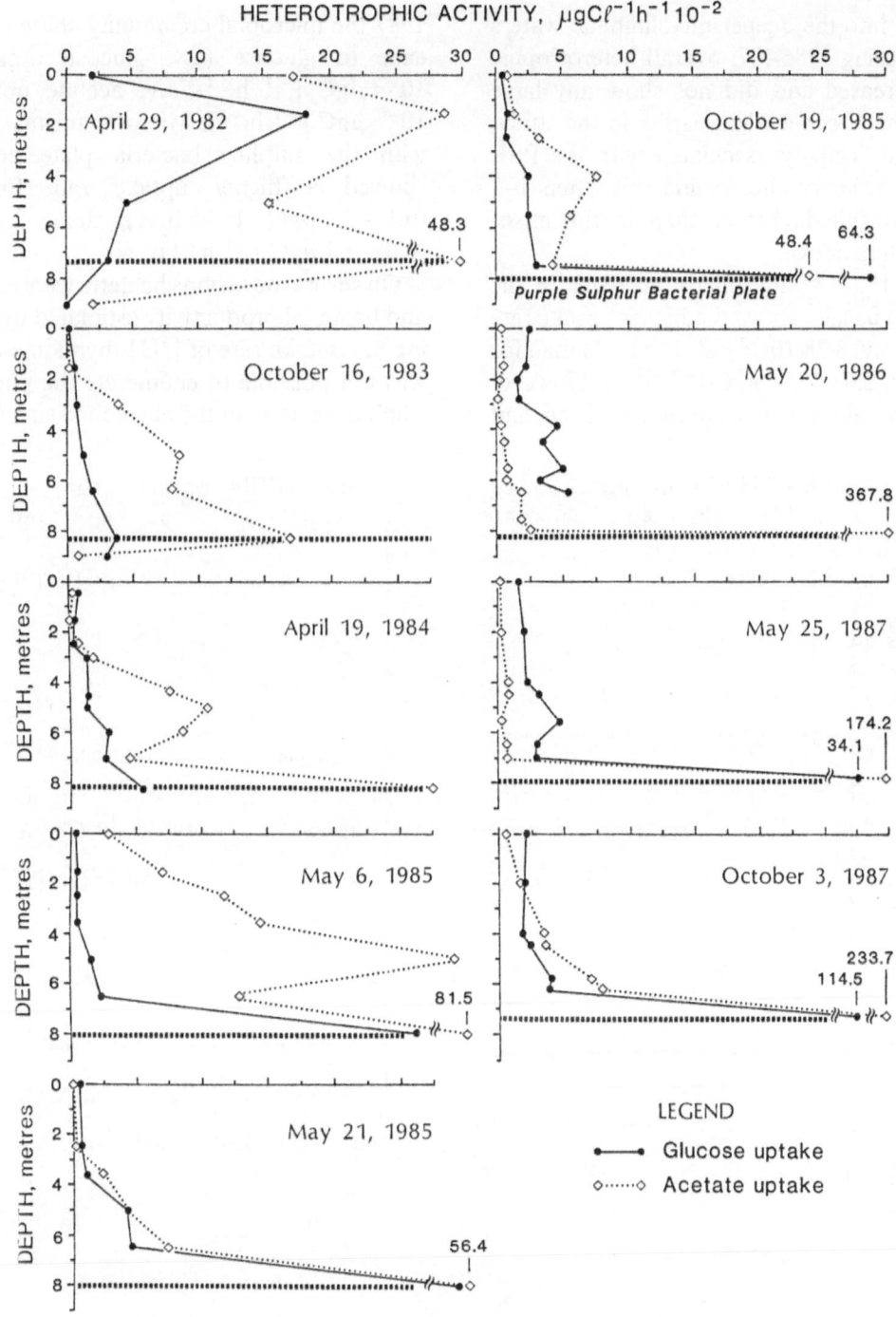

Fig. 2. Vertical distribution of net heterotropic assimilation of glucose and acetate (average values for $n = 2$ at each depth.

bial activity into the deeper mixolimnetic waters (4–6 m). During 1986–87, overall heterotrophic activity decreased and did not show any large peaks that had been evident earlier in the study. Heterotrophic activity associated with the PSB plate at the primary chemocline was often 1–2 orders of magnitude higher than in the mixolimnetic waters above.

Between 1982–85 the microorganisms in the mixolimnion usually showed a higher uptake rate for acetate (a.v. $8.28 \ 10^{-2} \ \mu gC \ l^{-1} \ h^{-1}$) than for glucose (a.v. $2.08 \ 10^{-2} \ \mu gC \ l^{-1} \ h^{-1}$). However as overall uptake rates decreased in 1986 and

1987 the microbial community shifted its preference to glucose (a.v. glucose uptake = $2.92 \ 10^{-2} \ \mu gC \ l^{-1} \ h^{-1}$; a.v. acetate uptake = $1.5 \ 10^{-2} \ \mu gC \ l^{-1} \ h^{-1}$). Microorganisms associated with the sulphur bacteria plate consistently showed a higher uptake rate for acetate ($0.17–3.7 \ \mu gC \ l^{-1} \ h^{-1}$) than for glucose ($0.03–1.1 \ \mu gC \ l^{-1} \ h^{-1}$).

On several occasions bacteria were enumerated and bacterial productivity estimated by determining the uptake rate of [^{3}H]-thymidine (Fig. 3). It was not possible to enumerate the phototrophic sulphur bacteria in the chemoline since the dense

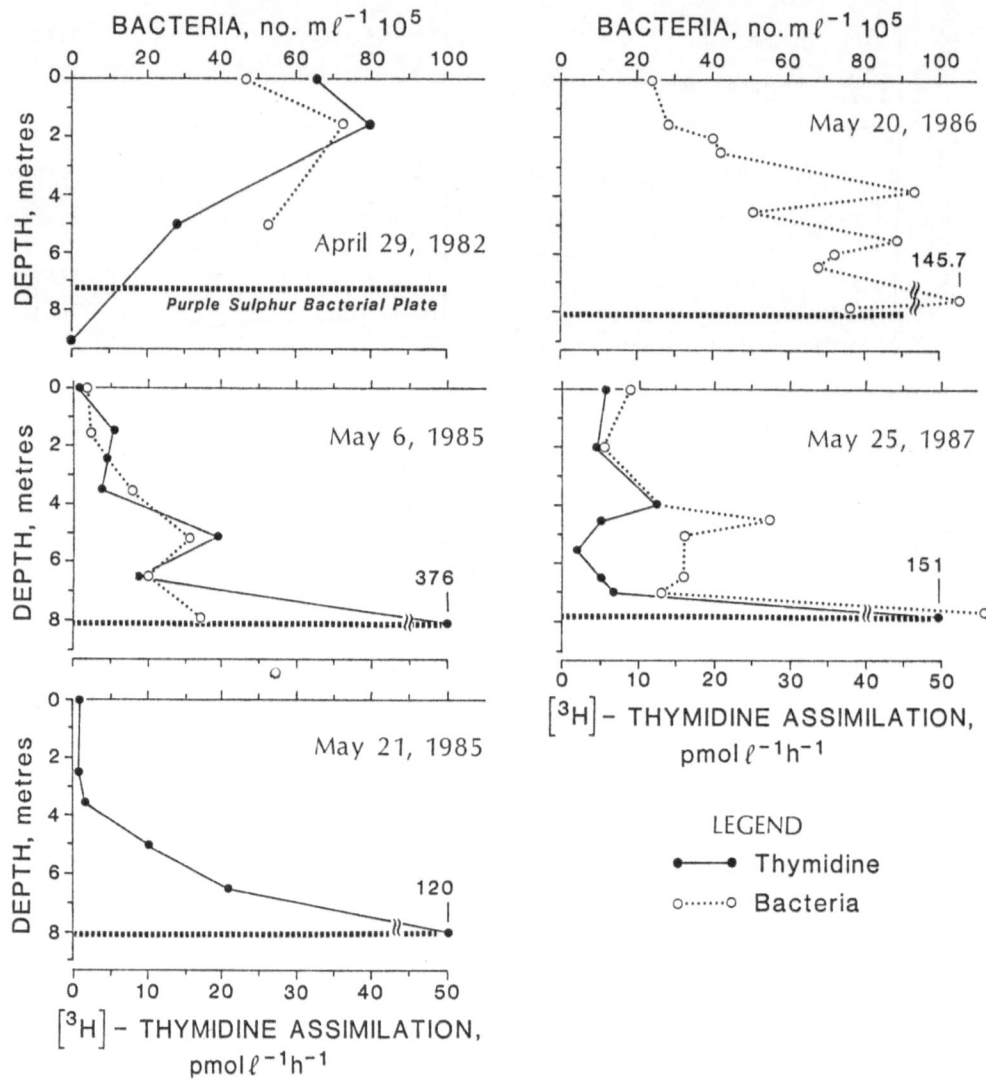

Fig. 3. Bacteria cell numbers and productivity (average values for $n = 2$ at each depth).

Table 4. Comparison of phytoplankton and bacterial production in Spring[1].

Date	Phytoplankton[2] Primary Production	Bacterial[3] Production	Production in Sulphur Bacteria Plate[4]	
			^{14}C-HCO_3^-	$^3[H]$-thymidine
3–6 May 1985	6.87	1.33	7.55	1.87
21–23 May 1985	25.98	1.32	13.92	1.05
25 May 1987	11.04	0.96	3.63	0.75

[1] All values as $mgC\,m^{-2}\,h^{-1}$.
[2] Vertical production in mixolimnion from surface to top of sulphur bacterial plate (0–7.5 m). Productivity of plate not included.
[3] Vertical production by bacterial tritated thymidine incorporation for same depth interval as phytoplankton production.
[4] Production for 0.25 m depth interval through plate using two different techniques.

clumps of cells could not be disaggregated. Bacteria abundance in the mixolimnion ranged between $5\cdot10^5$ and $145\cdot10^5$ cells ml^{-1}. Initially (April 1982) bacteria were numerous in the anoxic secondary chemocline established in the mixolimnion (Northcote & Hall, *MS*). Peak abundance shifted to deeper waters between May 1985 and May 1987 and at times showed considerable microstratification (20 May 1986).

Bacterial incorporation of labelled thymidine in the mixolimnion varied between 1 and 40 pMol $l^{-1}\,h^{-1}$ which represents bacterial production rates from 0.02 to 0.8 $\mu gC\,l^{-1}\,h^{-1}$. The highest bacterial productivity again was found in the anoxic secondary chemocline (April 1982) coincident with very high heterotrophic assimilation of glucose and acetate. Bacterial production associated with the plate varied from 2.4–7.5 $\mu gC\,l^{-1}\,h^{-1}$. Vertical integration of bacterial production in the mixolimnion and sulphur bacteria plate during May gave an average of 1.98 mgC $m^{-2}\,h^{-1}$ (range of 1.7–3.2 mgC $m^{-2}\,h^{-1}$ for $n=3$, Table 4). Bacterial production in the plate, which only occurred over a depth interval of 0.25 m, represented 44–58% of the total production. Therefore as much bacterial carbon was produced in the plate as occurred in the 7–8 m of water above.

Comparison of phytoplankton and bacteria

Using appropriate conversion factors from the literature (see methodology), it is possible to determine phytoplankton and bacterial carbon concentrations in the water column and to compare their relative production rates. Where comparative data were available, the average concentration of phytoplankton carbon was 163 $\mu gC\,l^{-1}$ (range 44–300 $\mu gC\,l^{-1}$). It was higher than found for bacterial carbon with an average of 90 $\mu gC\,l^{-1}$ (range 13–261 $\mu gC\,l^{-1}$). However, the bacterial carbon represented 14–57% of the total microbial carbon depending upon the depth interval in the mixolimnion over the three sampling periods considered, namely 29 April 1982, 3–6 May 1985 and 25 May 1987.

The production of phototrophic microorganisms (phytoplankton and purple sulphur bacteria) and bacterial productivity over the vertical water column are presented in Table 4 for three spring periods. Even though the bacterial carbon could represent up to 50% of the microbial carbon at some depths, integrated over the vertical water column, the bacterial production only represented 8–16% of the total cellular carbon fixed by microorganisms in the mixolimnion of Mahoney Lake. Primary productivity estimates in the PSB plate were considerably higher than were bacterial productivity measurements associated with the heterotrophic microorganisms in the bacterial plate.

Discussion

Early in the 1982 study, the distinct microstratification to form a secondary chemocline between

1.5–2 m in the spring provided a density discontinuity and high temperatures that favoured bacterial growth and activity (Northcote & Hall, 1983, *MS*). Similar microbial productivity and activity probably occurred during the spring of 1983 since conditions of shallow stratification and high temperatures developed again with a distinct anoxic zone between 2–3 m (Northcote & Hall, *MS*). The rapid increase in oxygen at 2 m from 10% saturation in April to over 200% saturation by mid-June (Northcote & Hall, *MS* – Fig. 3) in this strongly stratified region, provided evidence that the phytoplanktonic production was intense in these upper waters.

In the spring of 1984 an upper anoxic zone and a steep temperature gradient did not develop and peak microbial activity had shifted to the deeper waters (5 m). However, primary production was still relatively high at 2–3 m (Fig. 1) and oxygen was above 200% saturation (Northcote & Hall, *MS* – Fig. 3). In the spring of 1985 microbial production and activity were again concentrated in the deeper water of the mixolimnion and primary production and chlorophyll a peaks had shifted to this deeper water. Anoxic conditions were not observed in the upper mixolimnion and oxygen saturation was lower (<180%), shifting into the deeper water. In 1986 and 1987 when the shallow microstratification zone was not formed (Northcote & Hall, *MS*, Fig. 2), both microbial activity and primary productivity showed fairly uniform profiles in the water column above the chemocline. Thus over this six years study, phytoplankton and bacterial populations apparently responded to the gradual changes in the spring stratification pattern in the mixolimnion of Mahoney Lake.

Primary productivity by phytoplankton in the mixolimnion, over the study was comparable to that in an oligotrophic lake, although chlorophyll a concentrations were often in the mesotrophic lake range (Wetzel, 1975). Phosphorus rather than nitrogen apparently was one of the nutrients limiting phytoplankton productivity. However, under such hardwater conditions, iron would be very insoluble and also could limit phytoplankton growth as has been observed in other meromictic lakes (Priscu *et al.*, 1982). Adsorption of phosphorus during a calcium carbonate precipitation event has been observed in hardwater lakes near Mahoney (Murphy *et al.*, 1983). Total phosphorus in the upper waters of Mahoney Lake was 10 μg l^{-1} or less (Northcote & Hall, 1983). Visible cloudiness, evidence of calcium carbonate precipitation, was recorded in its water column especially during the warm summer months and varves of calcium carbonate were evident in a mid-lake sediment core, providing indirect evidence to support this mechanism of phosphorus removal from the water column.

The synchrony of vertical changes in microbial activity and primary productivity over the study period suggested that detrital decomposition by bacteria was probably essential for recycling of nutrients in the photic zone to stimulate phytoplankton in the water column. Early spring stratification with the anoxic conditions at the secondary chemocline indicated high microbial activity and respiratory activity was also stimulated by the elevated temperatures (Northcote & Hall, *MS*). This was subsequently followed by high primary productivity with high oxygen saturation values. However, it would be necessary to conduct kinetic uptake studies with radiolabelled phosphorus to help substantiate phosphorus limitation and recycling.

An extremely dense layer of purple phototrophic sulphur bacteria is present at the chemocline in Mahoney Lake (see Fig. 4 Northcote and Hall, 1983). The BChl a pigment concentration was found to exceed 7000 μg l^{-1} in the densest area of the plate which is higher than any concentration reported in the literature (Table 5). This concentration only occurred over a small depth interval (5 cm). Some of the highest pigment concentrations have been reported for BChl d in green sulphur bacteria (Lawrence *et al.*, 1978; Cromme & Tyler, 1984 – see Table 5).

It is difficult to compare the level of productivity by the PSB plate in Mahoney Lake to other meromictic lakes due to the wide variety of units used to report these data (Table 5). It is evident that there is considerable variability in the relative contribution of the PSB to overall productivity in

Table 5. Comparison of primary productivity and phototrophic sulphur bacteria in Mahoney Lake to other meromictic lakes.

Lake (Location)	Primary productivity			PSB Pigment[a]	Reference
	Phytoplankton	PSB	Total		
Mahoney (B.C.)	0.3–8[b] 54–232[c]	14–168[b] 29–314[c] (17–66%)[d]	115–476[c]	26–7090[h] 780[i]	Hall & Northcote (present study)
Waldsea (Sask.)	53–345[b] 38[e]				Hammer *et al.* (1975)
Waldsea (Sask.)		max. 1320[f] 398[c]		327–698[i] (BChl *d*)	Lawrence *et al.* (1978)
Deadmoose (Sask.)	90–493[b] 69[e]				Hammer *et al.* (1975)
Deadmoose (Sask.)		6.8–63.4[b]		1355[h]	Lawrence *et al.* (1978)
Green (New York)	<10–30[f] 51[e]	150–1628[f] (83%)[d]	2470[f] 290[e]		Culver & Brunskill (1969)
Medicine (S. Dakota)		(55%)[d]	190[c]		Hayden (1972)
Smith Hole (Indiana)		5700[f]			Wetzel (1973)
6 lakes (Michigan and Wisconsin)	276–1510[c]	1.98–59.8[c] (0.25–6.3%)[d]	357–1570[c]	11.1–630[h] (BChl *a* + *d*)	Parkin & Brock (1980)
Big Soda (Nevada)		58.9[c] (25%)[d]		9.3[h]	Priscu *et al.* (1982)
Big Soda (Nevada)		11.4[b]	500[e]	50[h] 300–480[i]	Cloern *et al.* (1983)
Vilar (Spain)				50–900[h]	Guerrero *et al.* (1985)
Cisó (Spain)		4.8–32.4[g]		40–450[h]	Guerrero *et al.* (1985)
Solar (Sinai)		(91%)[d]			Cohen *et al.* (1977)
Kisaratus (Japan)	4.7–9.2[b] 167–411[c]	0.3–154[b] 5–62[c] (3–13%)[d]		19.8–186[h]	Takahashi & Ichimura (1968)
Fidler (Tasmania)				250–>1500[h] (BChl *d*)	Croome & Tyler (1984)

PSB = phototrophic sulphur bacteria. a = BChl *a* unless otherwise stated.
Primary Productivity: b = μgC l^{-1} h^{-1}, c = mgC m^{-2} d^{-1}, d = % of total, e = gC m^{-2} y^{-1}, f = μgC l^{-1} d^{-1}, g = μgC mg cells^{-1} h^{-1}
PSB Pigment: h = μg l^{-1}, i = mg m^{-2}

meromictic lakes with a range of values from 1–91%. Mahoney appears intermediate in this range. Only a few lakes (Waldsea, Green and Smith Hole – Table 5) show daily primary productivity by the PSB which are similar to the high levels reported for Mahoney.

Light limitation and perhaps sulphide concentration probably were the main factors which limited primary productivity by the PSB in Mahoney Lake. van Gemerden & Beeftink (1983) reported a reduction in the specific growth rate of two species of phototrophic bacteria when sulphide concentrations exceeded 0.1 mM. Sulphide concentrations increased rapidly from 0.1 mM to over 2 mM over a 10 cm depth interval where the PSB plate showed its maximum density in Mahoney Lake. Different phototrophic bacteria show different responses both in affinity and in inhibition by sulphide, therefore specific experiments would be necessary to determine the most favorable sulphide conditions for the PSB in Mahoney Lake.

Several researches (Takahashi & Ichimura, 1970, Lawrence et al., 1978 and Parkin & Brock, 1980) reported the location of PSB in lakes where they received 1% or less of the surface radiation. The high transparency of Mahoney Lake allowed relatively high levels of light penetration (2–13%) to the top of the PSB plate, but the light was absorbed rapidly thereafter due to the high bacterial cell density. Laboratory experiments have shown that light intensities below about 90 μEinst m^{-2} s^{-1} are limiting to sulphur bacteria (Guerrero et al., 1985). Thus cells located in the dense area of the PSB plate would certainly be light limited unless they could regulate their buoyancy to move within the structure of the plate. van Gemerden et al. (1985) have shown that sulphur accumulates in the faster growing cells at the top of the plate which causes them to sink into deeper layers. This process may provide a mechanism for cell redistribution within the plate.

The dense PSB plate, concentrated at the chemocline in Mahoney Lake, could rapidly assimilate ^{14}C-acetate and was responsible for 10–90% of the areal acetate carbon assimilation

in the vertical water column. However, a comparison of acetate uptake (0.17–2.3 μg C l^{-1} h^{-1}) and phototrophic CO_2 assimilation (14–168 μg C l^{-1} h^{-1}) by the PSB in Mahoney Lake demonstrated the relative importance of CO_2 uptake to carbon assimilation by the bacterial plate. van Gemerden et al. (1985) could not show active assimilation of ^{3}H-acetate by sulphur bacteria in Lake Cisó and suggested that autotrophic CO_2 assimilation was the major process of carbon assimilation. Sulphur bacteria can certainly show a different affinity for acetate as has been demonstrated by Veldhuis & van Gemerden (1986) for purple Thiocapsa roseopersicina and brown Chlorobium phaeobacteroides isolated from Lake Kinneret (Israel).

Poly-β-hydroxybutyrate (PHB) is a common bacterial lipid polymer used for energy storage (Dawes & Senior, 1973). Studies in sewage systems have demonstrated a rapid incorporation of acetate into poly-β-hydroxybutyrate by microorganisms under anaerobic conditions which can be optimized to facilitate the biological removal of phosphorus from wastewaters (Comeau et al., 1987). The PSB in Mahoney Lake contained 0.7–1.1 mg l^{-1} of PHB in the dense area of the plate which is considerably higher than values (0.1–0.25 mg l^{-1}) reported by Guerrero et al., (1985). This difference may only reflect the high cell density in Mahoney Lake. Further studies are required to determine the importance of PHB to the sulphur bacteria.

Thymidine incorporation in the mixolimnion of Mahoney Lake was similar to levels found in the aerobic mixolimnion of Big Soda Lake (2–20 pMol l^{-1} h^{-1}, Zehr et al., 1987) and the surface waters of Lake Michigan in summer (6–18 pMol l^{-1} h^{-1}, Scavia & Laird, 1987). The exception was the high level (40 pMol l^{-1} h^{-1}) found in the anoxic secondary chemocline in the spring of 1982 where there was extremely high microbial activity. In the PSB plate thymidine incorporation was much higher (120–376 pMol l^{-1} h^{-1}) than has been observed in the zone of PSB in Big Soda Lake (Zehr et al., 1987) probably reflecting the high density of heterotrophic cells in Mahoney Lake.

Bacterial growth rates in the mixolimnion of Mahoney Lake varied from 0.05–0.55 d^{-1} using the conversion factor of 2 10^{18} cells mol^{-1} thymidine. These growth rates are in the range (0.07–0.59 d^{-1}) reported for Big Soda Lake (Zehr et al., 1987). Growth rates can be estimated from heterotrophic assimilation of labelled carbon solutes assuming that there is negligible isotope dilution due to unlabelled solute in the water column. The growth rate from acetate incorporation during the spring of 1982 and 1985 was 0.10 d^{-1} (n = 9) but this dropped to a very low value of 0.003 d^{-1} (n = 17) in the spring of 1986 and 1987 which probably reflected a change in the predominant species of microorganism rather than such a large change in overall microbial growth. The growth rate estimated from glucose net assimilation averaged 0.019 d^{-1} (n = 25) and was quite uniform over the four spring periods for which microbial abundance data were available. The growth rates were lower when calculated from labelled carbon solute assimilation than from thymidine incorporation similar to observations made by Zehr et al. (1987) for glutamate incorporation.

An estimate can also be made of the growth rate of bacteria in the plate from thymidine and labelled carbon (^{14}C-HCO$_3$ and ^{14}C-acetate) incorporation. The thymidine uptake rates provide an estimate of the heterotrophic and chemoautotrophic bacteria growth while labelled HCO$_3^-$ incorporation estimates the growth rate of the phototrophic sulphur bacteria. The growth rate from thymidine incorporation was 0.05–0.11 d^{-1} which on average was lower than observed from heterotrophs in the mixolimnion of Mahoney Lake. Zehr et al. (1987) found slightly higher growth rates in the anoxic mixolimnion where the PSB of Big Soda Lake are located. Growth rates based on phototrophic carbon dioxide incorporation in Mahoney varied from 0.1 to 1.1 d^{-1} which is approximately an order of magnitude faster than heterotrophic growth rates. Acetate can be utilized by heterotrophic bacteria and by PSB in the light. Growth rates based on the incorporation of acetate into PHB were low (0.003–0.04 d^{-1}) which suggested either that the overall importance of PHB was relatively minor in the energetics of bacteria associated with the plate or that the acetate was widely used in the cell for synthesis of other cell components or as an immediate energy source. No estimates were made of the respiration rate of the labelled organic solutes to evaluate their importance for rapid energy utilization.

Acknowledgements

We are grateful to Dr. M. A. Chapman and Dr. J. D. Green, University of Waikato, New Zealand for sampling assistance and demonstration of appropriate limnological attire during their visits to Canada. P. L. Wentzell, Department of Civil Engineering, U.B.C. enumerated the bacteria under the guidance of Dr. R. J. Daley, National Water Research Institute. Y. Comeau, Department of Civil Engineering, U.B.C. conducted the PHB analysis on the phototrophic sulphur bacteria. Chlorophyll a fluorescence measurements were made on instrumentation in the laboratory of Dr. P. J. Harrison, Department of Oceanography, U.B.C. Dra. E. Conejo enumerated the phytoplankton. Field laboratory facilities were provided by G. B. Northcote in Keremeos, B. C. Funding was obtained from NSERC grants 67-8935 (KJH) and 67-3454 (TGN).

References

American Public Health Association, American Water Works Association & Water Pollution Control Federation, 1985. Standard methods for the examination of water and wastewater, 16th ed. APHA, Wash. D.C. 1268 pp.

Anderson, G. C., 1958. Some limnological features of a shallow saline meromictic lake. Limnol. Oceanogr. 3: 259–270.

Braunegg, G., B. Sonnleitner & R. M. Lafferty, 1978. A rapid gas chromatographic method for the determination of poly-hydroxybutyric acid in microbial biomass. Biotechnol. 6: 29–37.

Cloern, J. E., B. E. Cole & R. S. Oremland, 1983. Autotrophic processes in meromictic Big Soda Lake, Nevada. Limnol. Oceanogr. 28: 1049–1061.

Cloern, J. E., B. E. Cole & S. M. Wienke, 1987. Big Soda Lake (Nevada). 4. Vertical fluxes of particulate matter: Seasonality and variations across the chemocline. Limnol. Oceanogr. 32: 815–824.

Cohen, Y., W. E. Krumbein & M. Shilo, 1977. Solar lake (Sinai) 2. Distribution of photosynthetic microorganisms and primary production. Limnol. Oceanogr. 22: 609–620.

128

Comeau, Y., W. K. Oldham & K. J. Hall, 1987. Dynamics of carbon reserves in biological dephosphatation of wastewater. Adv. in Water Pollut. Control. 39–55 IAWPRC Int. Conf. in Rome, Sept. 1987.

Croome, R. L. & P. A. Tyler, 1984. Microbial microstratification and crepuscular photosynthesis in meromictic Tasmanian lakes. Verh. int. Ver. Limnol. 22: 1216–1223.

Culver, D. A. & G. J. Brunskill, 1969. Fayetteville Green Lake v. Studies of primary production and zooplankton in a meromictic marl lake. Limnol. Oceanogr. 14: 862–873.

Dawes, E. A. & P. J. Senior, 1973. The role and regulation of energy reserve polymers in microorganisms. Adv. Microbial Physiol. 10: 135–266.

Fuhram, J. A. & F. Azam, 1982. Thymidine incorporation as a measure of heterotrophic bacterioplankton production in marine surface waters: Evaluation of field results. Mar. Biol. 66: 109–120.

Fuhram, J. A., H. W. Ducklow, D. L. Kirchman, J. Hudak, G. B. McManus & J. Kramer, 1986. Does adenine incorporation into nucleic acids measure total microbial production? Limnol. Oceanogr. 31: 627–636.

Griffiths, R. P., S. S. Hayasaka, T. M. McNamara & R. Y. Morita, 1977. Comparison between two methods of assaying relative microbial activity in marine environments. Appl. envir. Microbiol. 34: 801–805.

Guerrero, R., E. Montesinos, C. Pedros-Alio, I. Esteves, J. Mas, H. van Gemerden, P. A. G. Hofman & J. F. Bakker, 1985. Phototrophic sulfur bacteria in two Spanish lakes: Vertical distribution and limiting factors. Limnol. Oceanogr. 30: 919–931.

Hammer, U. T., R. C. Haynes, J. M. Heseltine & S. M. Swanson, 1975. The saline lakes of Saskatchewan. Verh. int. Ver. Limnol. 19: 589–598.

Hayden, J. F., 1972. A limnological investigation of a meromictic lake (Medicine Lake, South Dakota), M. Sc., Univ. of South Dakota, Vermillion.

Hobbie, J. E., R. J. Daley & S. Jasper, 1977. Use of Nuclepore filters for counting bacteria by fluorescence microscopy. Appl. envir. Microbiol. 33: 1225–1228.

Hudec, P. P. & P. Sonnenfeld, 1980. Comparison of Caribbean solar ponds with inland solar lakes of British Columbia. in A. Nissenbaum (ed.) Hypersaline brines and evaporitic environments. Elsevier, Amsterdam, 101–114.

Lawrence, J. R., R. C. Haynes & U. T. Hammer, 1978. Contribution of photosynthetic green sulphur bacteria to total primary production in a meromictic saline lake. Verh. int. Ver. Limnol. 20: 201–207.

Lovell, C. R. & A. Konopka, 1985. Seasonal bacterial production in a dimictic lake as measured by increases in cell numbers and thymidine incorporation. Appl. envir. Microbiol. 49: 492–500.

Murphy, T. P., K. J. Hall & I. Yesaki, 1983. Coprecipitation of phosphorus with calcite in a naturally eutrophic lake. Limnol. Oceanogr. 28: 58–69.

Northcote, T. G. & T. G. Halsey, 1969. Seasonal changes in the limnology of some meromictic lakes in southern British Columbia. J. Fish Res. Bd, Can. 26: 1763–1787.

Northcote, T. G. & K. J. Hall, 1983. Limnological contrasts and anomalies in two adjacent saline lakes. Hydrobiologia 105: 179–194.

Northcote, T. G. & K. J. Hall, MS. Vernal microstratification patterns in a meromictic saline lake: their causes and biological significance. Hydrobiologia

Parkin, T. B. & T. D. Brock, 1980. Photosynthetic bacterial production in lakes: The effect of light intensity. Limnol. Oceanogr. 25: 711–718.

Porter, K. G. & Y. S. Feig, 1980. The use of DAPI for identification and counting aquatic microflora. Limnol. Oceanogr. 25: 943–948.

Priscu, J. C., R. P. Axler, R. G. Carlton, J. E. Reuter, P. A. Arneson & C. R. Goldman, 1982. Vertical profiles of primary productivity biomass and physiochemical properties in meromictic Big Soda Lake, Nevada, USA. Hydrobiologia 96: 113–120.

Riemann, B., J. A. Fuhram & F. Azam, 1982. Bacterial secondary production in freshwater measured by ^{3}H-thymidine incorporation method. Microb. Ecol. 8: 101–114.

Scavia, D. & G. A. Laird, 1987. Bacterioplankton in Lake Michigan: Dynamics, controls, and significance to carbon flux. Limnol. Oceanogr. 32: 1017–1033.

Strickland, J. H. & T. R. Parsons, 1972. A practical handbook of seawater analysis, 2nd ed. Bull. Fish. Res. Bd, Can. 167.

Takahashi, M. & S. Ichimura, 1968. Vertical distribution and organic matter production of photosynthetic sulfur bacteria in Japanese Lakes. Limnol. Oceanogr. 13: 644–655.

Takahashi, M. & S. Ichimura, 1970. Photosynthetic properties and growth of photosynthetic sulphur bacteria in lakes. Limnol. Oceanogr. 15: 929–944.

van Gemerden, H. & H. H. Beeftink, 1983. Ecology of phototrophic bacteria. in J. G. Ormerod (ed.). The phototrophic bacteria, Studies in Microbiol. 4: 146–179, Univ. of Cal. Press, Berkley.

van Gemerden, H., E. Montesinos, J. Mas & R. Guerrero, 1985. Diel cycle of metabolism of phototrophic purple sulfur bacteria in Lake Cisó (Spain). Limnol. Oceanogr. 30: 932–943.

Veldhius, M. J. W. & H. van Gemerden, 1986. Competition between purple and brown bacteria in a stratified lake: Sulfide, acetate, and light as limiting factors. FEMS Microbial Ecol. 38: 31–38.

Wetzel, R. G., 1973. Productivity investigations of interconnected lakes 1. The eight lakes of the Oliver and Walters chains, northeastern Indiana. Hydrobiol. Stud. 3: 91–143.

Wetzel, R. G., 1975. Limnology, W. B. Saunders Co. Toronto. 743 pp.

Wetzel, R. G. & G. E. Likens, 1979. Limnological analyses. W. B. Saunders Co. Philadelphia. 357 pp.

Wood, L. W., 1985. Chloroform-methanol extraction of chlorophyll a. Can. J. Fish. aquat. Sci. 42: 38–43.

Zehr, J. P., R. W. Harvey, R. S. Oremland, J. E. Cloern, L. H. George & J. L. Lane, 1987. Big Soda Lake (Nevada). 1. Pelagic bacterial heterotrophy and biomass. Limnol. Oceanogr. 32: 781–793.

Hydrobiologia **197**: 129–138, 1990.
F. A. Comín and T. G. Northcote (eds), Saline Lakes.
© 1990 *Kluwer Academic Publishers.*

Autumnal mixing in Mahoney Lake, British Columbia

P.R.B. Ward[1], K.J. Hall[1], T.G. Northcote[2], W. Cheung[1] & T. Murphy[3]
[1]*Department of Civil Engineering and* [2]*Department of Zoology, University of British Columbia, 2075 Wesbrook Mall, Vancouver, B.C., V6T 1W5, Canada*; [3]*National Water Research Institute, Burlington, Ontario, L7R 4A6, Canada*

Key words: Saline lake, convection, meromixis, mixing, wind energy

Abstract

Salinity, water temperature and meteorlogical measurements were made over a 5 month period (June–Oct.) at Mahoney Lake, British Columbia to study autumnal mixing. The measurements were made during a time when the lake had been well stratified by a much larger than average runoff the previous spring. The potential energy of stratification decreased from 50 to 24 MJ, in the top 8 m of the water column, from mid August to mid October. Analysis of the energy available from wind shear on the water surface and from penetrative convection during the autumn cooling period was made. Winds were found to be weak (av. 2.17 m s^{-1}) at Mahoney Lake, and their average contribution to mixing energy during the study period was less than 30%. Penetrative convection from thermals descending from the cool surface contributed an average of 2.21 J m^{-2} d^{-1} to mixing which represented 72% of the energy available. An efficiency factor of 0.20 for the penetrative convection energy, larger than values previously reported in the literature, was found to fit the measured loss of potential energy of stratification during the period.

Introduction

Mahoney Lake is a meromictic saline lake in southern British Columbia (lat. 49° 17′ N long. 119° 35′ W, altitude 471 m above sea level). The unusual nature of the physical, chemical and biological properties of the lake were first documented by Northcote & Halsey (1969). Mahoney Lake receives a shallow ice cover for a period of 2 to 4 months every winter, and shows striking vertical variations in temperature due to the suppression of vertical mixing by strong salinity gradients.

The interesting nature of the physical limnology of meromictic lakes has been known for a long time, and calculations of the energy required for complete mixing of such lakes were made by Hutchinson (1957). Temperature and total dissolved solids data for several meromictic lakes in central Washington state (USA) have been published by Walker (1974) where he discussed the dual control of density by both temperature and chemical (salinity) effects on the water. Diurnal temperature changes measured near the surface of a shallow African salt pan, including data on total heat content and stability are reported by Ashton & Schoeman (1988).

A comparative limnological study of Mahoney

Lake, and a neighbouring dimictic lake (Green Lake) has been published (Northcote & Hall, 1983). The lakes were observed to be entirely different one from another in physical and geological characteristics and consequences to their limnology were presented. The possibility that differences in wind exposure caused significantly different physical stratification was postulated (Northcote & Hall, 1983).

During the spring and early summer of 1983, unusually high run-off caused a large layer of relatively fresh water to be introduced on the surface of Mahoney Lake. This layer was of lower salinity than the surface layer in preceeding or succeeding years (Northcote & Hall, 1990), and offered a good opportunity to study mixing mechanisms. Major changes in the potential energy of stratification in the upper 8 m of the lake were noted from August to November, and these changes are analyzed herein. Following Imberger (1979), an examination is made of the effects of wind driven circulation and penetrative convection induced by cooling, on mixing in the upper several meters of the lake. These calculations show whether wind or convective cooling or both are the main cause of autumnal mixing in Mahoney Lake.

Methods and mixing calculations

Methodology

Vertical temperature, conductivity and salinity readings were made with a YSI Model 33 meter at a sampling frequency of once per month, using depth intervals of 0.5 to 1.0 m. The probe was cleaned and the field instrument calibrated using laboratory instruments following Hall & Northcote (1986). The precision of temperature and conductivity measurements were estimated at ± 0.2 °C and $\pm 200 \mu$S cm^{-1} (high range scale) respectively.

Density was determined from the salinity readings from the YSI meter, using an oceanographic table which related density to salinity and temperature, Riley & Skirrow (1965). Calculations of density based on chemical composition and partial molal volumes, (MacIntyre & Melack, 1982),

showed very little difference between calculated density values and values estimated from seawater salinity tables. Thus the simpler approach, using the seawater density tables, was adopted.

During the summer and autumn of 1983, a data logger was operated at Mahoney Lake. Hourly measurements of wind velocity, water temperature at 4 depths (0, 2, 6, 12 m) and air temperature were recorded by a Hymet package (Plassey recorder), positioned on a raft at mid-lake. Data collection commenced on 19th June, and terminated on 6th November. Water temperature measurements had an accuracy of ± 0.1 °C within the range 0 to 35 °C. Wind measurements were made at 3 m above the water surface. Meteorological data for the complete study period were obtained from Penticton Airport station, 19 km north of Mahoney Lake, and were used to check on precipitation amounts during the study.

Bathymetric information for Mahoney Lake was taken from a 1 m depth contour map (scale 1 : 3000) with a Placom KP-90 digital planimeter and the area versus depth relationship calculated. Water surface levels relative to a bench mark on the lake shore were measured monthly.

Mixing calculations

An energy calculation may be carried out using the following equations (Imberger, 1979).

Velocity u_f of falling plumes:

$$u_f = \left[\frac{\alpha g h \tilde{H}}{\rho_w C_p} \right]^{1/3} \tag{1}$$

in which $\tilde{H}$ = rate of cooling at the water-air-
surface, W m^{-2}
h (depth of mixed upper layer) = 8 m
g (gravitational acceleration) = 9.8 m s^{-2}
C_p (specific heat of water) = 4182 J kg^{-1} °C^{-1}
ρ_w (density of water) = 1000 kg m^{-3}
α (thermal coefficient of expansion of water) = 1.8 × 10^{-4} °C^{-1}

Heat changes in the upper 8 m of the water column were calculated in order to determine the heat loss for the penetrative convection calculations. A temperature versus time profile for 4 m depth was synthesized from the monthly temperature readings. Using the water volumes in each 2 m thick layer, the weighted average water temperature for the top 8 m of the water column was calculated for each day, and the daily changes in temperature from the previous day determined. These temperature changes were converted to heat changes by multiplying by the depth times the specific heat of water, giving heat changes in $J\,m^{-2}\,d^{-1}$. These values were then converted to units of $W\,m^{-2}$.

The power per unit area available for stirring from cooling equals an efficiency C_k times the water density times the falling plume velocity cubed. C_k was found to be about 0.13 by Imberger (1979) in studies on Wellington Reservoir, Australia, and this value was in reasonable agreement with Sherman et al. (1978).

We used the product $\rho_w u^{*3}$ herein for calculating an efficiency of wind energy utilization following Turner (1973). The shear velocity (u^*) was determined from a coefficient of drag (C_d) for the air-water interface. For a medium range of wind velocity (4 to $11\,m\,s^{-1}$), C_d equals 0.0012 for winds measured at 10 m above the ground surface (Large & Pond, 1981). On this basis the shear velocity u^* is given by:

$$u^* = [0.0012\ u^2\ \rho_a/\rho_w]^{1/2} \qquad (2)$$

in which ρ_a = density of air, $kg\,mg^{-3}$
ρ_w = density of water, $kg\,m^{-3}$
u = wind velocity at 10 m, $m\,s^{-1}$

Using C^* as the efficiency of wind energy utilization, the wind power per unit area available for mixing is proportional to C^* times u^{*3}. In order to obtain an average wind speed representative of the wind power, the sums of the cubes of the hourly wind speeds were computed, and divided by 24 to determine the daily average values of the wind speed cubed. Values for C^* are still being researched, and a wide range of published values has appeared. Wind stress experiments by Wu (1973) yielded a value of C^* of 0.23, and values

reported by Imberger (1979) ranged from 0.03 to 0.23, when a remote station was used for wind data, depending on the amount of sheltering allowed for from surrounding hills. A value of 0.26 was used in studies on wind mixing in Babine Lake (Farmer & Carmack, 1981). Values used for modelling with the US Army Corps of Engineers model CE-QUAL-R1 are $C^* = 0.12$ and larger, dependent on the value of the Richardson number, Johnson & Ford (1981). We used a value of 0.23.

The potential energy of stratification is the difference between the potential energy of the stratified lake, and the potential energy of a hypothetical totally mixed lake of the same average temperature and the same average salinity. This difference in potential energy may be reduced to a small difference in height by dividing by the product total water mass times gravitational acceleration (Ward, 1980). The centre of gravity of the mixed lake is at a slightly greater elevation than that of the stratified lake. Change in the potential energy of stratification is equal to the amount of mixing energy that has been utilized. For lakes that show negligible changes in water surface level during the period of analysis, the calculations may be carried out from the surface downwards (using the water surface as datum). The integrations do not need to go for the full depth, and may be terminated at a depth deemed to show no change during the measuring period. The potential energy of stratification, Ps, in joules, is given by:

$$Ps = \int_0^{z_m} (\rho - \bar{\rho})gz\,A\,dz \qquad (3)$$

in which z = the depth measured from the surface of downwards, m
z_m = the depth of the level of no change (or bottom of the lake)
$A(z)$ = is the area of layer at depth z, m^2
$\rho(z)$ = is density of layer at depth z, $kg\,m^{-3}$
$\bar{\rho}$ = is mean density of layers above level of no change, $kg\,m^{-3}$
g = is gravitational acceleration, $m\,s^{-2}$.

Assuming that the power available from penetrative convection and wind energy are additive, then the time rate change of potential energy of stratification per unit area is:

Time rate change of potential = convective current + wind induced
energy per unit area   induced mixing   mixing

or: $\Delta Ps/(A\Delta t) = \rho_w C_k u_f^3 + \rho_w C^* u^{*3}$ (4)

The programme of field measurements enabled all the terms in equation 4 to be evaluated except the efficiencies. Data from the monthly field visits allowed the rate of change of potential energy of stratification to be found (left hand side of equation 4), and daily data on wind strengths and water temperatures were available from the data logger output and were used to determine the right hand side of equation 4, with assumed values for the efficiencies.

Results and discussion

The difference in power between the windiest day $(62 \text{ m}^3 \text{ s}^{-3}, u = 3.96 \text{ m s}^{-1})$ and the least windy day $(0.3 \text{ m}^3 \text{ s}^{-3}, u = 0.67 \text{ m s}^{-1})$ ranged over 200 times (Fig. 1). The average wind speed cubed for the whole period of record was $10.1 \text{ m}^3 \text{ s}^{-3}$, that is $u = 2.17 \text{ m s}^{-1}$. This is a low value, indicating the winds are light at Mahoney Lake.

Hourly temperatures at depths of 0 and 2 m increased each afternoon with heat input from the sun, by about $+0.5 \,^\circ\text{C}$ from the daily average

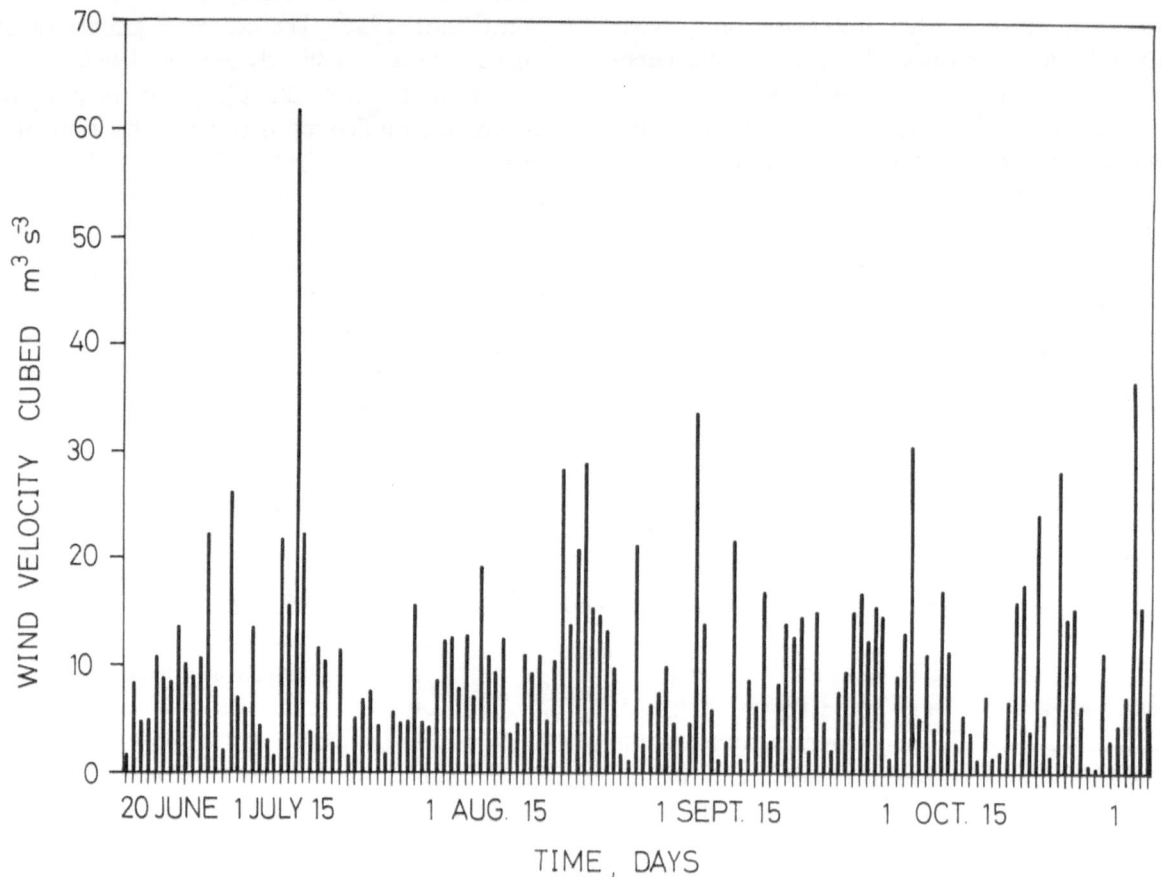

Fig. 1. Daily average wind speed 3 m above the surface of Mahoney Lake from June to October 1983 (Cubed values are presented to demonstrate the relative wind power available for mixing).

temperature. Daily average temperature values at 0 and 2 m were equal (Fig. 2). The temperatures in the well-mixed 0 to 2 m layer reached a maximum in early August. Temperature fluctuations with a period of 4 to 5 days, corresponding to synoptic meteorological weather pattern time scales, are seen in the surface layer data. Temperatures at 6 m depth become maximum much later (about mid October), and temperatures at 12 m depth display no change.

Figure 3 illustrates diagrammatically the well mixed upper layer with falling plumes of cool water eroding the stably stratified layers below. In meromictic lakes with small inflows this well mixed upper layer, the mixolimnion, is in the range 2 to approximately 8 m deep. In very sheltered meromictic lakes, such as crater bottom pans, the mixolimnion may be as shallow as 0.5 m (Ashton & Schoeman, 1988). During the autumn months,

the water adjacent to the surface is cooled by several processes (mainly evaporative losses and long wave radiation) and the cooling causes plumes of water to descend. The process is analogous to an inverted form of the convective currents that develop in a saucepan when water is heated on a stove.

The number of days when cooling occurred on Mahoney Lake is small up to mid August and then increases rapidly (Fig. 4). Bursts of cooling during 4 to 5 day synoptic cycles are clearly evident (Fig. 4). Several days when the cooling was at least 150 W m^{-2} are shown. This value, equal to 3.6 kWh m^{-2} d^{-1} is significant when compared to the incoming energy from solar radiation on a bright, sunny day (7 kWh m^{-2} d^{-1}). During mid August to mid September, the time of maximum cooling, the average daily value was 55 W m^{-2}.

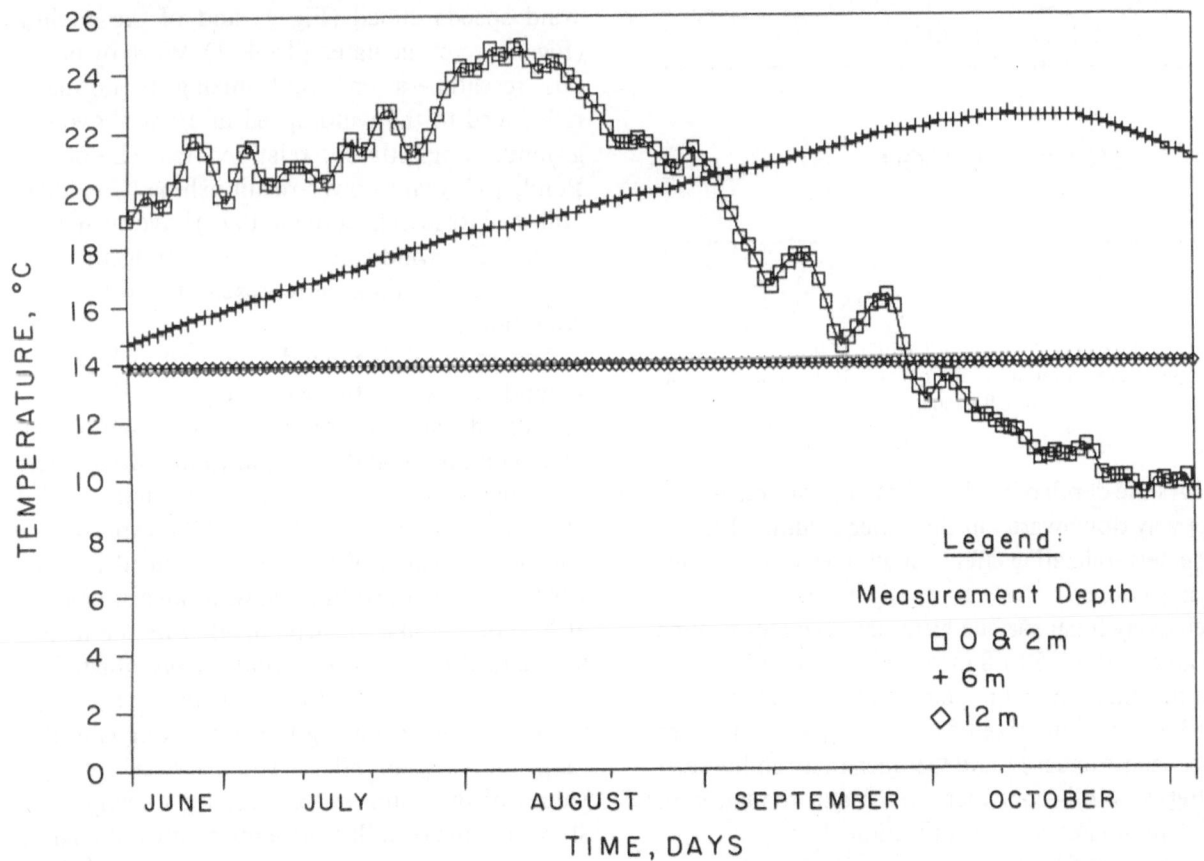

Fig. 2. Daily average water temperatures at 4 depths in Mahoney Lake from June to November 1983.

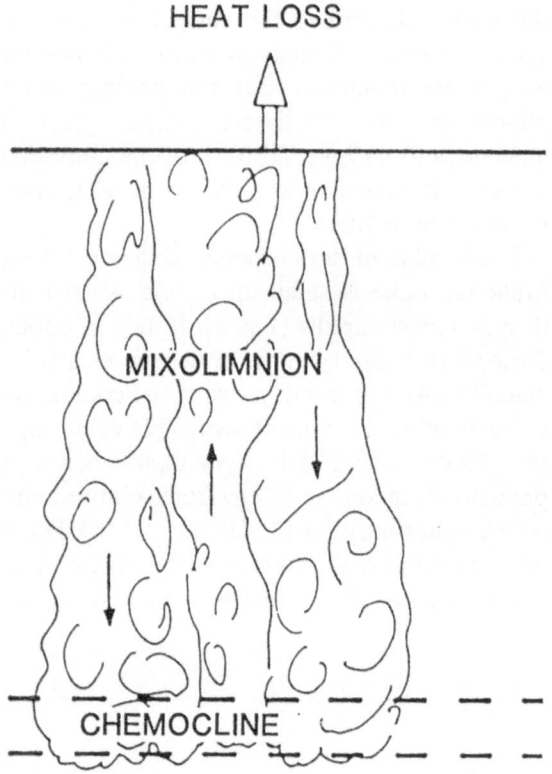

HEAT LOSS

MIXOLIMNION

CHEMOCLINE

MONIMOLIMNION

Fig. 3. A diagrammatic representation of mixing from penetrative convection in a meromictic lake (adapted from Fig. 6.9, Imberger, 1979).

As the depth of peak water temperatures works its way downwards in the water column (Fig. 5), the destabilizing effects of warmer water underlying cooler water causes significant mixing of relatively fresh surface water into the more saline water in the 2.5 to 8 m range. This is the period of the main annual mixing events in the lake. The reduction in the size of the density step at 2.5 m between August 11 and October 18 is readily seen (Fig. 5) and is equivalent to a major reduction in potential energy of stratification.

No significant changes in temperature or salinity were observed during the duration of the study below 8 m depth. Therefore the integration of equation 3 was evaluated for the range 0 to 8 m depth, for each of the 6 monthly data sets.

During the period of the study, the water surface height decreased between June 18 and October 18 by a negligible amount (0.11 m) and thus corrections for changing water surface levels were unnecessary. Values for the potential energy of stratification for the 0 to 8 m layer, in megajoules (MJ) are plotted in Fig. 6. An overall decrease of 37 MJ of potential energy of stratification occurred during a four month period, with the most significant decrease occurring between August 11 and October 18.

Efficiency calculations

For the analysis of the efficiency of utilization of energy for mixing, weekly averages of the daily wind speeds cubed (Fig. 1) and of the cooling (Fig. 4) were calculated (Table 1). Work by previous researchers on wind mixing is normally referenced to the wind speed at 10 m above the ground. Logarithmic relationships (Large & Pond, 1981) and power relationships (Canadian National Research Council, 1977) have been used to describe wind speed profiles. Judgement is involved in deciding on representative roughness conditions.

The ratio of the wind speed at 3 m above ground surface to the wind speed at 10 m was calculated using the power law distribution for open conditions and the logarithmic distribution for roughness height of 0.02 m. Results for the wind speed ratio were 0.84 and 0.80 respectively. An average value of 0.82 was assumed for the wind speed ratio, so that the wind measurements at Mahoney Lake must be divided by the factor 0.82 cubed (= 0.55) for wind energy utilization comparisons. This is done in Table 1. The contribution of wind mixing to total mixing is in the range of 10% to 60%. The 19-week average values of the wind and convective cooling data show that the overall contribution of wind mixing to total mixing is small (about one quarter). The results of balancing equation 4 using trial and

Table 1. Weekly average wind and convective cooling in Mahoney Lake

PERIOD		AVERAGE VALUE FOR THE PERIOD						Rate of change of potential energy[3] (J m^{-2} d^{-1})			Ratio of wind mixing to total
From	To	Wind speed cubed at 3 m (m^3 s^{-3})	Wind speed cubed at 10 m (m^3 s^{-3})	Shear velocity cubed[1] (m^3 s^{-3} × 10^{-5})	Heat loss by cooling/ unit area (W m^{-2})	Velocity of falling plumes[2] (m s^{-1})	Velocity cubed of falling plumes (m^3 s^{-3} × 10^{-6})	Wind	Convective	Total	
June 20–	26	6.73	12.2	0.0218	7.9	0.0030	0.027	0.43	0.46	0.89	0.49
27– July	3	10.82	19.6	0.0350	44.9	0.0053	0.152	0.70	2.62	3.31	0.21
July 4–	10	8.80	16.0	0.0284	28.6	0.0046	0.096	0.57	1.67	2.23	0.25
11–	17	21.10	38.3	0.0682	11.1	0.0033	0.038	1.36	0.65	2.00	0.68
18–	24	5.67	10.3	0.0183	12.1	0.0034	0.041	0.36	0.70	1.07	0.34
25–	31	5.92	10.7	0.0191	38.1	0.0050	0.129	0.38	2.22	2.60	0.15
Aug. 1–	7	11.56	21.0	0.0374	8.4	0.0031	0.028	0.74	0.49	1.23	0.60
8–	14	8.82	16.0	0.0285	27.9	0.0046	0.094	0.57	1.63	2.19	0.26
15–	21	16.95	30.7	0.0548	63.3	0.0060	0.214	1.09	3.69	4.78	0.23
22–	28	11.03	20.0	0.0357	35.7	0.0049	0.120	0.71	2.08	2.79	0.25
29– Sept.	4	5.60	10.2	0.0181	41.4	0.0052	0.140	0.36	2.41	2.77	0.13
Sept. 5–	11	11.47	20.8	0.0371	99.2	0.0069	0.335	0.74	5.79	6.52	0.11
12–	18	9.99	18.1	0.0323	44.5	0.0053	0.150	0.64	2.60	3.24	0.20
19–	25	7.86	14.3	0.0254	38.5	0.0051	0.130	0.50	2.25	2.75	0.18
26– Oct.	2	12.09	21.9	0.0391	91.5	0.0068	0.309	0.78	5.33	6.11	0.13
3–	9	13.23	24.0	0.0428	34.5	0.0049	0.116	0.85	2.01	2.86	0.30
10–	16	3.34	6.1	0.0108	36.4	0.0050	0.123	0.21	2.13	2.34	0.09
17–	23	10.78	19.6	0.0348	13.2	0.0035	0.044	0.69	0.77	1.46	0.47
24–	30	10.98	19.9	0.0355	41.5	0.0052	0.140	0.71	2.42	3.13	0.23
19 Week ave.		10.14	18.4	0.0328	37.8	0.0048	0.128	0.65	2.21	2.86	0.28

C* = 0.23, Ck = 0.20, Wind speed ratio = 0.82 [u_3/u_{10}].

[1] Calculated from equation 2, [2] calculated from equation 1, [3] calculated from equation 4.

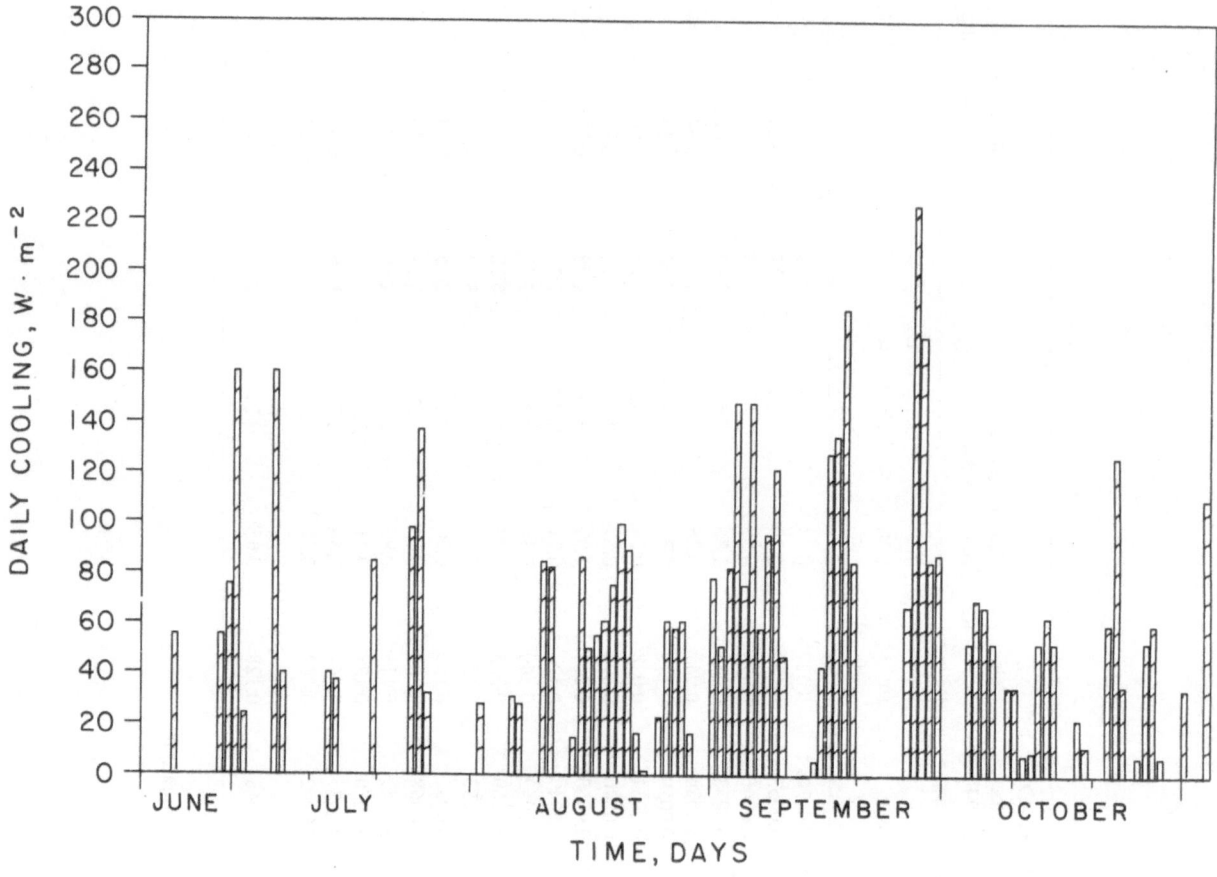

Fig. 4. Daily heat loss per unit surface area (0 to 8 m depth) in Mahoney Lake from June to November 1983. Only cooling changes are plotted, while changes associated with warming of the water column are shown as zero.

error values for the efficiency factors C^* (wind) and C_k (penetrative convection) are insensitive to the assumed values for C^*, but sensitive to the assumed values of C_k.

With an efficiency of wind energy utilization taken from previous work ($C^* = 0.23$) the efficiency of utilization of penetrative convection energy, C_k, may be determined using equation 4. Adjustments were made to the value of C_k, and the right hand side of equation 4 evaluated, until there was agreement with the rate of loss of potential energy of stratification (left hand side of equation 4). With a wind speed ratio u_3/u_{10} of 0.82, a good match was determined with $C_k = 0.20$. The result of this matching is shown in Fig. 6. A mean area of 110000 m² was assumed for the part of the lake between 0 and 8 m below

the surface. If the whole surface area of the lake were used for the calculation, then the value of C_k would be smaller and close to previously published results.

These physical measurements have demonstrated that wind is less important than penetrative convective cooling in autumnal mixing in Mahoney Lake. They also support earlier qualitative observations that poor exposure to the wind was an important factor in maintaining the stability and unusual microstratification in the mixolimnion of Mahoney Lake. A comparative analysis has been made with the meteorological conditions at nearby dimictic Green Lake (Ward *et al.*, 1989). The Mahoney Lake microstratification provided favourable experimental conditions to study gradual changes in the potential energy of

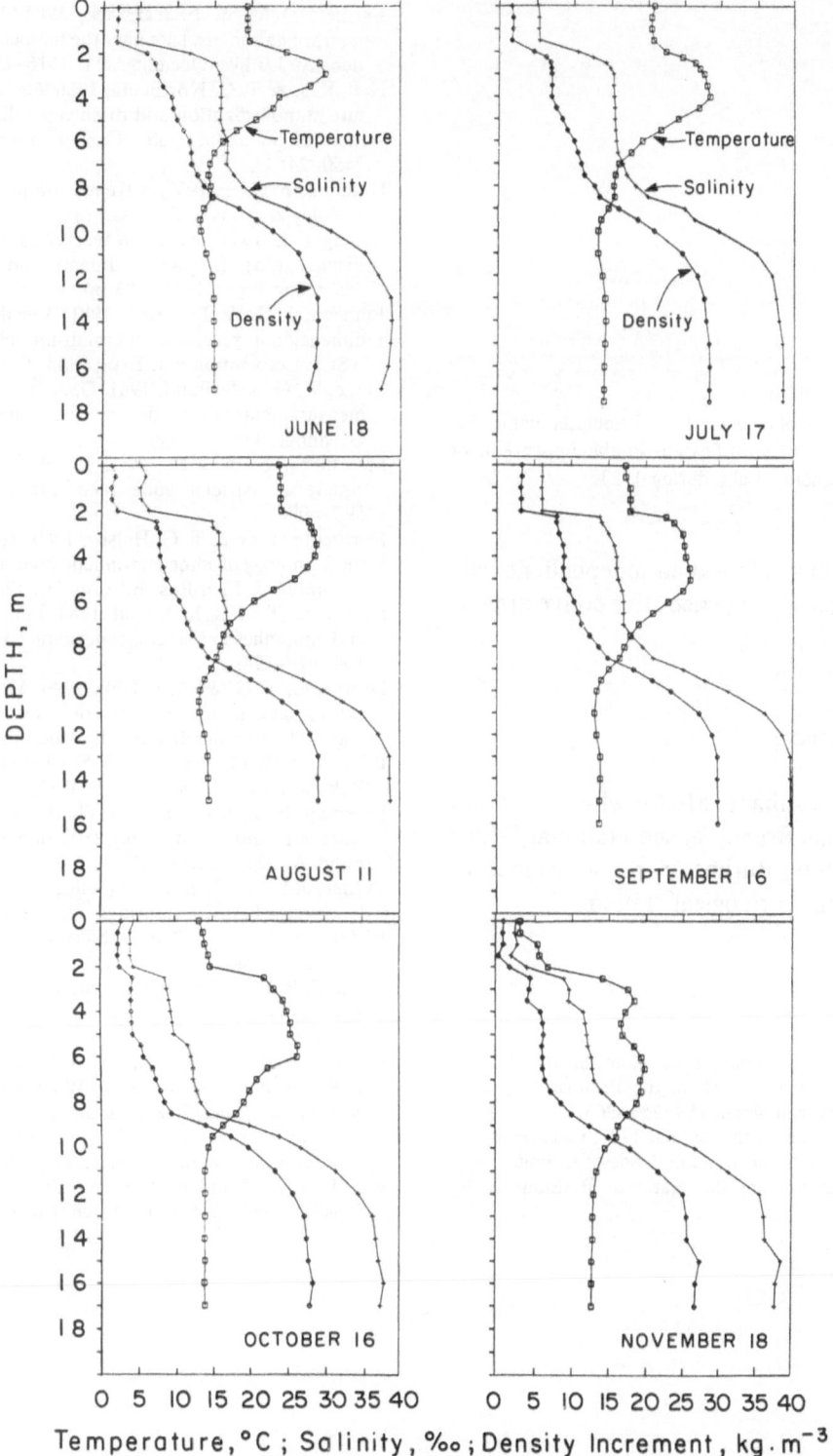

Fig. 5. Vertical salinity-temperature-density relationships in Mahoney Lake at approximately mid-monthly intervals between June and November 1983.

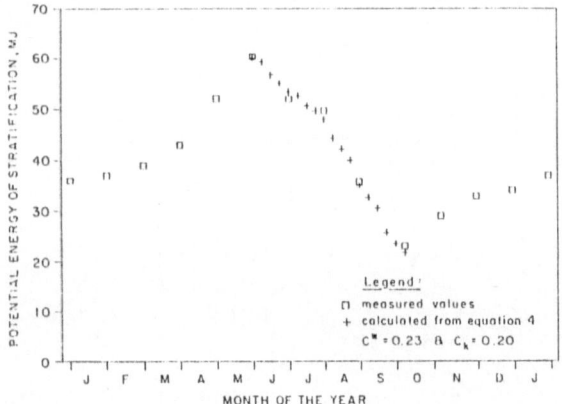

Fig. 6. A comparison of measured and calculated values for the potential energy of stratification in the upper 8 m of Mahoney Lake during 1983.

stratification which allowed an independent estimate of the efficiency of penetrative convection to be made.

Acknowledgements

We would like to thank M. Mawhinney of the Tech. Operations Branch at the National Water Research Institute, Burlington for assistance in setting up the meteorological station.

References

Ashton, P. J. & F. R. Schoeman, 1988. Thermal stratification and the stability of meromixis in the Pretoria Salt Pan, South Africa. Hydrobiologia 158: 253–265.

Canadian National Research Council, 1977. Commentaries on Part 4 of the National Building Code of Canada 1977. Associate Committee on the National Building Code, NRCC No. 15558.

Farmer, D. M. & E. Carmack, 1981. Wind mixing and restratification in a lake near the temperature of maximum density. J. Phys. Oceanogr. 11: 1516–1533.

Hall, K. J. & T. G. Northcote, 1986. Conductivity-temperature standardization and dissolved solids estimation in a meromictic saline lake. Can. J. Fish aquat. Sci. 43: 2450–2454.

Hutchinson, G. E., 1957. A treatise on limnology, Volume 1. J. Wiley & Sons, N.Y., 1015 pp.

Imberger, J., 1979. Mixing in reservoirs. Chapter 6 of H. B. Fischer (ed). Mixing in Inland and Coastal Waters, Academic Press, N.Y., 483 pp.

Johnson, L. S. & D. Ford, 1981. Vertification of a one-dimensional reservoir thermal model. Proceedings of ASCE Convention and Exposition, St. Louis, Mo. 29 pp.

Large, W. G. & S. Pond, 1981. Open ocean momentum flux measurements in moderate to strong winds. J. Phys. Oceanogr. 11: 324–336.

MacIntyre, S. & J. M. Melack, 1982. Meromixis in an equatorial African soda lake. Limnol. Oceanogr. 27: 595–609.

Northcote, T. G. & T. G. Halsey, 1969. Seasonal changes in the limnology of some meromictic lakes in southern British Columbia. J. Fish Res. Bd Can. 26: 1763–1787.

Northcote, T. G. & K. J. Hall, 1983. Limnological contrasts and anomalies in two adjacent saline lakes. Hydrobiologia 105: 179–194.

Northcote, T. G. & K. J. Hall, 1990. Vernal microstratification patterns in a meromictic saline lake: their causes and biological significance. Hydrobiologia 197: 105–114.

Riley, J. P. & G. Skirrow, 1965. Chemical oceanography, Volume 1. Academic Press, N.Y., 721 pp.

Sherman, F. S., J. Imberger & G. M. Corcos, 1978. Turbulence and mixing in stably stratified waters. Ann. Rev. Fluid. Mech. 10; 267–288.

Turner, J. S., 1973. Buoyancy effects in fluids. Cambridge University Press, Cambridge, UK. 368 pp.

Walker, K. F., 1974. The stability of meromictic lakes in central Washington. Limnol. Oceanogr. 19: 209–222.

Ward, P. R. B., 1980. Physical limnology. Chapter 3 of J. A. Thornton (ed.) Lake Macilwaine. Dr. W. Junk Publishers, The Hague. 23–41.

Ward, P. R. B., E. A. Cousins, K. J. Hall, T. G. Northcote & T. P. Murphy, 1989. Mixing by Wind and Penetrative Convection in Small Lakes. International Association for Hydraulic Research, 23rd. Congress, Ottawa, Canada. Proceedings, Technical Session D, pages D-331 to D-338.

Wu, J., 1973. Wind induced entrainment across a stable density interface. J. Fluid Mech. 61: 275–287.

Hydrobiologia **197**: 139–146, 1990.
F. A. Comin and T. G. Northcote (eds), Saline Lakes.
© 1990 *Kluwer Academic Publishers.*

Changes in lake levels, salinity and the biological community of Great Salt Lake (Utah, USA), 1847–1987

Doyle W. Stephens
U.S. Geological Survey, Salt Lake City, Utah 84104, USA

Key words: brine, salinity, brine shrimp, brine algae, halophiles

Abstract

Great Salt Lake is the fourth largest terminal lake in the world, with an area of about 6000 square kilometers at its historic high elevation. Since its historic low elevation of 1277.52 meters in 1963, the lake has risen to a new historic high elevation of 1283.77 meters in 1986–87, a net increase of about 6.25 meters. About 60 percent of this increase, 3.72 meters, has occurred since 1982 in response to greater than average precipitation and less than average evaporation.

Variations in salinity have resulted in changes in the composition of the aquatic biological community which consists of bacteria, protozoa, brine shrimp and brine flies. These changes were particularly evident following the completion of a causeway in 1959 which divided the lake. Subsequent salinities in the north part of the lake have ranged from 16 to 29 percent and in the south part from 6 to 28 percent.

Accompanying the rise in lake elevation from 1982 to 1987 have been large decreases in salinity of both parts of the lake. This has resulted in changes in the biota from obligate halophiles, such as *Dunaliella salina* and *D. viridis*, to opportunistic forms such as a blue-green alga (*Nodularia spumigena*). The distribution and abundance of brine shrimp (*Artemia salina*) in the lake also have followed closely the salinity. In 1986, when the salinity of the south part of the lake was about 6 percent, a population of brackish-water killifish (*Lucania parva*) was observed along the shore near inflow from a spring.

Introduction

Great Salt Lake, Utah, USA (Fig. 1), with a surface area of about 6200 km² at its historic high elevation, is the fourth largest terminal lake in the world. Between 1931 and 1976, the total annual inflow to the lake averaged about 2300 cubic hectometers (hm³) (Waddell & Barton, 1980). About 67 percent of the annual inflow to the lake is from the Bear, Weber, and Jordan river drainage systems. Precipitation on the lake surface contributes about 30 percent and ground water only 3 percent of the annual inflow. The size of the lake varies considerably depending on its surface elevation (Fig. 1), which varies with climatic changes. Because the lake is within a closed basin of nearly 90000 km², the elevation and volume of the lake reflect a dynamic equilibrium between the inflow and evaporation. Since its historic low elevation of 1277.52 meters (m) in 1963, the lake has risen to a new historic high elevation of 1283.77 m in 1986–87, a net rise of about 6.25 m. About 60 percent of this increase (3.72 m) has occurred since 1982 in response to greater than average precipitation and less than average evaporation.

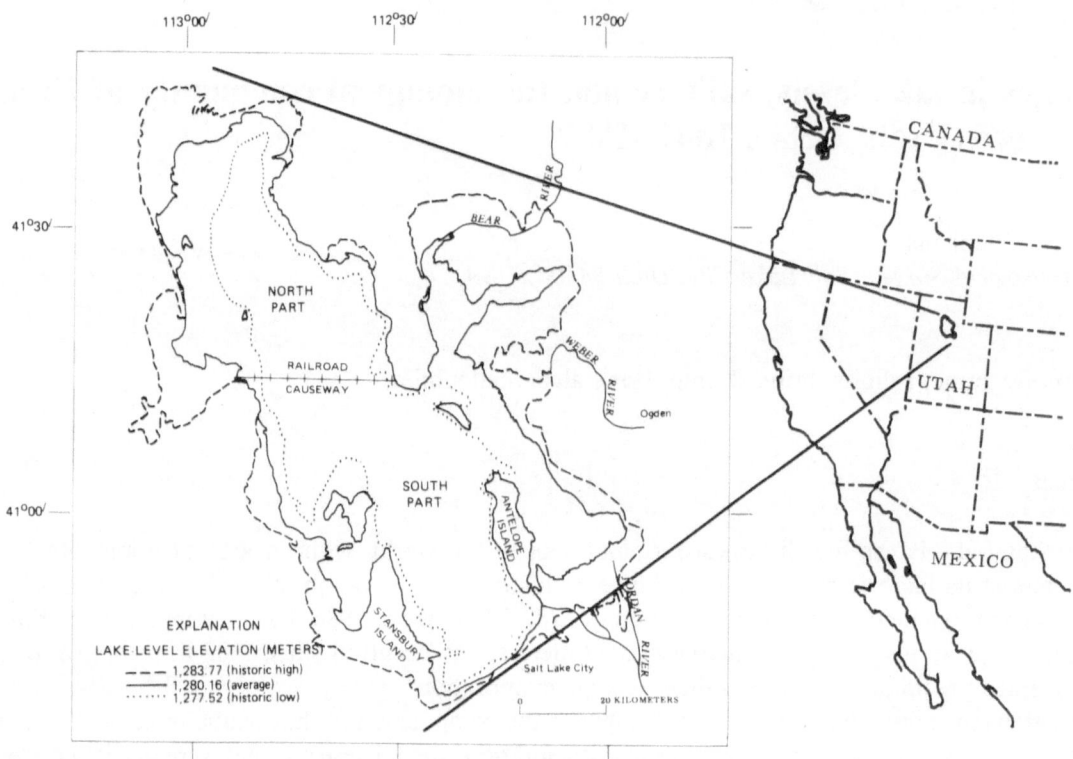

Fig. 1. Area covered by Great Salt Lake at historic high, low, and average elevations, 1847–1987. Adapted from Currey (1980).

The changes in lake level result in changes in the salinity of the brine, and consequently changes in the biological community of the lake. The brine has contained as much as 29 percent salt when the lake was at its historic low in 1963, but when the lake peaked in 1986, the salinity of the main southern part of the lake was about 6 percent. This paper shows how the major changes in lake elevation and salinity effect changes in the biological community of the lake. Additional information concerning lake elevation and salinity appears in Arnow (1984). The biology of the lake is reviewed in Stephens (1974), Post (1977) and Rushforth & Felix (1982).

Changes in lake elevation

When the Mormon pioneers arrived in Utah in 1847, the surface of Great Salt Lake was about 1280 m a.s.l. (Fig. 2). From 1862 to 1873 the lake

rose about 3.7 m to reach a historic high of about 1283.7 m. At this level, the lake covered about 6200 km². During the 31 years after the high level of 1873, the lake declined almost 5 m, and in November 1905 it was at a then historic low level of 1278.88 m. During 1906–52, the lake level fluctuated about 1.5 m but peaked at 1280.45 m in 1952. This was followed by another lengthy period of decline, and in October 1963 the lake had reached an all-time historic low level of 1277.52 m (Fig. 2). At that level, the lake covered only about 2500 km². From 1964 to 1982 the lake fluctuated over 3.4 m but by September 1982, the lake surface was at 1280.05 m, about the same level that it had been 135 years earlier when the pioneers arrived. Thus, the lake had fluctuated between 1277.52 and 1283.67 m but had no apparent net change of level.

The lake began to rise on September 18, 1982, in response to a series of storms earlier in the month. The total precipitation at Salt Lake City

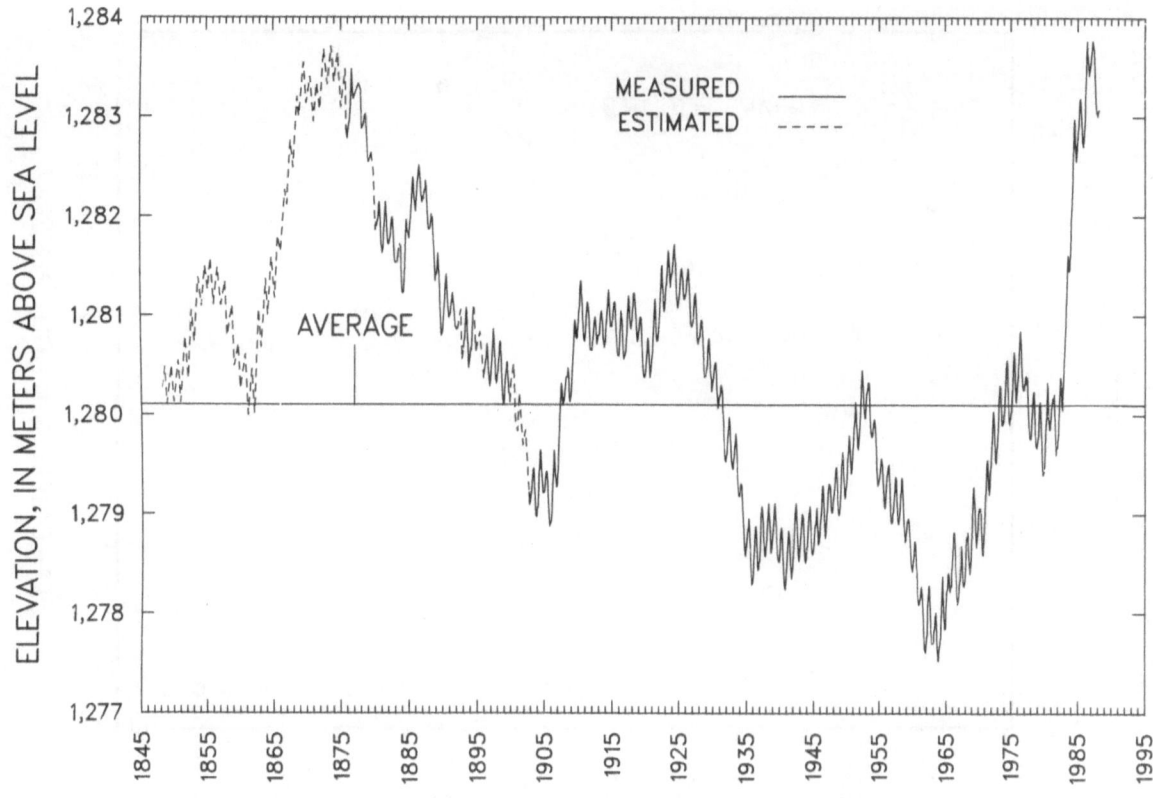

Fig. 2. Changes of water level of Great Salt Lake, 1847–1987.

during 1982 was 145 percent of the annual average of 40 cm. This was followed in 1983 and 1984 by precipitation which exceeded 135 percent of average. The unusually large precipitation resulted in unseasonably large inflow to Great Salt Lake, both from precipitation directly on the lake and from tributary surface streams. Historic peak flows were recorded for inflowing streams to Great Salt Lake in 1983 and 1984. The lake rose to 1283.19 m in 1985, the highest level it has reached since 1877.

Unusually large snowpack, ranging from 144 to 264 percent of average, accumulated within the drainage to the lake during the winter of 1985. This resulted in inflow to the lake from October through March 1986 of 4000 hm³, nearly equaling the 1983–84 record of 4500 hm³ during the equivalent period. On June 3, the lake peaked at 1283.77 m and covered about 6200 km². The net increase since the 1985 peak was greater than 2600 hm³.

A dryer climate prevailed during 1987 and precipitation at Salt Lake City was only 80 percent of average. The reduced inflow and the installation of a pumping system which transported water from the north end of the lake to an evaporation pond covering 2200 km² west of the Great Salt Lake, resulted in a peak level of 1283.77 m in 1987.

Changes in salinity

The lake has a dissolved-mineral content of almost 5 billion t (Sturm, 1980, p. 155). More than 2 million t of minerals enter the lake annually (Arnow & Mundorff, 1972, table 3), primarily from inflowing streams. More than one-half of this is calcium carbonate which precipitates to form oolitic sand and other sediments. The major dissolved ions in the brine are chloride, sodium, sulfate, magnesium and potassium. Chloride and

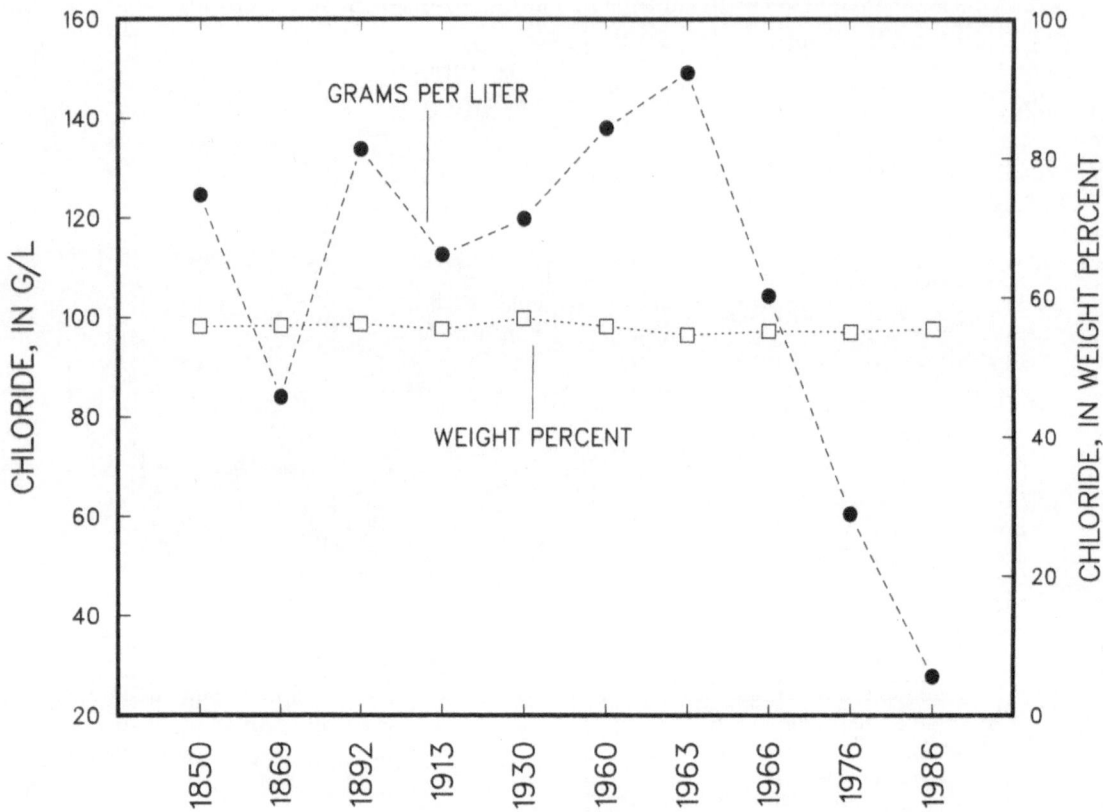

Fig. 3. Chloride in Great Salt Lake brine, in grams per liter and weight percent, 1850–1986. Data adapted from Hahl and Handy (1969, p. 14), and Utah Geological and Mineral Survey, unpublished.

sodium account for about 90 percent (by weight) of the dissolved ions. The brine contains small quantities of dissolved calcium, bicarbonate, bromium, lithium, boron, fluoride, silica, and other trace elements. While concentrations of the major dissolved ions in the brine has varied considerably as shown for chloride in Fig. 3, the relative composition indicated by weight percent has remained fairly constant throughout the recorded history of sampling on the lake.

Prior to completion of the railroad causeway that divided the lake in 1959, mineral concentrations in the lake probably were fairly uniform and salinity was dependent on lake volume. The railroad causeway is slightly permeable to water movement. It is constructed mostly of gravel and sand capped with boulder-sized riprap and is breached by two culverts, each 4.6 m wide and about 6 m high. (During the high water years of

1982–84, the culverts were filled with riprap.) The causeway separates the lake into two parts – about two-thirds of the lake is south of the causeway and about one-third is north of it. The southern part of the lake receives more than 90 percent of the freshwater inflow, whereas the northern part receives most of its water in the form of brine that moves through the causeway from the southern part. A dense monimolimnion, occupying only the deepest areas of the lake south of the causeway, is derived from dense brine in the north part. This brine was driven by a density gradient to flow southward through the lower parts of two culverts in the railroad causeway prior to 1984 when the culverts were filled with riprap.

These factors, in conjunction with restriction of flow by the causeway, have caused differences in salinity and in water elevations between the two

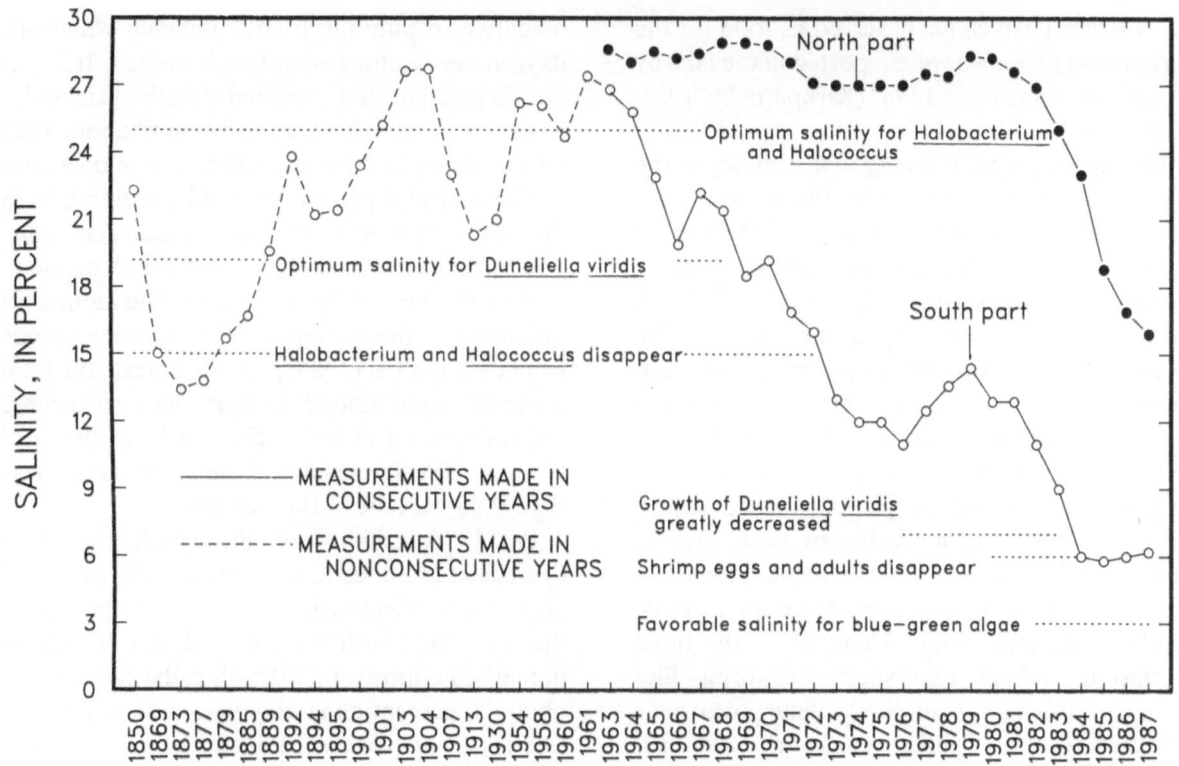

Fig. 4. Change in salinity of Great Salt Lake and its effect on the biota, 1850–1987.

parts of the lake (Fig. 4). The differences increased steadily throughout the 1960's. Since 1966, when measurements of the water elevation were begun in the northern part, the water elevation in the southern part has been consistently higher, and the difference reached a maximum of 1.1 m on July 1, 1984. Because of this elevation difference, the State of Utah recognized the northern part of the lake could store 2100 hm³ of additional water and, thereby, could decrease flooding that was occurring at the south end of the lake. The causeway was breached with a 91 m-wide opening on August 3, 1984, to provide flood relief to areas along the shore of the southern part of the lake. Within 2 months, the difference in water elevation across the causeway had decreased to 23 cm, and within 1 year, it had decreased to 15 cm.

The salinity north of the causeway remained relatively constant and near saturation, with a concentration of dissolved minerals of about 27 percent from 1959 to 1982, regardless of changes

in lake elevation. The concentration decreased however, during the large lake-elevation rises of 1983 and 1984, and a steady decrease continued after the breach of the causeway and as the lake continued to rise through May 1986. By June 1986 at the historic high elevation, the salinity north of the causeway had decreased to about 17 percent. South of the causeway it was only 6 percent.

Changes in biological community

During the late 1800's, lake salinity ranged from about 13 to 23 percent and brine shrimp (*Artemia salina* Leach) and a brine fly (*Ephydra gracilis* Packard = *E. cinerea* Jones) were reported inhabiting the lake (Verrill, 1869; Packard, 1871). Five different types of green and blue-green algae [*Aphanothece utahensis* Tilden, *Polycystis packardii* Farlow, *Dichothrix utahensis* Tilden, *Enteromorpha tubulosa* (Kütz.) Reinbold, and *Chara con-*

traria Braun], which likely served as food for the shrimp and flies, also were reported in the lake by Josephine Tilden in 1898 (Kirkpatrick, 1934, p. 2).

During the period 1900–1959 the salinity of the lake was relatively constant fluctuating from about 20 to 26 percent. Flowers (1934) reported six species of algae to be endogenous to the lake: *Aphanothece utahensis, Microcystis packardii* Farlow (Tilden), *Oscillatoria tenuis* var. *tergestina* (Kütz.), *O. tenuis* var. *natans* (Kütz.), *Chlamydomonas* sp., and *Tetraspora lubrica* var. *lacunosa* Chauv.. In addition, Kirkpatrick (1934) observed two species of *Navicula*, two species of *Chlamydomonas* and three ciliate protozoans (*Uroleptus packii* Calkins, *Prorodon utahensis* Pack, and an unidentified species), and one amoeba. Flowers & Evans (1966), using data collected prior to 1959, reported the lake biota to consist of the brine shrimp (*A. salina*), two species of brine flies (*Ephydra cinerea, E. hians* Say), ciliate protozoans [*Uroleptus packii* Calkins, *Chilophyra utahensis* (Pack), *Euplotes, Cyclidium, Pseudocohnilembus, Podophyra*], two unidentified amoebae, and at least two species of flagellates (*Tetramitus, Oikomonas*). The only algae present were *Coccochloris elabens* Drouet and Daily (formerly known as *Polycystis packardii, Microcystis packardii* and *Aphanothece utahensis*), *Entophysalis rivularis* (Kütz.) Drouet and two species of *Chlamydomonas. Ephydra cinerea* was the most common brine fly during this period (Jorgensen, 1956).

The completion of the railroad causeway in 1959 resulted in the creation of two ecologically distinct lakes due to differences in salinity. In addition to osmotic stress on the organisms, the large concentrations of dissolved salts which developed in the north part greatly decreased the ability of the water to absorb oxygen. As the salinity north of the causeway remained at about 28 percent and the salinity south of the causeway declined, populations of the bacteria *Halobacterium* and *Halococcus* greatly expanded in the north part. These bacteria contain bacteriorhodopsin, a pigment which allows the organisms to use light in an alternative metabolic pathway to oxidative respiration in environments where the oxygen needed for respiration is limited. It is this purple pigment that imparted a distinct red color to the north part that was visible until about 1982 when salinity in the north part began to decrease.

The only alga present in significant numbers in the north part after 1959 was *Dunaliella salina* Teodoresco, a chlorophyte with a red pigment. However, the red bacteria were the dominant organisms in the north part. Biomass calculations from data in Post (1980, p. 314), indicate the total bacterial population of the north part exceeded 2 million metric t in 1975. The weight of the algal biomass was 8 percent and the brine shrimp less than 0.05 percent of the bacteria.

By 1980, salinity of the south part had decreased to about 13 percent, and 20 species of algae were identified, including 14 species of diatoms (Rushforth & Felix, 1982, p. 160). This is in marked contrast to 1913 when the salinity was about 20 percent and only two genera of diatoms (*Navicula* and *Cymbella*) likely were residents of the lake (Daines, 1917). By 1980 the most common algae in the south part were the green alga *D. viridis* and the diatoms *Amphora coffeiformis* (Ag.) Kütz., *Navicula graciloides* A. Mayer, *N. tripunctata* (O. Müll.) and *Rhopalodia musculus* (Kütz.) O. Müll.. The biostrome-forming bluegreen alga *Coccochloris elabens* was quite rare as it prefers a salinity of about 25 percent (Felix & Rushforth, 1980, p. 306).

In response to large inflows of fresh water during 1982–86, salinity in the south part decreased from about 13 percent in 1981 to about 6 percent in 1986. The decrease in salinity has greatly affected the composition of the algal and shrimp communities. The filamentous blue-green alga, *Nodularia spumigena* Mertens, which was common in brackish water east of Antelope Island, but which appeared infrequently in the main south part of the lake during the early 1980's, was quite common throughout the south part by mid-1984. By 1987, there were species in the lake representing all major phyla of algae (Fig. 5).

The commercial harvest of adults and eggs of brine shrimp began about 1952 to provide food for tropical fish. By 1962, the salinity of the north

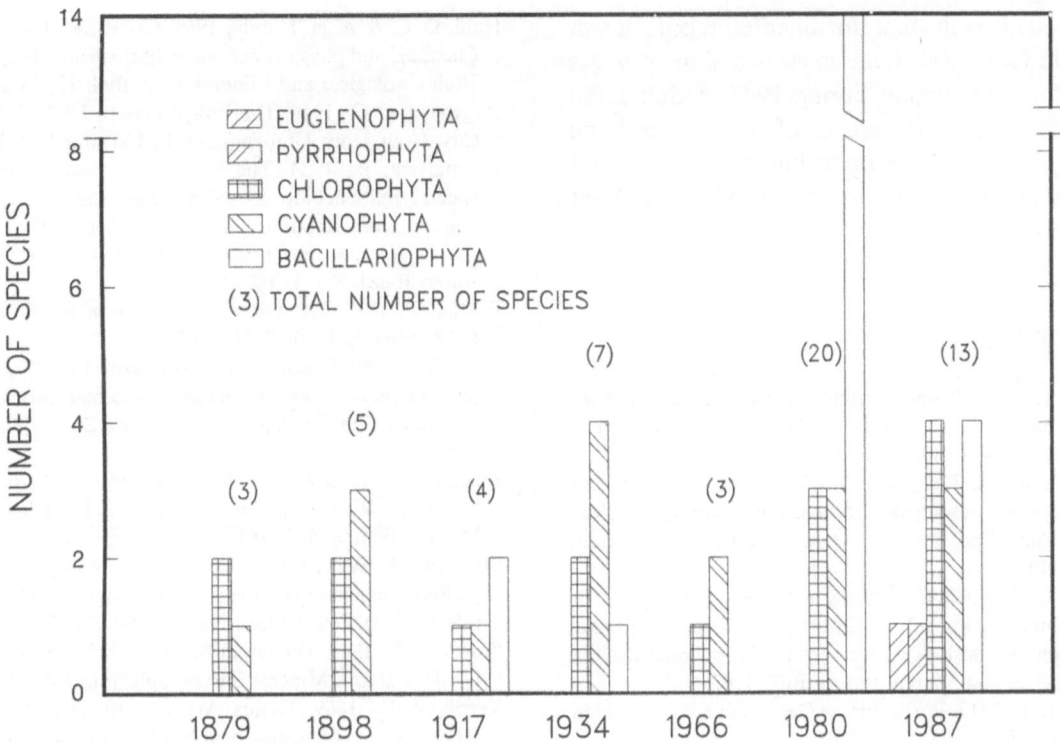

Fig. 5. Distribution of species in the major algal phyla in Great Salt Lake, 1879–1987. Data sources: 1879–1966, Felix & Rushforth (1980, table 2); 1980, Rushforth and Felix, (1982, p. 160); 1987, S.E. Rushforth, Brigham Young University, written commun., 1987.

part had increased to 28 percent and the brine shrimp industry moved to be south part. Harvest of brine-shrimp eggs, which during 1962–82 had been economically feasible only in the south part, decreased steadily from 77 metric t in 1965 to less than 8 metric t in 1981. The demise of the brine shrimp in the south part has been a result of decreasing salinity. As salinity decreases to 6 percent, the hard winter eggs produced by the shrimp sink to the bottom of the lake where they cannot hatch in response to inflow of fresh water in the spring. At a salinity of about 3 percent, adult shrimp fail to reproduce. The large inflows of freshwater, which began in 1982, also resulted in a decrease in salinity in the north part. By 1986, salinity there had decreased from about 28 to 17 percent. Harvests of eggs in the north part have increased from 15 metric t in 1982 to a record 1476 metric t in 1987 with a retail value of $26 million. The increased harvests are largely in

response to increased demand for the eggs for use as a food source in mariculture ponds.

Brine flies, the other major animal group in the lake, are tolerant of fairly high salinity, high temperatures, and low concentrations of dissolved oxygen. Of the two species of brine flies living in the Great Salt Lake, the smaller *E. cinerea* is more tolerant of increased salinity and in the past has been 100 times more abundant than *E. hians* (Felix & Rushforth, 1980, p. 309). With the decreasing salinity of the south part, *E. hians* is now becoming more dominant (Betinna Rosay, Salt Lake City Mosquito Abatement District, pers. commun., 1986).

During 1986, the decreased salinity of the south part allowed a breeding population of rainwater killifish, *Lucania parva* (Baird) to enter the lake near Stansbury Island. The 4-cm-long fish are members of the Cyprinodontidae family and are commonly found in brackish and slightly saline waters. This is the first documented report of fish living in Great Salt Lake in recorded history. As the total precipitation during 1987 at Salt Lake City was about 80 percent of average and the salinity of the southern part had increased to 7.2 percent by March 1988, the future of fish in Great Salt Lake is tenuous.

References

Arnow, T., 1984. Water-elevation and water-quality changes in Great Salt Lake, Utah, 1847–1983. U.S. Geological Survey Circular 913, 22 p.

Arnow, T. & J. C. Mundorff, 1972. The Great Salt Lake and Utah's water resources. Proceed. First Ann. Conf. Utah Sect. Am. Wat. Resources Ass., Salt Lake City, Utah, p. 29–40.

Currey, D. R., 1980. Coastal geomorphology of Great Salt Lake and vicinity. In J. W. Gwynn, (ed.), Great Salt Lake, a scientific, historical and economic overview. Utah Geological and Mineral Survey Bull. 116: 69–82.

Daines, L. L., 1917. On the flora of Great Salt Lake. Am. Nat. 51: 499–506.

Felix, E. A. & S. R. Rushforth, 1980. Biology of the south arm of the Great Salt Lake, Utah. In J. W. Gwynn, (ed.), Great Salt Lake, a scientific, historical and economic overview. Utah Geological Mineral Survey Bull. 116: 305–313.

Flowers, S., 1934. Vegetation of the Great Salt Lake Region. Bot. Gaz. 95: 353–418.

Flowers, Seville & F. R. Evans, 1966. The flora and fauna of the Great Salt Lake region, Utah. In H. Boyko, (ed.), Salinity and Aridity. W. Junk, the Hague. P. 367–393.

Hahl, D. C. & A. H. Handy, 1969. Great Salt Lake, Utah: Chemical and physical variations of the brine, 1963–1966. Utah Geological and Mineral Surv. Bull. 12, 33 p.

Jorgensen, E. C., 1956. The Ephydridae of Utah. Salt Lake City, Utah, Univ. Utah, unpublished Master's thesis, 62 p.

Kirkpatrick, R., 1934. The life of Great Salt Lake, with special reference to the algae. Salt Lake City, Utah, Univer. Utah, unpublished Master's thesis, 30 p.

Packard, A. S., Jr., 1871. On insects inhabiting salt water. Amer. Journ. Sci. 1: 100–110.

Post, F. J., 1977. The microbial ecology of the Great Salt Lake. Microb. Ecol. 3: 143–165.

Post, F. J., 1980. Biology of the north arm. In J. W. Gwynn, (ed.), Great Salt Lake, a scientific, historical and economic overview. Utah Geological and Mineral Survey Bull. 116: 311–321.

Rushforth, S. R. & E. A. Felix, 1982. Biotic adjustments to changing salinities in the Great Salt Lake, Utah, USA. Microb. Ecol. 8: 157–161.

Stephens, D. W., 1974. A summary of biological investigations concerning the Great Salt Lake, Utah (1861–1973). Great Basin Natural. 34: 221–229.

Sturm, P. A., 1980. The Great Salt Lake brine system. Utah Geological and Mineral Survey Bull. 116: 147–162.

Verrill, A. E., 1869. Genus Artemia. In Twelfth Annual Report of the U.S. Geological and Geographical Survey of the Territories of Wyoming and Idaho for the year 1878, Part I: 330–334, 1883, U.S. Government Printing Office, Washington, D.C., 809 p.

Waddell, K. M. & J. D. Barton, 1980. Estimated inflow and evaporation for Great Salt Lake, Utah, 1931–76, with revised model for evaluating the effects of dikes on the water and salt balance of the lake. Utah Div. of Water Resources Cooperative Invest. Rep. No. 20, 57 p.

Hydrobiologia **197**: 147–164, 1990.
F. A. Comin and T. G. Northcote (eds), Saline Lakes.
© 1990 *Kluwer Academic Publishers.*

Large-scale patterns of *Nodularia spumigena* blooms in Pyramid Lake, Nevada, determined from Landsat imagery: 1972–1986

David L. Galat,[1] James P. Verdin[2] & Lori L. Sims[3]

[1] *Department of Zoology, Arizona State University, Tempe, Arizona 85287, USA*; *Present Address: Missouri Cooperative Fish & Wildlife Research Unit, 112 Stephens Hall, University of Missouri, Columbia, Missouri 65211, USA*; [2] *Division of Research and Laboratory Services, U.S. Bureau of Reclamation, Denver, Colorado 80225, USA*; [3] *Advanced Sciences, Inc., Denver, Colorado 80225, USA*

Key words: Saline lakes, *Nodularia spumigena*, Pyramid Lake, phytoplankton periodicity, remote sensing

Abstract

Magnitude and long-term periodicity of summer-autumn blooms of the nitrogen-fixing cyanobacterium, *Nodularia spumigena*, were characterized for hyposaline Pyramid Lake, Nevada, from Landsat MSS band 3 film negatives. Predicted lakewide mean chlorophyll *a* concentrations for Landsat overpasses during the July–October *Nodularia* bloom season ranged from 27 to 72 mg m^{-3} with an overall average concentration of 32 ± 7 mg m^{-3} between 1972 and 1986.

Nodularia blooms were usually annual events. Blooms were not observed on Landsat images in only three of 15 years (1973, 1980, 1982) and midsummer calcium carbonate whitings occurred in two of these years (1973, 1980). Magnitude of *Nodularia* blooms was highly variable among years and 'very large' blooms, where maximum mean chlorophyll *a* concentration exceeded one standard deviation of the 15 year overall mean (> 39 mg m^{-3}), appeared in 1974, 1975, 1977, 1979, 1984, 1985 and 1986. Very large early-July blooms always occurred during or following years of above average fluvial discharge to Pyramid Lake (1984–1986) and were associated with meromixis produced by the large influx of freshwater.

Several problems arise using Landsat remote sensing to estimate magnitude and periodicity of scum-forming blue-green algal blooms. These complications may reduce accuracy and precision of phytoplankton biomass estimates made from Landsat images. Nevertheless, Landsat remote sensing enabled us to quantify relative bloom magnitude with limited collection of ground-based data and at a large-scale temporal and spatial resolution not possible using alternative methodologies.

Introduction

Seasonality of phytoplankton in lakes has attracted the attention of limnologists for over 50 years (Pearsall, 1932), and there are now numerous reports detailing phytoplankton periodicity for freshwater lakes (reviewed by Hutchinson, 1967; Reynolds, 1980; Harris, 1986; Fogg & Thake, 1987), and saline lakes (reviewed by Hammer, 1986). Determinants of phytoplankton seasonal succession may be grouped into external or allogenic factors such as basin morphometry, turbulence and climate, and internal or autogenic factors such as competition for light and nutrients, pathogens or predation by zooplankton (Reynolds, 1980; Harris & Trimbee, 1986; based on Tansley's, 1935, terminology). Allogenic and autogenic factors are strongly coupled in space and time with the result that a successional paradigm of phytoplankton periodicity in temperate and tropical lakes has been established (Lewis, 1978; 1986). Water-blooms whereby phytoplankton populations expand explosively within several days, are dramatic examples of these processes.

Research on phytoplankton periodicity has often been limited by spatial and temporal considerations. Typically, one or two sites within a lake are sampled for less than three years. Notable exceptions include the long-term studies of Lund (1965) on Lake Windermere, Edmondson & Lehman (1981) on Lake Washington, Munawar & Munawar (1986) on several of the North American Great Lakes, Likens (1985) on Mirror Lake, New Hampshire, Pollingher (1986) on Lake Kinneret, Israel, and Lewis (1986) on Lake Valencia, Venezuela. We extend the approach of long-term, multiple-site analysis of phytoplankton periodicity to a saline lake by assembling a 15-year record of large-scale spatial and temporal variability of a dominant bloom forming phytoplankton species in Pyramid Lake, Nevada USA.

Pyramid Lake is a large (446 km^2), deep (mean depth ca. 60 m) graben lake located in the western margin of the North American Great Basin Desert (Galat et al., 1981). It is a hyposaline (ca. 5 g l^{-1}), warm monomictic terminal lake receiving over 90% of its annual water influx from the Truckee River. Sodium, chloride and bicarbonate-carbonate are its principal ions, and alkalinity (ca. 23 meq l^{-1}) and pH (ca. 9.2) are high and stable.

Periodicity of phytoplankton during the growing season in the mixed layer of Pyramid Lake follows Lewis's (1978) generalized successional sequence. Diatoms, principally Cyclotella spp. and Stephanodiscus spp., dominate the epilimnion during early summer thermal stratification. They are followed by Chlorophytes (e.g. Crucigenia tetrapedia) and then Cyanobacteria, notably a filamentous nitrogen-fixing species, Nodularia spumigena. Diatoms (e.g., Chaetoceros elmori) and Chlorophytes (e.g., Chroococcus sp.) succeed Nodularia as thermal stratification erodes in autumn (Galat et al., 1981).

Our objective was to evaluate long-term periodicity and variability in abundance of one of these species, Nodularia spumigena Mertens. This nitrogen-fixing cyanobacterium is particularly important in Pyramid Lake because it produces massive water-blooms which are a major autochthonous source of organic matter (Galat, 1986) and nitrogen (Horne & Galat, 1985). Furthermore, decomposition of sedimented Nodularia has been causally associated with increased hypolimnetic oxygen consumption in Pyramid Lake (Hamilton-Galat & Galat, 1983) and also nearby saline Walker Lake (Cooper & Koch, 1984). Resource managers at Pyramid Lake are concerned about decreased hypolimnetic oxygen concentrations as a result of Nodularia blooms because of potential habitat loss to two of the lake's endemic fishes. One of these, the cui-ui (Chasmistes cujus), is federally listed as endangered (Sigler et al., 1985) and the other, Lahontan cutthroat trout (Salmo clarki henshawi), supports a world class trophy sport fishery vital to the economy of the Pyramid Lake Paiute Indian Tribe, whose reservation contains Pyramid Lake (Sigler et al., 1983). A second objective of this research was to document magnitude and periodicity of past Nodularia blooms as a reference against which to compare responses of future blooms to projected changes in anthropogenic nutrient loads to the Truckee River. Lastly, we discuss the applicability of remote sensing to quantify long-term periodicity of cyanobacteria blooms in large lakes where continuous field data are unavailable.

Nodularia spumigena is a cosmopolitan, euryhaline (<5 − >200 g l^{-1} salinity, Hammer, 1986) cyanobacterium producing massive blooms in a variety of saline waters including the Baltic Sea (Öström, 1976; Lindahl et al., 1980; Hübel & Hübel, 1980), Peel-Harvey Estuary, Western Australia (Huber, 1986; Lukatelich & McComb, 1986) and inland saline lakes of all non-polar continents (reviewed in Hammer, 1986). Blooms of Nodularia are not a modern phenomenon in Pyramid Lake. They were present in Pyramid Lake and its pluvial ancestor, Lake Lahontan, before settlement of the region by Anglo-Americans as Hattori (1982) reported Nodularia from a nearby archaeological site dated at 9540 ± 120 years Before Present. Although modern limnological investigations of Pyramid Lake were initiated by Hutchinson (1937) in 1933, he did not report *Nodularia*. LaRivers (1962) was the first to record a *Nodularia* bloom in Pyramid Lake during September 1951. Subsequent accounts of *Nodularia* blooms have been irregular (Koch *et al.*, 1976; Galat *et al.*, 1981; Hamilton-Galat & Galat, 1983; Horne & Galat, 1985; Galat, 1986) and, consequently, no continuous long-term record of their periodicity exists. An additional difficulty encountered documenting *Nodularia* periodicity is the large spatial heterogeneity characteristic of filamentous cyanobacteria blooms (Reynolds & Walsby, 1975). Results of a synoptic survey conducted during the peak of the 1979 *Nodularia* bloom at Pyramid Lake confirmed its horizontal distribution was also extremely patchy (Horne & Galat, 1985). Therefore, historic information on *Nodularia* abundance, based largely on collections from a few sites, by different investigators and using different techniques, provides unreliable evidence of variations in bloom magnitude. What was needed was an approach to furnish periodic, synoptic whole-lake coverage of phytoplankton biomass over extended time. This would enable us to integrate horizontal patchiness of contemporary *Nodularia* blooms and compare intra- and interannual variations in their occurrence and magnitude. We selected Landsat multispectral satellite imagery as a technique to supply this information. Landsat satellites have been in continuous operation since 1972 and successfully used to assess whole-lake phytoplankton biomass in freshwaters (Strong, 1978; Lillesand *et al.*, 1983; Verdin, 1985), saline lakes (Almanza & Melack, 1985), and also specifically to document magnitude of *Nodularia* blooms in the Baltic Sea (Oström, 1976; Ulbricht & Horstmann, 1980).

Methods

Radiance measured by the Landsat multispectral scanner (MSS) in the near-infrared wavelengths (band 3, 700–800 nm bandwidth) has been shown to be proportional to surface water chloro-phyll *a* concentrations and can, therefore, be used as an index of surface water phytoplankton biomass (Strong, 1978; Kirk, 1983; Verdin, 1985). *Nodularia*, like other filamentous blue-green algae, often forms dense surface scums which are highly reflective in MSS band 3 whose penetration depth in water is limited to about one meter (Hoffer, 1978). Landsats 1 through 5 have acquired imagery since 1972 on a schedule covering every point in the United States approximately once every 16 days at a spatial resolution of about 80-m pixel size (Lillesand & Kiefer, 1987). After reviewing Landsat film negatives and ground-based records of *Nodularia* occurrence in Pyramid Lake for all months over eight years, we chose a July though October time frame for appearance of *Nodularia* blooms. We restricted further analysis of archival Landsat images to this time interval extending from 1972 through 1986.

All film negatives during these intervals were acquired and redigitized using a scanning micro-densitometer to yield brightness histograms composed of land and water pixels. Image histograms were normalized for differences in photographic laboratory procedures, sun angle and atmospheric effects and registered to a series of master images selected to group scenes according to fluctuations in lake surface area (Galat & Verdin, 1987). Land and water pixels were segregated and land pixels masked. Fifteen of the 84 images had moderate cloud cover or data noise and were handled in a more subjective manner to assign an average density value, but no estimate of variance (see Galat & Verdin, 1987).

Conversion of film negative density to predictions of Pyramid Lake surface water chlorophyll *a* during the *Nodularia* bloom interval involved a two-step process. First, it was necessary to empirically relate surface water chlorophyll *a* concentrations during a *Nodularia* bloom to Landsat MSS band 3 radiance. This was accomplished by conducting a 32-site surface water chlorophyll *a* synoptic sampling during the peak of the 1986 *Nodularia* bloom on 20 July and concurrent with a flyby of Landsat 5. We used Landsat MSS band 3 digital tape data for this calculation rather than redigitized film negative data because digital tapes offer higher radiometric quality than scanned film negatives. Surface water chlorophyll a was predicted from digital tape data by pairing observed chlorophyll a concentrations (range included: 3–168 mg m^{-3}) with corresponding MSS band 3 tape radiance counts (range: 4–35) to the 85th percentile of the equivalent cumulative frequency distributions of both variables. Landsat band 3 tape radiance counts quantify radiance on a seven-bit, 0–127, scale and can be converted to physical units, mW cm^{-2} sr^{-2} μm^{-1}, using calibration data from Markham & Barker (1987). The least-squares linear regression model calculated to predict chlorophyll a concentration from tape radiance counts was:

$$CHL = -14.533 + 4.493\,(TRC), \qquad (1)$$

where CHL is chlorophyll a concentration in mg m^{-3} and TRC is Landsat 5 MSS band 3 tape radiance count (Galat & Verdin, 1988). The regression R^2 was 0.95 with a standard deviation (SD) about the regression line of 8 mg m^{-3} chlorophyll a ($P \le 0.001$, $N = 27$).

Before film density from archival Landsat film negatives of Pyramid Lake could be converted to chlorophyll a using this regression, we had to derive an empirical relationship between MSS band 3 scanned film density and tape image radiance. This step was performed by regressing film density counts for the 20 July 1986 image against log$_e$ transformed tape radiance counts for the same scene subset at equivalent five percent intervals also to the 85th percentile of the cumulative frequency histograms of both variables. The regression equation calculated was:

$$lnTRC = 1.787 + 0.015\,(FDC), \qquad (2)$$

where TRC is band 3 tape radiance count and FDC is band 3 film density count (Galat & Verdin, 1988). Coefficient of determination was 0.99 with a standard deviation about the regression line of 0.048 ($P \le 0.001$, $N = 17$). Film density values for individual pixels from all archival Landsat film negatives were converted to tape radiance counts using equation (2). Predicted radiances from over 90 000 lake pixels per film negative image were averaged for each image date and standard deviations calculated to evaluate horizontal patchiness.

Estimates of chlorophyll a are more uncertain than the ± 8 mg m^{-3} SD of equation (1) when also predicted from TRC using equation (2), rather than actual TRC. To evaluate this additional uncertainty we substituted values into equation (2) from the full range of TRC's used to develop equation (1). An additional ± 1 mg m^{-3} uncertainty was added to the chlorophyll a estimate at the low range while ± 7.7 mg m^{-3} chlorophyll a was added at the high range.

Equation (1) for estimating chlorophyll a from tape radiance counts is specific to *Nodularia* in Pyramid Lake using Landsat 5 MSS data for the scene conditions (sun angle, atmosphere) of 20 July 1986. We used it for historic estimation after histogram matching and converting film density counts to tape radiance counts with equation (2). This procedure standardized film density counts recorded on other dates to Landsat 5, 20 July 1986 tape radiance counts. Equation (1) is not appropriate for estimating chlorophyll a in other lakes, chiefly because algal species would be different, but also because background water color and other optical properties could be different.

Radiation reflected from natural waters and remotely sensed by Landsat MSS band 3 may be from other suspended particles in addition to phytoplankton. At Pyramid Lake these are primarily inorganic sediments of fluvial origin present during and following periods of high Truckee River discharge and particulate calcium carbonate present during lake whitings. All 84 Landsat film negatives were examined to judge the primary source of observed lake radiance. Based on published information and field records obtained from researchers investigating Pyramid Lake over the last 15 years, each image was placed into one of four categories: 1) calcium carbonate whiting present and likely the major source of observed radiance (see Berg et al., 1981; Galat & Jacobsen, 1985); 2) unusually high Truckee River

discharge preceded the image date and consequently suspended sediment of fluvial origin contributed to radiance (U.S. Geological Survey,
1972–1986); 3) *Nodularia* present (see earlier re-
ferences); and 4) no apparent *Nodularia* bloom,
whiting or high fluvial discharge, so radiance and
chlorophyll *a* were assumed to be from other phytoplankton. Tape radiances were calculated for
all film negatives, but because of possible contamination from non-phytoplankton sources, only radiances for Landsat images in categories 3)
and 4) were transformed to surface water chlorophyll *a* concentrations using the procedure outlined above. Lakewide mean chlorophyll *a* concentration for each image date was predicted from
lakewide mean tape radiance rather than tape
radiances for individual pixels because high
radiance pixels were occasionally above the upper
limit of radiance values used to derive the regression model. The image exhibiting the highest
predicted mean chlorophyll *a* concentration
among all category 3 and 4 scenes for each year
was defined as 'maximum mean chlorophyll *a*'
(MMCHL) between July and October for that
year.

Color coded maps of predicted surface water
chlorophyll *a* concentrations were prepared for
selected image dates to illustrate intra- and inter-
annual spatial heterogeneity of *Nodularia* blooms.
The total range of predicted chlorophyll *a* concentrations was partitioned into five intervals for
these maps. We designated the highest chlorophyll range as $> 150 \text{ mg m}^{-3}$ (red) because
168 mg m^{-3} was the highest observed chlorophyll *a* concentration used to calculate the
Landsat radiance-chlorophyll *a* regression.
Although surface water chlorophyll *a* concentrations in excess of 4 g m^{-3} were measured from
Nodularia surface scums during the 20 July 1986
synoptic survey, observations made with a hand-
held radiometer indicate radiance from these
scums becomes saturated at chlorophyll *a* concentrations above about 200 mg m^{-3} (Galat &
Verdin, 1989). This corresponds to the 85th
percentile used as the upper limit for the regression models outlined above. Consequently,
actual chlorophyll *a* concentrations in *Nodularia*

surface patches represented by red on these maps
are more on the order of grams per cubic meter.
Published and unpublished literature, field notes
and detailed visual inspection of enlarged film
negatives were used to aid our interpretation of
color maps. Predicted MMCHL concentrations
for each year, whether a *Nodularia* bloom was
present or not, were compared and ranked from
1 to 15, 1 being the lowest and 15 the highest
MMCHL.

Results

Mean lakewide radiances predicted from Landsat
MSS band 3 during the July through October
Nodularia bloom season for all processed Landsat
scenes were typically uniform within and among
years at about 10 (range possible: 0–127, Fig. 1).

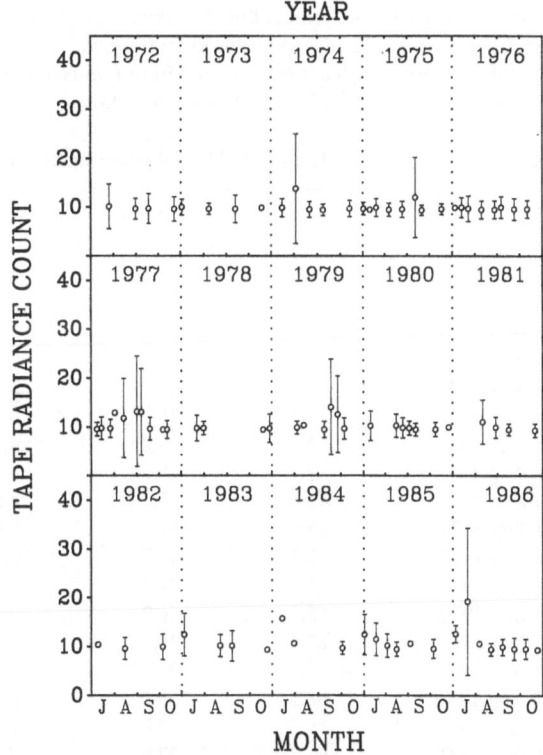

Fig. 1. Time series of predicted lakewide Landsat MSS
band 3 mean tape radiance counts of Pyramid Lake surface
water during the July–October *Nodularia spumigena* bloom
season, 1972–1986. Error bar represents ± one standard
deviation.

Variability of radiance over the lake surface for single dates was generally low, coefficients of variation (CV) averaged $\leq 5\%$ These patterns were punctuated in particular years by one or more dates of high mean radiance (≥ 12) and greater variance associated with the mean (CV's averaged 60%). Massive *Nodularia* blooms were responsible for these irregularities in nearly all cases.

Lakewide mean surface water chlorophyll *a* concentrations were predicted from mean MSS band 3 radiances only for dates where phytoplankton was considered the dominant source of radiance (categories 3 and 4). Chlorophyll *a* concentrations for these 73 cases ranged from 27 to 72 mg m^{-3} with a 15 year overall average ($\pm$ 1SD) of 32 ± 7 mg m^{-3}. Predicted lakewide mean chlorophyll *a* concentrations on a single date were seldom greater than the overall average except during *Nodularia* blooms. 'Very large' blooms were evidenced by, and defined *a posteriori* as exhibiting, average chlorophyll *a* concentrations greater than $+$ 1SD of the overall mean, or > 39 mg m^{-3}. 'Very large' blooms were recorded on 12 occasions using this criterion. Ten of these are illustrated in color maps (Figs. 2 and 3), and the remaining two are shown in black and white images because cloud cover interfered with production of maps (Fig. 4).

Temporal and spatial variability of *Nodularia* blooms between 1972 and 1986 is summarized below. Bloom periodicity is also illustrated in relation to calcium carbonate whitings and annual Truckee River discharge (Fig. 5).

1972. A distinct surface scum of *Nodularia* was evident at the south end of Pyramid Lake in the 25 July scene (rank 6, Fig. 2A, Table 1). No earlier images were available, and *Nodularia* was

Table 1. Proportion of Pyramid Lake's surface area predicted to occur in each of five chlorophyll *a* concentration ranges between 1 and > 150 mg m^{-3} and mean lakewide surface water chlorophyll *a* concentration for selected dates between July and October, 1972–1986. Chlorophyll *a* was predicted from Landsat MSS band 3 film negatives. Patterns of surface water chlorophyll *a* for most of these dates are illustrated in Figures 2–3.

Date	Range of chlorophyll *a* (mg m^{-3})					Mean chlorophyll *a* (mg m^{-3})
	1–30	31–50	51–80	81–150	> 150	
25 Jul 1972	84.8	11.2	1.9	1.4	0.7	31
02 Aug 1974	56.3	24.2	8.9	5.8	4.8	47
11 Sep 1975	59.8	26.4	8.1	3.9	1.8	40
20 Sep 1975	91.4	8.0	0.4	0.2	<0.1	28
05 Sep 1976	75.0	21.7	2.6	0.6	0.1	30
26 Jul 1977	79.3	19.3	0.7	0.6	<0.1	29
13 Aug 1977	65.1	24.0	4.7	3.5	2.7	39
31 Aug 1977	57.1	29.3	4.7	4.7	4.3	45
06 Sep 1977	45.6	37.7	8.3	5.1	3.3	44
18 Sep 1977	85.6	13.1	0.6	0.5	0.2	29
08 Sep 1979	87.1	11.7	0.7	0.4	0.1	28
17 Sep 1979	41.0	34.4	13.3	7.4	3.9	49
26 Sep 1979	55.5	26.2	9.1	6.7	2.5	42
05 Oct 1979	82.0	16.1	1.2	0.5	0.2	29
10 Aug 1981	57.0	34.5	4.8	2.9	0.8	35
28 Aug 1981	68.2	28.6	2.6	0.5	0.1	30
21 Aug 1983	59.6	37.0	2.7	0.6	0.1	31
06 Sep 1983	68.7	27.5	2.2	1.3	0.3	31
01 Jul 1985	40.4	32.9	24.1	2.4	0.1	42
17 Jul 1985	41.2	41.2	16.5	1.0	0.1	37
04 Jul 1986	7.7	87.4	4.2	0.4	0.2	42
20 Jul 1986	32.6	24.4	15.3	15.7	12.0	72

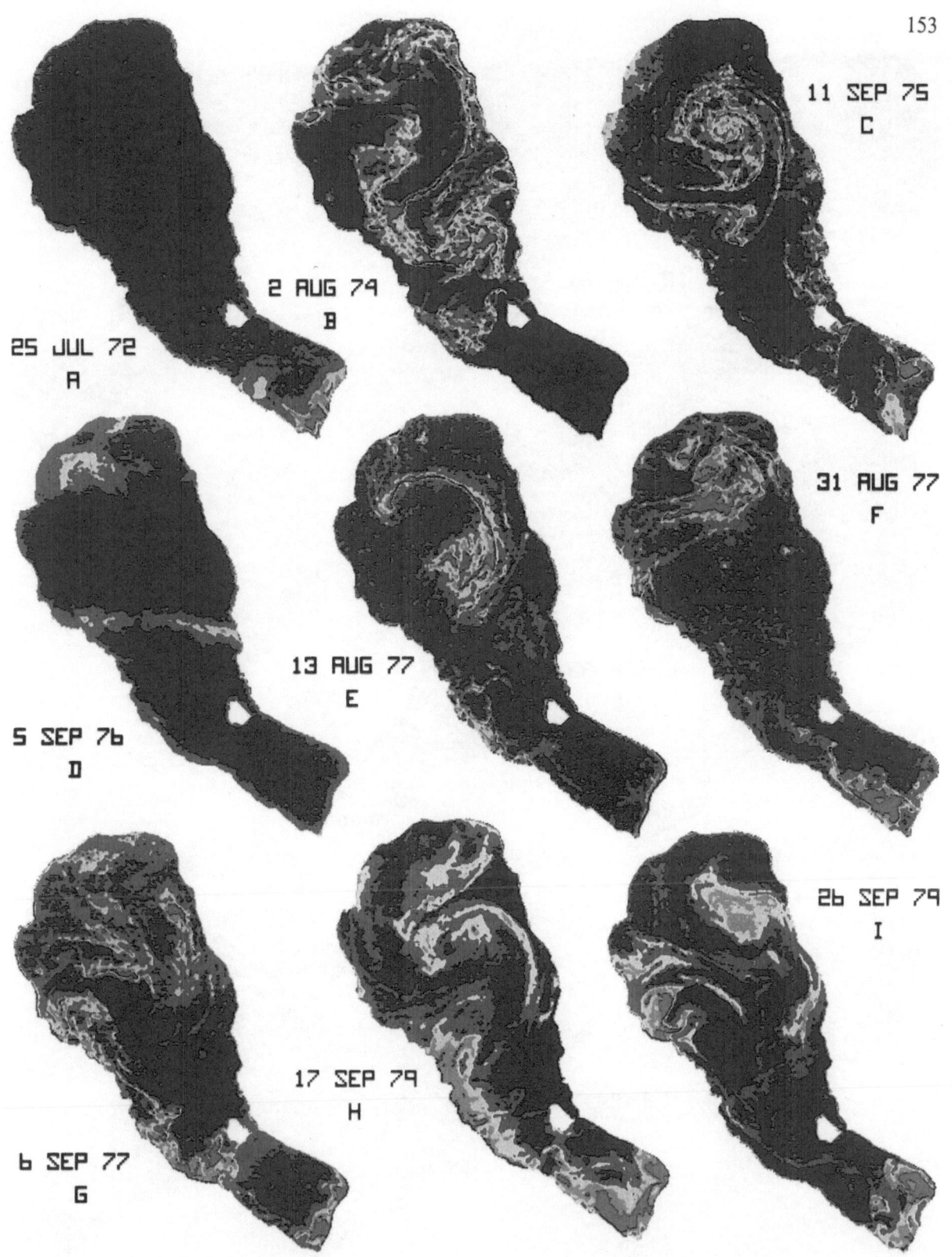

Fig. 2. Color image maps showing horizontal distribution of predicted surface water chlorophyll *a* concentrations in Pyramid Lake from July 1972 through September 1979. Key to colors: dark blue, ≤30; light blue, 31–50; yellow, 51–80; pink, 81–150; red, >150 mg m^{-3} chlorophyll *a*. White area within lake boundary is Anaho Island.

154

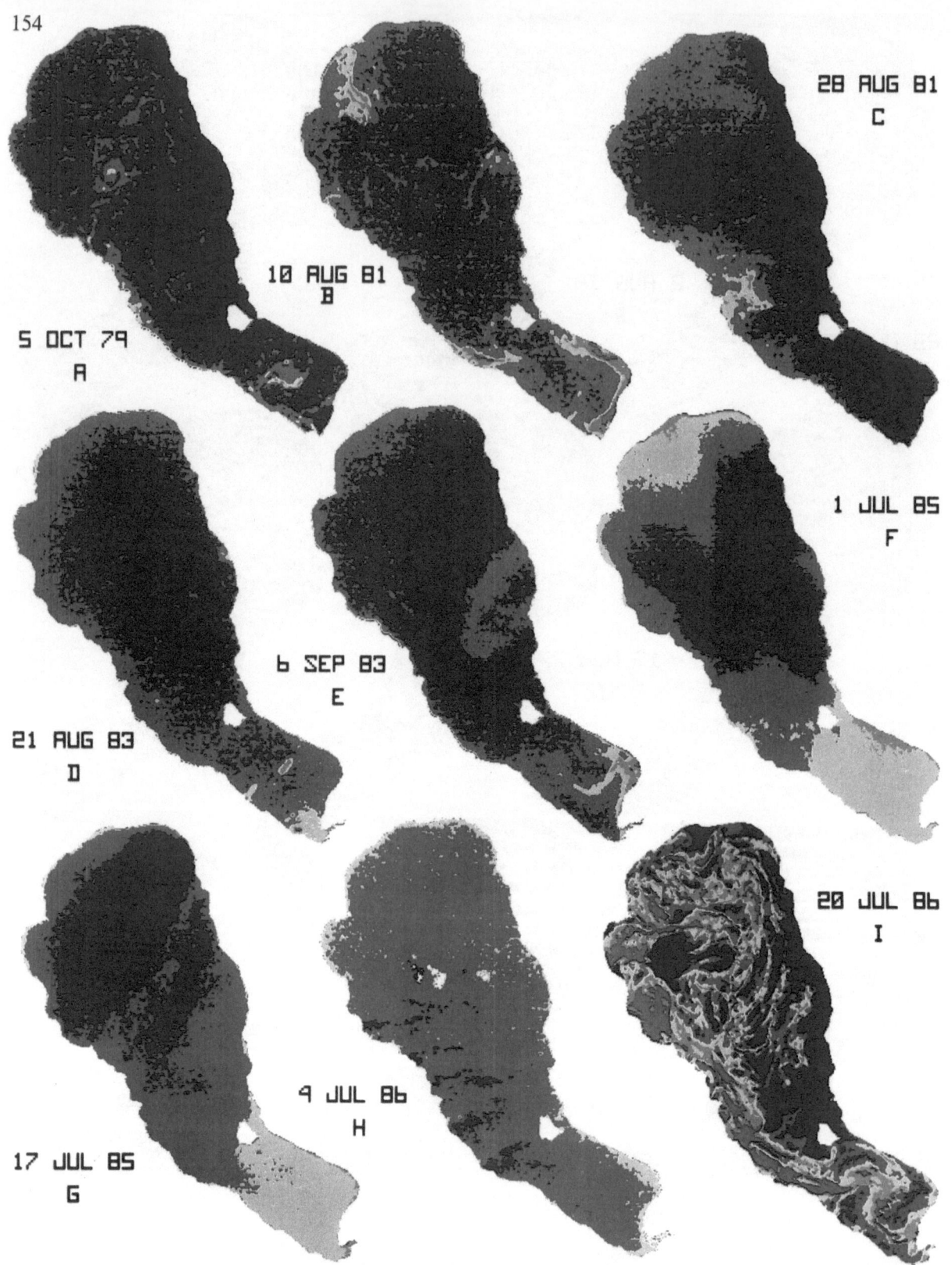

Fig. 3. Color image maps showing horizontal distribution of predicted surface water chlorophyll *a* concentrations in Pyramid Lake from October 1979 through July 1986. Unusual surface patterns of chlorophyll for 1 (F) and 17 (G) July 1985 and 4 July 1986 (H) are believed to be a result of wind dispersion. Refer to Fig. 2 for key to colors.

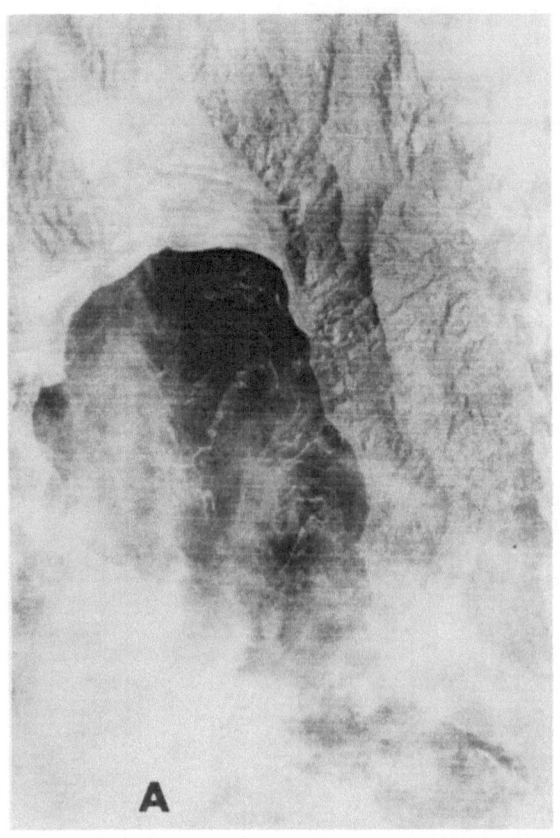 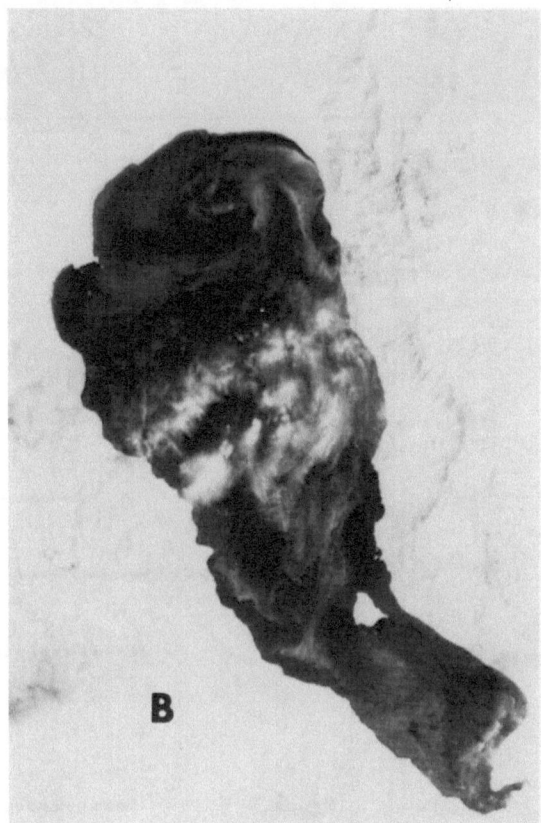

Fig. 4. Dense surface scums of *Nodularia spumigena* are largely obscured by cloud cover on this Landsat MSS band 3 black and white image of Pyramid Lake for 1 August 1977 (A). Another very large *Nodularia spumigena* bloom is clearly visible on 14 July 1984 beneath a band of high clouds extending southwest to northeast across the north-central lake basin (B).

not observed in the subsequent image (30 August). A late July bloom apparently occurred, but due to the small number of Landsat scenes available, we were unable to judge its actual size and duration. Hendrix & Deguire (1974) noted a *Nodularia* bloom in Pyramid Lake in 1972 but gave no dates of occurrence or other details.

1973. A *Nodularia* bloom was not observed and the MMCHL concentration of 30 mg m^{-3} chlorophyll *a* was therefore predicted from a category 4 image (rank 5). Galat and Jacobsen (1985) reported a calcium carbonate whiting in August and a large streak, typical of whitings, was visible at the north end of Pyramid Lake on the 7 August scene (Fig. 8 in Galat & Jacobsen, 1985).

1974. Truckee River discharge preceding the

bloom season was higher than average (299 m^3 × 10^6 during May–July, Fig. 6), and sediment appeared responsible for some of the radiance in the 15 July image. A 'very large' bloom was seen on the 2 August image as spirals and swirls characteristic of *Nodularia* surface scums (rank 12, Fig. 2B), and much of the lake's surface north of Anaho Island was affected. About 10% of the lake surface exhibited chlorophyll *a* concentrations greater than 80 mg m^{-3} (Table 1). *Nodularia* was not apparent on the 20 August or subsequent scenes.

Koch *et al.* (1976) sampled Pyramid Lake biweekly for *Nodularia* from June through early August 1974. Few filaments were counted from 29 May through 11 July, but large numbers were measured at the surface on 26 and 31 July.

156

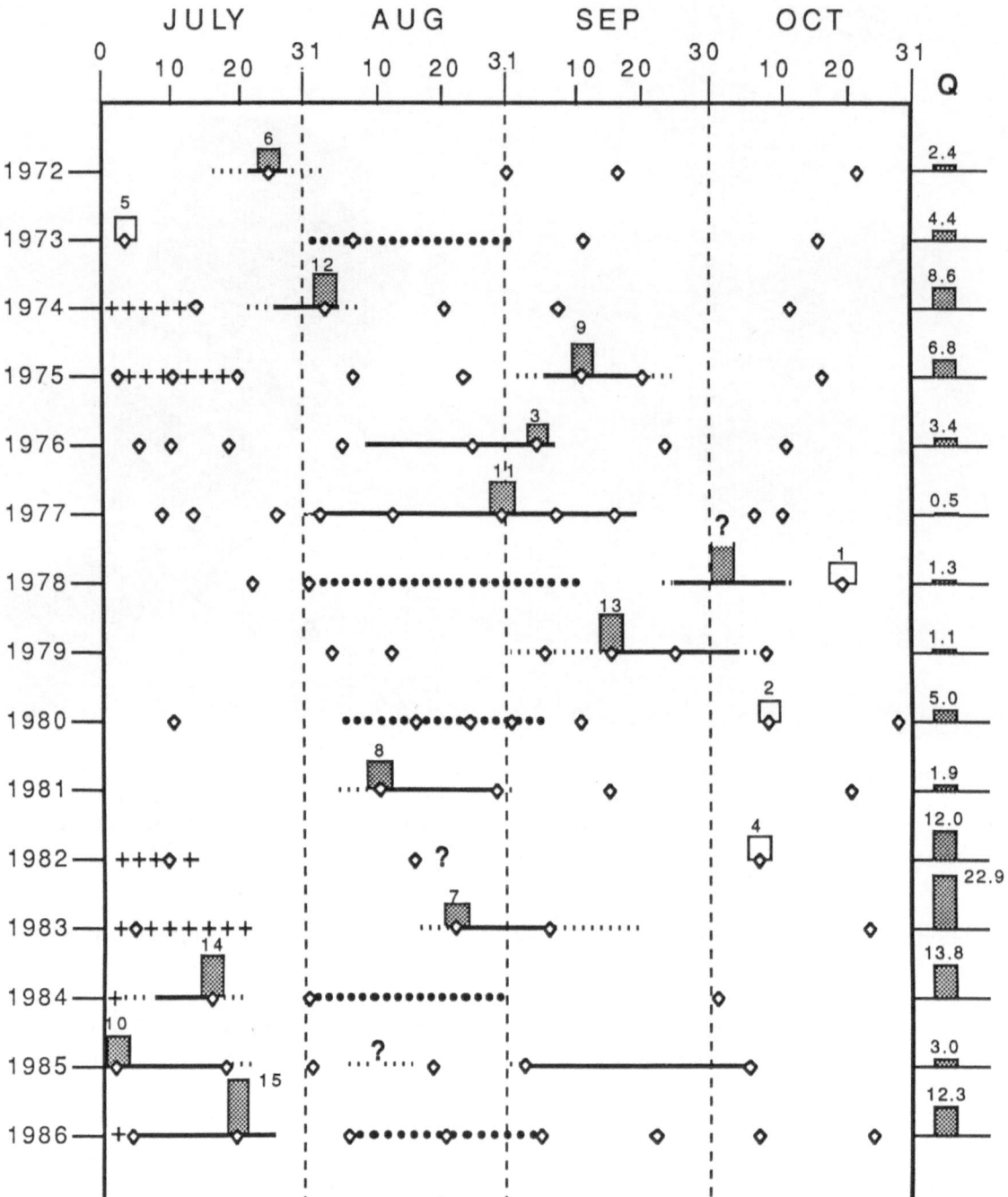

Fig. 5. Time series of *Nodularia spumigena* surface water blooms in Pyramid Lake, July–October 1972–1986. Diamonds show dates of Landsat images. Thick-solid horizontal lines illustrate duration of confirmed blooms and short-dashed horizontal lines represent probable blooms. Calcium carbonate whitings are indicated by large-dotted horizontal lines and periods of high suspended sediment concentrations are shown by pluses (+ + + + +). Vertical bars and numbers within figure axes represent relative size and ranking (1 being smallest) of maximum predicted mean chlorophyll *a* for each year. Shaded bars represent chlorophyll *a* primarily from *Nodularia spumigena* and open bars are chlorophyll *a* largely from other phytoplankton. Vertical bars and numbers to right of figure show total volume of November–October Truckee River discharge (Q, $m^3 \times 10^8$) to Pyramid Lake.

157

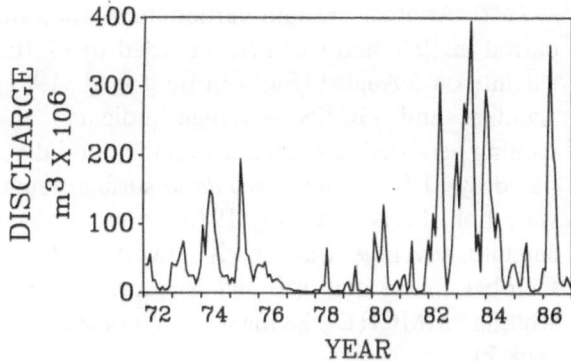

Fig. 6. Approximate total monthly discharge of Truckee River to Pyramid Lake measured at Nixon, Nevada, streamgage located 21 km upstream from Pyramid Lake (data from U.S. Geological Survey, 1972–1986).

Maximum number of filaments they counted were 1690 ml^{-1}, which is about 752000 cells ml^{-1} applying a filament to cell conversion factor from Galat & Verdin (1987). Numbers observed by Koch *et al.* (1976) were again low by 6 August and suggest the bloom lasted between 6 and 20 days.

1975. Spring Truckee River discharge was also above average in 1975 (331 m^3 × 10^6 during May and June, Fig. 6), and suspended sediment was considered the major cause of mean lake radiance on the three images from 1 to 19 July. The Landsat 2 scene for 11 September reveals another 'very large' *Nodularia* bloom as a major counterclockwise gyre in the north-central lake basin and smaller patches of high chlorophyll *a* concentrations north and southeast of Anaho Island (rank 9, Fig. 2C, Table 1). A distinct sediment plume is also visible at the terminus of the Truckee River at the southern-most point on the map. Only a few surface swirls of *Nodularia* remained by the 20 September flyby and predicted surface water concentrations of chlorophyll *a* for most of the lake surface were less than 30 mg m^{-3} (Table 1). Field notes (DLG) for 6 and 18 September confirmed a large *Nodularia* bloom. Duration of the 1975 bloom was at least 15 days.

1976. *Nodularia* was first seen on the 9 August film negative. Small patches were present on the 5 September image and only about three percent of the lake surface showed chlorophyll *a* concentrations exceeding 50 mg m^{-3} (rank 3, Fig. 2D,

Table 1). *Nodularia* was not apparent on Landsat scenes thereafter. Field notes (DLG) indicated small amounts of *Nodularia* at the north end of Pyramid Lake on 26 August, but a large bloom never developed because of subsequent cold, windy weather (Galat, 1986). Lider & Langdon (1978) sampled Pyramid Lake in 1976 for phytoplankton at three sites about every 10 days. Cell numbers were generally < 5000 ml^{-1} at the 1-m depth, and the highest density reported was about 21000 cells ml^{-1} on 10 August at the north end of the lake. Although a *Nodularia* bloom was verified for 1976, it was not large, but persisted for nearly one month.

1977. A thorough chronology of this year's bloom was recorded by Landsat. No surface patches of *Nodularia* were visible on the 26 July scene and chlorophyll *a* concentrations exceeding 50 mg m^{-3} were predicted over only a very small percent of the lake surface (Table 1). About 70% of Pyramid Lake was obscured by thin clouds on 1 August. Nevertheless, highly reflective swirls of *Nodularia* covering the entire lake surface can be seen beneath the overcast (Fig. 4A). Extensive surface scums continued to dominate the 13 and 31 August and 6 September scenes and yielded high predicted mean chlorophyll *a* concentrations (Table 1, Fig. 2E–G). By 18 September only a few indistinct wisps of *Nodularia* remained. This was a 'very large' bloom (rank 11) and discrete surface scums persisted for over 36 days. Galat (1986) measured phytoplankton photosynthesis at 11 sites on six dates in August and also reported this bloom to be very large; it contributed about 40% of Pyramid Lake's annual primary production. Lider & Langdon (1978) counted *Nodularia* once per month in August and September at three sites, but never measured cell numbers > 3000 ml^{-1}. The disparity between Lider & Langdon's results based on only a few samples per month and Galat's (1986) and ours emphasizes the necessity for frequent sampling at a large number of locations if reliable spatial and temporal data on size of heterogeneously distributed cyanobacterial blooms are to be gathered (c.f. Horne & Galat, 1985; Galat & Verdin, 1989).

1978. A calcium carbonate whiting was docu-

158

mented in summer 1978 (Fig. 3 in Berg et al., 1981; Galat & Jacobsen, 1985). It was first noticed on the 30 July Landsat scene and lasted through mid-September. Surface scums of *Nodularia* were not seen on any 1978 image. However, Galat (unpub. field notes) observed a small bloom at the north end of Pyramid Lake on 24 September and again on 3 and 10 October. Surface water peaks of primary production characteristic of *Nodularia* scums were also measured in late September, but were much smaller than recorded for either the 1977 or 1979 blooms (Galat, 1986). Landsat coverage was poor for 1978, and no scenes of acceptable quality were available for August or September. *Nodularia* was not evident on the 19 October film negative. The 19 October image was selected to represent predicted MMCHL (28 mg m^{-3} chlorophyll *a* from category 4, rank 1), although we recognize it underestimated the true magnitude of the 1978 bloom.

1979. Excellent records of *Nodularia* population dynamics were available for 1979 (Horne & Galat, 1985) as well as a comprehensive series of high quality Landsat images (Fig. 2H, I, 3A). *Nodularia* was first measured *in situ* in significant amounts on 15 August and on the 12 August image predicted mean lake chlorophyll *a* was assumed to be dominated by *Nodularia* although no surface scums were visible. Similar conditions existed for the 8 September Landsat scene. By 17 September a massive *Nodularia* bloom was evident (Fig. 2H, rank 13). Surface gyres in the deep northern basin were again noticeable as well as large areas of extremely dense *Nodularia* scums along the west shoreline and particularly in the southern basin. Chlorophyll *a* concentrations in excess of 80 mg m^{-3} were predicted over 11 percent of the lake surface (Table 1). Dense *Nodularia* surface scums continued to dominate on 26 September (Fig. 2I) and a faint counterclockwise gyre was still visible on the 5 October scene (Fig. 3A), although the predicted mean chlorophyll *a* concentration was low (Table 1). Duration of the 1979 bloom was estimated to be about 50 days because supporting field data were available. However, Landsat images showed distinct *Nodularia* surface patterns for only 19 days.

1980. Another calcium carbonate whiting occurred in 1980 and was first recorded by GOES satellite on 5 August (Fig. 4 in Berg et al., 1981). Landsat and GOES coverage indicated this whiting persisted through 2 September, and dissipated by 11 September. *Nodularia* surface scums were not observed on any 1980 scene, although satellite coverage was nearly biweekly. The 8 October image was selected to represent non-whiting MMCHL (28 mg m^{-3} chlorophyll *a*, rank 2).

1981. No Landsat images were available for July, but the 10 and 28 August scenes showed characteristic *Nodularia* surface patterns (Fig. 3B and C). *Nodularia* was not discernable on the 15 September image and this bloom appeared to be of moderate magnitude.

1982. Truckee River spring discharge was extremely high (495 m^3 × 10^6 during May and June, Fig. 6), and radiance in the 10 July scene appeared dominated by suspended sediment. Only three images were available for July through October and no patterns indicative of *Nodularia* were seen. However, we are not confident a bloom was absent. October 5 was selected by default to represent MMCHL unaffected by suspended sediment (30 mg m^{-3} chlorophyll *a*, rank 4).

1983. This was a year of record high Truckee River discharge to Pyramid Lake (704 m^3 × 10^6 during May and June, Fig. 6), and river flows remained high throughout the summer. High 4 July radiance (Fig. 1) was attributed to sediment and not *Nodularia* chlorophyll *a*, although we cannot conclude with certainty *Nodularia* was absent as no ground data were available. A small bloom was visible on images for 21 August and 6 September (Fig. 3D and 3E) and chlorophyll *a* concentrations less than 50 mg m^{-3} were predicted over about 95% of the lake surface (Table 1). Low numbers of *Nodularia* cells were counted on 26 September (< 500 cells ml^{-1}; Galat, 1984).

1984. May and June Truckee River discharge was again high (211 m^3 × 10^6), but declined to seasonally low levels by July (21 m^3 × 10^6, Fig. 6). A large *Nodularia* bloom was observed in late June and early July and was synoptically sampled on 8 July (Galat & Peacock, 1984). Maximum number of *Nodularia* vegetative cells and highest chlorophyll *a* concentration determined from surface to 2 m depth integrated samples was 38 000 and 18 mg m^{-3}, respectively. Results from this sub-surface collection cannot be quantitatively compared with results of surface grab samples (e.g. Horne & Galat, 1985; 23 September 1979) or surface to 0.5 m depth integrated samples (Galat & Verdin, 1989; 20 July 1986) because the positive buoyancy of *Nodularia* concentrates filaments in the upper few centimeters of lake water. We have been careful to report chlorophyll *a* in the present study as concentrations in *surface* water for this reason.

Complete lake coverage by a 'very large' bloom was observed on the 14 July film negative (Fig. 4B), although a band of patchy clouds prevented our analyzing magnitude and variability in the usual fashion. Predicted MMCHL for this scene was 56 mg m^{-3} (rank 14). A calcium carbonate whiting occurred approximately two weeks later (L. Carlson, Pyramid Lake Fisheries, pers. commun.) and was verified in October by analysis of calcium levels in the lake. Nineteen-eighty-four was the first year a confirmed whiting was preceded by a *Nodularia* bloom.

1985. After three years of atypically high Truckee River flows, 1985 was a year of below average discharge (Fig. 6). High mean radiances and predicted chlorophyll *a* concentrations for the 1 and 17 July images were representative of *Nodularia* blooms (Table 1), but color maps depict an unusual surface pattern, not characteristic of *Nodularia* surface scums seen in earlier scenes (Fig. 3F and 3G). Most images in which *Nodularia* is abundant show a complete range distribution of predicted chlorophyll *a* concentrations. Background chlorophyll levels in surface lakewater where dense *Nodularia* scums are absent appear on the color maps as dark blue. *Nodularia* patches typically appear on maps as a continuum of colors from light blue at the outside edge, to yellow, pink and finally grading to red at the center (Table 1, Figs. 2 and 3). In contrast, maps for 1 and 17 July 1985 show very large

uniform radiance areas, not graded patches of variable radiance. This was particularly evident in the 51–80 mg m^{-3} range of predicted chlorophyll *a* (Table 1). Clouds were not the cause, as they were absent from images on both dates. Low Truckee River discharge in 1985 eliminated suspended sediment as the source (Fig. 6). Finally, confirmed whitings in 1984 and 1986 made an event in 1985 highly unlikely (Galat & Jacobsen, 1985), although lakewater calcium was not measured to substantiate this conclusion. Furthermore, mean predicted tape lake radiance during confirmed whitings (e.g. 30 July 1978; 15 and 24 August 1980; 30 July 1984) never exceeded 10.6 and average radiances for 1 and 17 July 1985 were 12.5 and 11.5, respectively (Fig. 1). A large *Nodularia* bloom was likely present in July 1985, but it was more uniformly dispersed than normally observed. Horizontal wind advection is a possible, although unconfirmed, cause based on a somewhat similar surface pattern observed on the 4 July 1986 image (Fig. 3H).

Landsat images for 3 September (moderate cloud cover was present) and 5 October showed more typical, though faint, *Nodularia* surface spirals and *Nodularia* was confirmed present on 6 September (L. Carlson, Pyramid Lake Fisheries, pers. commun.). It appears the 1985 *Nodularia* bloom was either protracted (early July–early October), or consisted of two pulses, the first about two weeks long and the second nearly one month. Without any images for the 3 September to 5 October interval, we cannot judge magnitude of the second pulse. The latter proposition appears more tenable, however, as collapse of blue-green blooms is often induced by wind advection (Harris & Trimbee, 1986; see 1986).

1986. Truckee River discharge was again high during winter and spring, but characteristically low beginning in July (Fig. 6). Pyramid Lake Fisheries (L. Carlson, pers. commun.) kept us advised of *Nodularia* population changes on a weekly schedule beginning in June as this was the year of the coordinated Landsat flyby and *Nodularia* synoptic sampling.

Abundant surface scums were observed in early July, and predicted chlorophyll *a* on the 4 July

Landsat scene was high enough to register as a 'very large' *Nodularia* bloom (Table 1). However, background color of the map for this date was light blue (31–50 mg m^{-3}) rather than the usual dark blue (1–30 mg m^{-3}) indicating chlorophyll *a* concentrations in surface water were uniformly higher over the lake than ambient levels generally observed (Table 1). Furthermore, highest predicted chlorophyll *a* levels were present as uniformly colored bands along the north, east and southeast lake margins rather than the usually observed multiple-range chlorophyll swirls and patches in the limnetic zone. Finally, five parallel bands of low chlorophyll *a* extended east from the west-central shoreline (Fig. 3H). This pattern suggests west to east winds advected surface scums of *Nodularia* to the east shore and low chlorophyll *a* water upwelled along the western lake margin and flowed eastward. High winds on 4 July were confirmed from wind velocity measurements at Fallon, Nevada (150 km southeast of Pyramid Lake; National Climatic Data Center, 1986). The highest total wind recorded in July occurred on this date (209 km day^{-1}, mean for the month: 52 km day^{-1}).

No additional Landsat images were available until the 20 July synoptic sampling, and this scene exhibited the highest radiance and predicted chlorophyll *a* for the entire 15 years of Landsat coverage (rank 15, Fig. 3I, Table 1). Twelve percent of the lake surface exhibited chlorophyll *a* concentrations over 150 mg m^{-3}. We documented the abrupt wind-induced collapse of this largest of the 'very large' blooms between 20 and 28 July (Galat & Verdin, 1989). A calcium carbonate whiting was in progress by the 5 August Landsat overpass and largely responsible for the radiance observed on this and subsequent scenes until 22 September.

Review of 15 years of Landsat images enables us to present several generalizations about the periodicity and magnitude of *Nodularia* blooms and associated large-scale limnological events during summer and early autumn in Pyramid Lake (Fig. 5).

1. *Nodularia* blooms typically recur annually. Years without noticeable blooms were uncommon: 1973, 1980 and 1982, although we had too few Landsat scenes to rule out a bloom in 1982.

2. Calcium carbonate whitings are also recurrent events in Pyramid Lake (1973, 1978, 1980, 1984, 1986) and were regularly first observed in late July or early August. Whitings lasted about one month and never appeared in successive years. Whitings may influence size of subsequent *Nodularia* blooms, as blooms following whitings were small or absent (1973, 1978, 1980). Both *Nodularia* blooms preceding August whitings were large (1984, 1986).

3. Magnitude of *Nodularia* blooms was highly variable among years. 'Very large' blooms (ranks 9–15) infrequently appeared in successive years, although 1984–1986 was an exception when three very large blooms were observed. By assembling this long-term data set we were able to address the question: has the magnitude of *Nodularia* blooms shown any trend over the last 15 years? We tested for independence between MMCHL's for all 'very large' blooms (>39 mg m^{-3} chlorophyll *a*) and year of occurrence using Spearman's coefficient of rank correlation. Magnitude of 'very large' blooms was independent of year when all seven years of 'very large' blooms were included ($r_s = 0.50$, $Z = 1.23$, $P = 0.11$). However, examination of MMCHL's for 'very large' blooms plotted against year of occurrence illustrates a moderate trend of increasing bloom size during the past decade. The unusual bloom pattern registered for 1985 was the exception to this tendency (Fig. 7). Omitting this year yielded a significant positive correlation between MMCHL and year ($r_s = 0.83$, $Z = 1.85$, $P = 0.03$) and a positive slope significantly different from zero (Fig. 7). These results indicate that, while there appears to be a recent trend toward increasing magnitude of *Nodularia* blooms, it is not consistent. Longer-term monitoring of *Nodularia* blooms using remote sensing and ground truth is recommended to further evaluate this possibility.

4. *Nodularia spumigena* blooms were recorded during all months between July and October, nevertheless three generalities are evident from our results. First, early-July blooms were always

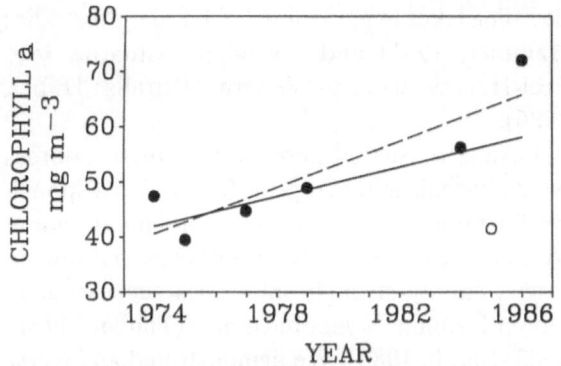

Fig. 7. Variations in predicted maximum mean surface water chlorophyll *a* concentration (MMCHL) for years when magnitude of *Nodularia spumigena* blooms exceeded one standard deviation of the 15 year average MMCHL ('very large' blooms, > 39 mg m^{-3}). The solid line is the slope of MMCHL for all years when very large blooms occurred and is not significantly different from zero ($N = 7$, $R^2 = 0.36$, $P \geq 0.05$). The dashed line illustrates a significantly positive slope when the unusual 1985 bloom was excluded (open circle, $R^2 = 0.80$, $P = 0.02$).

'very large' (1984-rank 14, 1985-rank 10, 1986-rank 15). Second, these 'very large', early-July blooms occurred only during or following years of above average Truckee River discharge to Pyramid Lake (> 1000 m$^3 \times 10^6$, or in 1985 which was preceded by three years of Truckee River flow over 1000 m$^3 \times 10^6$, Figs. 5 and 6). This does not imply high winter-spring Truckee River flows are always succeeded by 'very large', early-July *Nodularia* blooms (e.g. 1982, 1983), but does suggest 'very large', early-July blooms do not occur during or following years of average (≈ 500 m$^3 \times 10^6$ yr^{-1}) or low (≤ 300 m$^3 \times 10^6$ yr^{-1}) Truckee River discharge to Pyramid Lake. Lastly, blooms beginning in October were rare. The only confirmed *Nodularia* bloom first observed in October was in 1978 and it was preceded by a whiting. Its true size is unknown as no Landsat coverage was available, but moderate phytoplankton primary production in October 1978 (Galat, 1986) suggests it ranked less than 10.

The most probable season for *Nodularia* blooms in Pyramid Lake appears to be late-July through September (Fig. 5). Exceptions occur when Truckee River discharge has been above average the past year or more, and then an early-

July bloom is likely. Also, if no early-July *Nodularia* bloom develops, but a calcium carbonate whiting appears during midsummer, early-October blooms cannot be ruled out.

Discussion

Landsat remote sensing provided a historic window through which to review the yearly sequence of *Nodularia* blooms in Pyramid Lake. Its high spatial resolution enabled us to indirectly establish relative bloom magnitude and document large-scale spatial and temporal distribution patterns with limited collection of ground-based data. Qualitative information on patterns of surface currents was also supplied from observed configurations of *Nodularia* scums. Additionally, the importance of wind advection as an allogenic factor influencing distribution and duration of *Nodularia* blooms became apparent.

Remote sensing of scum-forming blue-green algal blooms is not without shortcomings, however. Time intervals between Landsat flybys were not consistent among years and the average 16 day period between overpasses may miss the peak of the bloom, or even the entire bloom if its duration is short (e.g. 1978). Other sources of radiance, including calcium carbonate whitings and suspended sediment, cannot always be differentiated from homogeneously mixed phytoplankton without collection of additional on-site information.

Dense surface scums of buoyant cyanobacteria may saturate reflectance with the result that vertical mats of variable thickness yield similar estimates of radiance, and thus similar predictions of phytoplankton biomass, when biomasses are actually quite different (Galat & Verdin, 1989). Estimates of total euphotic zone chlorophyll concentration are also imprecise using remote sensing as only the upper euphotic zone is registered in satellite images (Eppley *et al.*, 1985). Pyramid Lake has an average mid-summer euphotic depth of 8–10 m (Galat, 1986), and consequently only about 10% of the euphotic zone is recorded in the near-infrared MSS band 3, these wavelengths being quickly absorbed by water. *Nodularia* floats to the surface when wind speed is low and the lake surface is calm because of its positive buoyancy. The shallow registration depth of Landsat is not a problem under these circumstances, however, saturation of surface reflectance may be. In contrast, if the euphotic zone is well mixed, as it appeared to be on 1 and 17 July 1985 and definitely was on 4 July 1986, *Nodularia* will not be concentrated in the upper meter of the mixed zone, and surface saturation is no longer a concern. However, surface water chlorophyll *a* predicted from Landsat will continue to underestimate *Nodularia* biomass, because much of it is now mixed below the maximum depth of MSS band 3 penetration. Finally, it is not possible to differentiate *Nodularia* chlorophyll from chlorophyll of non-buoyant phytoplankton when the lake surface is well mixed because the distinctive patterns of *Nodularia* scums are absent. Duration of *Nodularia* blooms may, therefore, be underestimated using remote sensing unless recognizable surface scums dominate the image (e.g. 1979). These weaknesses of remote sensing will collectively yield underestimates of the true magnitude and duration of scum-forming filamentous cyanobacteria blooms. Although each of these complications may reduce accuracy and precision of remotely sensed estimates of phytoplankton biomass, nevertheless, remote sensing provides a large-scale spatial and temporal perspective not possible with any other conventional sampling strategy.

Results from this study confirm *Nodularia* blooms in Pyramid Lake are restricted to summer and autumn when the lake is thermally stratified. Although water-blooms of filamentous cyanobacteria often occur during summer thermal stratification in temperate lakes (Reynolds & Walsby, 1975), stable thermal stratification does not appear to be a prerequisite for bloom formation. Massive *Nodularia* blooms develop in Walker Lake, Nevada, in May and June prior to, or near the onset of, summer thermal stratification and may recur during summer or early-autumn when thermal stratification is stable (Cooper & Koch, 1984). *Nodularia* also blooms in shallow polymic-tic saline lakes (e.g. Corangamite Lake, Australia; Hammer, 1981) and polymictic estuaries (e.g. Peel-Harvey Estuary, Western Australia; Huber, 1986).

Results presented here enabled us to address what allogenic and autogenic factors are responsible for periodicity of *Nodularia* blooms (Galat & Verdin, 1988). Adequate phosphorus, but low nitrogen supply strongly influences succession to bloom-forming cyanobacteria (Tilman *et al.*, 1982; Smith, 1983). We demonstrated an inverse relation between July–October MMCHL concentrations predicted herein and total nitrogen concentrations in Pyramid Lake surface water the previous winter (Galat & Verdin, 1988). Adequate phosphorus, but limited nitrogen supply therefore appears to be an important antecedent of phytoplankton succession to *Nodularia* in Pyramid Lake.

Appearance of large *Nodularia* blooms in August and September (10 of 15 years) is consistent with Lewis's (1986) paradigm of phytoplankton periodicity, whereby the successional pattern to cyanobacteria is dictated by autogenic changes in nutrient availability following onset of thermal stratification. The unusually early *Nodularia* blooms in July 1984, 1985 and 1986 at first appear to be exceptions to this periodicity, but in fact illustrate how autogenic and allogenic processes interact to influence timing of successional sequences. The association of very large early-July *Nodularia* blooms with above average Truckee River discharge appears to be an allogenic effect of meromixis induced by the large influx of freshwater from 1982 through 1984. Incomplete winter circulation ensued, preventing relocation of hypolimnetic nitrogen to the epilimnion. Earlier nitrogen depletion via phytoplankton uptake occurred in the surface mixed layer and very large, early *Nodularia* blooms followed (Galat & Verdin, 1988). Hypersaline Mono Lake (ca. 80 g l^{-1}) is also located along the western margin of the Great Basin and became meromictic following exceptionally heavy snowfall and runoff in its watershed during 1982–1983. It is also a nitrogen limited saline lake and reduced vertical mixing decreased supply to surface waters and caused major changes in phytoplankton abundance (Mono Basin Ecosystem Study Committee, 1987) and primary production (Jellison & Melack, 1988). Although the causes of low surface water nitrogen were similar to those we observed at Pyramid Lake, specific changes in phytoplankton community structure and succession were not.

Our historic perspective necessitated we rely on the multiple-year Landsat MSS data base. Many limitations of Landsat MSS data have recently been overcome with the advent of more modern sensors with increased spatial, spectral and radiometric resolution (e.g. Landsat thematic mapper, Lathrop & Lillesand, 1986; Lillesand & Kiefer, 1987). Continued analysis of remotely sensed satellite data in conjunction with occasional synoptic ground surveys during *Nodularia* blooms should enhance our long-term objective of understanding the causal mechanisms for these blooms and applying this knowledge to develop reliable and accurate predictions of future blooms and their impact on Pyramid Lake water quality.

Acknowledgements

We are grateful for field assistance provided by L. Carlson, J. Davis, N. Vucinich and G. Wadsworth, Resources Department, Pyramid Lake Fisheries. Research was supported by Contract No. 68-03-6242 from the U.S. Environmental Protection Agency and funding from Pyramid Lake Fisheries who also contributed publication costs for color plates. We thank the Pyramid Lake Paiute Indian Tribe for permitting us to conduct research on their lake for the past decade.

References

Almanza, E. & J. Melack, 1985. Chlorophyll differences in Mono Lake (California) observable on Landsat imagery. Hydrobiologia 122: 13–17.

Berg, C. P., S. R. Schneider & D. L. Galat, 1981. Calcium carbonate precipitation in Pyramid Lake, Nevada, as monitored by satellite: 1979 and 1980. Proc. 15th. Int. Symp. Rem. Sen. Envir. 15: 721–731.

Cooper, J. J. & D. L. Koch, 1984. Limnology of a desertic terminal lake, Walker Lake, Nevada, U.S.A. Hydrobiologia 118: 275–292.

Edmondson, W. T. & J. T. Lehman, 1981. The effect of changes in the nutrient income on the condition of Lake Washington. Limnol. Oceanogr. 26: 1–29.

Eppley, R. W., E. Stewart, M. R. Abbott & U. Heyman, 1985. Estimating ocean primary production from satellite chlorophyll. Introduction to regional differences and statistics for the Southern California Bight. J. Plankton Res. 7: 57–70.

Fogg, G. E. & B. Thake, 1987. Algal cultures and phytoplankton ecology. The University of Wisconsin Press, Madison, 269 pp.

Galat, D. L., 1984. Pyramid Lake limnological survey – September 1983. Desert Research Institute, Reno, Nevada (unpublished report), 11 pp.

Galat, D. L., 1986. Organic carbon flux to a large salt lake: Pyramid Lake, Nevada, USA. Int. Revue ges. Hydrobiol. 71: 621–654.

Galat, D. L. & R. L. Jacobsen, 1985. Recurrent aragonite precipitation in saline-alkaline Pyramid Lake, Nevada. Arch. Hydrobiol. 105: 137–159.

Galat, D. L. & M. Peacock, 1984. Pyramid Lake limnological survey – July 1984. Bioresources Center, Desert Research Institute, Reno, Nevada (unpublished report), 14 pp.

Galat, D. L. & J. P. Verdin, 1987. Analysis of 1972–1986 *Nodularia spumigena* blooms in Pyramid Lake, Nevada using Landsat imagery. Department of Zoology, Arizona State University, Tempe, Arizona, 98 pp.

Galat, D. L. & J. P. Verdin, 1988. Magnitude of blue-green algal blooms in a saline desert lake evaluated by remote sensing: evidence for nitrogen control. Can. J. Fish. aquat. Sci. 45: 1959–1967.

Galat, D. L. & J. P. Verdin, 1989. Patchiness, collapse and succession of a cyanobacterial bloom evaluated by synoptic sampling and remote sensing. J. Plankton Res. 11: 925–948.

Galat, D. L., E. L. Lider, S. Vigg & S. R. Robertson, 1981. Limnology of a large, deep North American terminal lake, Pyramid Lake, Nevada, U.S.A. Hydrobiologia 82: 281–317.

Hamilton-Galat, K. & D. L. Galat, 1983. Seasonal variation of nutrients, organic carbon, ATP and microbial standing crops in a vertical profile of Pyramid Lake, Nevada. Hydrobiologia 105: 27–43.

Hammer, U. T., 1981. A comparative study of primary production and related factors in four saline lakes in Victoria, Australia. Int. Revue ges. Hydrobiol. 66: 701–743.

Hammer, U. T., 1986. Saline lake ecosystems of the world. Junk, Dordrecht, 616 pp.

Harris, G. P., 1986. Phytoplankton ecology: structure, function and fluctuation. Chapman & Hall, N.Y., 384 pp.

Harris, G. P. & A. M. Trimbee, 1986. Phytoplankton dynamics of a small reservoir: physical/biological coupling and the time scales of community change. J. Plankton Res. 8: 1011–1027.

Hattori, E. M., 1982. The archaeology of Falcon Hill, Winnemucca Lake, Washoe County, Nevada. Anthropological Papers No. 18, Nevada State Museum, Carson City, Nevada.

Hendrix, J. L. & M. F. Deguire, 1974. A study of the eutrophication of the surface waters of Pyramid Lake. Completion Report, Proj. A-047-NEV., Chemical and Metallurgical Engineering Dept., Mackay School of MInes, University of Nevada, Reno, Nevada, 18 pp.

164

Hoffer, R., 1978. Biological and physical considerations in applying computer-aided analysis techniques to remote sensor data. In P. Swain & S. Davis (eds.), Remote Sensing. McGraw-Hill, New York: 227–289.

Horne, A. J. & D. L. Galat, 1985. Nitrogen fixation in an oligotrophic, saline desert lake: Pyramid Lake, Nevada. Limnol. Oceanogr. 30: 1229–1239.

Hübel, H. & M. Hübel, 1980. Nitrogen fixation during blooms of *Nodularia* in coastal waters and backwaters of the Arkona Sea (Baltic Sea) in 1974. Int. Revue ges. Hydrobiol. 65: 793–808.

Huber, A. L., 1986. Nitrogen fixation by *Nodularia spumigena* Mertens (Cyanobacteriaceae). 1. field studies and contribution of blooms to the nitrogen budget of the Peel-Harvey Estuary, Western Australia. Hydrobiologia 131: 193–203.

Hutchinson, G. E., 1937. A contribution to the limnology of arid regions. Conn. Acad. Arts Sci. 33: 47–132.

Hutchinson, G. E., 1967. A treatise on limnology, 2. J. Wiley & Sons, New York, 1115 pp.

Jellison, R. & J. M. Melack, 1988. Photosynthetic activity of phytoplankton and its relation to environmental factors in hypersaline Mono Lake, California. Hydrobiologia 158: 69–88.

Kirk, J. T. O., 1983. Light and photosynthesis in aquatic ecosystems. Cambridge University Press, Cambridge, 401 pp.

Koch, D., L. Hoffman & J. Mahoney, 1976. Pyramid Lake: zooplankton distribution and blooms of the blue-green alga, *Nodularia spumigena*. Project Report No. 38, Water Resources Center, Desert Research Institute, Reno, Nevada, 46 pp.

LaRivers, I., 1962. Fishes and fisheries of Nevada. Nevada State Fish & Game Commission, Carson City, 782 pp.

Lathrop, R. G. & T. M. Lillesand, 1986. Use of Thematic Mapper data to assess water quality in Green Bay and Central Lake Michigan. Photo. Eng. Rem. Sen. 52: 671–680.

Lewis, W. M., Jr., 1978. Dynamics and succession of the phytoplankton in a tropical lake: Lake Lanao, Philippines. J. Ecol. 66: 849–880.

Lewis, W. M., Jr., 1986. Phytoplankton succession in Lake Valencia, Venezuela. Hydrobiologia 138: 189–203.

Lider, E. L. & R. Langdon, 1978. Plankton ecology. In W. F. Sigler & J. L. Kennedy (eds.) Pyramid Lake Ecological Study. W. F. Sigler & Associates, Logan Utah: 241–293.

Likens, G. E., (ed.), 1985. An ecosystem approach to ecology: Mirror Lake and its environment. Springer-Verlag, New York, 516 pp.

Lillesand, T. M. & R. W. Kiefer, 1987. Remote sensing and image interpretation. Wiley, New York, 721 pp.

Lillesand, T. M., W. Johnson, R. Deuell, O. Lindstrom & D. Meisner, 1983. Use of Landsat data to predict the trophic status of Minnesota lakes. Photo. Eng. Rem. Sen. 49: 219–229.

Lindahl, G., K. Wallström & G. Brattberg, 1980. Short-term variations in nitrogen fixation in a coastal area of the Northern Baltic. Arch. Hydrobiol. 89: 88–100.

Lukatelich, R. J. & A. J. McComb, 1986. Nutrient levels and the development of diatom and blue-green algal blooms in a shallow Australian estuary. J. Plankton Res. 8: 597–618.

Lund, J. W. G., 1965. The ecology of the freshwater phytoplankton. Biol. Rev. Camb. Philos. Soc. 40: 231–293.

Markham, B. L. & J. L. Barker, 1987. Radiometric properties of U.S. processed Landsat MSS data. Rem. Sen. Envir. 22: 39–71.

Mono Basin Ecosystem Study Committee, 1987. The Mono Basin Ecosystem, effects of changing lake level. National Academy Press, Washington, D.C., 272 pp.

Munawar, M. & I. F. Munawar, 1986. The seasonality of phytoplankton in the North American Great Lakes, a comparative synthesis. Hydrobiologia 138: 85–115.

National Climatic Data Center, 1986. Climatological data, Nevada. Vol. 101. U.S. Department of Commerce, National Climatic Data Center, Ashville, North Carolina.

Oström, B., 1976. Fertilization of the Baltic by nitrogen fixation in the blue-green alga *Nodularia spumigena*. Rem. Sen. Envir. 4: 305–310.

Pearsall, W. H., 1932. Phytoplankton in the English Lakes II. The composition of the phytoplankton in relation to dissolved substances. J. Ecol. 20: 241–262.

Pollingher, U., 1986. Phytoplankton periodicity in a subtropical lake (Lake Kinneret, Israel). Hydrobiologia 138: 127–138.

Reynolds, C. S., 1980. Phytoplankton periodicity in stratifying lake systems. Holarctic Ecol. 3: 141–159.

Reynolds, C. S. & A. E. Walsby, 1975. Water-blooms. Biol. Rev. 50: 437–481.

Sigler, W. F., S. Vigg & M. Bres, 1985. Life history of the cui-ui, *Chasmistes cujus* Cope, in Pyramid Lake, Nevada: a review. Great Basin Nat. 45: 571–603.

Sigler, W. F., W. T. Helm, P. A. Kucera, S. Vigg & G. W. Workman, 1983. Life history of the Lahontan cutthroat trout, *Salmo clarki henshawi*, in Pyramid Lake, Nevada. Great Basin Nat. 43: 1–29.

Smith, V. H., 1983. Low nitrogen to phosphorus ratios favor dominance by blue-green algae in lake phytoplankton. Science 221: 148–153.

Strong, A. E., 1978. Chemical whitings and chlorophyll distribution in the Great Lakes as viewed by Landsat. Rem. Sen. Envir. 7: 61–72.

Tansley, A. G., 1935. The use and abuse of vegetational concepts and terms. Ecology 16: 284–307.

Tilman, D., S. S. Kilham & P. Kilham, 1982. Phytoplankton community ecology: the role of limiting nutrients. Ann. Rev. Ecol. Syst. 13: 349–372.

Ulbricht, K. A. & U. Horstmann, 1980. Remotely sensed phytoplankton development in the Baltic Sea. In A. P. Cracknell (ed.). Coastal and Marine Applications of Remote Sensing. Remote Sensing Society, Reading, England: 69–76.

U.S. Geological Survey, 1972–1986. Water resources: Nevada. Water years 1972–1986. Water-data reports NV-72-1 to NV-86-1. U.S. Geological Survey, Carson City, Nevada.

Verdin, J. P., 1985. Monitoring water quality conditions in a large western reservoir with Landsat imagery. Photo. Eng. Rem. Sen. 51: 343–353.

Hydrobiologia **197**: 165–172, 1990.
F. A. Comin and T. G. Northcote (eds), Saline Lakes.
© 1990 *Kluwer Academic Publishers.*

The photosynthesis of *Dunaliella parva* Lerche as a function of temperature, light and salinity

C. Jiménez[1], F.X. Niell[1] & J.A. Fernández[2]

[1]*Departamento de Ecología, Facultad de Ciencias, Universidad de Málaga. Campus de Teatinos, 29071; Málaga (Spain)*; [2]*Departamento de Fisiología Vegetal, Facultad de Ciencias, Universidad de Málaga, Campus de Teatinos, 29071, Málaga. (Spain)*

Key words: *Dunaliella parva*, net photosynthesis, light, temperature, salinity

Abstract

The photosynthetic behaviour of *Dunaliella parva* Lerche from the athalassic lagoon of Fuente de Piedra (Málaga, Southern Spain) was studied experimentally at three NaCl concentrations (1, 2 and 3 M), five temperatures (15, 23, 31, 38 and 42 °C) and nine different irradiances between 82 and 891 mol m^{-2} s^{-1}. Results are analyzed to define the best growing conditions for the algae.

D. parva shows the highest photosynthetic rates at a NaCl molarity of 2 M, under a moderate light intensity (600 mol m^{-2} s^{-1}) at 31 °C. Above this light intensity a clear photoinhibition of the photosynthesis was found at 2 M and 3 M of NaCl. *D. parva* is a halotolerant and a thermoresistant species as evidenced by its net photosynthesis rate and positive values of oxygen evolution at 42 °C.

Two methods for modelling photosynthesis *vs.* irradiance curves are discussed. The first is a single model, based on third-order polynomial equations, and the second is double model, based on hyperbolical Michaelis-Menten type functions and negative exponential to define photoinhibition.

Introduction

A laboratory study was made of the photosynthetic performance of the flagellate *Dunaliella parva* Lerche from the athalassic lake of Fuente de Piedra (Málaga, Southern Spain). This lake has a natural cycle of draining each summer; that makes that its salinity fluctuates along the year within the range 31‰–350‰. *D. parva* and *D. salina* Teodoresco are the major species of spring phytoplankton. Neither species is present during fall and winter. *D. parva* is a halotolerant and thermoresistant species (Gimmler *et al.*, 1978) but is not a β-carotene accumulative species

(Ben-Amotz & Avron, 1983; Gimmler *et al.*, 1981; Borowitzka *et al.*, 1984).

We developed general models of the photosynthetic performance of *D. parva* based on its net photosynthetic rate and its pigment adaptation with respect to three variables: incident light intensity, temperature and salinity.

Material and methods

Photosynthetic rate was measured by means of the oxygen evolution method, using a closed transparent chamber of 60 ml of volume, to which a YSI oxygen probe model No. 5750 calibrated in air satured water with compensation for temperature was attached. Temperature was maintained constant for every treatment using a water flow system from a thermostatic bath to the water jacket of the incubation chamber.

D. parva cells used in the experiments were isolated from lake Fuente de Piedra (37° 06′ N, 04° 45′ W) and grown in axenic conditions in a chemostat, at a dilution rate of 0.25 day^{-1}.

Three salinities (1, 2 and 3 M NaCl), five temperatures (15, 23, 31, 38 and 42 °C) and nine light intensities (82, 94, 165, 214, 344, 412, 588, 691 and 891 μmol m^{-2} s^{-1}) have been used. The light source was a slide projector (ZEISS IKON model Perkeo Compact) with grey filters to obtain the different light intensities. The transmission spectra of the light source without filters and with the filters used for reducing light intensity are compared (Fig. 1). Light intensity was measured by means of a LICOR LI-1000 radiometer and light quality by means of a LICOR LI-1800 spectroradiometer (spectral sensitivity $\pm 4\%$) under the same light conditions as used for the experiments.

The different salinities have been achieved adding the appropriate amounts of NaCl to a basic medium (Johnson *et al.*, 1968). The cells were incubated at least for two weeks at each new salinity before the experiments.

Before and after each experiment, extending for twenty minutes in all cases, a 5 ml sample was taken from the incubation chamber and filtered through a WHATMAN GF/C filter. Chlorophyll extraction was made in neutralized acetone with $MgCO_3$ for 24 hours. Chlorophyll a (Chl a) content was estimated using the Jeffrey & Humphrey (1975) equations and expressed as a ratio between the final and the initial concentrations (Chl a_F/Chl a_I). The relation between Chl a and carotenoid pigments was estimated using the D_{430}/D_{664} ratio. The number of algal cells in the culture was estimated by means of a hematocytometer following the methods of Guillard (1973). Net photosynthesis rate was computed from the change in oxygen concentration within the incubation chamber. The data were plotted against light intensity (Fig. 2) and fitted to equation 1 in order to compute the maximum rate of photosynthesis and the light compensation point for every treatment.

$$P = P_{max}(1\text{-}LCP/L) \qquad (1)$$

where P and P_{max} are the photosynthetic rate and maximum photosynthetic rate, LCP is the light compensation point and L the irradiance. The equation was linearized by multiplying by L.

The data so transformed were fitted to equation 2 by means of the least squares method. The slope of this equation gives the maximum photosynthetic rate and the light compensation point is the Y-intercept divided by P_{max}.

$$P \cdot L = P_{max} \cdot L\text{-}P_{max} \cdot LCP \qquad (2)$$

The attempts to fit the experimental data of net photosynthesis *vs.* incident light to Jassby & Platt (1976), Platt & Jassby (1976) and Platt *et al.* (1980) equations were unsuccessful. The first part of the curves can be expressed by means of a Michaelis-Menten type equation, while photoinhibition can be fitted to a negative exponential expression; also, the whole process can be explained by a third order polynomial model. In

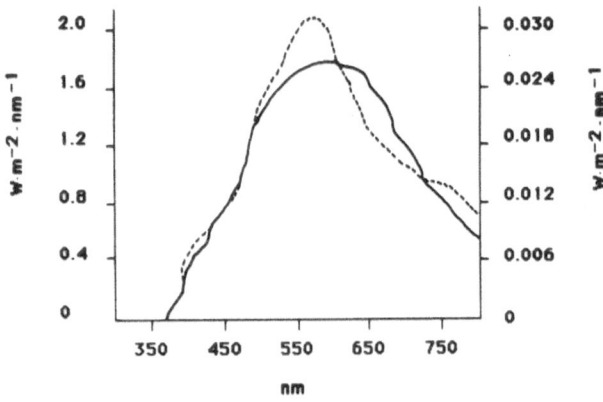

Fig. 1. Spectral irradiance of the light source without any filter (continuous line, left scale) and with a grey filter (broken line, right scale) (62 μmol m^{-2} s^{-1}) from 350 to 800 nm measured under the same conditions in which the experiments were performed.

all cases a significative value has been found. In order to compute the photoinhibition the slope of the negative exponential function has been used, comparable to β in Platt *et al.* (1980).

Results

Net Photosynthesis

Figure 2 shows the results of net photosynthesis for *D. parva* under the different treatments. The maximum photosynthetic rate is obtained at 31 °C at all the NaCl concentrations (Fig. 3), reaching values of 31.7 pg O_2 cell^{-1} h^{-1}. *D. parva* has its highest values of net photosynthesis at a NaCl concentration of 2 M; the differences between NaCl concentrations of 1 M and 3 M are only significant ($p < 0.001$) at 31 °C, and between 2 M and 3 M NaCl at 15, 23 and 31 °C. Beyond 38 °C there are no significant differences between the three tested NaCl concentrations (Fig. 3).

The results have been expressed in a hyperbolical Michaelis-Menten type function in the case of 1 M NaCl, because photoinhibition was

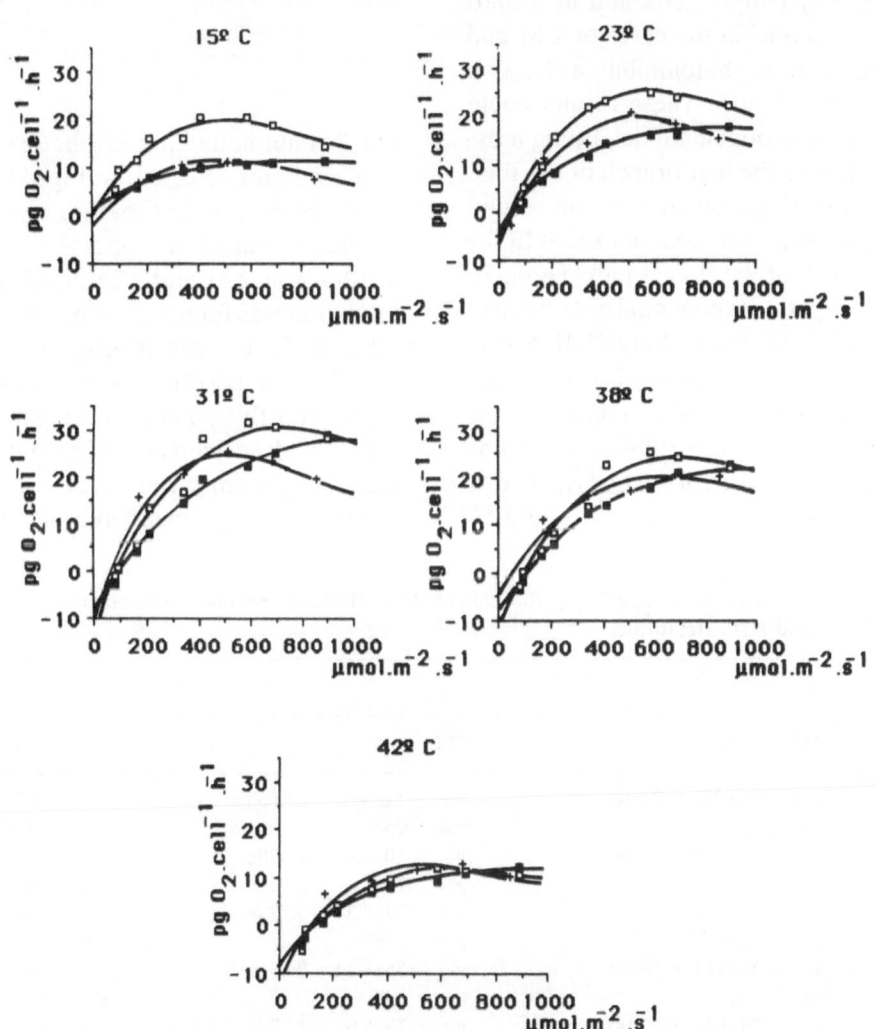

Fig. 2. Photosynthetic rates *vs.* light intensity at different treatments of temperature and salinity in *D. parva.* Solid squares, 1 M NaCl; empty squares, 2 M NaCl and crosses, 3 M NaCl.

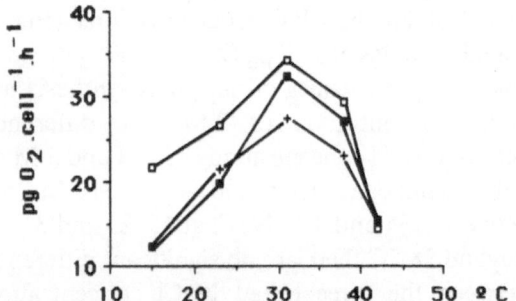

Fig. 3. Effect of temperature on maximum net photosynthetic rate at different salinity treatments. Symbols as in Fig. 2.

Table 2. Negative exponential functions fitting to the photoinhibition portion of *D. parva* photosynthetic rate *vs.* irradiance plots, at different temperatures and 2 M and 3 M NaCl concentrations.

T (°C)	NaCl	
	2M	3M
15	$y = 30{,}18 \cdot e^{-3{,}35.10^{-4}x}$ $r = 0{,}95$	$y = 20{,}95 \cdot e^{-4{,}85.10^{-4}x}$ $r = 0{,}96$
23	$y = 31{,}11 \cdot e^{-1{,}64.10^{-4}x}$ $r = 0{,}99$	$y = 32{,}66 \cdot e^{-3{,}75.10^{-4}x}$ $r = 0{,}99$
31	$y = 40{,}07 \cdot e^{-1{,}72.10^{-4}x}$ $r = 1{,}00$	$y = 35{,}79 \cdot e^{-2{,}91.10^{-4}x}$ $r = 0{,}99$
38	$y = 34{,}01 \cdot e^{-2{,}11.10^{-4}x}$ $r = 1{,}00$	$y = 37{,}68 \cdot e^{-3{,}53.10^{-4}x}$ $r = 0{,}94$
42	$y = 16{,}94 \cdot e^{-2{,}41.10^{-4}x}$ $r = 0{,}98$	$y = 22{,}81 \cdot e^{-3{,}78.10^{-4}x}$ $r = 0{,}96$

not detected at any temperature, and in a third order polynomial model in the cases of 2 M and 3 M NaCl where a clear photoinhibition effect on photosynthesis was found. These results could also be depicted on a double model, fitting a hyperbolical function to the first branch of the P *vs.* I plots and a negative exponential to the second one. Table 1 gives the linear equations that fit the hyperbolical branch of the P *vs.* I plots (Eqn. 2), and Table 2 the negative exponential ones fitted to the photoinhibition effects, at 2 and 3 M NaCl.

Although the highest values of net photosynthesis were obtained at a NaCl concentration of 2 M there was a clear photoinhibition of photosynthesis at high light intensity. *D. parva* had a similar net photosynthetic rate at 1 M and 3 M

NaCl. Photoinhibition of photosynthesis at 1 M NaCl was not detected. In Fig. 4 are depicted the results of the slopes (β) of the negative exponential functions that fit the photoinhibition of photosynthesis at 2 M and 3 M NaCl. Highest photoinhibition was found at the extreme temperatures, 15 and 42 °C, with β being almost constant at both NaCl concentrations between 23 and 38 °C. The curves fit significatively a parabolic function. Also, a clear increase of photoinhibition with increasing salinity was found.

The light compensation point (LCP) was esti-

Table 1. Linear equations fitting the first part (hyperbolical one; Eqn. 2) of *D. parva* photosynthetic rates *vs.* irradiance plots at different temperatures and NaCl treatments.

T (°C)	NaCl		
	1M	2M	3M
15	$y = -863.22 + 12.29x$ $r = 1.00$	$y = -1288.90 + 21.66x$ $r = 0.99$	$y = -733.21 + 12.29x$ $r = 1.00$
23	$y = -2136.65 + 19.19x$ $r = 1.00$	$y = -2056.92 + 26.70x$ $r = 1.00$	$y = -1505.21 + 21.37x$ $r = 0.99$
31	$y = -4544.71 + 32.37x$ $r = 0.99$	$y = -3713.33 + 34.24x$ $r = 0.99$	$y = -2132.91 + 27.39x$ $r = 0.99$
38	$y = -4588.01 + 26.93x$ $r = 1.00$	$y = -3568.14 + 29.36x$ $r = 0.99$	$y = -2170.87 + 22.90x$ $r = 1.00$
42	$y = -2973.58 + 15.38x$ $r = 1.00$	$y = -1855.05 + 14.33x$ $r = 0.99$	$y = -1809.30 + 15.43x$ $r = 1.00$

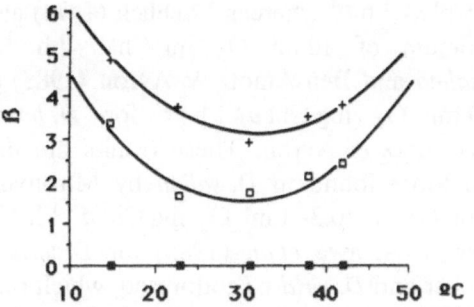

Fig. 4. Changes in photoinhibition effect, (β-coefficient of Platt *et al.*, 1980), with temperature at 1 M, 2 M and 3 M NaCl. Symbols as in Fig. 2.

mated according to a linear model of the hyperbolical branch of P *vs.* I plots (Eqn. 2). LCP decreases with increasing NaCl molarity for all temperatures (Fig. 5).

Pigment content

Chl *a* concentrations did not change in response to incident light intensity at 1 M and 2 M NaCl (Fig. 6). The ratio Chl a_F/Chl a_I fluctuated around 1 during the experiments developed in these two media. At 3 M NaCl, Chl *a* concentrations were always smaller at the end of the experiments and smaller than at 1 M and 2 M, probably due to the severity of the medium; also there was a decrease in the Chl a_F/Chl a_I ratio with increasing light intensity.

Increasing temperature caused a decrease in Chl *a* concentrations at any NaCl molarity in *D. parva*. The Chl a_F/Chl a_I index seemed to be

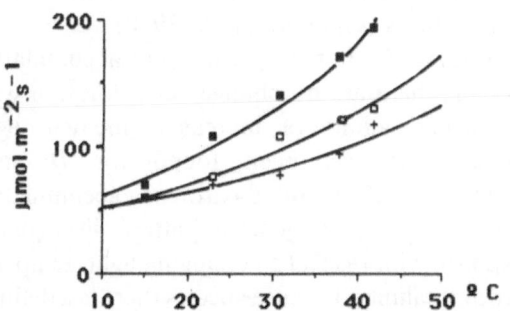

Fig. 5. Changes in light compensation point with temperature at 1 M, 2 M and 3 M NaCl. Symbols as in Fig. 2.

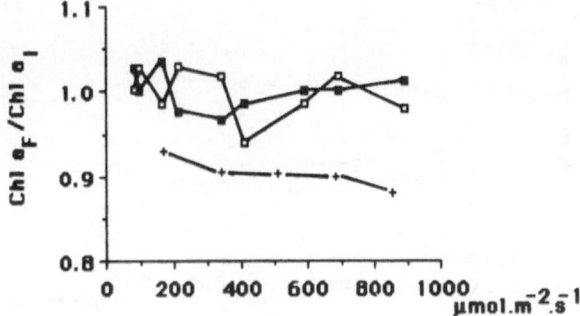

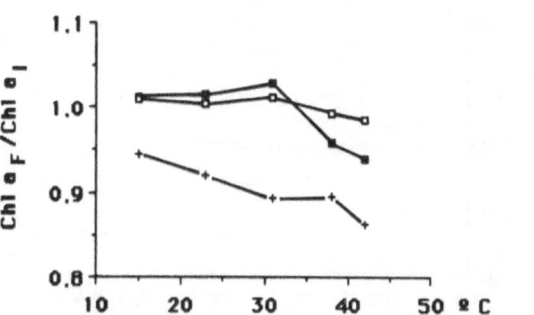

Fig. 6. Changes in the Chl a_F/Chl a_I ratio with light intensity (a) and temperature (b) at 1 M, 2 M and 3 M NaCl treatments. Symbols as in Fig. 2.

more affected by temperature than by the incident light. These conclusions agreed with the change in the D_{430}/D_{664} index (Fig. 7), which remained almost constant when plotted against increasing incident light but decreased with increasing temperature.

Changes in culture coloration with increasing NaCl concentrations were not found; *D. parva* could not be induced to accumulate carotenes in response to increasing salinity, as did *D. salina*. Our experimental data show a decrease in the D_{430}/D_{664} index with increasing chlorinity (Fig. 7), and the index seems to be very low, remaining between 1.87 and 2.16.

Discussion

The use of the Jassby & Platt (1976) and the Platt & Jassby (1976) hyperbolic tangent functions or the Platt *et al.* (1980) function introducing the β-parameter for explaining the photoinhibition of net photosynthesis, do not seem to be universally

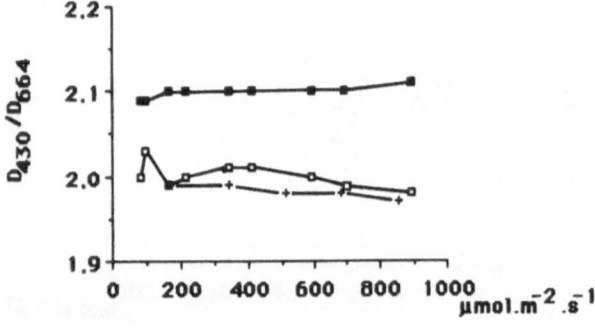

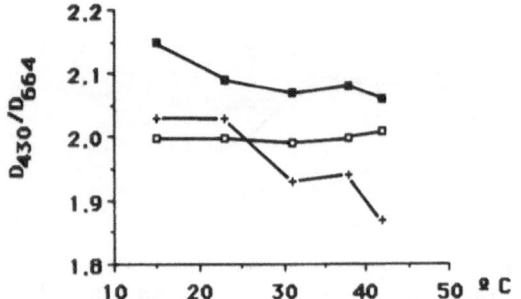

Fig. 7. Changes in the D_{430}/D_{664} index with light intensity (a) and temperature (b) at 1 M, 2 M and 3 M NaCl treatments. Symbols as in Fig. 2.

applicable. Peterson *et al.* (1987) introduced a simple method for describing *P* vs. *I* curves based on the use of exponential equations, but photo-inhibition was not included in the model. It is possible to model the changes of the photo-synthetic rate with increasing incident light according to a third-order polynomial function, or by using a different model for each one of the two branches of the curves. Neale & Richerson (1987) fitted the first one to a hyperbolical tangent function (Jassby & Platt, 1976; Platt & Jassby, 1976) and the second one, photoinhibition, to a simple negative exponential function. We found that a double model seemed to be very accurate, but the first branch of the curves can be fitted to a simple theoretical Michaelis-Menten type function (Eqn. 1), and the second one, agreeing with Neale & Richerson (1987), to a negative exponential equation, using the slope of the function as an estimator of photoinhibition (β).

Our maximum net photosynthetic rates seem to be very similar to those found in preceding papers. We found a maximum of 13.6 ml O_2 mg Chl a^{-1} h^{-1} whereas Loeblich (1974) gave a maximum of 10 ml O_2 mg Chl a^{-1} h^{-1} for *D. salina* and Ben-Amotz & Avron (1983) gave 13.4 ml O_2 mg Chl a^{-1} h^{-1} for *D. bardawil* Ben-Amotz & Avron. These results are higher than those found in *D. salina* by Mironyuk & Einor (1968) (0.3–1 ml O_2 mg Chl a^{-1} h^{-1}) or those of Aizawa *et al.* (1985) for *D. tertiolecta* Butcher and *D. viridis* Teodoresco, which ranged from 2.51 to 4.59 ml O_2 mg Chl a^{-1} h^{-1}.

The maximum rates were found at a NaCl molarity of 2 M, agreeing with the general pattern for *Dunaliella* species. Brock (1975), using a ^{14}C-labelled technique, gave the optimum for *Dunaliella* at a salinity of 10–15%, as did Loeblich (1982) for *D. salina*. Post *et al.* (1983) found that *D. salina* can grow in a wide range of NaCl concentrations, from sea salinity to more than 5 M NaCl, having the maximum rate at 2 M; Loeblich (1969) gives data for *D. salina* growth from 2% of NaCl to saturation, with the maximum rate of 10–15%. Our results agree with this pattern, from 15 to 31 °C, with the maximum photosynthesis rate at 2 M and the lowest rate at 1 M NaCl. *D. parva* from the athalassic lagoon of Fuente de Piedra could therefore be called a halotolerant species. The decrease of net photosynthetic rate with increasing salinity, up to 2 M NaCl, is explained on lower gas solubility (O_2 and CO_2) (Loeblich, 1970, 1972; Ginzburg & Ginzburg, 1985).

The similarity of photosynthetic rate at 38 and 42 °C and at the different salinities for *D. parva* could be explained by a lower availability of CO_2 during photosynthesis. Also there may be a general decrease in photosynthetic activity at high temperatures (Gimmler *et al.*, 1978).

D. parva is not a β-carotene accumulative species; there are no changes in coloration with increasing salinity or increasing incident light intensity as has been found for *D. salina* (Loeblich, 1974). In β-carotene accumulative species there is a general pattern in pigment response with respect to changing light, temperature and salinity. In these species there is a definite increase in Chl *a* concentrations with increasing salinity and temperature, but a decrease in response to increasing light intensity (Loeblich, 1969, 1970, 1974; Ben-Amotz & Avron, 1983). In *D. parva* we found a general decrease in Chl *a* concentrations with increasing temperature or salinity and an almost constant pattern with increasing light; these results agree with those found by Ben-Amotz & Avron (1983) for β-carotene non-accumulative species.

At least in *D. parva*, carotenoids do not prevent photoinhibition as has been suggested for *D. salina* (Loeblich, 1982) and *D. bardawil* (Ben-Amotz & Avron, 1983). In this work (Fig. 2) photoinhibition was found above $700 \text{ mol m}^{-2} \text{ s}^{-1}$, especially at low temperatures, in contrast to β-carotene accumulative species of the genus *Dunaliella* that do not show such decay in photosynthesis rate at high light intensities (Loeblich, 1982). Pigmentary responses of *D. parva* (Figs. 6 and 7) seem to point out that this species is not as well adapted to high light intensities as are β-carotene accumulative species. On the other hand, *D. parva* does not show a clear adaptation to high temperature (Fig. 4) because the lowest photoinhibition is found around 30 °C.

Note added in proof

After Borowitzka & Borowitzka (1988) we think that this species is in agreement with their description of *Dunaliella viridis* Theodoresco better than with *Dunaliella parva* Lerche.

Acknowledgements

This work has been supported by the C.A.I.C.Y.T. (1063/85) and by a fellowship of the 'Programa Nacional FPI/85, Plan Complementario'.

References

Aizawa, K., Y. Nakamura & S. Miyachi, 1985. Variation of phosphoenolpyruvate carboxylase activity in *Dunaliella* associated with changes in atmospheric CO_2 concentration. Pl. Cell Physiol. 26: 1199–1203.

Ben-Amotz, A. & M. Avron, 1983. On the factors which determine massive β-carotene accumulation in the halotolerant alga *Dunaliella bardawil*. Pl. Physiol. 72: 593–597.

Borowitzka, L. J., T. P. Moulton & M. A. Borowitzka, 1985. Salinity and the commercial production of beta-carotene from *Dunaliella salina*. In: W. J. Barclay & R. McIntosh (eds.), Algal biomass: an interdisciplinary perspective. J. Cramer Verlag, Verduz: 217–222.

Borowitzka, M. A. & L. J. Borowitzka, 1988. *Dunaliella*. In: Microalgal biotechnology. Eds. M. A. Borowitzka & L. J. Borowitzka. Cambridge University Press. 477 pp.

Brock, T. D., 1975. Salinity and the ecology of *Dunaliella* from Great Salt Lake. J. gen. Microbiol. 89: 285–292.

Bruff, 1968. Effects of salt on the halophilic alga *Dunaliella viridis*. J. Bact. 95: 1461–1468.

Gimmler, H., E. M. Kuhnl & G. Carl, 1978. Salinity dependent resistance of *Dunaliella parva* against extreme temperatures. I. Salinity and thermoresistance. Z. Pflanzenphysiol. 90: 133–153.

Gimmler, H., C. Weidemann & E. M. Moller, 1981. The metabolic response of the halotolerant alga *Dunaliella parva* to hypertonic shocks. Ber. Deutsch. Bot. Ges. Bd. 94: 613–634.

Ginzburg, B. Z. & M. Ginzburg, 1985. Studies of the comparative physiology of the genus *Dunaliella* (Chlorophyta, Volvocales). 1. Response of growth to NaCl concentration. Br. phycol. J. 20: 277–283.

Guillard, R. R. L., 1973. Division rates. In: Janet R. Stein (ed.), Handbook of phycological methods. Culture methods and growth measurements. Cambridge University Press: 289–311.

Jassby, A. D. & T. Platt, 1976. Mathematical formulation of the relationship between photosynthesis and light for phytoplankton. Limnol. Oceanogr. 21: 540–547.

Jeffrey, S. W. & G. F. Humphrey, 1975. New spectrophotometric equations for determining chlorophylls a, b, c 1 and c 2 in higher plants, algae and natural phytoplankton. Biochem. Physiol. Pflanz 167: 191–194.

Johnson, M. K., E. J. Johnson, R. D. MacElroy, H. L. Speer & B. S. Bruff, 1968. Effects of salts on the halophilic alga *Dunaliella viridis*. J. Bacteriol. 95: 1461–1468.

Loeblich, L. A., 1969. Aplanospores of *Dunaliella salina* (Chlorophyta). J. Protozool. 16: 22–23.

Loeblich, L. A., 1970. Growth limitation of *Dunaliella salina* by CO_2 at high salinity. J. Phycol. 6: (suppl) 9.

Loeblich, L. A., 1972. Studies on the brine flagellate *Dunaliella salina*. Ph.D. thesis, University of California, San Diego.

Loeblich, L. A., 1974. Action spectra and effect of light intensity on growth, pigments and photosynthesis in *Dunaliella salina*. J. Protozool. 21: 420.

Loeblich, L. A., 1982. Photosynthesis and pigments influenced by light intensity and salinity in the halophile *Dunaliella salina* (Chlorophyta). J. mar. biol. Ass. U.K. 62: 493–508.

Myronyuk, V. I. & L. O. Einor, 1968. Oxygen exchange and pigment content in various forms of *Dunaliella salina* Teod.

under conditions of increasing NaCl content. Gidro-biologichnii Zhurnal 4: 23–29.

Neale, P. J. & P. J. Richerson, 1987. Photoinhibition and the diurnal variation of phytoplankton photosynthesis. I. Development of a photosynthesis-irradiance model from studies of *in situ* responses. J. Plankton. Res. 9: 167–193.

Peterson, D. H., M. J. Perry, K. E. Bencala & M. C. Talbot, 1987. Phytoplankton productivity in relation to light intensity: a simple equation. Estuar. coast. & shelf Sci. 24: 813–832.

Platt, T. & A. D. Jassby, 1976. The relationship between photosynthesis and light for natural assemblages of coastal marine phytoplankton. J. Phycol. 12: 421–430.

Platt, T., C. L. Gallegos & W. G. Harrison, 1980. Photo-inhibition of photosynthesis in natural assemblages of marine phytoplankton. J. mar. Res. 38: 687–701.

Post, F. J., L. J. Borowitzka, M. A. Borowitzka, B. Mackay & T. Moulton, 1983. The protozoa of a Western Australian hypersaline lagoon. Hydrobiologia 105: 95–113.

Hydrobiologia **197**: 173–192, 1990.
F. A. Comin and T. G. Northcote (eds), Saline Lakes.
© 1990 Kluwer Academic Publishers.

Distribution and abundance of littoral benthic fauna in Canadian prairie saline lakes

U. T. Hammer, J. S. Sheard & J. Kranabetter
Biology Department, University of Saskatchewan, Saskatoon, Saskatchewan, S7N OWO, Canada

Key words: fauna, benthos, littoral, saline lakes

Abstract

The littoral benthos of 18 lakes in Alberta and Saskatchewan ranging in salinity from 3 to 126‰ (g l^{-1} TDS) were investigated twice, in the spring and in the summer of 1986. Multiple Ekman dredge samples were taken at water depths of about 0.5, 1.0 and 2 metres in each transect. Two to three transects were used in each lake according to its estimated limnological diversity for a total of 114 stations. A total of 76 species was present varying from 29–31 species in the three lakes of lowest salinity (means of 3.1–5.55‰) to only 2 species in lakes exceeding 100‰. Species richness decreased rapidly in salinities greater than 15‰.

Biomass maximum mean of 10.91 g m^{-2} dry weight (maximum 63.0 g m^{-2}) occurred in culturally eutrophic Humboldt Lake (3.1‰) but one third as great in other low salinity lakes. However, biomass again increased to about 4.5 g m^{-2} in two lakes of 15‰. As the salinity increased still further biomass declined steadily until a minimum of 0.0212 g m^{-2} was recorded in most saline Aroma Lake (mean 119‰). Summer biomass (11 lakes) was greater than spring biomass (4 lakes) because some groups such as amphipods, corixids and ostracods became more abundant in summer. Wet weight biomass averaged 15.8% of dry weight biomass.

Seasonality (spring or summer), sediment texture and organic matter content, water depth, pH, salinity (TDS) and the presence of aquatic plants (% plant cover) were considered in the matrix involving species dry weight biomass at each of 117 stations. TWINSPAN classification of the samples yielded a dendrogram with 18 indicator species. Successive dichotomies divided these indicator species into four main lake groups based on salinity, i.e., Group I: 3–10‰ (*Gammarus, Glyptotendipes* I, *Chironomus* cf. *plumosus*), Group II: 10–38‰ (*Hyalella, Enallagma, Bezzia*), Group III: 38–63‰ (*Hygrotus salinarius, Cricotopus ornatus*), Group IV: >63‰ (Dolichopodidae, *Ephydra hians*). Each of these main groups was subdivided into smaller groups of lakes based on factors such as pH, seasonality (spring or summer species dominance), % organic matter and % plant cover. Depth of samples played no apparent role.

Introduction

Saline lakes pervade endorheic regions on the Interior Plains of southern Saskatchewan an Alberta south of 54° N latitude and between 103°

and 114° W longitude. Rawson & Moore (1944) initiated saline lakes studies when they carried out a survey of some southern Saskatchewan lakes between 1938 and 1941. Detailed and specialized studies of saline lakes in both provinces have been carried out since 1970. These include a general introduction (Hammer *et al.*, 1975), chemistry (Hammer, 1978), physical aspects (Hammer & Haynes, 1978), algae (Hammer *et al.*, 1983), sedimentology (Last & Schweyen, 1983), benthic fauna (Timms *et al.*, 1986) and macrophytes (Hammer & Heseltine, 1988). More restricted studies were carried out on osmoregulation and ecology of selected species (Tones, 1978, 1983; Tones & Hammer, 1975), on bacterial primary productivity and ecology (Lawrence *et al.*, 1978; Parker & Hammer, 1983; Parker *et al.*, 1983) in Waldsea and Deadmoose lakes, and secondary production (Swanson & Hammer, 1983; Swift & Hammer, 1979) and sedimentology in Waldsea Lake (Schweyen & Last, 1983; Last & Schweyen, 1985). Hammer (1986) has reviewed in detail worldwide research on littoral saline lake fauna.

Rawson & Moore (1944) were the first to compare benthic fauna species and biomass in the survey which included 14 saline lakes. Many lakes were sampled only once. Identification of some groups (e.g., chironomids) was limited by the taxonomy of the day. Timms *et al.*, (1986) investigated the spring macrobenthos of 22 lakes in a band across Alberta and Saskatchewan between 51° and 53° N latitude. Sampling was restricted to sublittoral and profundal waters bare of benthic flora. Tones (1976) investigated the macrobenthos of the littoral of six Saskatchewan saline lakes (Basin, Big Quill, Deadmoose, Humboldt, Little Manitou, Waldsea) during 1974–75. The results remain unpublished but are reviewed in Hammer (1986).

The study reported here was an attempt to assess, in a broad spectrum of saline lakes, the littoral macrobenthos in terms of abundance and species and compare the results to those of Timms *et al.* (1986). Detailed studies of the fauna, including biomass, of the littoral zones of saline lakes have not been done elsewhere.

Saline lakes are defined as those equal to or exceeding 3‰ salts (Williams, 1964). The classification used herein is that of Hammer *et al.* (1983): hyposaline 3–20‰; mesosaline 20–50‰; hypersaline > 50‰.

Material and methods

The lakes chosen in our study were essentially the same as those studied by Timms *et al.* (1986) so that comparisons would be facilitated. However, we investigated no subsaline lakes. Opuntia, Landis and Whiteshore lakes were omitted and Deadmoose Lake (large and meromictic) was added. Relevant data on morphometry and physical features, and water chemistry for Saskatchewan lakes have been detailed in Hammer & Haynes (1978) and Hammer (1978) respectively and for Alberta in Hammer & Heseltine (1988). Timms *et al.* (1986) have additional information on some maximum depths.

Collections were made during spring and summer of 1986 to determine seasonal variations. Logistics prevented sampling all lakes at about the same time. The same order of sampling was maintained so that sampling intervals for different lakes would be similar. Two or three transects were established for each lake depending on limnological diversity. In each transect three stations were randomly selected at initial depths of about 50, 100 and 200 cm. The same stations were sampled in summer even though water depths may have changed. At each station triplicate bottom samples were taken with a 225 cm^2 gape, 15 cm high Ekman dredge. A sample of sediments was collected at each station in spring only. Particle size was categorized using the Wentworth Classification (Welch, 1948). Loss on ignition (roughly equivalent to organic content) was determined after 30 minutes ignition at 550 °C. Benthic vegetation was identified (except for algae) and plant ground cover estimated for each station during both seasons. The pH was measured *in situ* while surface water samples were returned to the laboratory for salinity estimates (Total Dissolved Solids, TDS, at 105 °C). Wind mixing precluded vertical differences over the upper 2 m except in meromictic lakes. There a conductivity meter was used to determined vertical changes and representative water samples were collected for salinity estimates. Table 1 lists the lakes, sampling dates and spring and summer salinities and pHs.

Table 1. Selected data for 18 Alberta and Saskatchewan lakes, 1986.*

Lake	No. of stations	Salinity (‰)		pH		Sampling dates	
		Spring	Summer	Spring	Summer		
1 Wakaw	9	2.75	3.42	8.1	8.4	May 6	June 23
2 Humboldt	6	3.2	3.0	8.7	8.8	May 16	July 16
3 Lenore	9	5.2	5.9	8.9	8.6	May 16	July 21
4 Killarney	6	5.6	5.7	9.6	9.4	June 2	July 28
5 Rabbit	6	8.5	8.7	8.7	8.7	May 30	August 12
6 Tramping	6	13.0	13.0	9.2	9.2	June 3	August 15
7 Arthur +	6	8.9	24.7	9.0	8.8	May 8	June 23
8 Sayer +	6	16.8	36.8	8.6	8.8	May 27	August 8
9 Redberry	9	22.2	23.8	8.8	8.7	May 30	August 12
10 Basin	6	22.4	26.0	8.9	8.9	June 7	August 4
11 Deadmoose +	6	24.1	28.3	8.7	9.0	May 27	August 4
12 Manito	6	27.1	27.8	9.3	9.3	June 2	July 28
13 Gooseberry	6	37.7	44.9	9.0	9.4	June 3	July 29
14 Big Quill	6	49.1	53.1	8.9	8.9	May 26	August 8
15 Marie +	6	9.8–91.8	58.4	8.8	8.8	May 8	June 23
16 Reflex (Salt)	6	61.6	83.3	9.6	9.2	June 2	July 28
17 Little Manitou	6	103.8	117.9	8.8	8.7	May 19	July 15
18 Aroma	6	111.3	126.3	9.0	9.1	June 17	August 15

* Data on lake morphometry and chemistry in Hammer (1978), Hammer and Haynes (1978), Hammer *et al.* (1983) and Timms *et al.* (1986).
+ These lakes are meromictic and data refer to mixolimnia except for Marie Lake on May 8 when the mean salinity was 63.3‰.

Bottom samples from each station were combined and screened in the laboratory. The minimum sieve pore size was 0.125 mm. The fauna were sorted into taxa, wet weighed, preserved in 70% ethanol and subsequently dried at 60 °C to constant weight. Mollusc shells were dissolved in HCl, caddis cases were removed and all species blotted of excess liquid before wet weighing. Mean numbers and wet and dry biomasses were determined for each taxon at each station. Seasonal and annual means were computed for each transect and for each lake.

The distribution and abundance of species was classified with a polythetic divisive classification procedure, TWINSPAN (Hill, 1979; Gauch, 1982). Pseudospecies cut levels of 0.0001, 0.01, 0.10, 1.00 and 10.00 produced a log transformation of the data. Sample and species classifications are produced simultaneously. The TWINSPAN analysis was based on species dry weight biomass at 117 stations during spring and summer.

Results and discussion

Environmental parameters

The lakes ranged in salinity from 2.75‰ in Wakaw Lake in spring to 126‰ in Aroma lake during summer sampling (Table 1). Lakes tended to increase in salinity from spring to summer particularly the most saline where increases ranged from 8–34‰. There was a little change in low salinity lakes and in others where local rainstorms maintained the status quo. Fresher waters from snow and ice melt superimposed over more saline lower layers were evident in early spring in meromictic Arthur, Marie and Sayer lakes but only in Marie Lake was there a distinct salinity gradient in the upper 2 metres.

Lake salinities tended to be higher in more saline lakes in 1986 as compared to 1985 (Timms *et al.*, 1986), an effect of 1985 summer – fall evaporation and low spring snowmelt runoff in 1986. Increases in spring salinities (Table 1) varied from 0.7‰ to 1.8‰ in Manito and Killarney lakes to 8‰ in Tramping Lake between 10 and 18‰ in Big Quill, Gooseberry, Reflex, Little Manitou and Rabbit lakes to 33‰ in Aroma Lake and 59‰ in Redberry Lake. (The variation in Redberry Lake seems to be inordinately high as this is a relatively deep lake). Local weather variations undoubtedly play geographic roles. Usually salinities increase over the summer and autumn so that surviving species must be tolerant of these higher salinities. In addition, an ice cover about 1 metre thick results in freezing out of salts increasing the salinity below the ice so that salinity is maximal during late winter (February, March). The effect is particularly magnified in shallow waters.

The pH values for surface waters changed only slightly if at all in these lakes. It was assumed from prior experience that there were no vertical pH differences over the upper 2 metres.

Sediment size range in most lakes varied over the whole range used (Table 2). Only in Tramping and Little Manitou lakes was the sediment consistently coarser (range > 0.25 mm, mean > 0.5 mm). Only finer sediments (< 0.5 mm) occurred in the transects examined in Deadmoose, Rabbit, Manito, Gooseberry and Reflex lakes.

Sediments tended to be finer in the pelagic zone (Timms et al., 1986) than in the littoral zone but five lakes also had very fine littoral sediments (Table 2). Sediments tend to grade from coarse at the margins to fine in the depths so the contrast was expected.

The organic content of the littoral sediments (Table 2) ranged from 0.01 to 29.5% at the 117 stations sampled. It was relatively low with means less than 5% in 6 lakes (Killarney, Rabbit, Tramping, Redberry, Manito, Big Quill) and ranges among stations in these lakes were also low. Five lakes (Wakaw, Arthur, Gooseberry, Marie, Aroma) exceeded 10% organic matter on average and the ranges for these lakes were broad.

Table 2. Sediment characteristics and vegetation at sampling stations in saline lakes of Saskatchewan and Alberta during spring and summer of 1986.

Lake	Sediment type[a]		% Organic		Vegetation type prevalence[b]
	Range	Mean	Range	Mean	(Number of stations in brackets)
Wakaw	1–5	3.3	6.6–19.2	11.6	0 (2); Chara (7); Myrio (1); Pota (1)
Humboldt	1–5	3.3	0.01–14.2	7.3	0 (1); algae (3); Myrio (1); Pota (4)
Lenore	1–5	3.8	1.5–19.7	8.3	0 (5); algae (7); Pota (5)
Killarney	1–5	3.5	0.7–2.2	1.3	0 (2); algae (3); Pota (2); Ruppia (1)
Rabbit	3–5	4.0	1.3–6.7	3.0	0 (2); Pota (3); Scirpus (2)
Tramping	1–3	1.5	0.05–3.3	1.9	algae (5); Pota (6); Ruppia (6); Scirpus (1)
Arthur	1–5	2.8	2.5–24.1	11.1	0 (1); algae (2); Pota (4); Ruppia (1)
Redberry	1–5	3.3	1.5–20.6	4.5	0 (4); Pota (5); Ruppia (1)
Basin	1–5	4.3	3.7–10.2	5.9	0 (1); algae (2); Pota (3); Ruppia (2)
Deadmoose	3–5	4.0	3.4–9.7	6.0	0 (1); algae (4); Ruppia (4)
Sayer	5	5.0	1.7–26.2	8.7	algae (6); Pota (1); Ruppia (6)
Manito	3–5	4.0	0.8–6.0	2.4	0 (1); algae (5); Ruppia (2)
Gooseberry	3–5	4.2	1.1–43.0	17.3	algae (2); Pota (1); Ruppia (6)
Big Quill	1–5	3.3	1.0–6.2	3.0	0 (4); algae (2)
Marie	1–5	3.0	6.6–27.7	13.9	algae (5); Ruppia (3)
Reflex	3–5	3.7	1.3–19.6	6.7	0 (1); algae (5)
Little Manitou	1–3	1.3	3.9–11.7	7.4	0 (4); algae (2)
Aroma	1–5	4.0	4.0–29.5	16.8	algae (6)

[a] Wentworth sediment classification: 1 > 1 mm; 2 0.5–1 mm; 3 0.25–0.5 mm; 4 0.125–0.25; 5 < 0.125 mm.

[b] 0-absence of vegetation; *Chara*- mix of *C. canescens* Desv. & *C. globularis* Thuill.; algae- one to several species (see Hammer et al., 1983); Myrio-*Myriophyllum exalbescens* Fernald; Pota-*Potamogeton pectinatus* L.; *Ruppia occidentalis* S. Wats.; *Scirpus americanus* Pers. (more than one vegetation type may be present at a station).

Organic matter was always higher in pelagic as compared to littoral sediments. These varied from slightly greater than 1 (Gooseberry, Little Manitou) to 8 to 10 times in Big Quill, Tramping, Killarney and Manito lakes. Presumably organic matter from the littoral is swept by wave action into the pelagic zone which also receives planktonic organic matter. Only littoral substrates had vegetation since Timms *et al.* (1986) deliberately avoided substrates with vegetation.

Only a few lakes had no stations without plants in either season (Tramping, Sayer, Gooseberry, Marie) and in no lakes were plants absent (Table 2). In Lenore, Little Manitou and Big Quill more than half the stations had no vegetation. *Chara* was only present in Wakaw Lake and was dense at several stations particularly along the first transect. Filamentous green algae were prominent and included *Cladophora glomerata*, *Rhizoclonium hieroglyphicum*, *Ctenocladus circinnatus* and *Enteromorpha* spp. (see Hammer *et al.*, 1983) but were not specifically identified in this study. They occurred over the entire range of lakes. Vascular macrophytes occurred into the hypersaline range but were most abundant in Lenore, Tramping, Basin and Sayer lakes. In some lakes the plant cover was not apparently much different seasonally but in Gooseberry, Arthur, Humboldt, Lenore, Marie, Rabbit, Tramping and Aroma lakes summer cover was obviously denser.

Species richness and distribution

Seventy-six different taxa or 'species' of benthic animals were identified in littoral dredge samples taken from 18 saline lakes (Table 3). Hyposaline lakes species numbers varied from 12 to 31 species per lake. In the least saline lakes (up to 6‰; Wakaw, Humboldt, Lenore) the number of species were very similar, i.e., 29 to 31. Although Killarney Lake was similar to Lake Lenore in salinity it only had 17 species and far lower densities of organisms. At higher salinities only Tramping Lake (13‰) had more species (18). Mesosaline lakes (20–50‰) species number

ranged from 5 (Manitou, Deadmoose) to 11 in Redberry and Sayer lakes. Hypersaline lakes (> 50‰) only had 2–5 species per lake with but 2 species per lake above 100‰. However, the Dolichopodidae are represented by two adult species (Little Manitou Lake, J. L. Hurley, pers. comm.) but these are not determinable at the larval stage. (Adults of *Ephydra hians* are abundant at Saskatchewan lakes of much higher salinities but no attempts have been made to collect larvae from their quicksand-like sediments). The number of species is graphed against the mean salinity of each lake in Fig. 1.

The dipteran family Chironomidae is represented by the most species with 23 different taxa recognized. A few species are very important numerically. *Chironomus* cf. *plumosus* (up to 2584 m^{-2} 100 cm deep in Wakaw in spring) and *Glyptotendipes* sp. I (up to 7572 m^{-2} in Humboldt Lake in summer) were very abundant and more or mess restricted to salinities below 6‰. *Criptopus ornatus* (up to 11659 m^{-2} in Marie Lake) appeared in most lakes (10) and occurred only above 10‰ but was very abundant between 25 and 75‰ salinity. *Tanypus nubifer* was only very abundant in Manito Lake (up to 12645 m^{-2} at 100 cm depth) but occurs in other highly alkaline lakes down to 13‰. Most other chironomid species are not abundant and probably play less important roles in the benthos. The ceratopogonid *Bezzia magnisetula* was present in 10 lakes over a salinity range of 3 to 45‰. Numbers were usually low but they reached 720 m^{-2} at 100 cm depth in Killarney Lake. Dolichopodid larvae were most abundant (up to 459 m^{-2}) in Reflex Lake at a salinity of 83‰ while *Ephydra hians* attained a maximum concentration of 387 m^{-2} when the salinity was 118‰ in Little Manitou Lake.

Amphipods are one of the more obvious faunal groups present. *Hyallela azteca* was abundant in all lakes up to about 25‰ salt and also in Sayer Lake during salinities of about 37‰. A maximum density of 24452 m^{-2} was reached in Sayer Lake then but in Humboldt Lake up to 57245 m^{-2} were present in summer. Both maxima occurred at 50 cm deep stations. *Gammarus lacustris* was

Table 3. Distribution and relative abundances of benthic invertebrates in 18 saline lakes in Alberta and Saskatchewan, 1986. *(mean number m⁻²: + = 1–9; ++ = 10–99; +++ = 100–999; ++++ = 1000–9999).

Species	Lake (abbr)	Waka	Humb	Leno	Kill	Rabb	Tram	Arth	Redb	Basi	Dead	Mani	Saye	Goos	BigQ	Mari	Refl	LMan	Aroma
ANNELIDA																			
Hirudinea																			
1.	*Erpobdella punctata* (Leidy)	+	+																
2.	*Glossiphonia complanata* (L.)	++	+																
3.	*Helobdella elongata* (Castle)		++																
4.	*Helobdella stagnalis* (L.)	+	++																
Oligochaeta																			
5.	*Limnodrilus hoffmeisteri* Claparede		+																
6.	*Limnodrilus profundicola* (Verrill)	++	+++																
CRUSTACEA																			
Amphipoda																			
7.	*Gammarus lacustris* Sars	+	+++	++		++++			+				++++						
8.	*Hyalella azteca* (Saussure)	+++	++++	+++++	+++	+++	+++++	+++++	+++	+++									
Ostracoda																			
9.	*Candona rawsoni* Tressler						+++												
10.	*Cyprinotus glaucus* Furtos	+++	+++	+	+++++	+++	+++		+++	+									
11.	*Megalocypris ingens* Delorme															++++			
ACARI																			
12.	*Arrenurus interpositus* Koenke	+																	
13.	*Eylais* sp.			+															
14.	*Hydrodroma despiciens* Muller	++	+	++	+		+												
15.	*Limnesia* sp.	++	++																
16.	*Piona carnea* (Koch)			++															
17.	*Piona* sp., *rotunda* group	++																	
18.	*Tiphys* sp.			++															
INSECTA																			
Ephemeroptera																			
19.	*Caenis simulans* McDunnough	++	+																
20.	*Callibaetis* sp.		+	+															
Coleoptera																			
21.	*Cercyon marinus* Thoms.												++						
22.	*Enochrous* sp.												+						
23.	*Haliplus strigatus* Roberts	++	++	+									+						
24.	*Haliplus immaculicollis* Harris		+																
25.	*Hygrotus punctilineatus* Fall		+					++											
26.	*Hygrotus salinarius* Wallis										+	++		++	+	++			
27.	*Ilybius* sp. (nr. *fraterculus* LeConte)		+																
28.	*Laccophilus maculosus* Say		++																
29.	*Potamonectes spenceri* Leech				++		+												
Diptera																			
Ceratopogonidae																			
30.	*Bezzia* nr. *magnisetula* Dow & Turner		+	++	+++	+++	++	++	++	++	++++	++	++	+					
Chironomidae																			
31.	*Chironomus* nr. *annularius* Kehl & Kehl	+++	+	++	+++	+++	++	+++	++	+++	++++	++	++	+					
32.	*Chironomus* cf. *plumosus* Meigen	+++	+++	++															
33.	*Cladotanytarsus* sp.								+	+									
34.	*Cricotopus ornatus* (Meigen)						++	++	++	++	+++	++++	++	++++	+++	++++	+		

	C1	C2	C3	C4	C5	C6	C7	C8	C9	C10	C11	C12	C13	C14	C15	C16	C17	C18	C19	C20
35. *Cryptochironomus* sp. a	+	++	++			+	+	+	++	+										
36. *Cryptochironomus* sp. b		++	+			+	+	+	++	+										
37. *Cryptotendipes* sp.										+++		++								
38. *Dicrotendipes* cf. *modestus* (Say)		+																		
39. *Einfeldia pagana* (Meigen)			++	+																
40. *Endochironomus nigricans* (Johannsen)	++	++++	++	+																
41. *Glyptotendipes* sp. I	++++	+++++	++		+															
42. *Glyptotendipes* sp. III		+++	++																	
43. Orthocladiinae (*Zalutschia* or *Chaetocladius*)			+																	
44. *Paratanytarsus* sp.	++					+	++		++											
45. *Polypedilum nubeculosum* Maschitz.	+		+	+	+++	+			+											
46. *Polypedilum* sp. IV	+																			
47. *Procladius freemani* Sublette	+	+	++	+		++	++	+		+++										
48. *Psectrocladius* sp.	+																			
49. *Pseudochironomus* sp.	+																			
50. *Tanypus nubifer* Coquillett	+			++		++	++		++	++++										
51. *Tanypus punctipennis* Meigen	+																			
52. *Tanytarsini* sp.	++			+																
53. *Zavelimyia* sp.																				
Culicidae																				
54. *Chaoborus* sp.					+															
Dolichopodae																				
55. *Hydatostega plumbea* (Aldrich)											+			+++						
Hydrophorus extrarius Aldrich																				
Ephydridae																				
57. *Ephydra hians* Say											+		++	++		+++				
Tabanidae																				
58. *Chrysops discalis* Williston				+							+									
Hemiptera																				
Corixidae																				
59. *Cenocorixa bifida* (Hungerford)	++				+	++		+	+	+					+					
60. *Cenocorixa dakotensis* (Hungerford)	++				+	++	++	++	+											
61. *Cenocorixa expleta* (Uhler)		++					++													
62. *Trichocorixa borealis* Sailer	++	++						+			+	+++	+++	++	++					
63. *Trichocorixa verticalis interiores* Sailer				+			++	+	++	+	+++	+++	+++	+++	++					
Notonectidae																				
64. *Notonecta kirbyi* Hungerford	+																			
Odonata																				
65. *Enallagma clausum* Morse	+			+		+	++	++	+		+	+++	+	++	+					
Trichoptera																				
66. *Agrypnia* sp. 1	++		++			++		+	++											
67. *Agrypnia* sp. 2	+		+			+														
68. *Limnephilus* nr. *labus* Ross	+	+																		
69. *Oecetis ochracea* (Curtis)			−+	+	+	+			+											
70. *Molanna flavicornis* Banks	+		−	+																
71. *Mystacides* sp.	+																			
72. *Triaenodes* sp. 1			+	+		+	+													
73. *Triaenoides* sp. 2			+	+			+													
MOLLUSCA																				
74. *Gyraulus deflectus* (Say)	+	++																		
75. *Limnaea stagnalis* L.	++	++	+																	
76. *Physa gyrina* Say	++	++																		
Total number of species	30	29	31	17	12	18	11	11	9	5	5	5	3	7	11	5	5	3	2	2

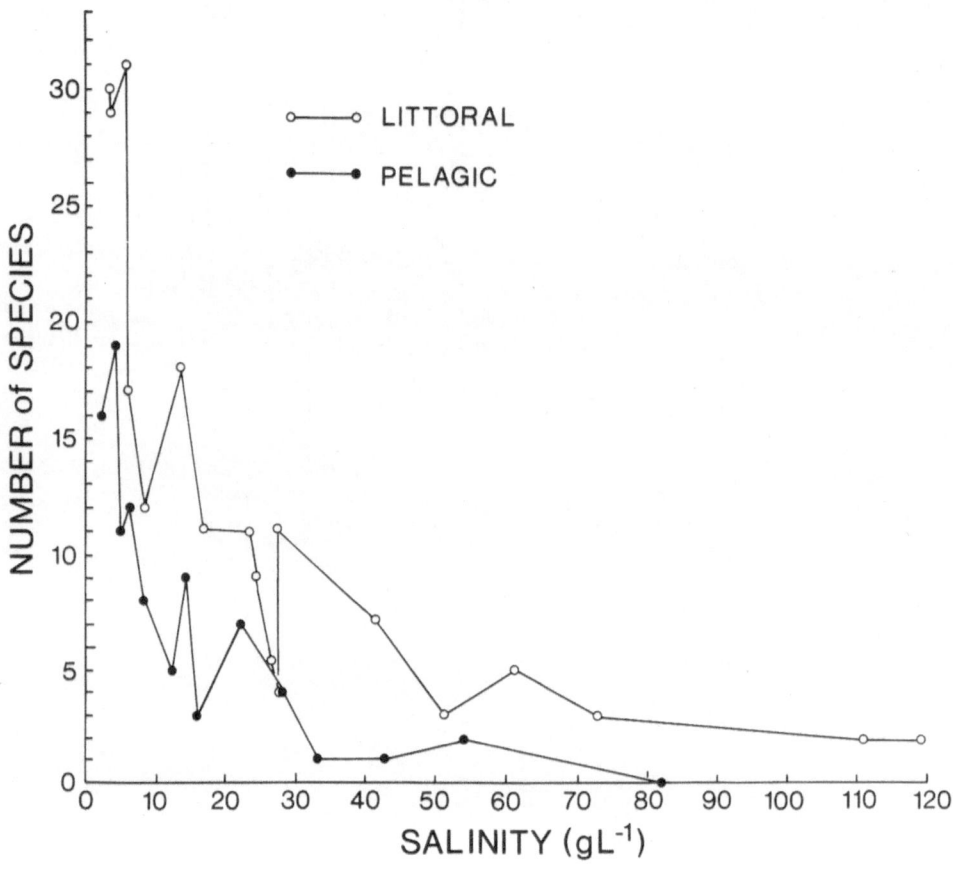

Fig. 1. A comparison of species richness in each lake according to its salinity in the littoral and pelagic (from Timms *et al.*, 1986) zones of saline lakes in Alberta and Saskatchewan.

much less abundant and reached highest concentrations in Humboldt (8088 m^{-2}) and in Rabbit (2612 m^{-2}) lakes but was not found above 24‰ salinity. It also was not present in high pH lakes such as Killarney and Tramping.

Only three species of ostracods were present. *Cyprinotus glaucus* (up to 3702 m^{-2} in Killarney Lake at 2 m depth in summer) was abundant (but not in all lakes) up to the salinity of Redberry Lake. *Megalocypris ingens* was very abundant (up to 2310 m^{-2} in summer) but occurred only in Marie Lake even though it has been abundant in Deadmoose Lake in previous years. It appears to prefer shallower waters less than 100 cm deep.

The corixid *Trichocorixa verticalis interiores* occurred over a salinity range of 3 to about 60‰ but

was most abundant in lakes of 28–60‰ salinity. Other corixids were limited by about 28‰ salt. The damselfly *Enallagma clausum* occurred over a spectrum of 5–37‰ salinity and was most abundant at the highest salinity, i.e., Sayer Lake.

Nine species of beetles were collected and most species were restricted to salinities of 13‰ or lower. The Hydrophilid *Cercyon marinus* (up to 517 larvae m^{-2}) was found only in Sayer Lake at salinities up to 37‰. *Haliplus strigatus* (present from 3–6‰) and *Enochrus* sp. were also present in this lake when the salinity was 16.8‰. *Hygrotus salinarius* occurred in salinities of 28‰ (Deadmoose) to 58‰ (Marie) with maximum numbers reaching 86 m^{-2} in Marie Lake. Numbers of all species were usually low.

Leeches and oligochaetes were only found in the two lakes with the lowest salinities, ca. 3‰. The water mites (Acari) were restricted to waters of 13‰ or less. Most of the caddisfly larvae were also found in this range but *Agrypnia* sp. 1 was found in Arthur Lake and Basin Lake at 22.4‰ while *Oecetis ochracea* occurred in Redberry Lake at 22.2‰. Gastropods were only found in Humboldt and Lenore lakes. All the species in these groups were usually present in low numbers but their relatively large sizes contributed considerably to biomass when they were present.

It should be noted that some aquatic invertebrates, e.g., corixids, amphipods, beetles, water mites, tend to be nektonic as well as benthic, and therefore Ekman samples underestimate their abundance.

Numerically, the highest concentration of fauna (74138 organisms m^{-2}) occurred in Humboldt Lake on July 16 at a depth of 50 cm. It involved 14 species but was dominated by amphipods and *Gylptotendipes* I.

The littoral zones were occupied by 76 different taxa (Table 3) compared to only 42 species in pelagic substrates of saline lakes (including Humboldt and Wakaw lakes) (Timms *et al.*, 1986). Presumably the more diverse habitats of the littoral zones are partly responsible since 16 of 18 lakes are the same in the two studies. The added presence of Deadmoose and Sayer lakes resulted in the addition of 2 hydrophilid species which occurred only in Sayer Lake. Thirty of the 42 pelagic species were also present in our littoral samples. (It is presumed that *Chaoborus*, *Haliplus*, *Hygrotus*, *Bezzia*, *C. plumosus* and *Limnephilus* are the same species reported by Timms *et al.*, 1986). The two abundant littoral species of ostracods (*Cyprinotus glaucus*, *Candona rawsoni*) were not mentioned in the pelagic samples although they were present. The sieves used then did not adequately retain them so they were not considered in the pelagic studies (B. V. Timms, pers. comm.). The abundance of corixids in the littoral and their complete absence from the pelagic zone is not surprising. The amphipods *Hyallela azteca* and *Gammarus lacustris* and the damselfly *Enallagma* were much more abundant in littoral habitats

while the chironomid species in common tended to be more abundant in pelagic substrates.

Only Redberry Lake had the same number of species (11) present in the littoral and pelagic (Timms *et al.*, 1986) zones but only 5 species were common to both zones. Figure 1 illustrates that there is a marked increase of species in the same lakes (16 to 18) as one proceeds from the pelagic to the littoral zones. We found about twice as many species in the littoral of Humboldt, Lenore, Arthur and Little Manitou lakes, three times as many in Tramping and Big Quill lakes and seven times as many in Gooseberry lake. In Aroma Lake we found two species while none were found in the pelagic zone. In other lakes there were 25–50% more species in the littoral zone. One of the major reasons for this difference in species numbers is the lack of summer sampling of the pelagic zone. A number of species were added when Timms *et al.* (1986) also sampled some lakes in summer.

When one compares littoral species from spring and summer sampling, in all lakes except two (Wakaw, Sayer) more additional species appear in the summer than disappeared from spring samples. The additional species varied from one (Big Quill, Reflex, Aroma) to 11 in Humboldt Lake and 14 in Lenore Lake. In lower salinity waters (6–17‰) 4–6 additional species were present during summer while in higher salinities the usual range was 1–2 additional species. New species appearing in the summer tend to be corixids, gastropods, leeches, ostracod and water mites. *Ephydra hians* larvae are only prominent in the summer. Species which tend to disappear by summer are mainly chironomid species which have emerged.

Tones (1976) investigated the taxonomy and ecology of the littoral fauna of six saline lakes 6–7 times each during 1972. More species were found than in our study (maximum salinity): Little Manitou 7 species (85‰); Big Quill 5 species (55‰); Deadmoose 24 species (24‰); Waldsea 30 species (22‰); Basin 14 species (18.5‰); Humboldt 35 species (3.5‰). The species we found were also present then but many more beetle species were present. Timms & Hammer

(1988) also found many more beetle species present when sampling with nets. No biomass measurements were made. In Deadmoose and Waldsea lakes, sites with high plant cover had many more species and higher numbers compared to sites with low or no plant cover. Tones used a much larger sampling unit and this undoubtedly made possible more species caught including some which are more nektonic.

Rawson & Moore (1944) found *Gammarus lacustris* abundant in Redberry Lake (14.2‰ TDS) but the 1985 and 1986 studies showed only small numbers in the littoral. Rawson & Moore (*loc. cit.*) also found various gastropod species in Stoney Lake (Humboldt) at 7.87‰ TDS and *Stagnicola palustris* in the littoral of Redberry Lake. They attributed the higher tolerance there to wave action. None were found in our studies in lakes exceeding 5.9‰ TDS. Although Rawson & Moore (*loc. cit.*) reported leeches (*Hemiclepsis occidentalis*) in the littoral of Stoney and Redberry lakes, we found other leech species restricted to salinities less than 4‰ (Wakaw, Humboldt). Tubificids were also restricted to these low salinity lakes in 1985 and 1986 although some were found in Redberry Lake in 1941. Water mites were not found in waters exceeding 3‰ TDS by Rawson & Moore (*loc. cit.*). We found *Hydrodroma despiciens* in Tramping Lake at 13‰ but most mite species were restricted to salinities below 6‰. Rawson & Moore reported the presence of caddisfly larvae common in salinities up to $14.2 \, \text{g} \, \text{l}^{-1}$ but do not specify the species found there. Beetles such as the dytiscid *Coelambus salinarius* (= *Hygrotus salinarius*?) were confined to saline lakes (Redberry, Big and Little Quill) while the hydrophilid *Enochrus diffusus* was found from freshwater to Little Manitou Lake. *Enochrus* was only found on one occasion in our survey. The corixid *Trichocorixa verticalis* was only found in 'more saline lakes'. Ephydrid larvae were the chief benthic organisms in Little Manitou Lake $(118 \, \text{g} \, \text{l}^{-1})$ while dolichopodids made up the other 20% of the fauna.

Hammer (1986) discusses the littoral fauna of world saline lakes in great detail. Much information on taxonomy and ecology is available but biomass data is virtually absent. *Ephydra hians* occurs in Mono Lake, California, and has the ability to withstand high salinities and high alkalinities (Winkler, 1977). Presumably this latter species occupies still higher salinity lakes on the Canadian prairies as well.

Standing crop of littoral benthos

The spring mean dry weight biomass ranged from 0.0001 in Aroma Lake to $6.071 \, \text{g} \, \text{m}^{-2}$ in Tramping Lake while the summer values ranged from 0.0424 in Aroma Lake to $16.41 \, \text{g} \, \text{m}^{-2}$ (maximum of $63 \, \text{g} \, \text{m}^{-2}$ at 50 cm) in Humboldt Lake (Table 4). Except for Killarney, Tramping, Redberry, Deadmoose and Manito lakes summer biomass exceeded spring biomass (Fig. 2). Combined values give an annual range from 10.91 in hyposaline Humboldt Lake to $0.0212 \, \text{g} \, \text{m}^{-2}$ dry weight biomass in hypersaline Aroma Lake. Variations in dry weight biomass are compared to lake salinities in Fig. 3. A large increase is notable around 15‰ salinity. Other hyposaline lakes (except Killarney Lake) had biomass dry weights an order of magnitude higher than mesosaline and hypersaline lakes with three exceptions. In the mesosaline range Manito and Sayer lakes had standing crops similar to most hyposaline lakes. The most saline Aroma Lake is the only lake where the standing crop was three orders of magnitude lower than Humboldt Lake. Mean annual wet weight ranged from $0.1209 \, \text{g} \, \text{m}^{-2}$ in Aroma Lake to $83.09 \, \text{g} \, \text{m}^{-2}$ in Humboldt Lake (Table 4). Dry weight as a percentage of wet weight was usually close to 16% (Table 4) with a range of 11.0 to 23.2% and a mean of 15.8% ± 0.6 S.E.

Table 5 provides information on the relative proportions of the dry weight biomass contributed by various taxonomic groups. The Crustacea (amphipods) contributed the major portion in the less saline lakes (below 27‰) except for Wakaw, Lenore and Deadmoose. At higher salinities only Marie Lake has a crustacean, the ostracod *Megalocypris*, as a major contributor. Chironomids contributed most to the littoral biomass in only

Table 4. Dry weight biomass (per transect – average of 3 stations, lake seasonal and annual means) and wet weight biomass (annual mean) in Saskatchewan and Alberta saline lakes during spring and summer of 1986. Percentage mean dry weight of mean wet weight is given.

Lake	Transect	Dry weight (g m^{-2})		Transect mean	Lake mean	Lake mean wet weight (g m^{-2})	Dry weight % of wet weight
		Spring	Summer				
Wakaw	1	4.157	6.659	5.408			
	2	1.075	0.9521	1.014			
	3	0.2770	2.171	1.224			
	mean	1.836	3.261		2.549	23.22	11.0
Humboldt	1	4.010	2.650	3.330			
	2	6.790	30.18	18.49			
	mean	5.400	16.41		10.91	83.09	13.1
Lenore	1	0.6774	3.689	2.183			
	2	1.140	2.332	1.736			
	3	1.126	2.924	2.025			
	mean	0.9811	2.982		1.982	12.53	15.8
Killarney	1	0.3985	0.2195	0.3090			
	2	0.9812	0.5137	0.7475			
	mean	0.6899	0.3666		0.5283	3.599	14.7
Rabbit	1	1.211	3.654	2.433			
	2	0.3197	1.811	1.065			
	mean	0.7654	2.724		1.745	11.71	14.9
Tramping	1	3.123	1.787	2.455			
	2	9.020	3.862	6.441			
	mean	6.072	2.825		4.448	27.04	16.4
Arthur	1	2.567	2.743	2.655			
	2	2.544	10.23	6.387			
	mean	2.556	6.486		4.521	30.13	15.0
Redberry	1	0.6097	0.0969	0.3533			
	2	0.0892	0.5227	0.3060			
	3	0.2912	0.2115	0.2513			
	mean	0.3300	0.2770		0.3035	1.809	16.8
Basin	1	0.0785	0.2448	0.1616			
	2	0.3097	0.2210	0.2654			
	mean	0.1941	0.2329		0.2135	1.143	18.7
Deadmoose	1	0.2109	0.0368	0.1160			
	2	0.1294	0.0177	0.2144			
	mean	0.1702	0.0273		0.0987	0.6762	14.5
Manito	1	0.1878	0.3791	0.2835			
	2	2.934	0.9441	1.939			
	mean	1.561	0.6616		1.111	9.151	12.1
Sayer	1	0.2601	2.176	1.218			
	2	0.3425	1.960	1.151			
	mean	0.3013	2.068		1.185	7.725	15.3
Gooseberry	1	0.1997	0.5381	0.3689			
	2	1.052	0.8534	0.9527			
	mean	0.6259	0.6958		0.6608	4.832	13.7
Big Quill	1	0.0543	0.5732	0.3138			
	2	0.0037	0.6280	0.3159			
	mean	0.0290	0.6006		0.3148	1.859	16.9
Marie	1	0.0814	0.8339	0.4577			
	2	0.4809	0.1845	0.5827			
	mean	0.2812	0.7592		0.5202	3.068	17.0
Reflex	1	0.2243	0.3836	0.3040			
	2	0.0428	0.4593	0.2511			
	mean	0.1336	0.4215		0.2776	1.194	23.2
Little Manitou	1	0.0626	0.6524	0.3575			
	2	0.1199	0.4535	0.2867			
	mean	0.0913	0.5530		0.3221	1.758	18.3
Aroma	1	0.0000	0.0507	0.0254			
	2	0.0001	0.0340	0.0171			
	mean	0.0001	0.0424		0.0212	0.1209	17.4

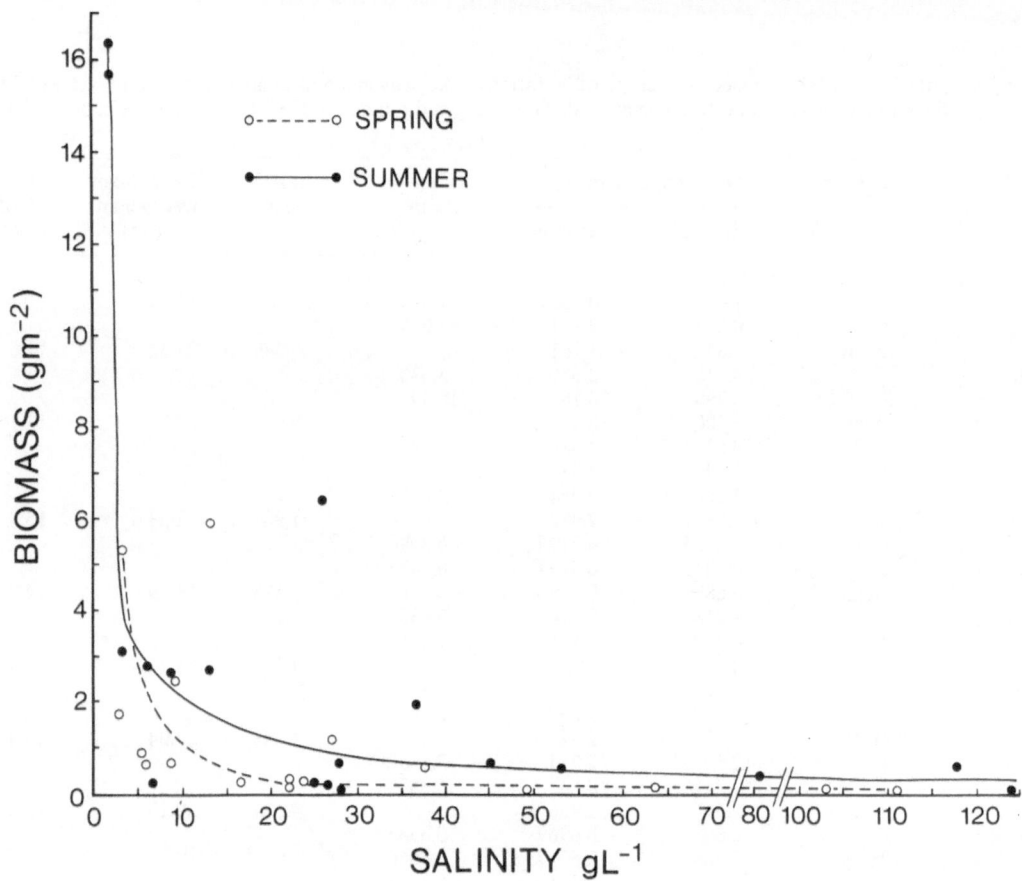

Fig. 2. Seasonal variations in littoral dry weight biomass as related to lake salinity in Saskatchewan and Alberta saline lakes, 1986.

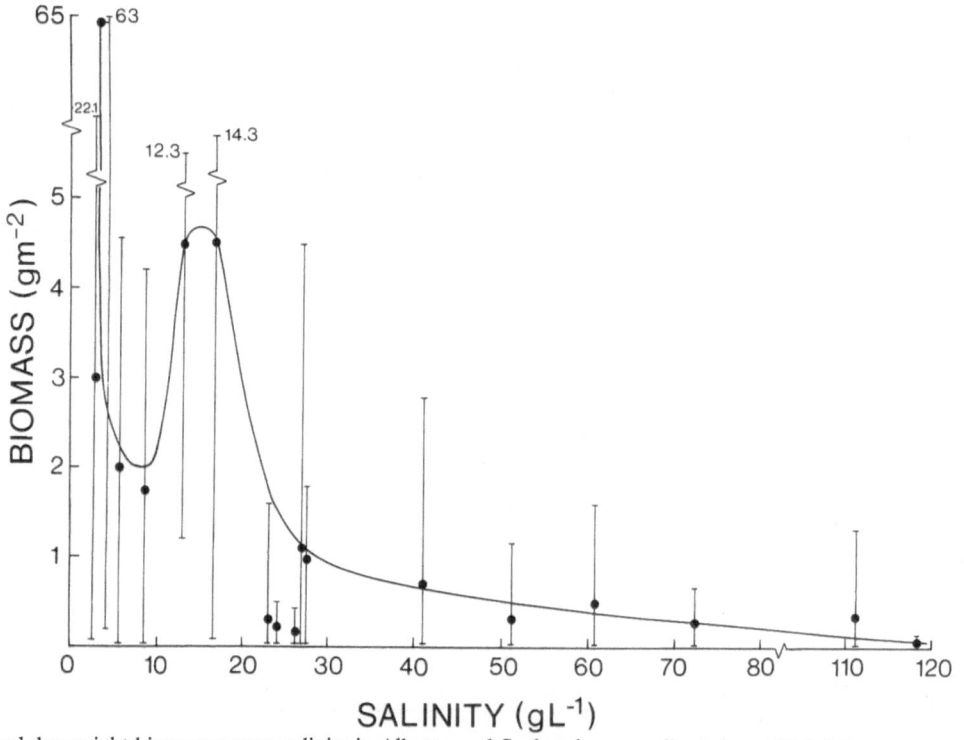

Fig. 3. Littoral dry weight biomass versus salinity in Alberta and Saskatchewan saline lakes, 1986. (Means are represented by points while ranges are represented by bars)

Table 5. Percentage contribution to littoral faunal dry weight biomass by major taxonomic groups in 18 Saskatchewan and Alberta saline lakes.

Lake	Salinity % TDS	Taxonomic groups									
		CRUS	ACAR	CERA	CHIR	DOLI	EPHY	ODON	COLE	CORI	TRIC
Wakaw	3.1	18.6	01.	0	61.7	0	0	0.5	2.4	0.7	3.3
Humboldt	3.1	51.0	0.1	0.0	45.7	0	0	0	0.4	0.7	0.1
Lenore	5.55	31.0	1.7	0.1	16.9	0	0	0.3	0.5	2.4	7.0
Killarney	5.65	86.0	0.0	4.9	3.8	0	0	0.8	0.5	0.1	1.2
Rabbit	8.6	95.7	0.0	0.1	2.0	0	0	0	0	0.6	0.1
Tramping	13.1	94.4	0.0	0.0	1.4	0	0	0.0	1.3	2.4	0.3
Arthur	16.8	90.5	0	0.0	6.1	0	0	2.1	0	0.8	0.5
Redberry	23.0	72.1	0	1.4	8.8	0	0	10.1	0	1.2	0.1
Basin	24.2	58.9	0	2.9	13.4	0	0	24.8	0	0	0.4
Deadmoose	26.3	0	0	0	77.7	0	0	6.9	3.4	12.0	0
Sayer	26.8	64.0	0	0.1	1.7	0	0	22.4	4.8	7.0	0
Manito	27.4	0	0	0	91.9	0	0	0.6	1.0	6.4	0
Gooseberry	41.3	0	0	0.1	53.9	2.7	0.6	0	18.5	22.9	0
Big Quill	51.1	0	0	0	6.4	0	0	0	2.7	91.0	0
Marie	60.8	48.0	0	0	44.0	0	0	0	5.2	2.8	0
Reflex	72.5	0	0	0	0.5	79.3	20.2	0	0	0	0
Little Manitou	111	0	0	0	0	14.8	85.1	0	0	0	0
Aroma	119	0	0	0	0	40.2	59.8	0	0	0	0

Taxonomic groups key: CRUS – Crustacea; ACAR – Acari; CERA – Ceratopogonidae; CHIR – Chironomidae; DOLI – Dolichopodidae; EPHY – Ephydridae; ODON – Odonata; COLE – Coleoptera; CORI – Corixidae; TRIC – Trichoptera.

four lakes (Wakaw, Deadmoose, Manito, Gooseberry) but were a close second in Humboldt and Marie lakes. Ephydrids dominated the biomass in two lakes (Little Manitou, Aroma) and were important in Reflex lakes. Dolichopodids were most important in Reflex Lake but were also prominent in the other two highly saline lakes. Big Quill Lake was dominated by the corixid *Trichocorixa verticalis interiores* which is the only corixid which overwinters in the egg stage (Tones & Hammer, 1975). The damselfly nymph *Enallagma* made a major contribution to the faunal dry weights of Redberry, Basin and Sayer lakes. The beetle *Hygrotus salinarius* and corixid *T. v. interiores* were important elements of the biomass of Gooseberry Lake while corixids were also important in Deadmoose Lake. Other major taxonomic groups such as the Hirudinea, Oligochaeta, Ostracoda, Acari, Ephemeroptera, Coleoptera, Trichoptera and Gastropoda usually contributed less than 5% of the biomass in these lakes.

Dry weight biomass tended to be greater in the summer than in the spring except for five lakes in the 5–28‰ salinity range (Table 4, Fig. 2). The major reason for the summer increase is related to the kinds of species which predominate then and the increase in numbers and biomass from spring to summer. *Glyptotendipes* I is a dominant contributor to summer biomass in Wakaw and Humboldt lakes. *Hyallela* and *Gammarus* make up a major portion of the summer biomass in Humboldt and Lenore lakes. *Gammarus* is a major contributor to the Rabbit Lake summer biomass while *Hyallela* assumes that role in Arthur, Basin and Sayer lakes. *Enallagma* plays a major role in the shift to summer maxima in Basin and Sayer lakes. *Trichocorixa* supplies much of the summer biomass in Gooseberry and Big Quill lakes while the dolichopodids and *Ephydra* mature later in the summer in the three most saline lakes. In lakes with higher spring maxima *Hyallela* plays the major role in Killarney. Tramping and Redberry lakes when these lakes were sampled around June 1. In Deadmoose and Manito lakes *Cricotopus ornatus* and *Tanypus nubifer* contribute most of the biomass and their emergence occurs before summer.

Figure 3 illustrates seasonal changes in biomass related to salinity. Although the pattern tends to be a parabolic relationship, a large increase around 15‰ alters the pattern. This shift is related to the ranges as well as the means. Timms *et al.* (1986) found a similar peak at that salinity when the mean dry weight biomass of Redberry lake was 7.08 g m⁻² (Fig. 4) although other lakes with similar salinities had much more lower biomass. Similarly, in Australian lakes an increase occurred but at a lower salinity of about $7 \, g \, l^{-1}$ (Timms, 1983). The overall pattern in Australian saline lakes is also similar to that of Canadian lakes although the maxima are lower. Maximum biomass dry weights in both littoral and pelagic zones in Canadian lakes occurred in Humboldt Lake (10.91 and 9.12 g m⁻², respectively) which is highly fertilized by sewage from the town of Humboldt. The littoral biomass of Redberry Lake is contrast to the pelagic biomass was very low, i.e., 0.30 *vs* 7.08 g m⁻². Tramping and Arthur lakes had much higher littoral biomass, about 4.5 g m⁻². Between 3 and 28‰ salinity there was considerable variability between littoral and pelagic biomass whereas littoral biomass was always greater than pelagic biomass in more saline lakes. Rawson & Moore (1944) found pelagic biomass of 0.53 g m⁻² in Stoney Lake (Humboldt), 0.32 in Redberry, 0.47 in Manito and 0.18 g m⁻² in Little Manitou Lake. Biomass was higher only in Little Manitou Lake compared to present day values.

Percentage biomass composition in the pelagic zones was dominated by the Chironomidae (Timms *et al.*, 1986, Table 4) with the exception of the two most saline lakes where other insect larvae dominated and Arthur Lake where chironomids were similar to crustaceans in biomass. In the littoral, crustaceans dominated the lakes having salinities below 25‰ and Marie Lake (ostracods) while chironomids dominated only Deadmoose, Manito and Gooseberry lakes (Table 5). Other dipteran larvae dominated the pelagic and littoral substrates of the most saline lakes.

Statistical analyses

Spearman coefficients of rank correlation were calculated (based on logarithms of values) to

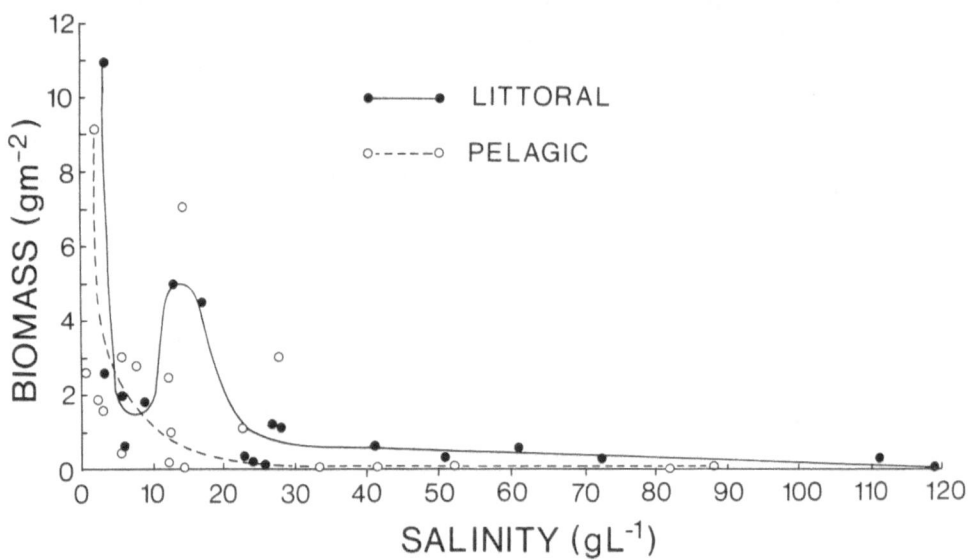

Fig. 4. A comparison of mean dry weight biomass of pelagic (from Timms *et al.*, 1986) benthos (1985) and littoral benthos (1986) versus salinity in saline lakes of Alberta and Saskatchewan.

Table 6. Spearman coefficients of rank correlation (r_s) and significance (p) relating the logarithms of lake salinities ($g\,l^{-1}$), number of species and dry weight biomasses ($g\,m^{-2}$) seasonally distributed (spring, summer). (n = 18; *** $p \leq 0.001$, ** $p \leq 0.01$, * $p \leq 0.05$).

	Spring		Summer	
	Species	Biomass	Species	Biomass
Spring salinity	− 0.907 ***	− 0.791 ***	−	−
Summer salinity	−	−	− 0.935 ***	− 0.529 *
Spring species	−	0.736 ***	0.916 ***	−
Summer species	0.916 ***	−	−	0.657 **

ascertain relationships between spring and summer salinities and the numbers of species and dry weight biomass in each of the seasons in 18 lakes (Table 6). All were significant at the $p \leq 0.001$ level except summer salinity *vs* summer biomass

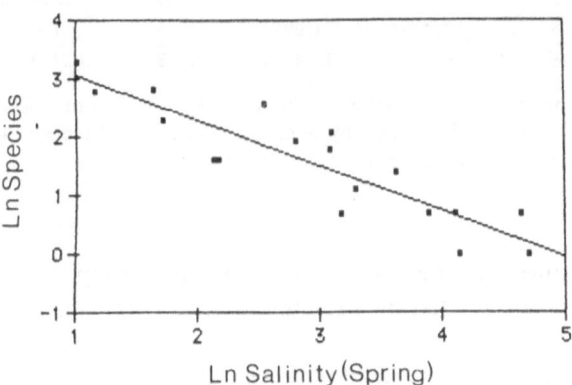

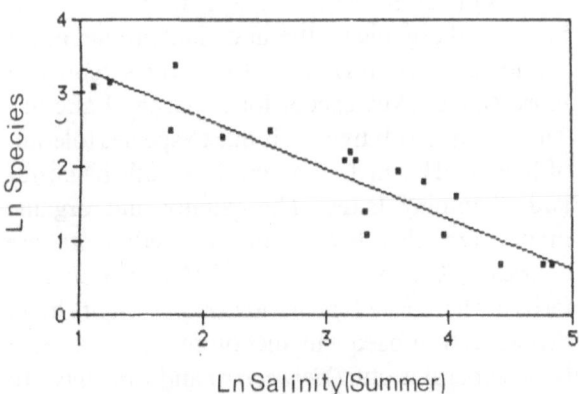

Fig. 5. The relationships between the logarithms of species richness and salinities in 18 Alberta and Saskatchewan saline lakes in 1986.

where the level of significance was only at $p \leq 0.05$. Very high levels of significance were also characteristic of the relationships between number of species and the salinity during the spring or summer season. Figure 5 illustrates these relationships. The decrease in number of species with increasing salinity is obvious and similar in the two seasons. Timms *et al.* (1986) found a similar relationship regarding pelagic species richness and salinity. Numbers of spring and summer species were significantly related when the 18 lakes were compared. The *t* test showed that there is no difference between numbers of species on a seasonal basis at a level of $p \leq 0.05$ ($t = 2.512$ d.f. 17).

Lake classification

TWINSPAN classification sequentially subdivided the original samples, on the basis of indicator species (Fig. 6), into four major groups of lakes which appear to be closely related to salinity (TDS $g\,l^{-1}$). The first dichotomy split off species (Dolichopodidae (55), *Ephydra hians* (57)) characteristic of the three most saline lakes, i.e., Reflex, Little Manitou, Aroma, or range of above 63‰ salinity (lake Group IV). On the left *Hyallela azteca* (8) is an indicator species for a broad range of salinities (Fig. 7). A second division produced lake Group III including lakes (10–15) (Deadmoose to Marie) from about 28 to 63‰. The indicator species for this group are *Hygrotus salinarius* (26) and *Cricotopus ornatus* (34). A third dichotomy produced Groups I and II typifying

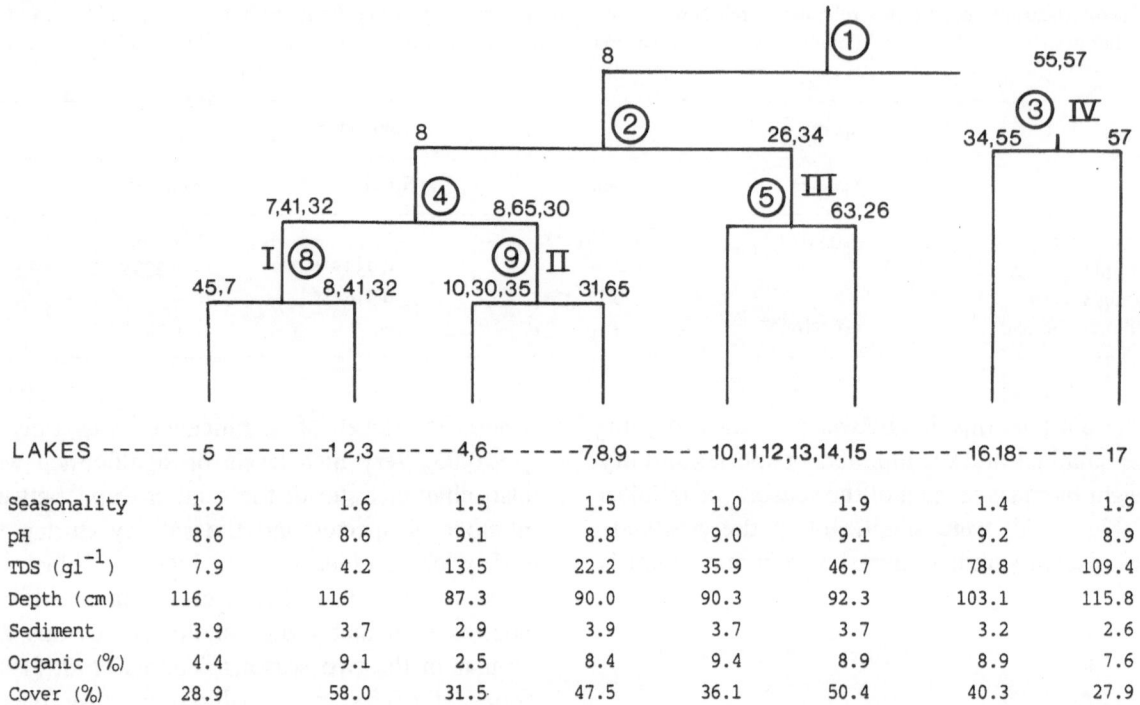

Fig. 6. Classification of faunal species according to TWINSPAN. Successive dichotomies are indicated by encircled numbers while indicator species (Table 3) are shown at each dichotomy. Lakes are designated by numbers (Table 1) and means of environmental factors are given at the base of the dendrogram. Roman numerals suggest the four lake groups.

lakes of lower salinities. The indicator species for Group I lakes are *Gammarus* (7), *Glyptotendipes* I (41) and *Chironomus* cf. *plumosus* (32) while Group II lakes are characterized by an abundance of *Hyallela* (8), *Enallagma* (65) and the ceratopogonid *Bezzia* (30). The Group II indicator species are representative of broad environmental ranges and overlap two lake groups (Fig. 7).

Further subdivisions of each Groups (at different levels) separated the lakes according to more environmental information. Eighteen indicator species were selected by TWINSPAN but we rejected two (*Caenis simulans*, *Candona rawsoni*) since they were of very limited occurrence and abundance. Mean values for seven environmental factors are given for the eight derived subdivisions of lakes (Fig. 6). Seasonality (spring or summer), salinity, organic matter and plant ground cover are the main environmental factors separating the species of Group I into two categories. The three

lower salinity lakes (1, 2, 3) in Group I are characterized by an abundance of *Glyptotendipes* I and *Chironomus* cf. *plumosus* while for Rabbit Lake (5) the indicator species are *Gammarus* and *Polypedilum nubeculosum*. This latter lake is split off from others in Group I since the indicator species occur earlier in the season, the salinity is higher and organic matter and plant ground cover are lower. Group II indicator species represent more saline lakes except for Killarney Lake (4). The left hand subdivision is due to species tolerant of higher pH which characterizes both Killarney and Tramping lakes. The salinity and organic matter are also lower and the sediments are coarser. *Chironomus* nr. *annularius* (31) characterizes the more saline lower pH right hand division. A subsequent dichotomy of Group III lakes separates out *Trichocorixa* and *Cricotopus* to the right on the basis of abundant occurrence in samples, other environmental values are fairly similar. All three indicator species in Group III

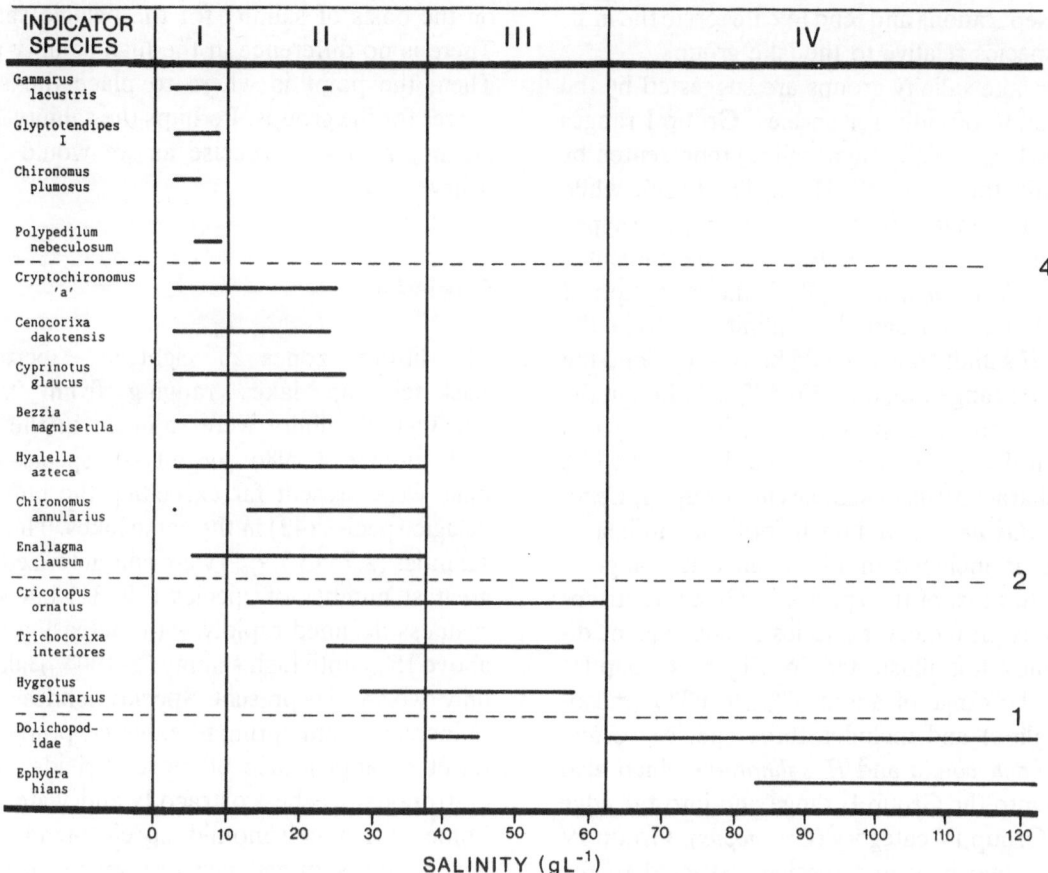

Fig. 7. The salinity ranges of the indicator species and their distribution according to the successive dichotomies generated by TWINSPAN. The four salinity groups are designated in Roman numerals.

also have a broader tolerance and also occur in Group II lakes (Fig. 7). Group IV includes lakes with the two species which are tolerant to high salinity. In a subsequent dichotomy the ephydrids (57) prefer lower pH and occur later in the season in Little Manitou Lake (17) compared to the dolichopodids (55) in Reflex and Aroma lakes (16, 17). *Cricotopus* also appears in this latter group of samples in Reflex Lake (16). Other environmental factors are quite similar in their values.

Timms *et al.* (1986) considered depth in the pelagic zone (2–21 m) to be a discriminating factor in separating species (and salinity groups). By contrast depth variations in the littoral are very limited (up to 200 cm) and means for the 8 subdivisions (Fig. 6) vary little.

Six of 11 indicator species obtained by Timms

et al. (1986) were also members of the 18 found in our study, i.e., *Hyallela, Cryptochironomus* a, *Chironomus* nr. *muratensis* = *C.* cf. *plumosus, C.* nr. *annularius, Cricotopus ornatus* and Dolichopodidae, while five other species we determined to be indicator species were preferential species in pelagic habitats. Two indicator species in the pelagic study (*Procladius freemani, Tanypus nubifer*) are essentially spring species since they emerge then.

Figure 7 gives the salinity tolerance ranges of the indicator species. The four lake groups are superimposed upon these distributions and pictorially suggests that the indicator species are not restricted to the groups based on salinity and that other environmental factors are involved. The sequential dichotomies are represented by the horizontal separations and tend to categorize the indicator species relative to the lake groups.

Four lake salinity groups are suggested by the distribution of indicator species. Group I ranges from 3–10‰ salinity (hyposaline) represented by four indicator species (7, 41, 32, 47) (Fig. 7) while Group II ranges from 10 to 38‰ (hypomesosaline) and has only one true indicator species, *C.* nr. *annularius*. With the exception of Sayer Lake on August 8 (salinity 36.8‰) the Group II salinity range would be 10–30‰ and the Group III range would be 30–63‰. Although the other five indicator species (8, 10, 30, 35, 65) also occur in the lower salinity range they tend to be more characteristic of samples in Group II. *Cenocorixa dakotensis*, although not an indicator species, is included in Fig. 7 since its range is similar to most of the species in Group II. It appeared as an indicator species in subsequent dichotomies not illustrated in Fig. 6. Group III covers the range of about 38‰ to 63‰ (mesohypersaline) and includes three species (*Cricotopus*, *Trichocorixa* and *H. salinarius*) which also extend into the Group II range and into the edge of the Group IV category (*Cricotopus*). Group IV contains two indicator species restricted to the high salinity range > 60‰ or hypersaline.

Timms *et al.* (1986) also generated four faunal groups but one of these was restricted to lakes with salinities < 3‰. However, Wakaw Lake (3.2‰) was included in this group as well as Humboldt Lake which is always saline (50 years) at least for the most of the year. The other three faunal groups encompassed salinity ranges of 3–20‰, 20–50‰ and > 50‰. These lower ranges are an effect of sampling lakes in the spring when salinities were much lower with only two lakes exceeding 50‰.

When we compare the lakes represented in four salinity groups generated for the littoral and pelagic species they are almost identical. The exceptions are Lenore and Rabbit lakes which we placed in hyposaline Group I. Perhaps Group I should include subsaline (but not fresh water ≤ 0.5‰) and lakes up to 10‰. Arthur and Basin lakes are interchanged between groups II and III although it is difficult to see why this is so

on the basis of salinity for the pelagic samples. There is no difference in the high salinity group. Then, the point is where to place the salinity ranges for the groups. Perhaps the salinity groups are not nearly as precise as we would like to believe.

Conclusion

The littoral zones of eighteen Alberta and Saskatchewan lakes ranging from 2.75 to 126.3 g l^{-1} salinity were sampled in the spring and summer of 1986. Seventy-six species of animals were present far exceeding the number of pelagic species (42) in the same lakes. The lowest salinities (2.75 to 5.9‰) were characterized by the greatest number of species (29–31) but species richness declined rapidly with increasing salinity above 15‰ until high salinity (> 100‰) lakes had only two species present. Species richness tended to increase from spring to summer, primarily the result of appearance of more corixids, beetles, gastropods, leeches, ostracods and water mites. Some spring chironomid species tend to disappear by summer because emergence takes place. Higher species richness in the littoral zone can be attributed to sampling both in spring and summer and probably greater habitat diversity including the presence of vegetation compared to pelagic habitats.

Biomass also tends to decrease with increasing salinity but a large increase is apparent around 15‰ salinity. The highest biomass occurred in culturally eutrophic Humboldt Lake. In most lakes biomass was greater in summer than in spring samples. This was primarily related to increasing numbers and biomass of amphipods (*Hyallela*, *Gammarus*), corixids (*Trichocorixa*), damselfly nymphs (*Enallagma*), a chironomid species (*Glyptotenpides* I) and other dipterans (Dolichopodidae, *Ephydra hians*). Biomass in the littoral zone tended to be higher than that of the pelagic zone in the same lakes in 1985 and in some of these lakes in 1983–1941.

TWINSPAN analysis of the biomass of species yielded a dendrogram with 18 indicator species.

Successive dichotomies permitted these species to characterize particular environmental ranges. Salinity was one of the main related factors resulting in four main salinity groups: 3–10‰, 10–38‰, 38–63‰ and >63‰. In spite of the higher salinity ranges, most of the same lakes and some of the same indicators species were found by Timms *et al.* (1986) in four lower salinity range categories. Therefore there seems to be some flexibility with respect to species preferences because species continue to live in a lake even if environmental factors such as salinity change temporally. Other factors such as pH, seasonality, and perhaps % organic matter and % plant cover played a determinant role with regard to distribution of indicator species. Depth of water did not, apparently, play a role in species distribution in the littoral zone.

The relative importance of the littoral zone is a consideration. Wetzel (1983) argues that the littoral zone is very important on the basis that most world lakes are small. Other factors determining the extent of the littoral zone are the relative shallow area of the basin and depth of light penetration. Most lakes in our study have relatively large shallow areas and therefore large littoral areas. The only real exceptions are Manito and Redberry lakes and even in them plants grow to considerable depths, e.g., Redberry Lake to 8 metres (Hammer & Heseltine, 1988) due to high light penetration. Thus littoral zones are very important to most Alberta and Saskatchewan saline lakes in terms of relative benthos production in each lake.

Acknowledgements

This investigation was supported by NSERC Grant No. 1412 to UTH. Special thanks are extended to Dr. R. L. Randall who carried out the statistical analyses. We thank Dennis Dyck for drawing the figures and Kathy Meeres for the computer analyses of the data for TWINSPAN. We are grateful to the following for providing identifications: I. M. Smith (Acari), R. A. Cannings (*Enallagma*), W. W. Wirth (Ceratopogonidae), R. Brinkhurst (Oligochaeta), M. E. Dillon, D. R. Oliver, O. A. Saether, W. F. Warwick (Chironomidae), D. Larson, L. LeSage, A. Smetana (Coleoptera), D. Delorme (Ostracoda), R. L. Hurley (Dolichopodidae adults), R. S. Zack (Ephydridae), R. W. Davies (Hirudinea), H. J. Teskey (Tabanidae), D. Parker, E. Whiting (Ephemeroptera).

References

Gauch, H. G., 1982. Multivariate Analysis in Community Ecology. Cambridge University Press, Cambridge, 298 pp.

Hammer, U. T., 1978. The saline lakes of Saskatchewan. III. Chemical characterization. Int. Revue ges. Hydrobiol. 63: 311–335.

Hammer, U. T., 1986. Saline Lake Ecosystems of the World. Junk, The Hague, 602 pp.

Hammer, U. T. & R. C. Haynes, 1978. The saline lakes of Saskatchewan. II. Locale, hydrogeography and other physical effects. Int. Revue ges. Hydrobiol. 63: 179–203.

Hammer, U. T., R. S. Haynes, J. M. Heseltine & S. Swanson, 1975. The saline lakes of Saskatchewan. Verh. int. Ver. Limnol. 19: 589–598.

Hammer, U. T. & J. M. Heseltine, 1988. Aquatic macrophytes in saline lakes of the Canadian prairies. Hydrobiologia 158: 101–116.

Hammer, U. T., J. Shamess & R. C. Haynes, 1983. The distribution and abundance of algae in saline lakes of Saskatchewan, Canada. Hydrobiologia 105: 1–26.

Haynes, R. C. & U. T. Hammer, 1978. The saline lakes of Saskatchewan. IV. Primary production by phytoplankton in selected saline ecosystems. Int. Revue ges. Hydrobiol. 63: 337–351.

Hill, M. O. 1979. TWINSPAN-A FORTRAN Program for Averaging Multivariate Data in an Ordered Two-Way Table by Classification of the Individuals and Attributes. Cornell University, Ithaca, 90 pp.

Last, W. M. & T. M. Schweyen, 1983. Sedimentology and geochemistry of saline lakes of the Great Plains. Hydrobiologia 105: 245–264.

Last, W. M. & T. H. Schweyen, 1985. Late Holocene history of Waldsea Lake, Saskatchewan, Canada. Quater. Res. 24: 219–234.

Parker, R. D. & U. T. Hammer, 1983. A study of the Chromatiaceae in a saline meromictic lake in Saskatchewan, Canada. Int. Revue ges. Hydrobiol. 68: 839–851.

Parker, R. D., J. R. Lawrence & U. T. Hammer, 1983. A comparison of phototrophic bacteria in two adjacent saline meromictic lakes. Hydrobiologia 105: 53–62.

Rawson, D. S. & J. E. Moore, 1944. The saline lakes of Saskatchewan. Can. J. Res., D. 22: 141–201.

Schweyen, T. & W. Last, 1983. Sedimentology and paleohydrology of Waldsea Lake, Saskatchewan. In M. D. Scott (ed.) Canadian Plains Proc. 11, Canadian Plains Research Center, Regina: 45–59.

Swanson, S. M. & U. T. Hammer, 1983. Production of *Cricotopus ornatus* (Meigen Dipera: Chironomidae) in Waldsea Lake, Saskatchewan. Hydrobiologia 105: 155–164.

Swift, M. C. & U. T. Hammer, 1979. Zooplankton production dynamics and *Diaptomus* production in Waldsea Lake, a saline meromictic lake in Saskatchewan. J. Fish. Res. Bd Can. 36: 1431–1438.

Timms, B. V., 1983. A study of benthic communities in some shallow saline lakes of western Victoria, Australia. Hydrobiologia 105: 165–178.

Timms, B. V., U. T. Hammer & J. W. Sheard, 1986. A study of benthic communities in some saline lakes in Saskatchewan and Alberta, Canada. Int. Revue ges. Hydrobiol. 71: 759–777.

Tones, P. I., 1976. Factors influencing selected littoral fauna in saline lakes in Saskatchewan. Ph. D. Thesis, University of Saskatchewan, Saskatoon. 185 pp.

Tones, P. I., 1978. Osmoregulation in adults and larvae of *Hygrotus salinarius* Wallis (Coleoptera: Dytiscidae). Comp. Biochem. Physiol. 60: 247–250.

Tones, P. I., 1983. *Megalocypris ingens* Delorme (Ostracoda) in Saskatchewan saline lakes: Osmoregulation and abundance. Hydrobiologia 105: 133–136.

Tones, P. I. & U. T Hammer, 1975. Osmoregulation in *Trichocorixa verticalis interiores* Sailer (Hemiptera: Corixidae) – an inhabitant of Saskatchewan saline lakes. Can. J. Zool. 53: 1207–1212.

Welch, P. S., 1948. Limnological methods. McGraw-Hill, Toronto. 381 p.

Wetzel, R. G., 1983. Limnology. 2nd Ed. W. B. Saunders Company Toronto. 767 p.

Williams, W. D., 1964. A contribution to lake typology in Victoria, Australia. Verh. int. Ver. Limnol. 15: 158–163.

Winkler, D. W. (ed), 1977. An ecological study of Mono Lake, California. Inst. Ecology Publ. No. 12, University of California, Davis. 184 pp.

Hydrobiologia **197**: 193–205, 1990.
F. A. Comin and T. G. Northcote (eds), Saline Lakes.
© 1990 *Kluwer Academic Publishers.*

Distribution and abundance of the alkali fly (*Ephydra hians*) Say at Mono Lake, California (USA) in relation to physical habitat

David B. Herbst
Sierra Nevada Aquatic Research Laboratory, University of California, Star Route 1, Box 198, Mammoth Lakes, CA 93546, USA

Key words: *Ephydra*, life cycle, development, distribution, Mono Lake, substrate

Abstract

The distribution and abundance of larval, pupal, and adult stages of the alkali fly *Ephydra hians* Say were examined in relation to location, benthic substrate type, and shoreline features at Mono Lake. Generation time was calculated as a degree-day model for development time at different temperatures, and compared to the thermal environment of the lake at different depths.

Larvae and pupae have a contagious distribution and occur in greatest abundance in benthic habitats containing tufa (a porous limestone deposit), and in least abundance on sand or sand/mud substrates. Numbers increase with increasing area of tufa present in a sample, but not on other rocky substrates (alluvial gravel/cobble or cemented sand). Standing stock densities are greatest at locations around the lake containing a mixture of tufa deposits, detrital mud sediments, and submerged vegetation. Shoreline adult abundance is also greatest in areas adjacent to tufa. The shore fly (ephydrid) community varies in composition among different shoreline habitats and shows a zonation with distance from shore.

The duration of pupation (from pupa formation to adult eclosion) becomes shorter as temperature increases. The temperature dependence of pupa development time is not linear and results in prolonged time requirements to complete development at temperatures below 20 °C. About 700 to 1000 degree-days are required to complete a generation. Degree-days of time available in nature declines by 10 to 50% at depths of 5 and 10 metres relative to surface waters (depending on the extent of mixing), resulting in fewer possible generations. Essentially no growth would be expected at 15 m, where temperature seldom exceeds the developmental minimum. It is concluded that reduced substrate availability and low temperatures may limit productivity of the alkali fly at increasing depths in Mono Lake.

Introduction

Ephydrids are common in saline aquatic habitats throughout the world, and are often the most abundant inhabitants of the benthic and shoreline portions of these ecosystems. Ephydrids display a remarkable capacity for physiological adap-tation to extreme physicochemical conditions (Nemenz, 1960; Wirth & Mathis, 1979; Barnby, 1987; Herbst *et al.*, 1988). In addition to living in inland salt lakes and ponds varying from perma-nent to seasonally ephemeral, ephydrids inhabit alkaline and acidic thermal springs, mud flats and playa seeps, tidal splash pools, and coastal salt marshes. As consumers of benthic algae and detritus, and in turn being a major food resource to a variety of migratory water birds, these flies occupy a key trophic position in transforming these harsh environments into important wildlife habitats. Despite the abundance, varied habitats, and potential ecological role of halobiont ephydrids, we know little about how the physical environment may influence their distribution and abundance.

Ephydra Fallen is the main genus of brine flies found in saline waters of both the New World (Wirth, 1971), and Old World (Wirth, 1975). The genus is absent only in Australia, where it is apparently replaced by the endemic *Ephydrella* Tonnoir and Malloch.

The account of Aldrich (1912) on the distribution and biology of the alkali fly, *Ephydra hians* Say, has essentially been the sole source of information on this insect. Most of the lakes where *E. hians* has been found are alkaline, thus the common name. The alkali fly was the first North American species of *Ephydra* described (Say, 1830). Wirth (1971) provided distribution records from museum specimens, and an excellent review of the natural history of the genus. Simpson (1976) described larvae and pupae of *E. hians* and *E. cinerea* Jones. Studies by Herbst *et al.* (1988) have shown that *E. hians* survives longer and osmoregulates more effectively in natural alkaline carbonate waters than in sea water, indicating that this species is alkali-adapted.

In studies of *E. cinerea* from Great Salt Lake (Utah), Collins (1980a) found that larvae tend to emigrate from sand substrates, and occur in greatest densities on bioherm reef-rock or shallow water mud substrates. Both this species and *E. hians* have clawed larval prolegs used in crawling over and clinging to bottom substrates, and pupae that attach to submerged objects. At Mono Lake, where sand, mud, and rock substrates occur, larval and pupal densities were expected to be greatest on the rocky substrates used for pupal attachment, and least on sand.

The objective of this paper is to describe the distribution of the alkali fly in relation to benthic habitat and substrate associations, and the influence of temperature on development and generation time. These studies are based on field sampling of the population from Mono Lake (California, USA).

Life cycle of the alkali fly and ecological setting

All life stages except eggs have been collected throughout the year at Mono Lake. This suggests, as Collins (1980a) reported for *E. cinerea* from Great Salt Lake (Utah), that there is no developmental diapause stage for this species. Low temperatures slow the development of all life stages, and prolong the lifespan. Adults apparently do not oviposit during winter months, and show no vitellogenic activity, indicating possible reproductive diapause during this time (Herbst, 1988). In contrast to *E. cinerea*, which deposits its eggs on the water surface, where they may settle into unfavourable habitat, *E. hians* may choose oviposition sites. Adult females are able to go underwater by walking down partially submerged objects into the water. An air bubble forms around the bodies of the flies due to presence of a dense covering of hairs and bristles, and acts as a physical gill. Flies may then search over the lake bottom, using their long tarsal claws to hold on. Females often place their eggs in mats of algae, substrate presumably favorable to larval development. Some oviposition may occur on the water surface, and has been observed on floating algal mats. Although most submerged flies are found to be females, males too are sometimes observed, suggesting that this behavior may not be exclusively related to egg laying. Close observation reveals that adults spend much time feeding, rasping the film of algae off rock surfaces. Food quality is critical to reproductive success in *E. hians* (Herbst, 1986) and benthic epilithic algae and algal mats probably represent a superior food source to that onshore (except possibly on detrital mud flats). Submergence may thus have evolved originally as an adult feeding habit. This is consistent with the fact that oviposition may occur without feeding (autogenously) in *E. cinerea* (Collins, 1980b), whereas feeding is essential for

E. hians. Eggs hatch in 1–2 days and develop through 3 larval instars. Larvae also feed on epilithic and sediment surface algal mats, composed mainly of diatoms, filamentous green, and blue-green algae, and the microbial-detrital component of the sediments (benthic algal species are listed in Herbst, 1988). Prolonged recruitment by oviposition over much of the year, and variable larval development rates result in overlapping generations, without clearly distinguishable cohorts.

Morphology and behavior of larvae and pupae appear to provide safeguards from mortality caused by storm-related dislodgement from lake bottom habitats. Clawed prolegs permit the clinging larvae to retain their hold in the wave-swept rocky littoral shallows along the open, windy shores of desert lakes. Larvae have a caudal respiratory siphon, as do related ephydrids, but they respire primarily by diffusion of dissolved gases across the cuticle and seldom rise to the surface to breathe. The siphon probably serves as a tracheal gill in this insect. In addition, mature larvae usually select the protected undersides of rocks to pupate, and the coarctate pupae remain attached by means of the clamp-like terminal prolegs. Larvae cast adrift are subject to resettling in unfavorable regions (such as profundal sediments or sandy deposits), and dislodged pupae will invariably float to the surface and be cast on the shore in windrows where they are subject to desiccation, parasitism, and predation. Pupae may be subject to attack by the pteromalid parasitoid *Urolepis rufipes* (Ashm.) (Essig, 1926), with those cast ashore in windrows being especially vulnerable (Herbst, unpubl. data).

Mono Lake is located on the eastern side of the Sierra Nevada mountains in east-central California, at an elevation of almost 1950 meters. Mono Lakes becomes thermally stratified during the summer at a depth usually between 10 and 15 meters. Below the thermocline, the profundal lake bottom consists primarily of fine detrital sediments of autochthonous origin (anaerobic much of the year), and fine alluvium. Localized around sublacustrine springs and seeps (found mostly along the western, NW and SW lake margins), are extensive formations of the main hard substrate of the littoral region, porous limestone deposits known as tufa (Dunn, 1953). On the south and southeast lakeshore, the bottom is covered by windblown volcanic sands. East and northeast shores are composed mainly of soft detrital muds, with some areas having a surface veneer of sand.

Material and methods

Benthic sampling

The densities of larvae and pupae at different sites around Mono Lake, and on different substrate types, were estimated from samples taken with a 15 cm ($0.018 \, m^{-2}$) diameter coring tube (Wilding or stovepipe sampler type, Merritt & Cummins, 1984). Samples were taken in spring, summer, and autumn of 1986 at 15 sites in locations representing each of the major geologic substrata of the shallow littoral region (Fig. 1). These are deposits of tufa, sand, alluvial gravel and mud. In addition, combinations of these substrates with detritus deposits, submerged terrestrial vegetation, pumice, cemented sands, and algal mats produce varied habitat around the lake.

Samples were collected at depths between 20 and 50 cm by working the core sampler into the substrate and removing the stirred contents at the base of the tube using a bilge pump and/or fine mesh hand net. Five samples at each site were taken both in early July and late October (corresponding to early and late in peak seasonal abundance, Herbst, 1988). Each sample was selected at random along a 100 m transect parallel to shore at a marked location. Any rock material larger than a centimeter was removed for processing and measuring of surface area. Samples were taken to the laboratory for processing within 12–24 hours after collection (refridgerated). Samples were put into a large volume of saturated sodium chloride to float out all insect life stages in the sample. These were filtered off and the remaining debris examined microscopically for any remaining larvae or pupae. These samples were preserved for counting. Surface area of rock substrate in

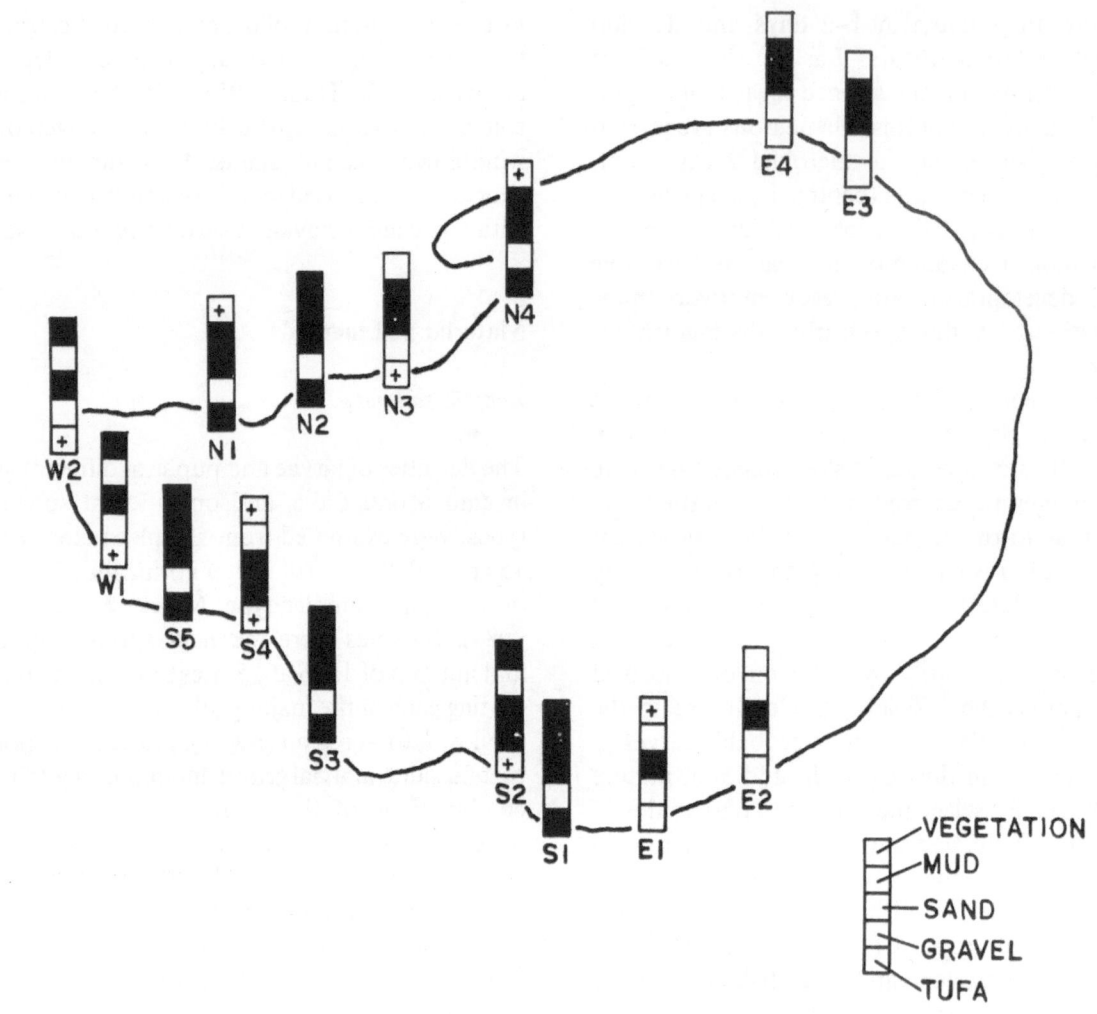

Fig. 1. Substrate composition of benthic samples sites at Mono Lake. Site prefix refers to compass sector of location. Boxes are filled if half or more of the samples taken contained that substrate type. Boxes with a plus (+) show a substrate was present but in less than half the samples.

samples was estimated by outlining their contours on graph paper and measuring the area on the grid (2-dimensional area). Salinity was measured from surface to bottom along the transect as specific gravity (using a hydrometer and translucent cylinder). Temperature was recorded at the same time using a standard mercury thermometer. Substrate type in each sample was scored as presence or absence of each of the five categories listed in Fig. 1.

First, second, and third larval instars and pupae were counted in each sample and the frequency distributions of larval and pupal abundance examined for conformity to a contagious or clumped distribution. This was done by a chi-square test of the dispersion coefficient (k) to a negative binomial distribution (Elliott, 1977). The value of k was obtained using the maximum-likelihood equation and this was used to determine sample size requirements.

Substrate preference was examined by plotting the frequency distribution of densities for samples containing (1) tufa (2) gravel alluvium (3) consolidated or cemented sand (4) unconsolidated sand or sand/mud mixture and (5) submerged vegetation present on sand or sand/mud mixtures.

The relationship of hard substrate area available in a sample to density of larvae and pupae was plotted to compare suitability of different hard substrates (categories 1, 2 and 3 above).

The relative abundance of adult *Ephydra hians* at different locations and on different shore substrates around the lake shoreline was evaluated by coating 25 × 75 mm slides with sticky-trap (trademark), and setting them out for 15 minute intervals. Zonation away from the shoreline, and species composition of the shore insect community was observed from these samples.

Temperature Effects on Development

Temperature effects on development were measured using the duration of the pupal stage as an indicator. Field-collected 3rd instar larvae from Mono Lake were held in the laboratory in the lake water at 20 °C, and within several hours after pupariation (puparium formation), were transferred to incubation chambers at 5, 10, 15, 20, and 25 °C, on a L : D 14 : 10 photoperiod. Eclosions were recorded daily. Larval development time at 25 °C was also determined, with laboratory cultured algae (mainly diatoms and filamentous blue-green algae) provided regularly as food, at a salinity of $100 \, g \, l^{-1}$ (near present lake salinity). Larval development at other temperatures was assumed to change in proportion to that observed for the effect of temperature on pupation time. Pupation time is independent of environmental factors such as food and salinity (Herbst, 1986) and is used here as a standardizing factor that reduces natural variability in larval development not due solely to temperature. Total degree-days from egg to adult was estimated from these data at each experimental temperature. Degree-days in the field (above a developmental temperature minimum) were calculated from data on temperature at the water surface, 5 m, 10 m and 15 m depth (using data from pelagic monitoring stations, courtesy G. Dana & R. Jellison). These were compared to the laboratory data on development times to estimate potential generations completed during the growing season within

the epilimnion (the thermocline usually forms between 10 and 15 meters depth).

Results

Habitat associations and Ephydra hians distribution at Mono Lake

The map of dominant substrate types present in samples from study sites (Fig. 1) show some general patterns:
(1) East shore sites are composed almost exclusively of sand substrate except in the north, where some detrital mud deposition occurs. In addition, consolidated or cemented sand provides a hard substrate in some locations, and patches of submerged vegetation and a few scattered tufa formations also occur.
(2) West shore sites (sites with W, S or N prefixes) are a heterogenous mix of substrates composed of tufa, sand, mud, detritus, alluvial gravel, and submerged vegetation, each of which may be locally dominant.

Shallow tufa shoals in certain areas (N1, N2, S1, S5) also promote stabilization and protection of detrital mud deposits, permitting the development of a benthic algal-microbial mat in these locations. Submerged vegetation has its origin in the growth of salt grass (*Distichlis spicata*) onto lakeshore exposed by dropping lake level that has since become re-submerged by rising water. The submerged salt grass is a dead rhizomatous mat, and is an erodable, impermanent substrate feature. This root mass has an important stabilizing effect on sand/mud substrates, and extends around much of the lake as a band parallel to the lakeshore (Burch *et al.*, 1977).

A preliminary set of 50 benthic samples were taken in May 1986 from the mixed substrates of the west bay area (S5-W1-W2) to determine the spatial dispersion of larvae and pupae, and the sampling intensity required to estimate mean density with a 20% SE. Maximum-likelihood values calculated for the index of dispersion k (Elliot, 1977) showed significant conformity to a negative binomial distribution by a chi-squared goodness-

198

of-fit test. Smaller values of k for pupae relative to larvae indicated pupae had an even more clumped distribution than larvae. A sample size of 29 was calculated to obtain 40% precision in estimation of the 95% CI for the mean density. Since aggregated distributions of benthos are often due to substrate preferences (behavior and morphology of *E. hians* larvae and pupae clearly suggest this), sampling was expanded to incorporate other locations and substrate types (sample $n = 75$).

The distribution map of larval and pupal abundance (Fig. 2) shows large variations among sample sites, but that densities within sites are similar between summer and autumn samples. The most densely occupied locations around the lake are associated with areas of mixed tufa deposition, detrital mud sediments, and submerged vegetation. Sand was poorly inhabited unless tufa and/or vegetation was also present. Substrate preferences are examined in the frequency distribution of densities for different substrate categories (Fig. 3). Densities are highest on tufa (mean = $14050 \, m^{-2}$) and vegetated

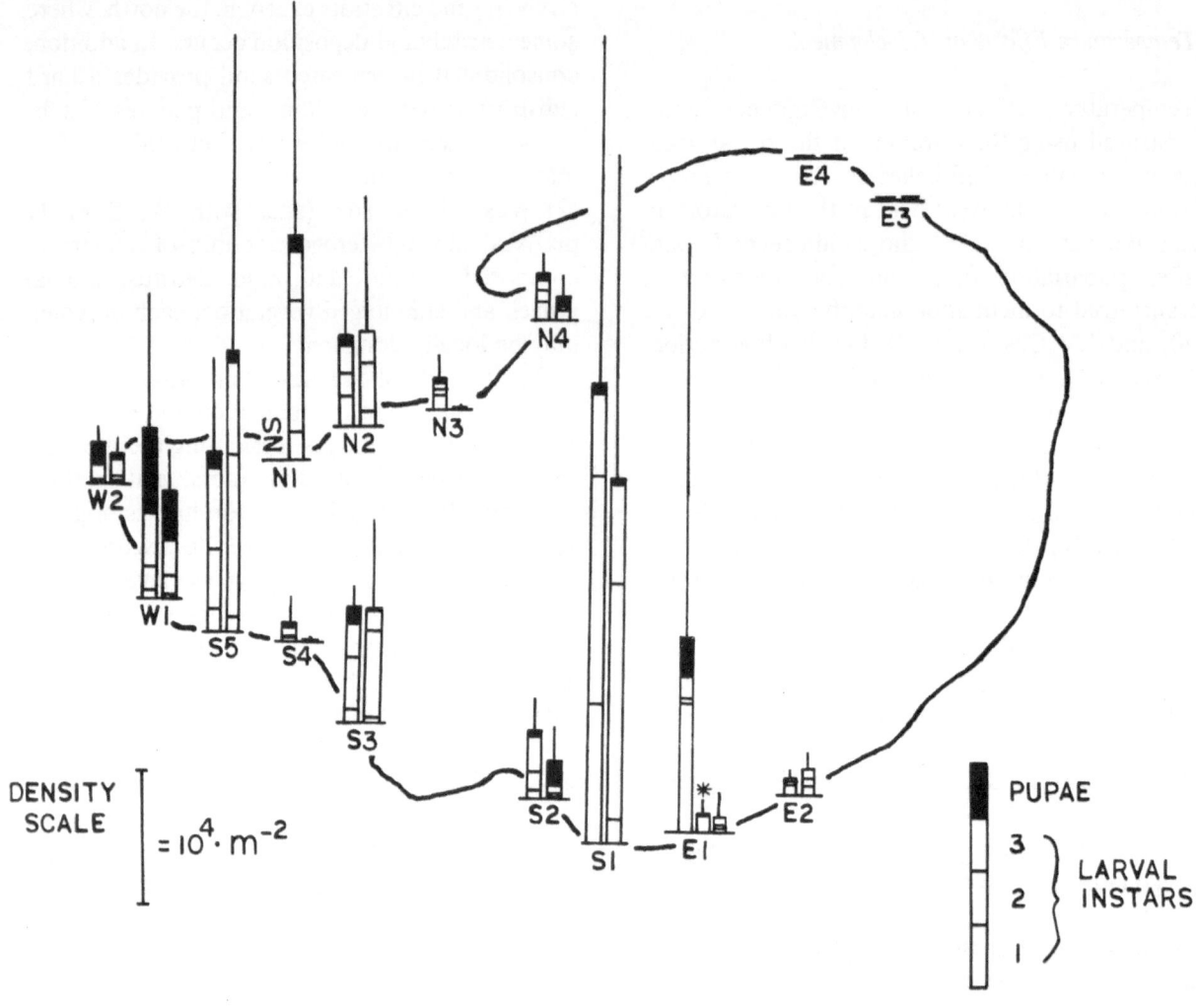

Fig. 2. Density of larval instars and pupae at sample sites in summer (left bar at each site, 1–9 July) and autumn (second bar at each site, 20–30 October) of 1986. N = 5 samples each, SD shown as line at top of each bar. Note that N1 was not sampled (NS) in summer, and middle bar (*) of E1 is the summer collection with 1 sample removed that contained a branch of drifted vegetation covered with attached larvae and pupae.

199

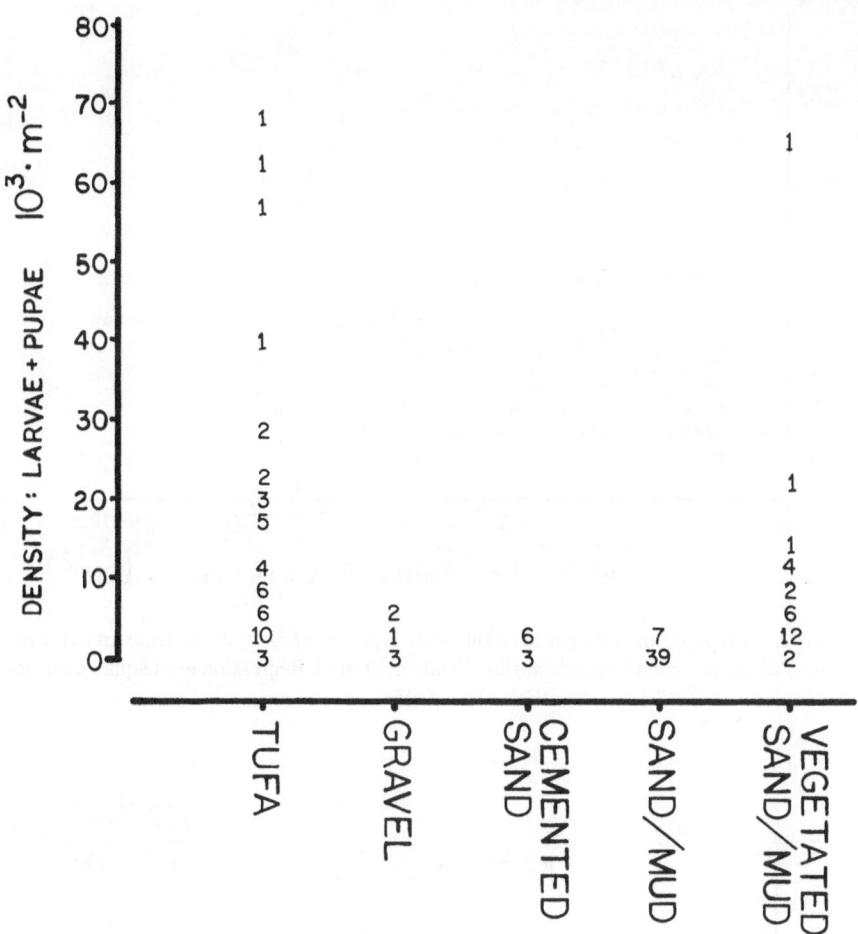

Fig. 3. Frequency distribution of abundance (larvae + pupae m^{-2}) on different substrate types for summer and autumn samples pooled. Number on graph is the frequency of samples in each density class, drawn from a total of 145 benthic collections. Tufa, gravel and cemented sand (hard substrates) were samples in which only one of these substrates was present, but sand, mud, or vegetation could also be present. Sand/mud was samples consisting exclusively of sand or sand/mud mixtures; or vegetated sand/mud if submerged vegetation was present.

sand/mud (8100 m^{-2}), intermediate on gravel (2400^{-2}) and cemented sand (1850 m^{-2}), and lowest on sand or sand/mud deposits (550 m^{-2}). Over the 0–150 cm^{-2} size range of hard substrates sampled, the abundance of larvae and pupae increases significantly with the area of tufa present, but not on gravel or consolidated sand substrate (Fig. 4).

To examine changes in population age structure from summer into autumn, changes in the proportion of each life stage were compared as matched-pairs within sites (to control for location and substrate effects) using Wilcoxon's signed-ranks test (Fig. 5). This showed significantly fewer first instars and more second instars in autumn.

Adult abundance along Mono Lake shores is greatest adjacent to littoral zones where tufa substrate is present (Table 1). Adult density also decreases rapidly with distance from shore, and most flies are usually found in a band less than a meter wide along the shoreline. Other ephydrid species occur back from the shoreline (1–3 meters), out of the zone where *E. hians* is most abundant. In addition, although the alkali fly is abundant at a nearby saline pond, *Mosillus bidentatus* and *Paracoenia bisetosa* are also com-

200

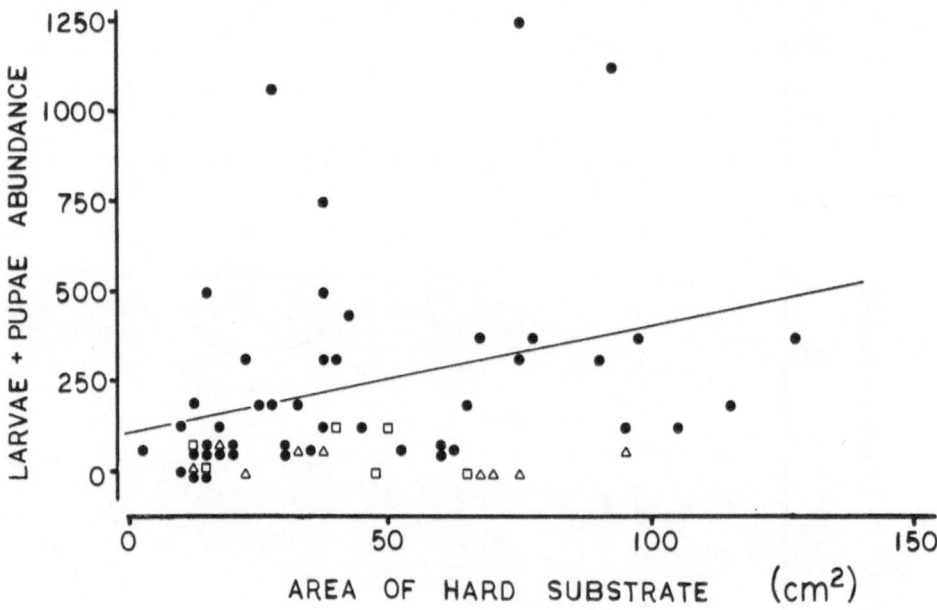

Fig. 4. Abundance of larvae and pupae on different hard substrate types in relation to substrate area (for areas up to 150 cm²). Filled circles = tufa; open squares = gravel, open triangles = cemented sand. Regression least squares line for tufa only: $r = 0.32$ ($p = 0.03$).

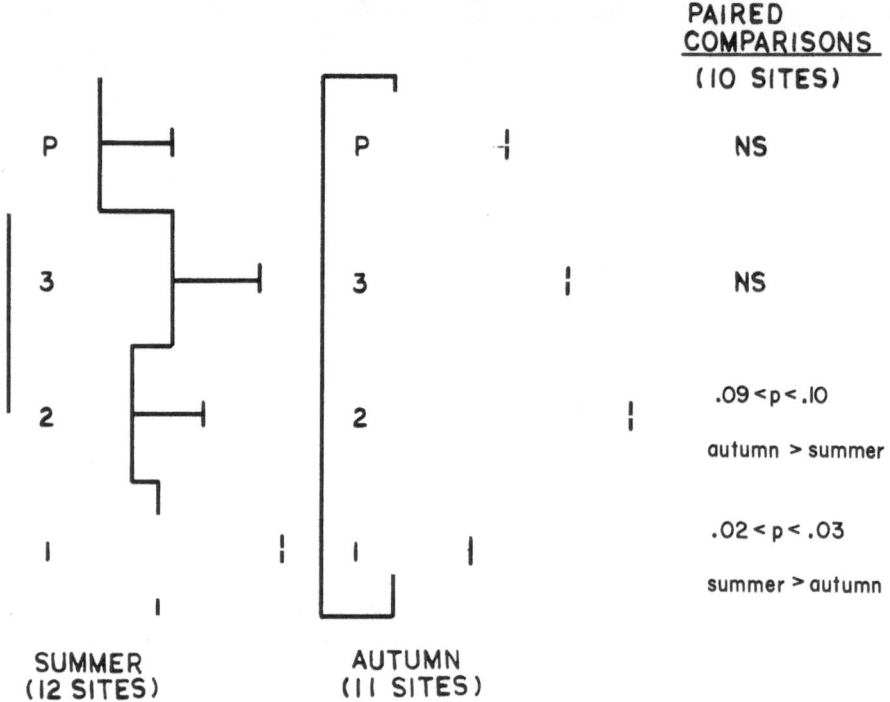

Fig. 5. Population age structure: overall average proportions of larval instar and pupa stages in summer and autumn samples. Compared within sites using Wilcoxon's matched-pair signed-ranks test (2-tailed), for differences between summer and autumn sample periods. Mean for sites calculated using only samples with $n > 20$ larvae + pupae ($n = 10$ paired sites). Error bars are SD of proportions.

Table 1. Lakeshore habitat and adult shore fly distribution at Mono Lake and nearby ponds in July/August 1980.

Substrate and habitat type	N	Distance from shore	Ephydrid (species)				
			Eh	Mb	Pb	Lc	Po
Detrital mud							
Tufa present	7	0 m	***********	0	0	0	+
	7	1 m	*	0	0	0	+
	4	3 m	0	0	0	+	+
Tufa absent	3	0 m	*	0	0	0	0
	3	1 m	+	+	0	0	0
	3	3 m	+	+	0	0	*
Sand							
Tufa present	4	0 m	********	0	0	0	0
	4	1 m	*	0	0	+	+
Tufa absent	6	0 m	+	0	0	0	0
	6	1 m	0	0	0	*	+
Saline pond (2 km from ML)							
Mud	3	0 m	********	+	+	0	0
	3	3 m	*******	*****	***	0	0
	3	15 m	0	0	0	0	**
Sand	3	0 m	0	***	0	0	*
	3	2 m	0	0	0	0	**
Freshwater seep pool (20 m from ML)							
Mud	7	0 m	0	+	*	0	0

Key: N = sample size, each * represents an average of 10 flies per sample, each + indicates a species is present, but fewer than 10 flies in a collection. Coefficient of variation = 30–50% for most samples. Tufa presence or absence refers to submerged nearshore tufa habitat. Species abbreviations: Eh = *Ephydra hians* Say, Mb = *Mosillus bidentatus* (Cresson), Pb = *Paracoenia bisetosa* (Coquillett), Lc = *Lamproscatella cephalotes* Cresson, Po = *Ptilomyia occidentalis* Sturtevant and Wheeler. Other species commonly encountered onshore:
Hydrophorus plumbeus Aldrich (Dolichopodidae)
Saldula arenicola (Scholz) (Saldidae)
Lispe sp. (Anthomyiidae)
Cicindela fulgida Say (Cicindelidae)

mon, and these species are the only ones found in samples at a freshwater seep adjacent to the lake.

Temperature and generation time

Pupa development time for *E. hians* is temperature-dependent (Fig. 6), and is more rapid in comparison to *E. cinerea* from Great Salt Lake (Collins, 1980a). This is a useful indication of seasonal effects on growth rate, independent of food and salinity, because pupae are non-feeding, and sequestered from the aquatic environment (Herbst, 1986). The most rapid change in development rate occurs between 10–20 °C for both species. This corresponds to the temperature range found in spring and autumn, coincident with periods of population growth and decline.

202

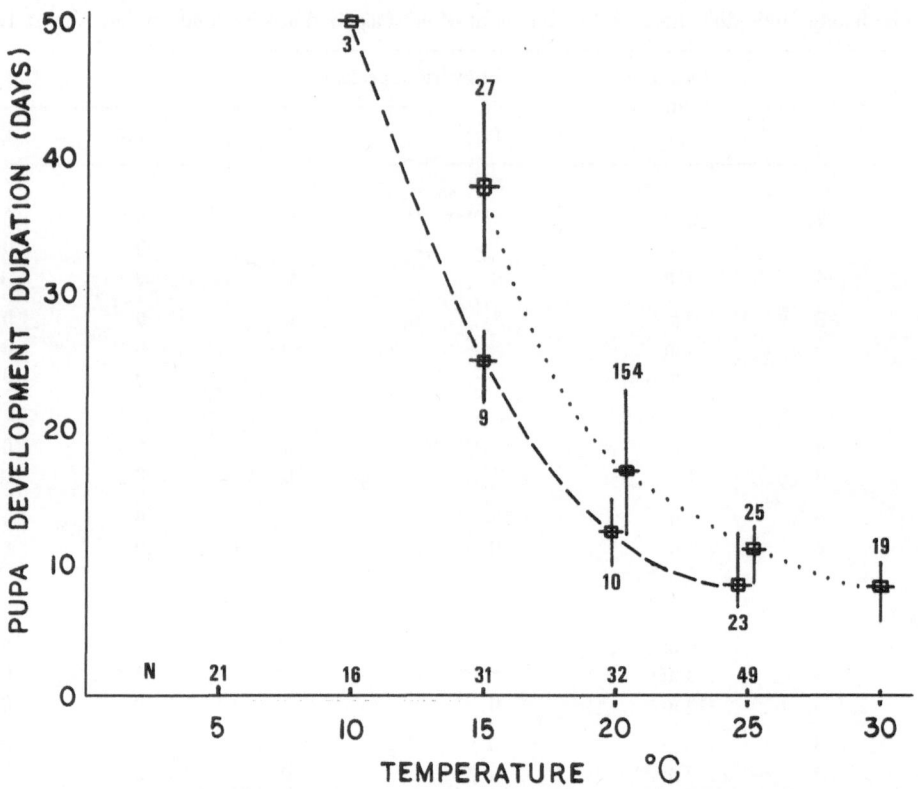

Fig. 6. Influence of temperature on pupa development time in *Ephydra hians* (dashed curve) compared to *E. cinerea* (dotted curve; data taken from Collins, 1980a). Vertical bars are the SE of the mean (horizontal lines), and vertical lines are the range. N above X axis refers to the total number of *E. hians* pupae exposed to that temperature.

The relation of temperature to development is non-linear and has a Q-10 of 2.1 over 20–25 °C, but is 4.1 over the 10 to 20 °C range.

The minimum temperature for development is estimated as 10 °C since no adults emerged from pupae exposed to 5 °C, but a few emerged at 10 °C. Based on laboratory experiments of larval development at 25 °C, egg hatching to pupa formation requires 40 to 60 days.

Combining pupation times with the ranges estimated for larval development times at each temperature provides a measure of egg to adult generation time in degree-days (Table 2). Using 10 °C as the developmental minimum, the estimates agree at all temperatures that 700–1000 degree-days are required to complete a generation (excluding adult reproduction time).

Degree-days of development time available in nature vary with depth and location (Fig. 7). In surface waters, between 1200 and 1400 degree-days are accumulated above the 10 °C minimum temperature. In a warm year (e.g. 1984), when the epilimnion was well-mixed, degree-day time was reduced by only 10 to 20% at 5 and 10 meters depth, respectively (relative to the surface). in colder years (1982 and 1983), degree-days were reduced by 20 to over 50% at the same depths. Greater variability at the surface between east and west sides of the lake (separated by a large island) suggests that in any year, the thermal environment is more constant in deep water. The relative thermal inertia of deep water is likely to be even more pronounced in the benthic littoral region due to solar radiation inputs to substrate in shallow water. Herbst (1988) observed large variations in shallow benthic temperatures within and between locations studied in 1983 and 1984. In 1984, mean degree-days accumulated above 10 °C was 1650,

Table 2. Degree-days generation time at different temperatures, using pupation time as a standard.

Temperature °C	Puration Time (d)	Larval Development Time (d)*	Estimated Sum** Degree-Days Required for Egg to Adult Generation Time
25	9	40–60	735–1035
20	13	60–85	730– 980
15	25	110–170	675– 975
10	50 +	200–300	–

* Larval development time at 20, 15, and 10 °C calculated from the range found at 25 °C, proportional to the increase in pupation time.

** Degree-days calculated as degrees above 10 °C times total days of larval + pupal development time.

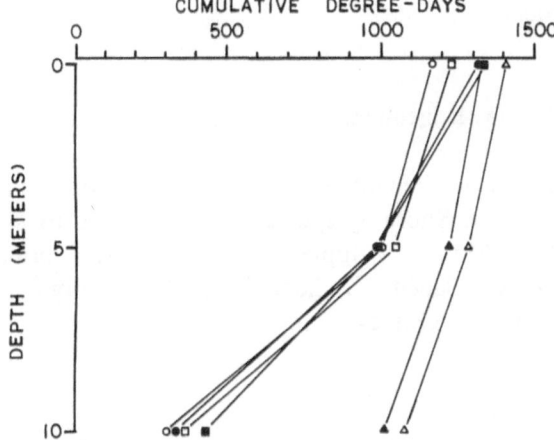

Fig. 7. Cumulative degree-days of development time available at different epilimnion depths for 1982 (circles), 1983 (squares) and 1984 (triangles). Calculated from pelagic temperature profiles using 10 °C as the minimum developmental temperature of *Ephydra hians*. Filled symbols from profiles taken on west side of lake, open symbols from east side of lake.

and the maximum annual sum (at site N1) was 3150. Using the minimum generation time, between 2.4 and 4.5 generations per year would be possible under these conditions. Applied to the degree-day data of Fig. 7, between 1.7 to 2.0 generations could be completed in shallow surface waters, 1.4 to 1.8 at 5 m, and from less than 1 to 1.5 at 10 m. Only in 1984 did temperatures at 15 m reach 10 °C or above, and only 80–120 degree-days were accumulated.

Discussion

Tufa is apparently the most suitable substrate for aggregation and persistence of the immature stages of the alkali fly, unless soft or unconsolidated substrates are otherwise stabilized and protected from wave action (Figs. 2, 3 and 4). Collins (1980a) also showed that abundance of *E. cinerea* was greater on the rocky, reef-like bioherms of Great Salt Lake than in sandy areas. At Abert Lake (Oregon), another western Great Basin lake where *E. hians* is common, larvae and pupae occur mainly among the rocky fragments of porous basalt on the eastern shore (Herbst, 1988). Tufa provides an attachment site for pupae, and is actively grazed by larvae since it serves as an epilithic growth site for diatoms. The association of adults with shorelines adjacent to offshore tufa (Table 1), and the observation of adult feeding and oviposition on partially submerged tufa, is also consistent with the importance of this rocky substrate in the distribution of the alkali fly at Mono Lake. The other hard substrates that occurred in samples (alluvial gravel/cobble, and cemented sands) did hold higher densities than unconsolidated sand or sand/mud, but apparently the surface of these substrates was not as useful for attachment, because area was not correlated with larval and pupal abundance (Fig. 4). Submerged vegetation also serves as an attachment site, but the value of the habitat it occurs in for larval development probably depends more on food availability and quality.

Declining lake levels at Mono Lake, resulting from the diversion of inflowing streams, has become a major conservation concern in California (National Academy of Sciences, 1987). Among the potential problems are that exposed lakebed will result in the elimination of submerged tufa habitat. Geomorphic studies of the benthic topography of Mono Lake indicate that the distribution of tufa is restricted mainly to shallow water (Stine, 1988). Approximately 60% of presently submerged tufa would be exposed by a 3 meter decline in lake elevation (5–10 years from present, at the current diversion rate). The results of the present study show that tufa is preferred substrate of alkali fly larvae and pupae, suggesting that lower lake levels will restrict habitat availability, and thus limit the distribution and abundance of this benthic insect.

Although densities within sites remain consistent between early and late season samples, there is a significant decline in the proportion of first instars present, and increase in second instars (Fig. 5). This change in age structure matches the autumn decline in adult reproductive condition and ovipositional activity previously noted by Herbst (1988). Fewer eggs were observed in autumn collections as well, indicating that little or no new recruitment is occurring at this time, and that slow growth into later instars and pupae should characterize the overwintering, non-diapausing population.

Temperature effects on growth are a useful way of modeling and comparing the population dynamics of aquatic insects (Sweeney, 1984). The more rapid development rates for *E. hians* than for *E. cinerea* (Fig. 6) are indicative of an intrinsic difference that should permit earlier seasonal activity and greater potential productivity (growing season) for the alkali fly, under comparable ecological conditions. Collins (1980a) reports that probably no more than 1 to 2 generations of *E. cinerea* occur annually at Great Salt Lake. In contrast, Mono Lake *E. hians* can probably complete 2 to 3 or more generations in shallow habitat, depending on local heating conditions. In deeper

littoral waters there may be much less potential for growth because of lower temperatures (Fig. 7). However, in warm years, with deeper mixing, there may be only a slight reduction in development time available (see 1984 in Fig. 7). Although there are certainly other factors influencing development (e.g. food and salinity: Herbst 1986), further studies of seasonal and local variations in the thermal environment would be useful in understanding annual trends in population growth and life history traits. In addition, surveys of the depth distribution of larvae and pupae in Mono Lake should be conducted to substantiate the conclusion that productivity will be limited in deep water by decreased substrate habitat availability and less degree-day time available for development.

Acknowledgements

This work benefited from discussions with T. J. Bradley. Shore fly species were identified by W. N. Mathis. Supported by NSF grant DCB-8608664. Dedicated to the memory of David A. Gaines.

References

Aldrich, J. M., 1912. The biology of some western species of the dipterous genus Ephydra. J. N.Y. Ent. Soc. 20: 77–99.

Barnby, M. A., 1987. Osmotic and ionic regulation of two brine fly species (Diptera: Ephydridae) from a saline hot spring. Physiol. Zool. 60: 327–338.

Burch, J. B., J. Robbins & T. Wainwright, 1977. Botany. In D. W. Winkler (ed.), An ecological study of Mono Lake, California. University of California at Davis, Institute of Ecology, Publ. no. 12: 114–142.

Collins, N. C., 1980a. Population ecology of *Ephydra cinerea* Jones (Diptera: Ephydridae), the only benthic metazoan of the Great Salt Lake, USA, Hydrobiologia 68: 99–112.

Collins, N. C., 1980b. Developmental responses to food limitation as indicators of environmental conditions for *Ephydra cinerea* Jones (Diptera). Ecology 61: 650–661.

Dunn, R. C., 1953. The origin of the deposits of tufa at Mono Lake. J. Sed. Petrol. 23: 18–23.

Elliot, J. M., 1977. Some methods for the statistical analysis of samples of benthic invertebrates. 2ne edition. Freshwater Biological Association Scientific Publication No. 25, Ambleside, England. 160 pp.

Essig, E. O., 1926. Insects of western North America. The Macmillan Co. New York, 1050 pp.

Herbst, D. B., 1986. Comparative studies of the population ecology and life history patterns of an alkaline salt lake insect: *Ephydra* (*Hydropyrus*) *hians* (Diptera: Ephydridae). PhD thesis, Oregon State University, Corvallis. 206 pp.

Herbst, D. B., 1988. Comparative population ecology of *Ephydra hians* Say (Diptera: Ephydridae) at Mono Lake (California) and Abert Lake (Oregon). Hydrobiologia 158: 145–166.

Herbst, D. B., F. P. Conte & V. J. Brookes, 1988. Osmoregulation in an alkaline salt lake insect, *Ephydra* (*Hydropyrus*) *hians* Say (Diptera: Ephydridae), in relation to water chemistry. J. Insect Physiol. 34: 903–909.

Merritt, R. W. & K. W. Cummins (eds), 1984. An introduction to the aquatic insects of North America. 2nd edition. Kendall/Hunt Publishing Co., Dubuque, IA. 722 pp.

National Academy of Sciences, 1987. The Mono Basin ecosystem: effects of changing lake level. National Academy press, Washington, D.C.: 272 pp.

Nemenz, H., 1960. On the osmotic regulation of the larvae of *Ephydra cinerea*. J. Insect Physiol. 4: 38–44.

Say, T., 1830. Descriptions of North American dipterous insects. J. Acad. Nat. Sci. Philadelphia. 6: 183–188.

Simpson, K. W., 1976. The mature larvae and puparia of *Ephydra* (*Halephydra*) *cinerea* Jones and *Ephydra* (*Hydropyrus*) *hians* Say (Diptera: Ephydridae). Proc. Ent. Soc. Wash. 78: 263–269.

Stine, S., 1988. Geomorphic and geohydrographic aspects of the Mono Lake controversy. In D. B. Botkin *et al.* (eds), The future of Mono Lake. University of California Water Resources Center, Riverside. Report No. 68. Appendix D: 1–135.

Sweeney, B. W., 1984. Factors influencing life history patterns of aquatic insects. In Resh, V. H. & D. M. Rosenberg (eds), The ecology of aquatic insects. Praeger Publishers, N.Y.: 56–100.

Wirth, W. W., 1971. The brine flies of the genus *Ephydra* in North America (Diptera: Ephydridae). Ann. ent. Soc. Am. 64: 357–377.

Wirth, W. W., 1975. A revision of the brine flies of the genus *Ephydra* of the Old World (Diptera: Ephydridae). Ent. Scand. 6: 11–44.

Wirth, W. W. & W. N. Mathis, 1979. A review of the Ephydridae (Diptera) living in thermal springs. In D. L. Doenier (ed.), First symposium on the systematics and ecology of Ephydridae (Diptera). North American Benthological Society publ. Lawrence (KS): 21–45.

Hydrobiologia **197**: 207–220, 1990.
F. A. Comin and T. G. Northcote (eds), Saline Lakes.
© 1990 *Kluwer Academic Publishers.*

Distribution patterns of ostracods in iberian saline lakes. Influence of ecological factors

Angel Baltanás[1]; Carlos Montes[2] & Paloma Martino[2]
[1]*Museo Nacional de Ciencias Naturales (C.S.I.C.), c/José Gutierrez Abascal, 2. 28006 Madrid, Spain*;
[2]*Departamento de Ecología, Universidad Autónoma de Madrid, 28049 Madrid, Spain*

Key words: ecological factors, Iberian saline lakes, ostracods, zoogeography

Abstract

A limnological survey of the Iberian mesosaline and hypersaline lakes allowed study of biogeographical and ecological aspects of their ostracod populations.

Eucypris aragonica Brehm & Margalef; *Eucypris mareotica* (Fischer) and *Heterocypris barbara inermis* (Gauthier) are the only species found in waters with salinities between 10 and 100‰. Each species tends to be restricted to one of the three large Iberian Tertiary depressions: *E. aragonica* in the Ebro River basin; *E. mareotica* in the Guadalquivir River basin (South Spain) and *H. barbara inermis* in the tableland of La Mancha (Central Spain). Both *E. mareotica* and *H. barbara inermis* are distributed in inland waters of regions around the Mediterranean Sea, while *E. aragonica* is only known from Spain. Historical factors have been traditionally used to account for the distribution of several crustacean species and recent faunas have been seen as the remnat of those inhabiting Tertiary Depressions around ancient Mediterranean Sea.

Ecological factors, mainly ion composition, account for their observed distribution pattern in Spanish lakes. *E. mareotica* typically inhabits high chloride waters while *H. barbara inermis* 'prefers' lakes with high sulphate. *E. aragonica* seems to be restricted to chloride waters with a high sulphate content and very irregular (or aperiodic) hydrological regimes. Presence of parthenogenetic populations and waterfowl exchange between different saline-lake areas in the Iberian Peninsula facilitate ostracod dispersion.

Introduction

Ostracods are known to be important in controlling community structure (Diner *et al.*, 1986). They act as herbivores grazing on algae, and as detritivores or generalists in a great variety of habitats, including such extreme environments as hot springs (Wickstrom & Castenholz, 1973) and highly mineralized waters. Ostracods are an important component of the fauna in saline lakes (Hammer, 1986); and, possibly, have dominated these waters since the Triassic period (McKenzie, 1981). Several studies in taxonomy and ecology of ostracods in saline lakes have been published recently (see De Deckker, 1981 for a review).

Data on this subject are still scarce for the Iberian Peninsula, despite the fact that it has a large number of athalassic saline lakes (Montes & Martino, 1987). Major sources of information are taxonomic and regional limnology studies (Brehm

208

& Margalef, 1948; Margalef, 1947, 1953, 1956; Marazaof, 1967; Armengol *et al.*, 1975; Comín *et al.*, 1983; Marín, 1983).

This paper deals with species distribution and related ecological and biogeographical considerations on ostracods collected from a large number of inland Iberian saline lakes; i.e., those in which halobiont and halophilic communities occur almost exclusively. Such lakes have concentrations of total dissolved solids (TDS) greater than 10‰ (Bayly, 1972; Williams, 1981) and may be defined as hypo-mesosaline and hypersaline waters sensu Por (1980) and Hammer *et al.* (1983).

Description of the sites studied

On the Iberian Peninsula, due to its peculiar climatic, geological and hydrological characteristics the number of shallow saline lakes is high. They are located in the arid or semiarid regions of the three sedimentary Tertiary basins (Fig. 1A): the marine Tertiary depression of the Guadalquivir in the South (Andalucia); and two continental Tertiary depressions in Central (the natural region of La Mancha) and North-Eastern Spain (the Ebro River depression (Aragón)). Another endorheic area situated in the Duero basin (NW Spain) has not been considered here because of the low salinities of its waters.

There are approximately 65 Spanish saline lakes with salinity values (TDS) at the highest water level greater than 10‰. Most mesosaline and hypersaline Spanish lakes are temporary (3 to 10 months with water), shallow at maximum recorded depths (5–70 cm) and of small area (0.02–1.92 km^2). Salinity ranges between 10–430‰. The dominant chemical composition of the lake waters is usually magnesium-sulphate Mg-(Na)-SO$_4$-(Cl) for most lakes in La Mancha (originating from continental evaporites); sodium-chloride Na-(Mg)-Cl-(SO$_4$) for most Andalusian lakes (marine evaporites); and mixed Mg-Na-Cl-SO$_4$ for most Aragonese lakes (marine and continental evaporites). More information on limnological aspects of Iberian saline lakes can be found in Montes & Martino (1987) and Comín & Alonso (1988).

Material and methods

Sixty-two lakes were sampled in the winter (January or February) and spring (March or May) of 1986 and 1987. At each locality a surface water sample was collected close to the shoreline. Samples were analyzed in the laboratory between 1 and 30 days after collection. Major ions concentrations were measured following standard methods (see APHA, 1980). Some other variables (depth, temperature, pH, conductivity) were measured *in situ*.

Ostracod fauna was collected with a hand plankton net (60 μm mesh size). Samples were preserved with 4% formalin.

Results

Geographical distribution

Table 1 provides a list of the lakes studied and ostracod species found in them.

Eucypris mareotica (Fischer, 1855) Gauthier, 1928 syn.: *Eucypris inflata* (Sars, 1903) fide Gauthier, 1928.

In this paper we follow Gauthier's criterion, revised by Martens (1984), and use *Eucypris mareotica* as a senior synonym instead of *Eucypris inflata*, more widely used in the literature.

In the Iberian Peninsula this species is distributed mainly in the southern parts (amphigonic populations), the Guadalquivir valley. There are several separate distribution patches (parthenogenetic populations) in Central and Northeastern Spain: lakes from Pueblo (Ciudad Real), Manjavacas (Cuenca) and Gallocanta (Zaragoza) (Fig. 1A). Records in this paper are the first ones for the Iberian Peninsula and Western Europe.

Morphological descriptions can be found in Gauthier (1928). Fig. 2 shows key anatomical details of this species.

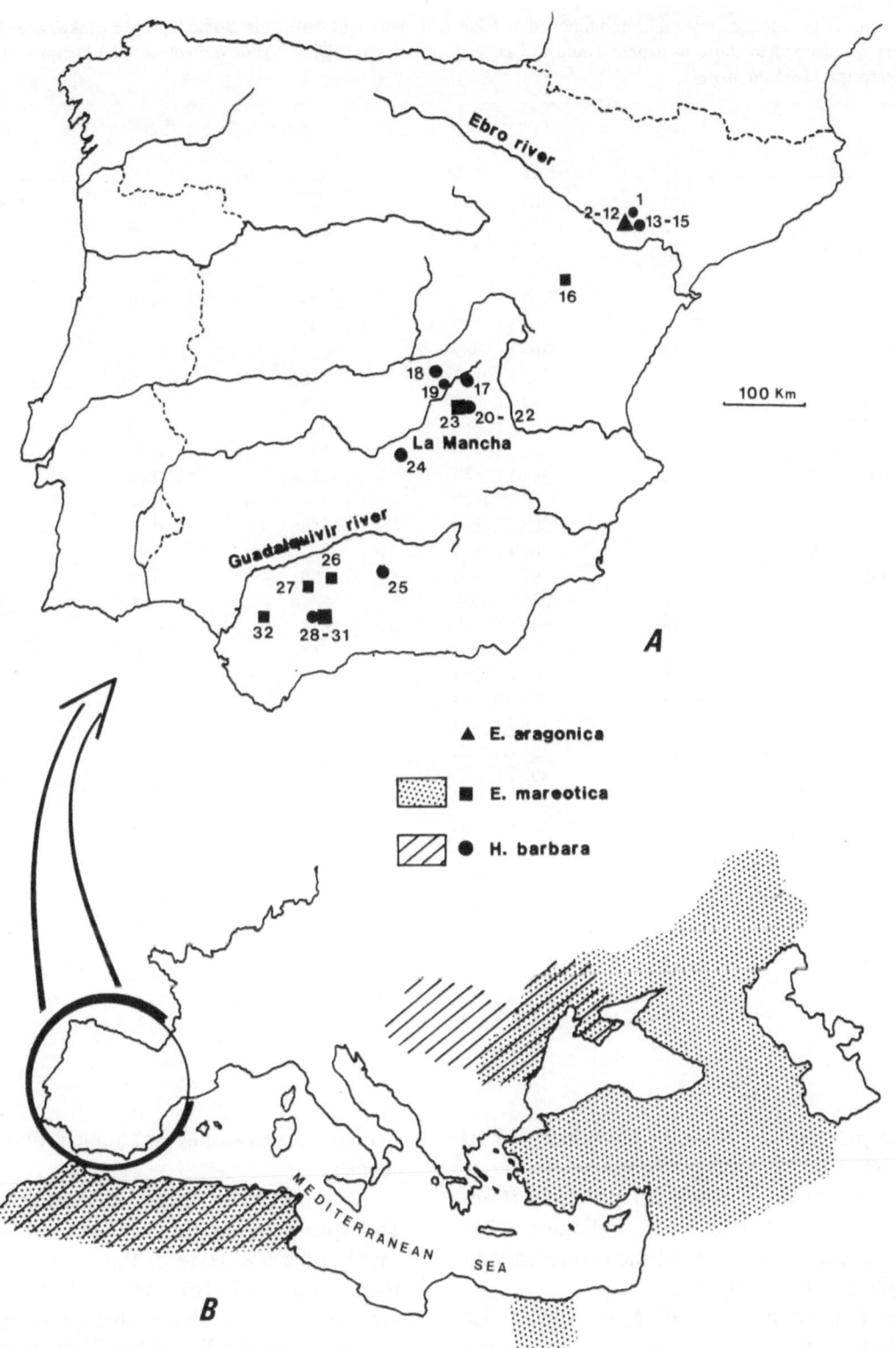

Fig. 1. Iberian (A) and world (B) distribution of the three ostracod species which inhabit Iberian athalassic saline lakes.

Table 1. List of the Iberian mesosaline and hipersaline lakes with halobiont ostracods. Some hiposaline lakes are also included. Numbers correspond to those in figures 1 and 5. Species symbols are: (■) *Eucypris mareotica*; (▲) *Eucypris aragonica* and (●) *Heterocypris barbara inermis*.

Lake	N°	Location	Altitude (m)	Surface (km²)	Species
El Salobral	1	30TYL3898	325	0.08	●
El Saladar	2	30TYL3797	322	0.44	▲
Salina de la Playa	3	30TYL3488	324	1.92	▲
Pueyo	4	30TYL3688	326	0.18	▲
Pito	5	30TYL3788	328	0.40	▲
Guallar	6	30TYL3187	336	0.08	▲
Salina del Piñol	7	30TYL2987	339	0.10	▲
Salina de la Muerte	8	30TYL2887	330	0.10	▲
Salina del Camarón	9	30TYL2687	326	0.29	▲
Salina del Rollico	10	30TYL2685	322	0.30	▲
Salada del Pez	11	30TYL2984	330	0.06	▲
Salina del Rebollón	12	30TYL2584	316	0.11	▲
Amarga I	13	30TYL4687	332	0.12	●
Amarga II	14	30TYL4888	350	0.14	●
Hoya de los Aljeces	15	30TYL3183	335	0.16	●
Gallocanta	16	30TYL2636	1000	13.30	■
Hito	17	30SWK2613	830	2.91	●
Altillo	18	30SVJ7394	680	0.14	●
El Salobral	19	30SVH4997	667	0.08	●
Sánchez-Gómez	20	30SWJ1364	680	0.50	●
Dehesilla	21	30SWJ3063	680	0.17	●
Manjavacas	22	30SWJ1162	670	1.06	●■
Pueblo	23	30SWJ0563	654	0.38	■
Pozuelo	24	30SVJ2707	620	0.46	●
Conde	25	30SVG9359	420	0.46	●
Salada de Tiscar	26	30SVG3844	180	0.11	■
Ballestera	27	30SVG0738	150	0.25	■
Ratosa	28	30SVG4918	450	0.17	■
Fuente de Piedra	29	30SVG4208	410	12.19	■
Salada	30	30SVG3600	460	0.13	■
Cerero	31	30SVG3800	460	0.04	■●
Zarracatin	32	30STG5002	50	0.55	■

Heterocypris barbara inermis (Gauthier, 1928) Margalef, 1948
This species occurred mainly in La Mancha (Central Spain) but also in three Aragonese lakes (Salobral and Amargas I & II) and in one Andalusian lake (Cerero) (Fig. 1A).

Cerero (Andalucia) and Manjavacas (La Mancha) are the only lakes where *H. b. inermis* has been found with another ostracod species, *E. mareotica*.

Figure 3 shows key anatomical details of this species.

Eucypris aragonica Brehm & Margalef, 1948.
This species has so far only been found in a very small endorheic zone in the Ebro valley of the Iberian Peninsula (Fig. 1A). Due to an unusual very wet year, we were able to collect several samples from the Bujaraloz-Sástago lakes complex and found many fullgrown specimens of *E. aragonica*.

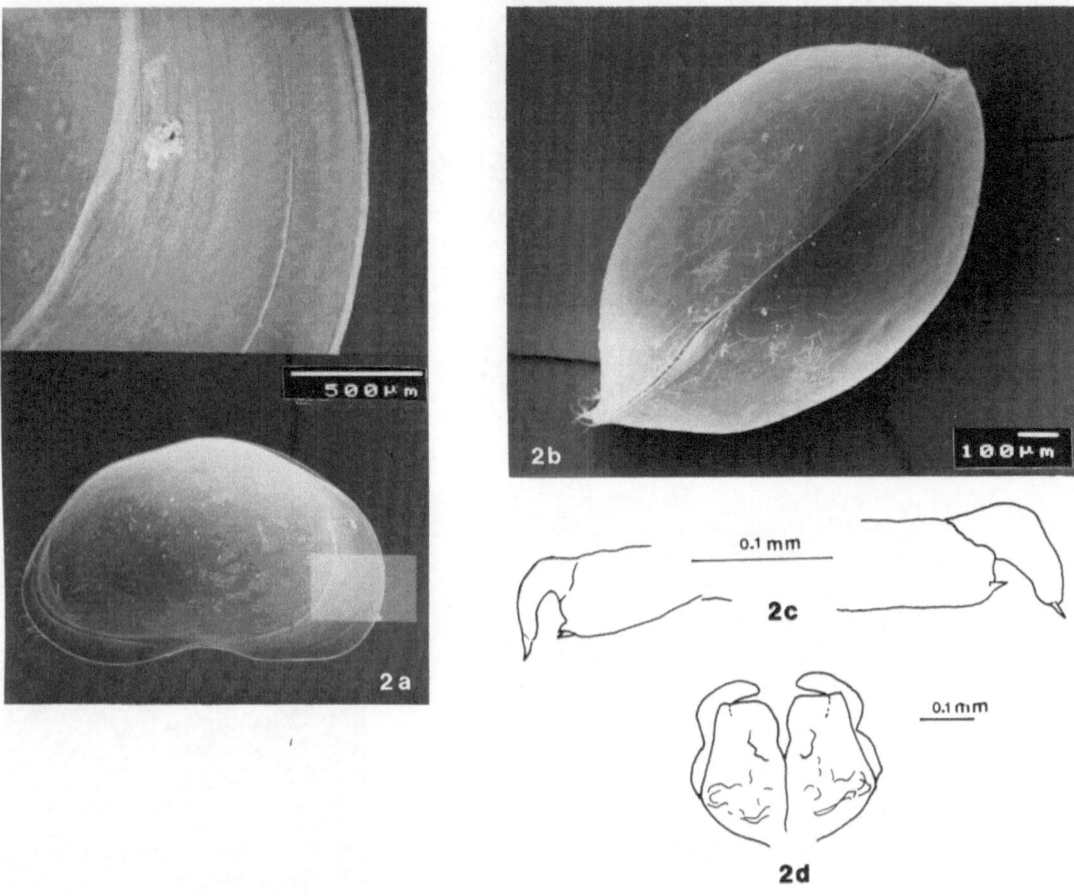

Fig. 2. Eucypris mareotica. a) Lateral view of the left valve and a detail of the inner lamella, b) dorsal view of the carapace, c) maxillae (male), d) copulatory organ (male).

Morphological descriptions are given in Brehm & Margalef (1948). Figure 4 shows key anatomical details of this species.

All these ostracods inhabit temporary waters, although *H. b. inermis* has produced several generations in laboratory cultures without dessication (unpublished data). They all are benthonic but because of the very low depth of these lakes (see Table 2) they also are often found in the water column.

Ecological factors

The three species occur in overlapping salinity ranges: 12.9–80.9‰ for *E. mareotica* and 2.6–116.6% for *H. b. inermis* (Table 2). *E. ara-gonica* has a narrower range (10–51‰); this agrees with Marín (1984) who found this species only in salinities below 85‰.

Instead of salinity *per se*, the two anions, Cl^{-1} and SO_4^{2-}, seem to be important in determining occurrence of the species. *E. mareotica* is strongly associatedpresents with sodium-chloride waters; *H. b. inermis* 'prefers' sulphate-rich waters and *E. aragonica* inhabits exclusively waters rich in chlorides but also with a high SO_4^{2-} content.

In order to identify quantitatively the factors that may explain this pattern, a stepwise discriminant analysis (Dixon, 1983) was performed on water chemical features summarized in Table 2. Groups were defined by the presence of each of the three ostracod species considered. Only one

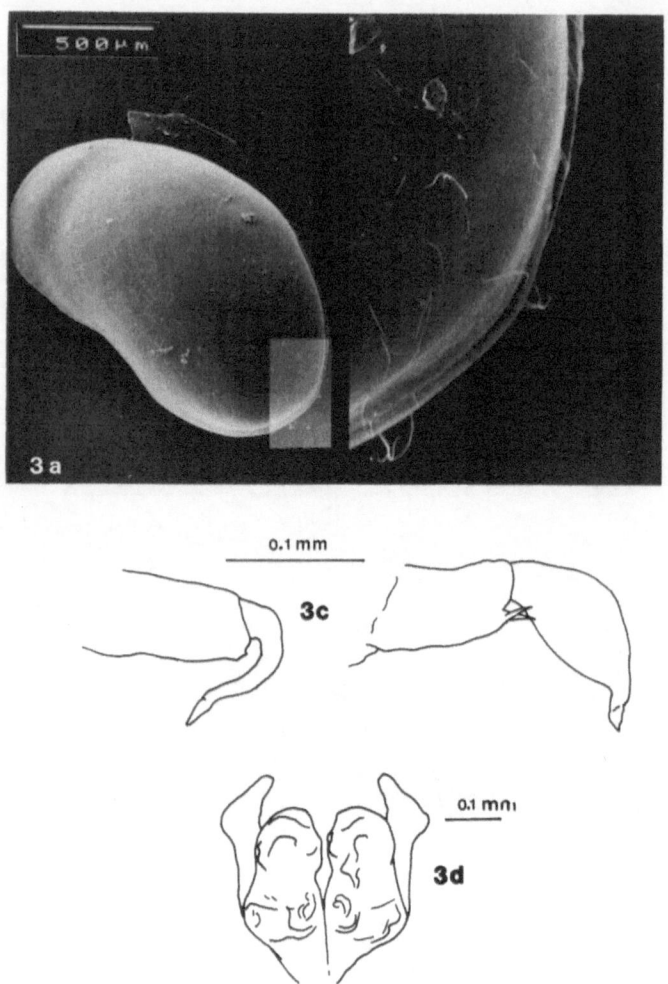

Fig. 3. Heterocypris barbara inermis. a) Lateral view with detail of the posterior part of the left valve, b) ventral view of the carapace, c) maxillae (male), d) copulatory organ (male).

variable, the Cl^-/SO_4^{2-} ratio, had a significant relationship ($p < 0.001$) F-values (16.7). Chemical water composition may explain at a first level, the disjunct distribution pattern among the three ostracod species. In fact, lakes with quite different chemical features, because of their origins, in the same endorheic basins sustain ostracod populations according to their chemical characteristics and not with their geographical location. Thus is in the Guadalquivir basin *H. barbara inermis* inhabits Conde lake (sulphate rich waters) instead of *E. mareotica*, which inhabits the sodium-chloride lakes in that area. Similarly *E. mareotica* occurs in Pueblo Lake (La Mancha, Central

Spain) instead of *H. b. inermis* which occupies the major lakes with high sulphate in that basin. More examples can be found in the Aragonese region; the Monegros basin, a small endorheic zone inhabited by the endemic *E. aragonica*, has lakes rich in sulphates (Salobral, Amargas, Hoya de los Aljeces) in which *H. b. inermis* replaces the former species. Near Monegros is located Gallocanta Lake, characterized by a high chlorides and consequently inhabited by *E. mareotica*.

Some Iberian saline lakes were misclassified by the discriminant analysis. Two of them, Manjavacas Lake (La Mancha, Central Spain) and Cerero Lake (Guadalquivir basin) have very balanced

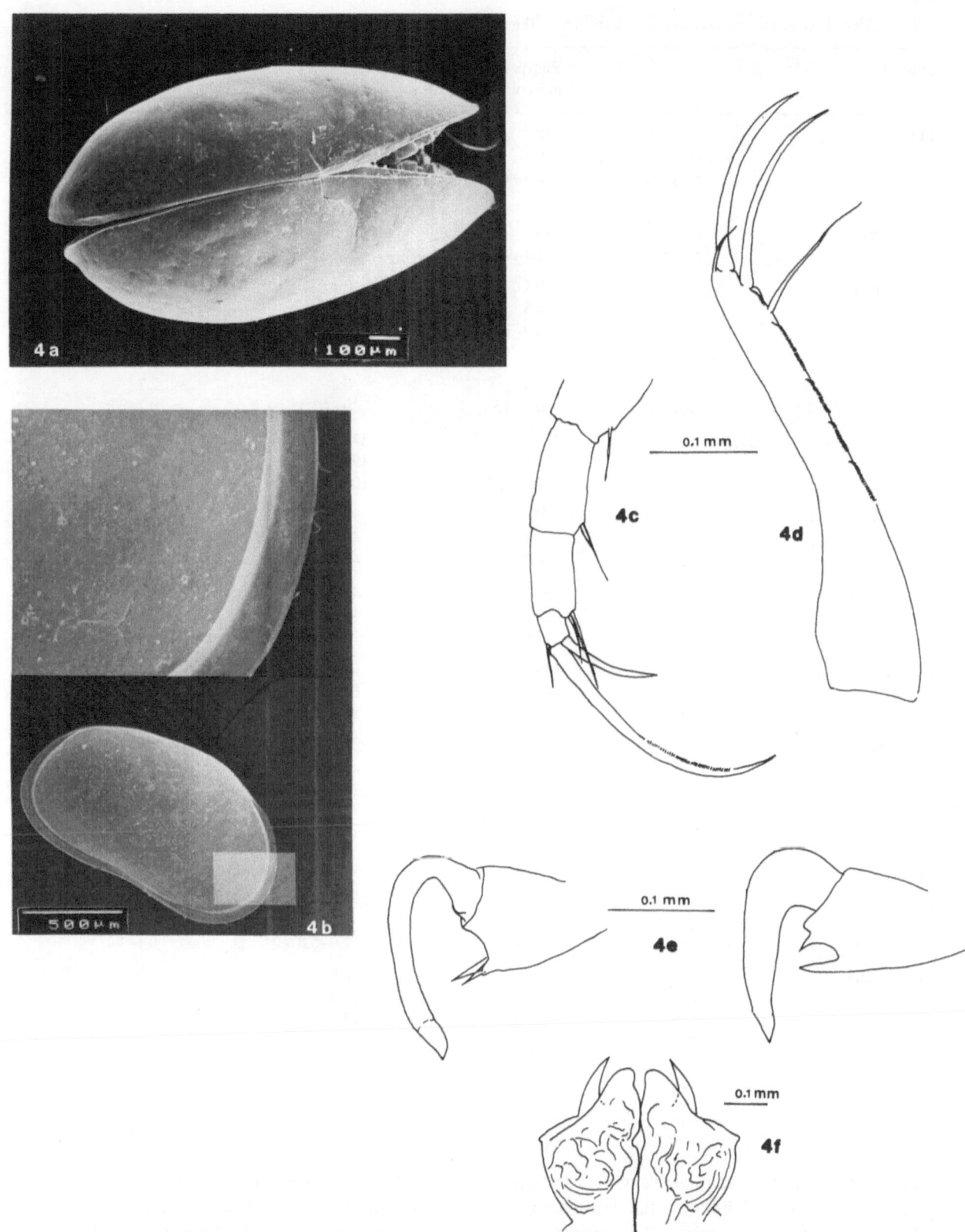

Fig. 4. Eucypris aragonica. a) Dorsal view of the carapace, b) lateral view of the left valve with detail of the inner lamella, c) first thoracopod, d) furcal ramus, e) maxillae, f) copulatory organ (male).

Table 2. Variation in salinity and ion concentration in the water samples collected in a number of three ostracod habitats.

Species		Eucypris mareotica	Eucypris aragonica	Heterocypris barbara
TDS (%)	x	38.2	28.2	39.8
	s	20.3	11.8	31.7
	range	12.9–80.9	10.1–50.5	2.6–116.6
Conductiv. ($mS\ cm^{-1}$)	x	50.0	33.4	35.3
	s	28.9	13.0	22.7
	range	11.3–95.0	12.6–56.1	2.6–84
CO_3^{2-} ($meq\ l^{-1}$)	x	0.1	0.1	2.3
	s	0.3	0.12	2.9
	range	0.0–1.1	0.0–0.38	0.0–1.7
HCO_3^{-} ($meq\ l^{-1}$)	x	1.9	1.9	3.4
	s	0.9	0.8	2.7
	range	0.7–4.5	0.8–3.3	0.7–8.3
CL^{-} ($meq\ l^{-1}$)	x	479.1	378.2	220.2
	s	282.5	192.8	190.6
	range	115.6–956.0	107.0–806.3	3.1–654.2
$SO4^{2-}$ ($meq\ l^{-1}$)	x	79.8	99.7	318.2
	s	34.4	43.4	288.0
	range	46.4–135.8	40.6–193.6	37.9–895.3
Ca^{2+} ($meq\ l^{-1}$)	x	53.2	45.1	29.4
	s	22.2	12.0	8.9
	range	17.6–99.8	27.7–59.9	10.9–45.1
Mg^{2+} ($meq\ l^{-1}$)	x	144.2	91.0	422.1
	s	101.4	67.9	466.8
	range	46.4–290.5	16.9–263	22.8–1485.9
Na^{+} ($meq\ l^{-1}$)	x	483.5	342.1	185
	s	286.4	140.8	131.9
	range	53.1–957	137.0–604.6	35.2–435
K^{+} ($meq\ l^{-1}$)	x	2.6	6.3	1.5
	s	1.7	3.6	2.1
	range	1.4–5.1	0.8–10.9	0.02–6.1
Cl^{-}/SO_4^{2-}	x	6.1	3.8	0.7
	s	2.9	1.3	0.4
	range	1.4–12	2.6–5.8	0.1–1.1
Depth (cm)	x	35.5	26.9	26.4
	s	14.9	11.9	16.2
	range	35–60	15–42	4–60
Water persistence (months)		4–9	4–7	2–8

values for the chloride-sulphate ratio (1.8 and 2.6, respectively). These lakes are the only ones in which two ostracod species, *H. b. inermis* and *E. mareotica*, co-occur. Two lakes from the Monegros (Ebro basin) also were misclassified, Rollico Lake and Salada del Pez lake, which have high chloride concentration. *E. aragonica* inhabits these lakes but the discriminant analysis asigns *E. mareotica* to them.

E. aragonica has never been collected with another ostracod species.

Discussion

Species distribution

Only three species of ostracods belonging to two genera from the non-marine family *Cyprididae* were collected in mesohaline and hyperhaline Spanish waters. Absence of cytherid species, unable to produce dry resistant eggs, is probably due to the ephemeral character of the lakes considered. The low ostracod species richness in these habitats agrees with known data from other geographical regions. As De Deckker (1983) points out only 10 of the 44 species that inhabit athalassic environments in Europe are truly halobiont. In Canada (Delorme, 1969a, 1969b) just four species are known to live in TDS values above 10‰. In this sense Australia is an exception having a large number of halobiont ostracods, many of them endemic (De Deckker, 1981, 1983).

Eucypris mareotica is a common inhabitant of North African and Central Asian saline waters (Fischer, 1955; Sars, 1903; Gauthier, 1928; Hartmann, 1964; Schornikov, 1964; Löffler & Danielopol, 1978; De Deckker, 1981) (Fig. 1B). It follows the pattern of the Mediterranean East-West disjunction (Brehm, 1947; Miracle, 1982). It typically occurs in waters with salt concentrations up to 110‰ (Löffler, 1961) and it also can resist wide salinity ranges (Bronstein, 1947).

Eucypris aragonica was described from specimens reared from dried mud from Piñol Lake (Monegros, Zaragoza) (Brehm & Margalef, 1948). Afterwards it was recorded by Marín (1983, 1984), who dealt with some aspects of the growth and development of this species reared in the laboratory. This species has never been found outside the endorheic area of Bujaraloz-Sástago (Monegros, Zaragoza). It is characteristic of temporary and saline waters (Brehm & Margalef,

1948; Marín, 1984) but nothing else is known about its natural habitats. Only amphigonic populations have been found.

H. barbara inermis was described by Gauthier (1928) from North-African saline lakes, whereas *H. barbara barbara* (Gauthier & Brehm, 1928) seems to inhabit freshwaters. *H. b. inermis* lacks the characteristic tubercules on the margin of the right valve. Other features are well described both in Gauthier & Brehm (1928) and in Gauthier (1928). Males of *H. b. inermis* have so far not been described but the ones found in Iberian saline lakes were similar to the males described for *H. b. barbara*, apart from the fact that they lack the ventral tubercules (Fig. 3). Its distribution is similar to that of *Eucypris mareotica*, a typical East-west Mediterranean disjunct (Gauthier & Brehm, 1928; Gauthier, 1928b; Löffler & Danielopol, 1978). In the Iberian Peninsula the species has been reported mainly from La Mancha (Central Spain) (Margalef, 1948; Armengol *et al.*, 1972). It has also been recorded in Doñana (Southwestern Spain) (Bigot & Marazaof, 1965), a saltmarsh not included in this paper because of its thalassic nature. This species is a typical saline ostracod which prefers temporary (Gauthier, 1928b) and sulphate-rich waters (Margalef, 1953). Amphigonic and parthenogenetic populations are known to live in salinities up to 88% (Löffler, 1961).

Comín *et al.* (1983) reported *Heterocypris salina* from Gallocanta lake during September-December 1980. Instead, we confirm the presence of an amphigonic population of *Eucypris mareotica* in one sample from the same lake collected in July 1980. The former species could be considered as uncertainly classified because of the high salt content of the lake waters (minimum of 32‰) which seems inappropriate for the maintenance of a stable population of this species according to the laboratory experiments carried out by Ganning (1971).

In the Monegros area, Marín (1984) recorded *Heterocypris salina* from Salobral Lake where we have found *H. b. inermis*. He raised both males and females from mud cultures but males of *H. salina* are not known in Europe nor in North

Africa. Description and drawings of Marín's specimens do not allow us to confirm his identification so we also consider it as dubious.

Bigot & Marazanof (1965) collected *H. barbara* in Doñana National Park (specimens determined by Margalef). Afterwards, Marazanof (1967) found no specimens of *H. barbara* but instead *Heterocypris exigua* (material examined and identified by G. Hartmann). *H. exigua* (Gauthier & Brehm, 1928) is another Northafrican species easily distinguishable from the former on the basis of the shape of segment II of the mandibular palp. In the paper of Marazanof (1967), *H. barbara* was only found in the Fuentedepiedra Lake, a place where we have collected *E. mareotica*.

Hammer (1986) assumes the presence of *Eucypris virens* in high saline habitats (10–100%) based on data from Armengol *et al.* (1975). Nevertheless, this cosmopolitan species is well known to inhabit freshwaters and slightly saline waters (Margalef, 1953; De Deckker, 1981) but not waters with high salt content. Lakes mentioned in Armengol *et al.* (1975) can be included in the hyposaline type but not in the meso- or hypersaline ones. We consider Hammer's (1986) statement to be incorrect.

Importance of ecological and historical factors

Iberian-Pontocaspian distribution can be observed in many floristic and faunistic elements in the area and has been explained on the basis of historical factors (Brehm, 1947; Miracle, 1982; Alonso, 1985b). Recent steppic lakes are considered as the remains of a Tertiary steppic area, around the Mediterranean Sea, that was fragmented during glacial periods. On the other hand, *E. aragonica* is an Iberian endemic limited to a small endorheic zone in the Ebro valley, which is a relatively well isolated river basin.

Nevertheless, it is far from proven that steppic conditions have persisted since the end of the Tertiary until now (Pérez-González, 1982). Consequently it is difficult to establish whether or not the ostracod fauna inhabiting Iberian saline lakes is the remains of Tertiary communities or is due

to later colonization (Margalef, 1947). In elucidating this question paleontological data would be of major interest. As far as we know there are no information on ostracod fossils for the endorheic basins considered here. Despite this, we could consider look the possibility that the patterns observed may have arisen by recent colonization (or by a series of recurrent colonizations). But first we would have to account for the dispersion abilities of the species found in the Iberian saline lakes.

Cypridid ostracods are well known to produce resistant eggs that can survive dessication and be available for easy dispersal. Several records of the passive dispersal of ostracod eggs (and ostracods themselves) are documented in literature: by moist wind (McKenzie, 1971); by rice seeds (Fox, 1965; Moroni & Ghetti, 1970); fishes (Kornicker & Sohn, 1971; Victor *et al.*, 1979) or insects (Fryer, 1953); inside the alimentary canal of birds (Proctor, 1964; Proctor & Malone, 1965; Frith, 1967; Frith *et al.*, 1969) and attached to the body of birds (Sandberg & Plusquellec, 1974; De Deckker, 1977).

The Iberian Peninsula is well known to be a 'land bridge' for migratory birds on their way down to Africa and back again to Europe in summer. With waterfowl migrants, lakes become of major importance because they act like stepping-stones where that birds can feed and rest. Bird-ringing and census data (SEO, 1985) support this view. A great faunal exchange occurs between La Mancha (Central Spain) and the Guadalquivir basin lakes (Southern Spain); Gallocanta (Ebro valley) lake is also important for waterfowl migrations. There is also an exchange between the Iberian Peninsula and North Africa with several waterfowl species involved: *Anas strepera*, *A. platyrhynchos*, *A. clypeata*, *Marmorenetta angustirostris*, *Aythya ferina* were ringed in Spanish lakes and recaptured in North Africa (or viceversa) (SEO, 1985). If not only ducks but waders and herons are considered as well, the existence of recent colonizations of an ostracod fauna via bird migrations seems to be highly probable, as it was also suggested for Euphyllopoda (Alonso, 1985a). This implies that historical fac-

217

tors may have been obscured by the dispersal abilities of these organisms. However, faunal exchange cannot explain for observed distribution patterns in the Iberian Peninsula (Fig. 1A), especially in the case of *Eucypris aragonica*. Other ecological factors must be considered.

Despite their capability to colonize widely separate lakes, *E. mareotica* and *H. b. inermis* exhibit a markedly disjunct distribution that can be attributed to their water chemistry requirements, but not to salinity. A similar conclusion was reached by Forester (1983) studying the mutually exclusive occurrence of *Limnocythere sappaensis* and *L. staplini* in North America: '...solute composition and not salinity is the most critical factor that control their occurrence'. All these characteristics added to the strong geographical differences in the water chemistry of the Iberian saline lakes (caused by the different genesis of each of

the endorheic basins) may account for the distributions observed.

Eucypris aragonica distribution could not be explained as clearly as desired on the basis of the arguments used above, because there is some overlap with *E. mareotica* in its chemical requirements (Fig. 5). This is a very special case because of endemic nature of the species and the highly unpredictable hydrological cycle, both in seasonal and inter-anual aspects, of the lakes in which it inhabits. *E. aragonica* has amphigonic populations only. Parthenogenesis is known to be a good strategy for improving dispersion abilities and is widely used among the ostracods (McKenzie, 1971) and other entomostracan groups like anostracans (Brown & McDonald, 1982). In fact, we have found several parthenogenetic populations of both *E. mareotica* and *H. b. inermis* and many other saline and fresh-

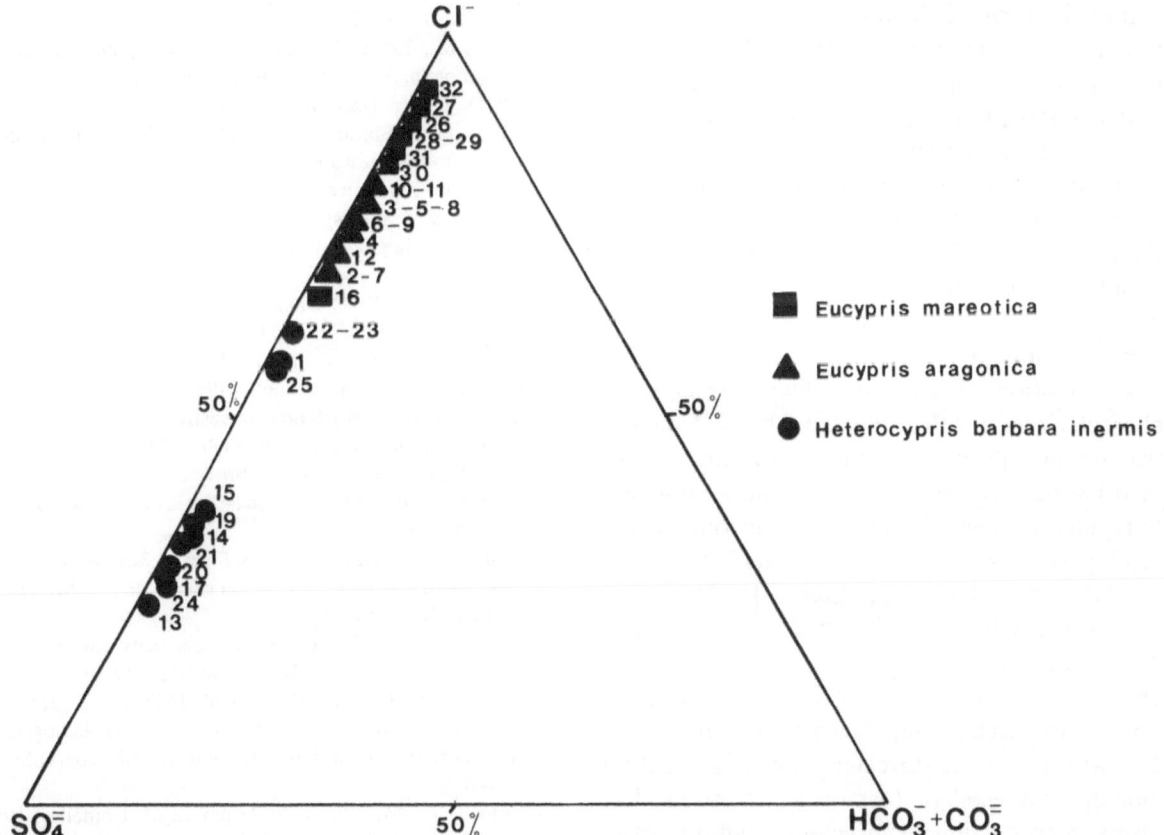

Fig. 5. Ternary diagram showing the distribution of the ostracod species on Iberian saline lakes in relation to ion composition.

218

water ostracod species have parthenogenetic populations in their European range.

Thus, the absence of parthenogenesis in *E. aragonica*, the scarcity of waterfowl on the lakes which it inhabits, the peculiar chemical composition of the lakes (rich both in chlorides and sulphates) contribute to isolate this species in the endorheic basin of Monegros. Probably, this species is so well adapted to the very special conditions of its habitat that can prevent the occurrence of other species (e.g. *E. mareotica*) which theoretically, as the discriminant analysis suggests, are capable of living in those lakes. In order to test our hypothesis we aim to make several experimental approaches in the laboratory.

Concluding remarks

A water salinity (TDS) of 10‰ seems to be a good indicator to delimit the presence of halobiont ostracod fauna. Only three species have been found in this kind of waters in Spain. They tolerate a wide range of salinity above that level.

It is known (De Deckker, 1983) that stressed habitats like saline lakes sustain very simplified communities made of highly specialized organisms, both animals and plants. In the Iberian Peninsula saline ostracod communities are mainly monospecific, except for the more stable lakes (deeper and with mixed chloride-sulphate water chemistry).

Distribution of Iberian saline ostracods is characterized by different geographical elements: two steppic species with an Ibero-Pontocaspian distribution and one, as far as we know, endemic. Sympatry in their distribution is accounted for vagility (dispersal ability via waterfowl movements) and ecological requirements. Historical factors, which have been invoked as determinants by several authors (Miracle, 1982; Alonso, 1985b) in explaining the distribution patterns of several crustacean groups, seems not to be important with ostracods. Greater knowledge of their biology and ecology (osmorregulatory mechanisms, reproduction requirements, salinity tolerance, etc.) would permit a wider evaluation of both historical and ecological factors influencing ostracod distribution in Iberian saline lakes.

Acknowledgements

Dr.K. Martens, Dr.G. Valdecasas and two anonymous referees made useful comments and suggestions. Mr. Soriano (M.N.C.N.-C.S.I.C.) provide us the samples from Gallocanta lake (July-1980). S.E.M. photographs were done by Miguel Jerez (Real Jardín Botánico de Madrid-C.S.I.C.). This research programm was supported by C.O.N.A.I. (Diputación General de Aragón) and Dirección General de Medio Ambiente (M.O.P.U.). One of us (A. Baltanás) was supported by C.A.I.C.Y.T. project PR84-0181.

References

Alonso, M., 1985a. A survey of the Spanish Euphyllopoda. Misc. Zool. 9: 179–208.

Alonso, M., 1985b. Las lagunas de la España Peninsular: Taxonomía, Ecología y Distribución de los Cladóceros. Ph.D. Thesis. Univ. Barcelona.

APHA, 1980. Standard Methods for the examination of water and wastewater. 15th Edition.

Armengol, J., M. Estrada, A. Guiset, R. Margalef, D. Planas, J. Toja & F. Vallespinós, 1975. Observaciones limnológicas en las lagunas de La Mancha. Bol. Est. C. Ecol. 4: 11–27.

Bayly, I. A. E., 1972. Salinity tolerance and osmotic behaviour of animals in athalassic saline and marine hypersaline waters. Annu. Rev. Ecol. Syst. 3: 233–268.

Bigot, L. & F. Marazanoff, 1965. Considerations sur l'écologie des invertebres terrestres et aquatiques des Marismas du Guadalquivir. Vie et Milieu 16: 411–473.

Brehm, V., 1947. Reflexiones sobre relaciones zoogeográficas de la fauna de agua dulce de la Península Ibérica. P. Inst. Biol. Apl. 4: 53–74.

Brehm, V. & R. Margalef, 1948. *Eucypris aragonica* nov. sp., nuevo ostrácodo de una laguna salada de Los Monegros. P. Inst. Biol. Apl. 5: 5–9.

Bronstein, Z. S., 1947. Ostracodes des eaux douces. Faune de l'URSS. Crustacea II (in Russian), 339 pp.

Browne, R. A. & G. H. Macdonald, 1982. Biogeography of the brine shrimp, *Artemia*: distribution of parthenogenetic and sexual populations. Journal of Biogeography 9: 331–338.

Comín, F. A., M. Alonso, P. López & M. Comelles, 1983. Limnology of Gallocanta Lake, northeastern Spain. Hydrobiologia 105: 207–221.

Comín, F. A., M. Alonso, 1988. Spanish salt lakes: Their chemistry and biota. Hydrobiologia 158: 237–245.

De Deckker, P., 1977. The distribution of the 'giant' ostracods (family: Cyprididae Baird 1845) endemic to Australia. In: Löffler, H. & D. L. Danielopol (eds.). Aspects of Ecology and Zoogeography of Recent and Fossil Ostracoda. Junk, The Hague.

De Deckker, P., 1981. Ostracods of athalassic saline lakes. Hydrobiologia 81: 131–144.

De Deckker, P., 1983. Notes on the ecology and distribution of non-marine ostracods in Australia. Hydrobiologia 106: 223–234.

Delorme, L. D., 1969a. Ostracods as Quaternary paleoecological indicators. Can. J. Earth Sci. 6: 1321–1426.

Delorme, L. D., 1969b. On the identity of the ostracode genera Cypriconcha and Megalocypris. Can. J. Zool. 47: 271–281.

Diner, M. P., E. P. Odum & P. F. Hendrix, 1986. Comparison of the roles of Ostracods and Cladocerans in regulating community structure and metabolism in freshwater microcosms. Hydrobiologia 133: 59–63.

Dixon, W. J. (Ed.), 1983. BMDP Statistical Software. University of California Press.

Fischer, S., 1855. Beitrag zur kenntnis der Ostracoden. Abhandlungen der Matematisch-Physikalischen classe der Koeniglich-Bayerischen Akademie der Wissenschaften 7: 635–666.

Forester, R. M., 1983. Relationship of two lacustrine ostracod species to solute composition and salinity: Implications for palaeohydrochemistry. Geology 11: 435–438.

Fox, H. M., 1965. Ostracod Crustacea from ricefields in Italy. Memorie dell'Istituto Italiano di Idrobiologia 'Dott. Marco de Marchi', Pallanza 18: 205–214.

Frith, H. J., 1967. Waterfowl of Australia. Angus & Robertson, Sydney.

Frith, H. J., L. W. Braithwaite & J. L. McLean, 1969. Waterfowl in inland swamp in New South Wales. II. Food. CSIRO Wild Res. 14: 17–64.

Fryer, G., 1953. Notes on certain freshwater crustaceans. Naturalist, Hull 846: 101–109.

Ganning, B., 1971. On the ecology of Heterocypris salinus, H. incongruens and Cypridopsis aculeata (Crustacea: Ostracoda) from brackish-water rockpools. Mar. Biol. 8: 271–279.

Gauthier, H., 1928a. Ostracodes et Cladoceres de l'Afrique du Nord (Iière note). Bull. Soc. Hist. nat. Afr. N. 19: 10–19.

Gauthier, H., 1928b. Rècherches sur la faune des eaux continentales de l'Algerie et de la Tunisie. Lechevalier, Paris, 420 pp.

Gauthier, H., 1931. Faune aquatique du Sahara centrale. Récoltes de MLG Seurat au Hoggar en 1928. Bull. Soc. Hist. nat. Afr. N. 22: 350–400.

Gauthier, H. & Brehm, V., 1928. Ostracode et Cladoceres de l'Algerie et de la Tunisie (3ième note) Bull. Soc. Hist. nat. Afr. N. 19: 114–121.

Hammer, U. T., 1986. Saline lakes ecosystems of the world. Junk Publishers. Lancaster. 616 pp.

Hammer, U. T., J. Shames & R. C. Haynes, 1983. The distribution and abundance of algae in saline lakes of Saskatchewan, Canada. Hydrobiologia 16: 1–26.

Hartmann, G., 1964. Asiatische Ostracoden, systematische und zoogeographische Untersuchungen. Int. Revue ges. Hydrobiol., Syst. beihefte 3, Akademie-Verlag, Berlin, 155 pp.

Kornicker, L. S. & I. G. Sohn, 1971. Viability of ostracod eggs egested by fish and effect of digestive fluids on ostracod shell- ecologic and paleontologic implications. Bull. Centre. Rech. Pau SNPA 5 Suppl. 207–237.

Löffler, H. H., 1961. Beiträge zur Kenntnis der Iranischen Binnengewasser II. Int. Revue ges. Hydrobiol. 46(3): 309–406.

Löffler, H. H. & D. D. Danielopol, 1978. Ostracoda (Cyprididae). In: Illies, J. (Ed.) Limnofauna Europaea. G. Fisher Verlag. Stuttgart.

Marazanof, F., 1967. Ostracodes, Cladoceres, Heteropteres et Hydracariens nouveaux pour les marismas du Guadalquivir (Andalousie) donnes ecologiques. Annls. Limnol. 3: 47–64.

Margalef, R., 1947. Estudios sobre la vida en las aguas continentales de la región endorreica manchega. P. Inst. Biol. Apl. 4: 5–51.

Margalef, R., 1953. Los Crustáceos de las Aguas Continentales Ibéricas. Biología de las Aguas Continentales 10: 1–243.

Margalef, R., 1956. La vida en las aguas de elevado residuo salino de la provincia de Zamora. P. Inst. Biol. Apl. 24: 123–137.

Marín, J. A., 1983. Las lagunas atalasohalinas de los Monegros (zona de Bujaraloz-Sástago) como ambiente natural de los ostrácodos Eucypris aragonica y Heterocypris salina. Tesis de Licenciatura. Univ. de Barcelona, 114 pp.

Martens, K., 1984. Anotated checklist of the non-marine ostracods from african inland waters. Musee Royal de lÿfrique Centrale, Documentation Zoologique, n. 20, 51 pp.

McKenzie, K. G., 1971. Paleozoogeography of freshwater Ostracoda. Bull. Centre Rech. Pau SPA 5 Suppl.: 207–237.

McKenzie, K. G., 1981. Paleobiogeography of some salt lake faunas. Hydrobiologia 82: 407–418.

Miracle, M. R., 1982. Biogeography of the freshwater zooplanktonic communities of Spain. Biogeogr. 9: 455–467.

Montes, C. & P. Martino, 1987. Las lagunas salinas españolas. Seminario sobre Bases Científicas para la Protección de los Humedales en España. Real Academia de Ciencias Exactas, Físiscas y Naturales: 95–145.

Moroni, A. & P. F. Ghetti, 1970. Popolamente di Ostracodi di risaie Italienne. Note de Ecologia Bull. Zool. 37: 1–2.

Pérez-González, A., 1982. Neógeno y Cuaternario de la llanura manchega y sus conexiones con la cuenca del Tajo. Tesis Doctoral. Univ. Complutense de Madrid.

Por, F. D., 1980. A classification of hypersaline waters, based on trophic criteria. Mar. Biol. 1: 121–131.

Proctor, V. W., 1964. Viability of crustacean eggs recovered from ducks. Ecology 45: 656–658.

Proctor, V. W. & C. R. Malone, 1965. Further evidence of the

220

passive dispersal of small aquatic organisms via the iternal tract of birds. Ecology 46: 728–729.

Sandberg, P. A. & P. L. Plusquellec, 1974. Notes on the anatomy and passive dispersal of Cyprideis (Cytheracea, Ostracoda). Geoscience and Man 6: 1–26.

Sars, G. O., 1903. On the Crustacean fauna of Central Asia. Part 3: Copepoda and Ostracoda. Anuaire du Musee Zoologique de l'Academie Imperiale des Sciences de St. Petersbourg, 8(2): 195–264.

Schornikov, E. I., 1964. Ecological questions on Azov-Black Seas Ostracods. Biologia Moria 26: 53–88.

S.E.O., 1985. Estudio sobre la biología migratoria del orden Anseriformes (Aves) en España. Monografías No. 38, I.C.O.N.A.

Victor, R., G. L. Chan & C. H. Fernando, 1979. Viability of ostracods passing through the gut of *Catostomus comersoni* (Lacepede) (Pisces, Catostomidae). Can. J. Zool. 57: 1745–1747.

Wickstrom, C. E. & R. W. Castenholz, 1978. Association of *Pleurocaspa* and *Calothrix* (Cyanophyta) in a thermal spring. J. Phycol. 14: 84–88.

Williams, W. D., 1981. The limnology of asaline lakes in Western Victoria. A review of some recent studies. Hydrobiologia 82: 233–259.

Hydrobiologia **197**: 221–231, 1990.
F. A. Comín and T. G. Northcote (eds), Saline Lakes.
© 1990 Kluwer Academic Publishers.

Anostraca, Cladocera and Copepoda of Spanish saline lakes

M. Alonso
Departament d'Ecologia, University of Barcelona, Spain; Present address: LIMNOS, S.A., Bruc 168, entlo. 2a., 08037, Barcelona, Spain

Key words: anostraca, cladocera, copepoda, saline lakes

Abstract

A study of 102 samples from almost all salt water bodies in Spain has allowed the preparation of a comprehensive list of anostracans, cladocerans and copepods living in such extreme environments. Among the 26 species recorded, 9 are halobionts, but 17 can exist in less saline waters. Of the halobionts, several are widely distributed throughout arid areas around the Mediterranean (*Arctodiaptomus salinus, Cletocamptus retrogressus, Branchinectella media, Branchinella spinosa, Daphnia mediterranea, Moina salina*); *Branchinecta orientalis* (= *B. cervantesi*) only appears in Guadiana watershed and toward the east of Hungary, and the *Alona* belonging to the *A. elegans* complex is a Spanish endemic. In the second group are many typically freshwater species which also appear occasionally in saline waters, and colonizers of wetlands in steppes, characteristically adapted to a wide range of salinity; one of the formers, *Diaphanosoma* cf. *mongolianum*, deserves closer study. The Spanish halobiontic fauna seems to be very old judging by the existence of some isolated species, e.g. *B. orientalis* may be a Tertiary relic. Persistence through time could have resulted from the continuous aridity of some Iberian localities during the Pleistocene and the ecological constancy of wetlands maintained by regional groundwater discharges.

Introduction

Anostracans, cladocerans and copepods are often the most abundant invertebrates of inland temporary saline lakes. Although biological adaptations, types and species richness in such lakes are similar around worldwide, the taxa are regionally different and form characteristic regional groups. The Mediterranean saline lake fauna was one of the earliest studied (e.g. Gauthier, 1928; Beadle, 1943; Margaritora, 1971, 1972; Margalef, 1947). Later papers of significance include those on saline lakes of Morocco (Thiery, 1986), Rumania (Negrea, 1983) and Spain (Armengol *et al.*, 1975; Alonso & Comelles, 1981, 1984; Comín & Alonso, 1988). The present paper provides further information on species in Spanish saline lakes of salinities > 5‰, with special reference to their autoecology and geographical distribution.

Material and methods

Collections were made between 1978 and 1986 from 99 temporary saline lakes distributed over the main Spanish endorheic areas (Fig. 1). A few of the lakes were sampled several times giving a

222

Table 1. Saline lakes sampled in the five endorheic areas of Spain grouped according their relative anion concentration (ACS = total alkalinity > chloride > sulphate, CAS = chloride > total alkalinity > sulphate, CSA = total alkalinity > sulphate > chloride > total alkalinity, SCA = sulphate > chloride > total alkalinity). Bracketed numbers are salinities as ‰ (*data from Martino, 1988). Dates of sampling are also indicated.

Chemical type	Watershed				
	Duero	Guadiana	Guadalquivir	Segura	Ebro
ACS	Caballo Alba, Jul 79 (15)				
CAS	Bodón Blanco, Mar 80 (6)				
CSA	Bodón Blanco, Mar 83 (17) Las Eras, May 80 (6) Las Eras, Jul 79 (29)	Retamar, Apr 79 (8) Cambronera, Apr 79 (14) Petrola Pequeña, Apr 81 (20) Salobral, Apr 79 (20) Altillo I Apr 79 (22) Albardiosa, Apr 79 (38) Larga Villacañas, Apr 79 (39) Zacatena, Apr 82 (39) Saladar, Apr 79 (119) Petrola Grande, Apr 81 (125) Peña Hueca, Apr 79 (139)	Amarga, Apr 82 (8) Ballestera, Apr 79 (13) Benamejí, Apr 82 (14) Salada Puentegenil, Apr 82 (20) Zarracatin, Apr 79 (21) Conde, Feb 86 (23)* Gosque, Feb 86 (25)* Tiscar, Feb 86 (26) Zarracatin, Feb 86 (27) Salada Campillos, Apr 79 (30) Gosque, Apr 79 (31) Ralosa, Feb 86 (38)* Fuentedepiedra, Apr 79 (42) Conde, Apr 82 (47) Palmar de Troya, Nov 78 (59) Gosque, Nov 78 (59) Zarracatin, Nov 78 (62)	Corral Rubio, Apr 82 (28)	Santed, Mar 81 (9) Carralogroño, Mar 86 (23)* Gallocanta, Apr 79 (26) Carravalseca, Mar 86 (47)* Grande Alcañiz, Mar 84 (58) Salineta, May 80 (437)
SCA	Boecillo, Mar 83 (17)	Camino Villafranca, Nov 78 (6) Grande Villafranca, Nov 78 (6) Hito, Apr 79 (6) Laguna del Pueblo, Apr 79 (7) Pajares, Jan 86 (15)* Longar, Nov 78 (16) De la Sal, Apr 79 (20) Longar, Apr 79 (21) Grande Villafranca, Apr 79 (23) Dehesilla, Mar 86 (24) Navarredonda, Apr 79 (25) Sanchez Gómez, Mar 86 (27)* Altillo I, Nov 78 (39) Salobrejo, Apr 79 (40) Navalafuente, Apr 79 (41) Larga Villacañas, Nov 78 (43) Alabardiosa, Nov 78 (72) Grande de Quero, Nov 78 (101)	Medina, Nov 78 (15) Zorrilla Salada, Nov 78 (19)	Hoya Rasa, Apr 82 (28) Atalaya Ojicos, Apr 82 (325)	Chica Alcañiz, Mar 84 (8) Magallón, Apr 80 (12) Bujaraloz, May 80 (15) Salada Chiprana, Feb 86 (84)* Salada Chiprana, Jun 80 (105) Gallocanta, Mar 81 (112)

			Sariñena, Mar 80

		Grande de Quero, Apr 79 (108)	
		Cambronera, Nov 78 (117)	
		Tirez, Nov 78 (149)	
		Tirez, Apr 79 (149)	
		Peña Hueca, Nov 78 (214)	

With no chemical information	Bodón Blanco, Jul 79	Alabardiosa, May 85	
	Las Eras, Nov 77	Alcahozo, May 85	
		Altillo I, Apr 84	
		Altillo II, May 84	
		Camino Villafranca, Nov 78	
		Chica Villafranca, Jul 84	
		Dehesilla, Apr 84	
		Grande Quero, Apr 84	
		Grande Villafranca, Jul 84	
		Hijosa, Jun 84	
		Hito, Apr 84	
		Honda, Feb 86	
		Laguna del Pueblo, Mar 86	
		La Playa, Mar 86	
		Manjavacas, Apr 34, May 85, Mar 86	
		Petrola Grande, May 85	
		Retamar, Apr 84	
		Salicor, Mar 86	
		Salobral, Apr 84, May 84	
		Salobrejo, May 85	
		Sanchez Gómez, Apr 84	
		Tirez, Jun 84	

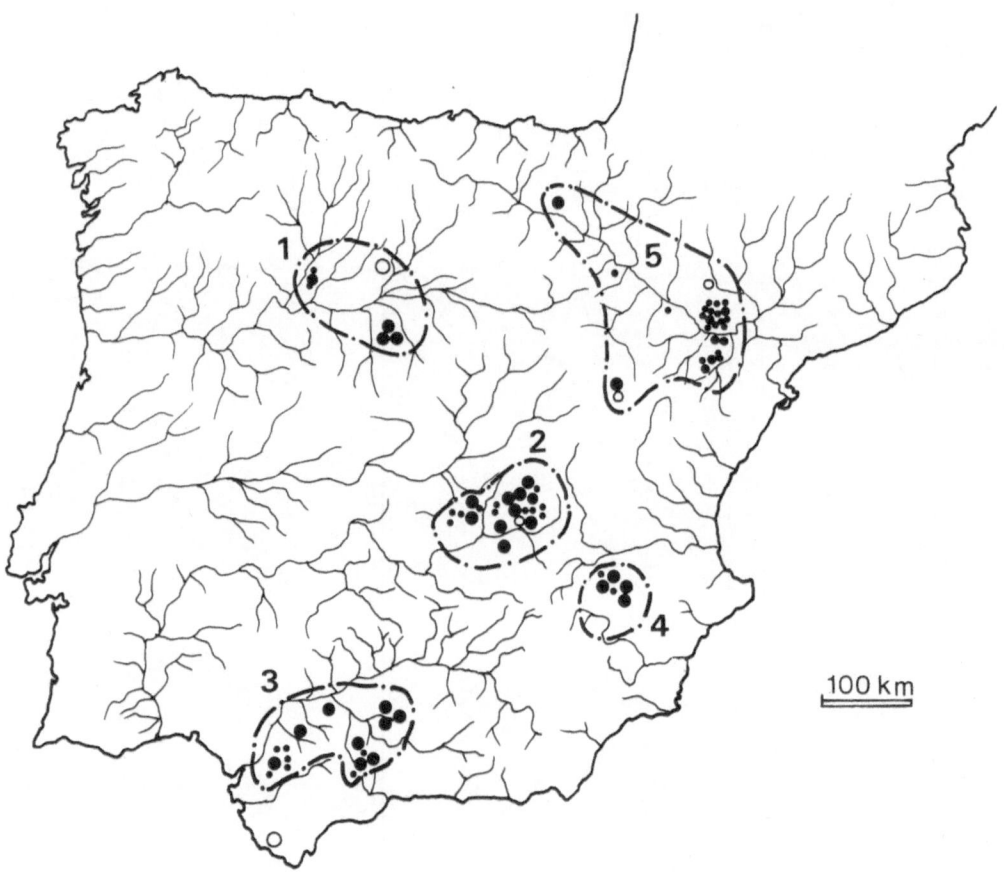

Fig. 1. Distribution of saline lake groups in Spain; 1) Duero watershed, 2) Guadiana watershed, 3) Guadalquivir watershed, 4) Segura watershed, 5) Ebro watershed. Lake size is indicated by solid circle size: lakes $< 100 \ m^2$ = small circles; lakes to 100 m^2 to 1 km^2 = large circles. The location of some dessicated or drained salt lakes are also shown by white circles. (Adapted from Comín & Alonso, 1988).

total of 102 samples. The study lakes are listed in Table 1 along with some details about chemical characteristics and sampling dates.

Samples were obtained from representative environments in each lake (littoral areas, open water, among vegetation, etc.) using a hand net of mesh size 100 μm. Specimens were preserved in 4% formaldehyde.

Results

Species composition

Twenty six species were recorded in the samples. (Table 2). Thirteen were cladocerans, 4 anostracans, and 9 copepods.

The most abundant cladocerans are *Alona* sp., *Daphnia* (*C.*) *mediterranea*, *D.* (*C.*) *magna* and *Moina salina*. The *Alona* sp. is close to *Alona elegans* and *A. elegans lebes*, but its taxonomic status cannot be defined at present without a complete revision of this species complex. Morphological differences between these three species are given in Table 3 and Fig. 2 to aid future studies. Note that until recently *Daphnia* (*C.*) *mediterranea* was confused with *D.* (*C.*) *dolichocephala*. The latter does not live in saline waters (see Alonso, 1985a). Characteristically, *D. magna* colonizes a wide range of environments, and populations in saline lakes seem to have the typical form. However, the allometry of the shell spine growth is modified with increasing salinity, so that this spine becomes very small in adults found at

Table 2. Anostraca, Cladocera and Copepoda in Spanish temporary saline lakes. Numbers reflect percentage of lakes where the species was recorded. Number of samples = 102.

Anostraca	
Branchinectella media Schmankewitsch	16
Branchinella spinosa Milne-Edwards	7
Artemia salina (L.)	2
Branchinecta orientalis Sars	2
Cladocera	
Alona sp.	18
Daphnia (C.) *mediterranea* Alonso	15
Daphnia (C.) *magna* Straus	13
Moina salina Daday	6
Moina brachiata Jurine	5
Daphnia (C.) *atkinsoni* Baird	3
Simocephalus exspinosus Koch	3
Pleuroxus letourneuxi Richard	3
Macrothrix hirsuticornis Norman & Brady	3
Ceriodaphnia reticulata Jurine	1
Dunhevedia crassa King	1
Diaphanosoma gr. *mongolianum*	*1*
Macrothrix laticornis Fischer	1
Copepoda	
Arctodiaptomus (Rh.) *salinus* Daday	58
Cletocamptus retrogressus Schmankevitsch	37
Metacyclops minutus Claus	5
Acanthocyclops sp.	3
Diacyclops bicuspidatus Claus	2
Diacyclops bisetosus Rehberg	1
Arctodiaptomus (A.) *wierzejskii* Richard	1
Mixodiaptomus incrassatus Sars	1
Megacyclops viridis Jurine	1

higher salinities (Alonso, 1985b). The Spanish *Moina salina* corresponds to the neotype established by Negrea (1984) from Rumania, although some minor differences could be noticed in the number of lateral ciliated processes of the postabdomen (7–13 in Spanish populations, 7–10 in Rumanian). Other species of Table 2 are weakly represented in saline environments, but their presence is not rare since they are typical colonizers of lacustrine environments in steppes. Of them, only *Diaphanosoma* cf. *mongolianum* deserves additional comments because of its aberrant morphology (its head is small; Korovchinsky, in litt.) and occurrence.

The anostracans *Branchinectella media* and *Branchinella spinosa* are also very common in Spanish saline environments. *Artemia salina*, occurs frequently in artificial salt pans but is scarce in natural salt athalassohaline waters where it appears only in permanent localities such as Laguna Salada de Chiprana. *Branchinecta orientalis* is also rare; it has only been found in the Hito lagoon where it has a stable population. *Branchinecta cervantesi* Margalef 1947 is a synonimy of *B. orientalis* Sars 1901 (Brtek, in litt. and original observations); Spanish literature (Alonso, 1985c) and more general works (Hammer, 1986) refer to *B. orientalis* as *B. cervantesi*.

Of the copepods, *Arctodiaptomus salinus* and

Table 3. Morphological differences between *Alona* sp. from Spain, *A. elegans* Kurz 1875 from Wurzburg (sample 4498 of Dr. Frey's collection) and *A. elegans lebes* Dumont & Van de Velde (from Rio de Oro, paratype sample 4500 of Dr. Frey's collection).

	Alona sp.	*Alona elegans*	*A. elegans lebes*
Ocellus	Faintly pigmented; smaller than the eye	Heavily pigmented; about the same size as the eye	Heavily pigmented; bigger than the eye
Postabdomen	Anal groove as long as the postanal part	Anal groove longer than he postanal part	Anal groove longer than the postanal part
Shell sculpture	Five longitudinal ridges each 50 µm. Distance between them is twice their width	Ten-twelve fine longitudinal ridges each 50 µm; distance between them is six times their width	Five longitudinal ridges each 50 µm; distance between them is three times their width

226

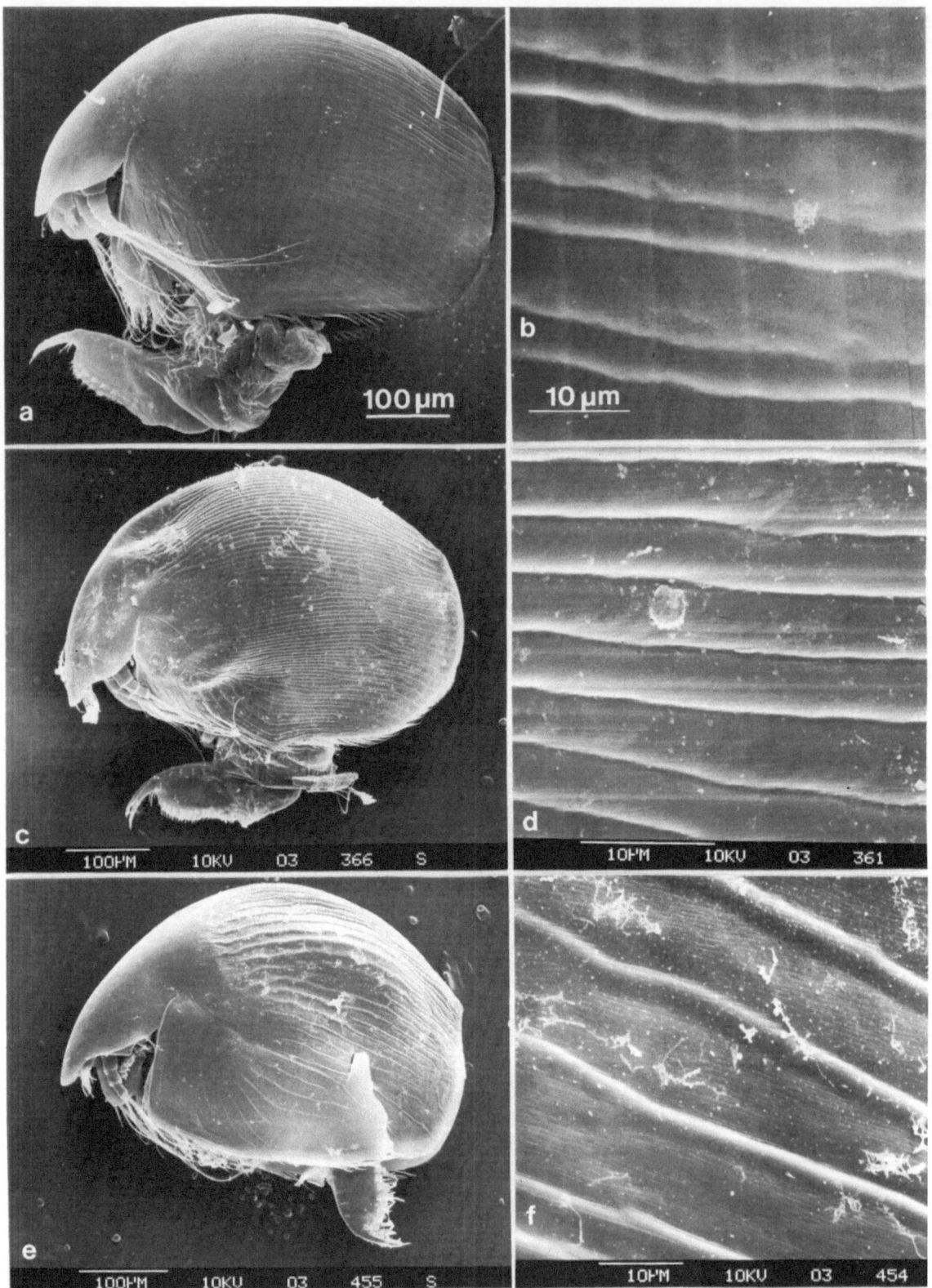

Fig. 2. Alona sp., parthenogenetic female from Laguna de las Eras; a) lateral view, b) shell sculpture. *Alona elegans* Kurz 1875 from Wurzburg, sample 4498 of Dr. Frey's Collection, c) lateral view, d) shell sculpture. *Alona elegans lebes* Dumont & Van de Velde, e) lateral view, f) shell sculpture.

227

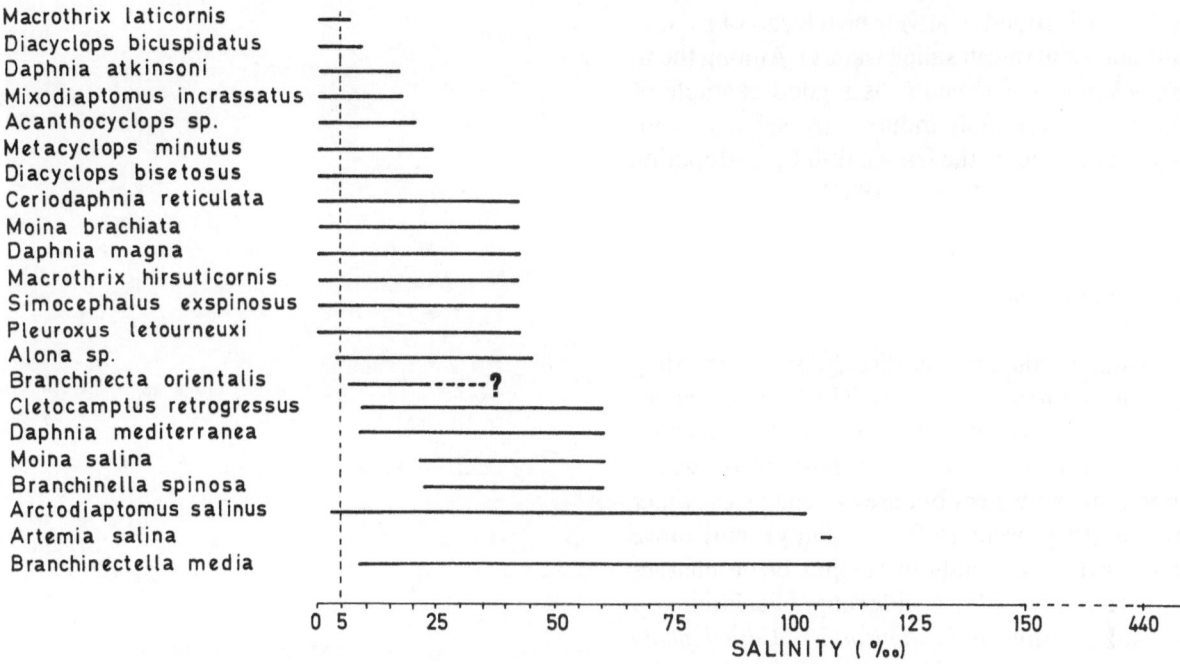

Fig. 3. Distribution of species in Spanish temporary salt lakes according to salinity.

Cletocamptus retrogressus are the most common of these crustaceans in the salt lake communities. Surprisingly, *A. salinus* also occurs in two karstic freshwater lakes in Spain (Banyoles, Miracle, 1976; Tobar, Alonso, 1985b). *Acanthocyclops* sp.

forms a complex in Spain, and detailed studies are needed to elucidate whether there is a highly variable euryoic species or a number of closely related species adapted to different environments. Other copepods are very common in waters in steppes

Table 4. Distribution of halobionts in Spanish saline lakes from five watershed areas (see Fig. 1). Salinity ranges and chemical types of each group are also given. Numbers for species are the number of localities in which the taxon was recorded.

	Watershed				
	Duero	Guadiana	Guadalquivir	Segura	Ebro
Number of samples	8	59	19	3	13
Salinity range (%)	6–29	6–214	8–26	28–325	8–437
Chemical type*	ACS, CAS, CSA, SCA	SCA, CSA	CSA, SCA	SCA, CSA	SCA, CSA
Crustaceans absent		10	1	2	2
Alona sp.	5	9	3		3
Cletocamptus retrogressus		18	13		6
Daphnia (C.) *mediterranea*		11	3		1
Moina salina			5		1
Branchinella spinosa		1	7		1
Arctodiaptomus salinus	5	34	13	1	8
Artemia salina					2
Branchinectella media		13	2		2
Branchinecta orientalis		1			

* See text for explanation.

and can withstand relatively high levels of salinity (although still rare in saline waters). Among them, *Mixodiaptomus incrassatus* is a good example of phenotypic variation induced by salinity; with increasing salinity, the female fifth leg endopodite becomes shorter (Alonso, 1984).

Species distribution

Two major groups can be distinguished according to salinity tolerance (Fig. 3). The first comprises 17 species with limits between 5 and 40‰ salinity. Here, two types of species coexist: those which appear in salt waters but are typically freshwater species (they occur to 25‰ salinity), and those which live in wetlands in steppes being characteristically adapted to a wide range of salinities up to 40‰. *Macrothrix laticornis*, *Ceriodaphnia reticulata*, *Acanthocyclops* sp. and *Simocephalus exspinosus* are examples of the first type; all others fall into the second type.

The second group comprises 9 species with salinity ranges between 5 and slightly above 100‰. The lower limit of 5‰ seems suitable to define this group, but species such as *B. orientalis* and *Alona* sp. were absent at salinities over 45‰. Although tolerance to high salinities cannot be demonstrated for these two species, they evidently do not live in fresh waters, which seems to be more important. Moreover, the data of Margalef (1953) on *B. orientalis* of $3.9\,g\,Cl^-\,l^{-1}$ and $14.6\,g\,SO^{-2}\,l^{-1}$ suggest that the salinity tolerance of this species may be higher than 18‰.

For other species, field data show that *Branchinectella media*, *Artemia salina* and *Arctodiaptomus salinus* are the most tolerant species. The former occurs in the most saline sample found with live crustaceans (108‰ salinity, Laguna Grande de Quero, April, 1979). *Cletocamptus retrogressus*, *Daphnia mediterranea*, *Moina salina* and *Branchinella spinosa* appear in waters with a salinity from 10 to 60‰.

The distribution of halobionts according to relative anionic dominance shows minor differences (Fig. 4). Most species are placed along the $Cl^- - SO^{-2}$ axis where, in any event, occur

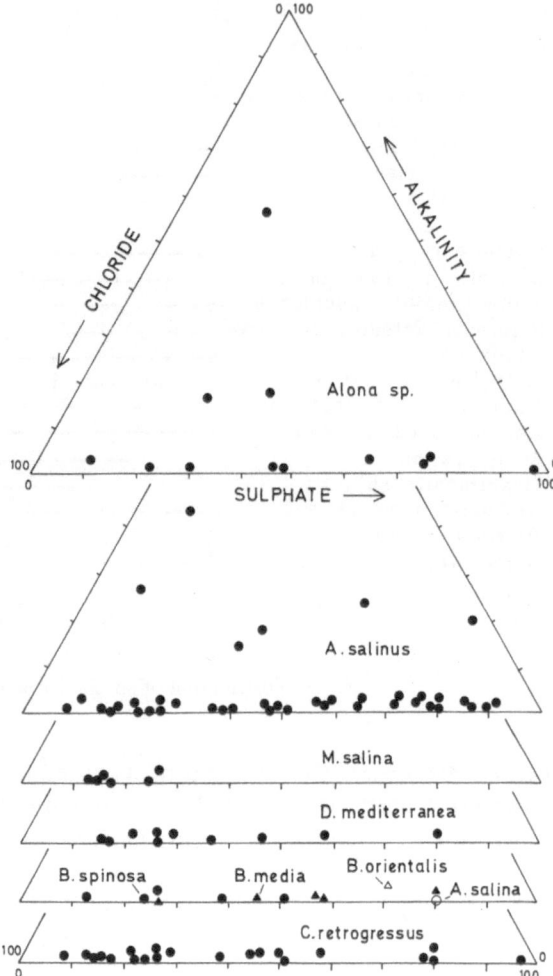

Fig. 4. Distribution of halobiont species in triangular diagrams of major anions in Spanish saline lakes.

most Spanish saline lakes (Comín & Alonso, 1988). *Moina salina* and *Branchinella spinosa* could be related to chloride waters; *Alona* sp. and *Arctodiaptomus salinus* may also occur in waters with a higher proportion of alkalinity. Other species seem unaffected by anionic differences.

Geographical distribution of halobiont species

The halobiont fauna appears richer in the saline environments of the Ebro, Guadalquivir and Guadiana watersheds (Table 4). This may be because these areas were more intensively sampled, but could also reflect a greater environmental diversity in those areas.

All of these areas have almost the same fauna, although there are some minor differences; for example, *Moina salina* does not appear in the Guadiana watershed, *Branchinella spinosa* is more abundant in the Guadalquivir than in the Guadiana and Ebro watersheds, and *Branchinecta orientalis* seems to be restricted to the Guadiana watershed. In contrast, the fauna of the Segura and Duero watersheds is very poor: *Arctodiaptomus salinus* occurs in the former, and *A. salinus* and *Alona* sp. in the latter. *A. salina* is widely distributed but it appears mainly in artificial salt pans. These were not studied. The only permanent natural environment where *A. salina* was found is Chiprana lagoon, in the Ebro watershed.

Discussion

The fauna of Spanish saline lakes is similar to that in saline lakes of Algeria and Tunisia (Gauthier, 1928; Beadle, 1943), Morocco (Ramdani, 1986; Thiery, 1986), Sardinia (Margaritora, 1971; Cottarelli & Mura, 1983; Stella, 1970) and Rumania (Negrea, 1983). Thus it can be considered as typically Mediterranean. However, some minor differences distinguish the Spanish halobiontic fauna, especially the taxa *Alona* sp. and *Branchinecta orientalis*. The former has not been recorded elsewhere although a revision of more material, especially of *Alona rectangularis* from Algeria (Beadle, 1943) and *A. elegans lebes* from Morocco (Ramdani, 1986), is necessary to determine whether the distribution of this species is really restricted to Spain. *B. orientalis*, also, does not appear to occur in the Mediterranean region but does have a wide distribution from Central Europe to Mongolia (Brtek *et al.*, 1984).

The remaining species are apparently the same in all arid regions in the Mediterranean area. However, additional taxonomic studies are needed for: 1, the identity of *Moina salina* and *M. mongolica* (Forró, 1985); 2, the identity of the Spanish and Sardinian *Daphnia* (*C.*) *mediterranea*, *D. C.* cf. *mediterranea* from Rumania, and *D.* (*C.*) *dolichocephala* from Morocco (Ramdani, 1986); 3, the relationships between *Diaphanosoma brachyurum* in Algeria (Beadle, 1943), *D.* gr. *mongolianum* in Morocco (Ramdani, 1986) and the Spanish *D.* gr. *mongolianum*.

The salinity tolerances of the Spanish species follow the pattern observed in other arid and semi-arid areas of the World. The upper limit of 25‰ established here for the freshwater species found in salt waters is near the 20‰, limit fixed by Williams (1978) for salt-tolerant freshwater species in Australia and by Hammer *et al.* (1983) for hyposaline waters (although euryhaline species from lacustrine environments in steppes can survive to 40‰ salinity). The second group, true 'salt water species', tolerates from 5 to more than 100‰. Some of them, mainly *Alona* sp. and *Branchinecta orientalis*, would correspond to the halophilic group of Williams (10–60‰), whilst the rest would correspond to the hypersaline water fauna of Hammer *et al.*; and typical halobionts according to Williams (50–100‰).

The salinity ranges proposed by Williams (1978) and Hammer *et al.* (1983) may well apply in Spain. However, due to extreme variability in salinity, the lower limit in mesosaline lakes should be reduced until the salinity at which the 'true halobionts' appear, namely *ca.* 5‰. Thus, the classification proposed for Spain is: *hyposaline lakes* (with salt-tolerant freshwater species or euryhaline species from waters in steppes, in the range < 5–25 (– 40)‰ salinity); *mesosaline lakes* (halophile species in the range 5–40‰ salinity); and *hypersaline lakes* (halobiont species in the range 5–110‰ salinity).

The present distribution of species depends on ecological and historical factors. Apparently, relative anion composition is the only ecological factor affecting some of the distributions, but more data are required on this subject. Two examples, *Moina salina* and *Branchinella spinosa*, occur preferentially in chloride waters, and apparently are more abundant in the Guadalquivir watershed which is characteristically rich in this type of water. Interestingly, both species also occur in coastal saline lakes (Cotarelly & Mura, 1983;

Margaritora, 1985), in contrast to the athalassic preference of other halobionts referred to in this paper.

The history of colonization of saline lakes may have started during the Tertiary, when aridity was widespread. As Beadle (1959) noted, halobionts probably originated from freshwater species and, for this reason it would be interesting to determine whether the freshwater populations of *Arcto-diaptomus salinus* in Banyoles and Tobar karstic lakes correspond to the primitive form. Another important question concerns the temporal persistence of these species. Margalef (1947) suggested that aridity was maintained during Pleistocene climatic fluctuations, and thus the associated fauna of wetlands in steppes. This is probable because of the incomplete glaciation of the Iberian peninsula in the Pleistocene, and because of the particular hydrological situation of Spanish saline lakes. Most are fed by groundwater discharges which are characterized by high salt concentrations and chemical anomalies (Custodio & Llamas, 1983; González-Bernáldez *et al.*, 1987) and thus ensuring stability through time. Some groundwater discharges maintain a relic and endemic biota, as in springs in arid areas (Cole, 1968), or the Australian mound springs (Ponder, 1986). For example, the only known population of the charophyte *Tolypella salina* is in Bodón Blanco (Duero watershed, Comelles, 1986), and *Branchinecta orientalis* is a relic restricted to the Hito lagoon (Guadiana watershed). These occurrences are very interesting since both localities are often maintained by stable groundwater discharges.

Acknowledgements

This work was supported by the University of Barcelona. I am grateful to C. Montes and P. Martino for the samples taken in several saline lakes in 1984, 1985 and 1986. I am also very grateful to Dr. Frey for material of *Alona elegans* and *A. elegans lebes*, and to Dr. Turner for scanning electron microscopy.

References

Alonso, M., 1984. The genus *Mixodiaptomus* Kiefer 1932 (Copepoda Diaptomidae) in Spain. Hydrobiologia 118: 135–146.

Alonso, M., 1985a. *Daphnia (Ctenodaphnia) mediterranea*. A new species of hypersaline waters long confused with *D. (C.) dolichocephala* Sars 1895. Hydrobiologia 128: 217–228.

Alonso, M., 1985b. Las lagunas de la España peninsular: Taxonomía, ecología y distribución de los cladóceros. Ph.D. Thesis, University of Barcelona. 795 pp.

Alonso, M., 1985c. A survey of the Spanish Euphyllopoda. Misc. Zool. 9: 179–208.

Alonso, M. & M. Comelles, 1981. Criterios básicos para la clasificación limnológica de las aguas continentales de pequeño volumen de España. Act. I Congr. esp. Limnol., p. 35–42.

Alonso, M. & M. Comelles, 1984. A preliminary grouping of the small epicontinental water bodies in Spain and distribution of Crustaceans and Charophyte. Verh. int. Ver. Limnol. 22: 1699–1703.

Armengol, J., M. Estrada, A. Guiset, R. Margalef, M. D. Planas, J. Toja & F. Vallespinos, 1975. Observaciones limnológicas en las lagunas de la Mancha. Bol. Est. Centr. Ecología 8: 11–27.

Beadle, L. C., 1943. An ecological survey of some inland saline waters of Algérie. J. linn. Soc. Zool. 41: 218–242.

Beadle, L. C., 1959. Osmotic and ionic regulation in relation to the classification of brackish and inland saline waters. Arch. Oceanogr. Limnol. 11(Suppl.): 143–151.

Brtek, J., L. Forró & E. Ponyi, 1984. Contributions to the knowledge of the Branchiopoda (Crustacea) fauna of Mongolia. Annls hist.-nat. Mus. natn. hung. 76: 91–99.

Cole, G. A., 1968. Desert limnology. In G. W. Brown Jr. (ed.), Desert Biology Vol. 1. Academic Press; New York: 423–486.

Comelles, M., 1986. *Tolypella salina* Corillion 1960, nueva carofícea para España. An. Jard. Bot. Madrid 42(2): 293–298.

Comín, F. & M. Alonso, 1988. Spanish salt lakes: Their chemistry and biota. Hydrobiologia 158: 237–245.

Cottarelli, V. & G; Mura, 1983. Anostraci, Notostraci, Concostraci (Crustacea: Anostraca, Notostraca, Conchostraca). Guide per il riconoscimento delle specie animale delle acque interne italiane, 18, AQ/1/194. Consiglio Nazionale delle Ricerche. 71 pp.

Custodio, E. & M. R. Llamas, 1983. Hidrología subterránea. 2a ed., 2 vols. Ed. Omega, Barcelona.

Forró, L., 1985. A new species of *Moina* from Australia (Crustacea: Cladocera). Acta zool. hung. 31: 111–118.

Gauthier, H., 1928. Recherches sur la faune des eaux continentales de l'Algérie et de la Tunisie. Minerva, Alger, 420 pp.

González-Bernáldez, F., P. Herrera Moreno, A. Sastre Merlin, J. M. Rey Benayas & R. Vicente Lapuente, 1987.

Comparación preliminar de los ecosistemas de descarga de aguas subterráneas de las cuencas del Duero y del Tajo. IV Simposio de Hidrología (Palma de Mallorca), pp. 19–34. Asociación Española de Hidrología Subterránea.

Hammer, U. T., J. Shamess & R. C. Haynes, 1983. The distribution and abundance of algae in saline lakes of Saskatchewan, Canada. In U. T. Hammer (ed.) Saline Lakes. Proc. 2nd Int. Symp. Athalassic (Indland) Saline Lakes, Dev. Hydrobiol. 16, Junk, The Hague: 1–26.

Hammer, U. T., 1986. Saline lakes. Ecosystems of the World. Monographic biologicae, Vol. 59. Junk Publishers. 616 pp.

Margalef, R., 1947. Estudios sobre la vida en las aguas dulces de la región endorréica manchega. Publ. Inst. Biol. Apl., 4: 5–51.

Margalef, R., 1953. Los crustáceos de las aguas continentales ibéricas. Min. Agricultura, Inst. For. Inv. Exp. Madrid. 300 pp.

Margaritora, F., 1971. Sulle presenze di *Moina mongolica* Daday 1901 nella Sardegne occidentale. Riv. Idrobiol. 10: 5–9.

Margaritora, F., 1972. Sulle presenze di *Ctenodaphnia* Dybowski e Grochowski 1895 nelle acque astatiche delle Sardegna. Istituto Lombardo (Rend. Sc.) B 106: 36–49.

Margaritora, F., 1985. Cladocera. Fauna d'Italia, 399 pp.

Martino, P., 1988. Limnologia de las lagunas salinas españolas. Ph.D. Thesis, Autonomous University of Madrid. 264 pp.

Miracle, M. R., 1976. Distribución en el espacio y en el tiempo de las especies del zooplancton del lago de Banyoles. ICONA Monografías, 5: 1–270.

Negrea, S., 1983. Fauna Republicii Socialiste România. Crustacea. Cladocera. Vol. 4, fasc. 12, Fauna Republicii Socialiste România. Ed. Acad. Republ. Soc. România. 399 pp.

Negrea, S., 1984. Redescription de *Moina salina* Daday, 1888 (Cladocera, Moinidae) d'après des exemplaires trovés en terre typica. Crustaceana 47(1): 83–97.

Ponder, W. F., 1986. Mound springs of the Great Artesian Basin. In P. De Deckker & W. D. Williams (eds.), Limnology in Australia. Dr. W. Junk Publishers; Boston: 403–420.

Ramdani, M., 1986. Ecologie des crustacés (Copepodes, Cladoceres et Ostracodes) des Dayes Marocaines. Thèse, Université de Provence (Aix-Marseille I). 226 pp.

Stella, E., 1970. Diaptomidi della Sardegna. Istituto Lombardo (Rend. Sc.) B 104: 69–87.

Thiery, A., 1986. Les Crustacés Branchiopodes (Anostraca, Notostraca et Conchostraca) du Maroc occidental. I. Inventaire et répartition. Bull. Soc. Hist. Nat., Toulouse, 122: 145–155.

Williams, W. D., 1978. Limnology of Victorian salt lakes, Australia. Verh. int. Ver. Limnol. 20: 1165–1174.

Hydrobiologia **197**: 233–243, 1990.
F. A. Comin and T. G. Northcote (eds), Saline Lakes.
© 1990 Kluwer Academic Publishers.

Artemia monica cyst production and recruitment in Mono Lake, California, USA

Gayle L. Dana, Robert Jellison & John M. Melack
Marine Science Institute, University of California, Santa Barbara, CA 93106, USA

Key words: Artemia monica, cyst production, recruitment

Abstract

Annual egg production was determined for *Artemia monica* in Mono Lake, California, from 1983 to 1987. Annual oviparous (overwintering cyst) production was 3 and 7 million cysts $m^{-2} yr^{-1}$ in 1986 and 1987, respectively, as measured by *in situ* sediment traps. Cyst production for the entire five year period was calculated using *Artemia* census data and inter-brood duration derived from mixolimnetic temperature. These estimates ranged from 2 to 5 million cysts $m^{-2} yr^{-1}$. This method underestimated annual production by 30%, when compared to estimates using sediment traps. Cyst production was similar during 1983–1986 and showed a significant increase in 1987, which was due primarily to a larger reproductive population later in the year. Recruitment into the adult populations of the following spring ranged between 1.4 to 3.2%. Overall abundance of this generation reflected the patterns in annual cyst production. Compensatory effects must operate on the second generation of each year, since summer populations were similar in all years despite differences in cyst production.

Introduction

Egg production in zooplankton is used as a measure of total reproductive output and is an important factor in recruitment. The focus of most studies has been to determine the relationship of egg production rates to environmental factors. Food and temperature have been shown repeatedly to affect egg production rates (Edmondson *et al.*, 1962; McLaren, 1965; Weglenska, 1971; Durbin *et al.*, 1983; Uye, 1981) both during development and after maturity (Checkley, 1980). While it is important to establish these causal relationships, few studies have evaluated year-to-year variation in annual egg production (Edmondson, 1964) and recruitment into the subsequent generation.

Variation in year-to-year population size results from interactions between several population processes leading to differences in recruitment. In Mono Lake, *Artemia monica* hatch from overwintering cysts in early March and mature to adulthood by late May. This generation gives rise to a second generation ovoviviparously. This second generation produces predominantly cysts. The first generation population size of adult *Artemia monica* Verill in Mono Lake, California was relatively stable from 1984 to 1987, but significantly increased in 1988. The second generation was very similar in all years from 1983 to 1988. In this study we examine the variation of oviparous (cyst) production, and assess its contribution to the observed population size of *Artemia*.

Estimates of annual cyst production in 1986 and 1987 were calculated from cyst deposition data obtained from *in situ* sediment traps in Mono Lake. Annual cyst production was also calculated for 1983 to 1987 using life-history data obtained from the field. Recruitment was calculated using estimates of annual cyst production and observed first generation adult abundances of the following year. Factors important to annual cyst production in *Artemia* from 1983–1987 are identified and their relative importance is discussed.

Description of site studied

Mono Lake is a large (150 km^2), moderately deep (mean depth, 17 m), alkaline, saline lake. It occupies a tectonic basin on the western edge of the North American Great Basin just east of the Sierra Nevada, California (latitude 38° N 119° W; elevation 1942 m). The lake was meromictic during the period of the study with a persistent chemocline lying below the seasonal thermocline. It begins to thermally stratify in late March or early April and remains stratified until November (Mason, 1967; Melack, 1983). In the offshore areas, mean mixolimnetic temperatures in the summer reach ca. 20 °C; minimum winter temperatures are near 0 °C. Approximately half of the sediments lie within the monimolimnion (non-mixing layer) and are anoxic.

The planktonic community of Mono Lake includes few species as is typical of hypersaline waters. The phytoplankton are dominated by a coccoid chlorophyte tentatively identified as *Nannochloris*, (pers. commun., B. Javor) and cyanobacteria with several bacillarophytes, mainly *Nitzschia* spp. (Mason, 1967; Lovejoy & Dana, 1977; J. Melack, unpublished data.). An endemic brine shrimp species, *Artemia monica* is the major zooplankter (Lenz, 1982). *A. monica* hatch from overwintering cysts as early as January and continue to hatch into May. Hatching is preceded by an obligate dormant period of three months in cold (<5 °C) water (Dana, 1981). Dormancy may occur in anoxic conditions but hatching requires the presence of oxygen (Dutrieu

& Chrestia-Blanchine, 1966). By mid-May the first adult *Artemia* are observed. The *Artemia* exert heavy grazing pressure on the phytoplankton during this time, which contributes to a rapid decrease in the algal levels in the epilimnion (Jellison & Melack, 1988). For approximately one month *Artemia* females bear live young (ovoviviparity), which rapidly mature in the warm epilimnetic water. Starting in June, females switch to oviparous (cyst) reproduction. The diapause cysts lie dormant on the bottom of the lake until the following winter. The adult population reaches its peak in the summer when densities are 6–8 adults l^{-1} in the pelagic region, and 15–17 adults l^{-1} in the nearshore region (Conte *et al.*, 1988). During summer the lake is separated into an upper region with sparse phytoplankton and abundant *Artemia* and a deep region with cold temperatures, very low light, no oxygen, very few *Artemia*, and high concentrations of phytoplankton. In autumn the phytoplankton bloom in the surface waters as thermal stratification weakens and as the *Artemia* begin to decline in numbers. Adult *Artemia* are nearly absent in December and none overwinter.

Materials and methods

Sediment trap methodology

Deposition of *Artemia* cysts was measured biweekly with *in situ* sediment traps. The traps were sampled in 1986 and 1987 for the duration of the cyst producing season. Each trap consisted of four replicate cylinders, each with an inner diameter of 7.3 cm and a height of 23.3 cm, resulting in an aspect ratio (height/inner diameter) of ca. 3 to 1. Gardener (1980) tested a variety of sediment trap configurations and found that cylinders yield accurate measurements of vertical flux in non-turbulent waters if the aspect ratio is between 2 and 3. A properly proportioned cylinder collects about 95–100% of the actual sedimentation (Bloesch & Burns, 1980).

One trap each was set in the east, west, and south sectors of Mono Lake at a depth of 24 m,

3–15 m above the bottom. Bloesch & Burns (1980) state that 1 to 3 m above the mud surface is sufficient to prevent resuspension into the traps. However, in winter of 1986 low numbers of cysts were collected in the west sector trap (3 m above the bottom), even though cyst production had stopped before this time. We assumed this was due to resuspension, and on 27 February 1987 we moved this trap to a deeper west sector location, 7 m above the bottom. All traps were situated below the thermocline and chemocline during the period of study.

Annual cyst production was determined by averaging cyst deposition from the three traps and summing cyst totals for each sampling interval (usually two weeks) over the entire reproductive season.

Calculated cyst production estimates

Annual cyst production was calculated for the years 1983–1987 using female population and brood size data, and inter-brood duration based on temperature.

Lakewide female *Artemia* abundance was based on biweekly samples from ten pelagic stations. Duplicate (through mid-1983) or triplicate (mid-1983 to 1987) samples of *Artemia* were taken with a plankton net (1 m × 0.30 m diam., 120 μm Nitex mesh) towed vertically through the oxygenated portion of the water column. Samples were preserved with 5% formalin in lake water and individuals were counted under a stereomicroscope (6× magnification). Depending on the density of shrimp, counts were made of the entire sample or of subsamples made with a Folsom plankton splitter. Between 150 to 200 animals were counted from each subsample, and instars and reproductive state were determined as in Lenz (1984). Shrimp counts were corrected for net size and net efficiency, 0.70 (Lenz, 1984).

Live females were collected for brood size analysis. Shrimp were collected with 20 m vertical net tows and kept cool (< 15 °C) and in low densities during transport to the laboratory. Approximately 100 females were randomly selected, isolated in individual vials, and preserved. Brood size was determined by counting the number of eggs in the ovisac, and egg type (oviparous or ovoviviparous) was noted. In 1983, 30 females were measured from each of three index stations. Since there was no significant difference between mean fecundity at the south and east index stations in 1983, (ANOVA, $p > 0.05$) the south station was dropped in 1984 and 50 females were counted from the other two index stations. To obtain better estimates of lakewide fecundity, ten females were counted from each of the ten pelagic stations from 1985 to 1987.

Inter-brood duration of zooplankton is highly dependent on temperature (Edmondson, 1960). Since the empirical relationship of inter-brood duration to temperature has not been determined for *A. monica*, we used the data of Wear *et al.* (1986) on a closely related species, *A. franciscana*. In this study, Wear *et al.* present reproductive data (total number of broods produced and length of reproductive life) at 14, 20 and 26 °C. From these data, we calculated an average inter-brood duration for each temperature and fitted an exponential curve to the data. The regression equation describing this relationship is: $Y = 66.5\,e^{-0.1027X}$, where Y is the inter-brood duration and X is the temperature in degrees centigrade ($r^2 = 0.99$). The inter-brood durations calculated from this equation at 20 and 25 °C are 8.5 and 5.2 days, respectively. This is in close agreement with the 9 (s.d., 2; Dana & Lenz, unpublished data) and 5 days (s.d., 1; G. Starrett, pers. commun.) determined at 20 and 25 °C from laboratory experiments on *A. monica*. In addition, inter-brood duration derived from the sediment trap and *A. monica* census data for the summers of 1986 and 1987 was 8.5 days (s.d., 2.5), when epilimnetic temperatures reached 20 °C.

Mean daily cyst production rates were calculated for each sampling date from 1983–1987 by multiplying ovoviviparous female abundance with cyst brood size and dividing by the inter-brood duration calculated from the regression equation and mean mixolimnetic temperature from each sampling date. Daily rates were linearly interpolated between dates and summed over all days to obtain annual production. Female specific rates were calculated for each sampling date by dividing brood size by the inter-brood duration. An average female specific rate was calculated by linearly interpolating between sampling dates, and then averaging over all days. Female-days is the sum of the daily oviparous abundances.

A single estimate of annual cyst production was obtained for each year, 1983–1987. The confidence intervals associated with these estimates were obtained using Monte Carlo simulations of the uncertainties associated with each of the factors involved in the estimate of cyst production. One thousand pseudo-replicates were obtained by drawing random variates from a normal distribution with the standard deviation of the individual factors: number of oviparous females,

brood interval, and brood size. The standard deviations of the number of oviparous females and brood size were based on replicate samples from field surveys. The standard deviation associated with inter-brood duration was based on the regression equation on temperature and was constant during the entire year. When the standard deviation of an individual component in the calculation was large relative to the mean, sampling from a normal distribution will sometimes result in negative values. In these cases, negative values were set to zero. The mean of the distribution from which the individual variates were drawn was shifted so that the mean of the pseudo-replicates would equal the observed mean after the negative values were set to zero. The random variates for each date were then used to calculate the total annual cyst production. Ninety-five per-

Table 1. Artemia cyst (oviparous) production, population size, and recruitment, 1983–1988.

	1983	1984	1985	1986	1987	1988
Annual cyst production (cysts m^{-2} yr^{-1} × 10^6)	4.0	3.5	4.1	2.4 (3.4)[1]	5.1 (7.3)	–
Female specific Avg. daily cyst prod. (cysts female^{-1}d^{-1})	3.6	2.6	3.1	3.2	3.1	–
Length of cyst producing season (d)	212	191	194	196	272	–
No. female-days (× 10^5)	12.3	9.5	10.1	7.7	14.4	–
Ratio of summer to first gen. peak	1.4	1.3	1.4	0.8	2.3	0.5
[3]Prop. of sediments oxygenated (> 1 mg l^{-1})	–	0.49	0.47	0.42	0.42	0.44
[4]Cysts in oxygenated sediments (cysts m^{-2} × 10^6)	–	1.6	1.7	1.7	1.0	2.2
Peak first generation adults (ind. m^{-2} × 10^3)	5.3	30.7	23.3	33.0	24.0	71.0
[5]Percent recruitment into first generation	–	1.9	1.4	1.9	2.4	3.2

[1] Numbers in parentheses are derived from sediment trap data.
[2] See methods for calculation
[3] (Mean area of oxygenated sediments, March-May) ÷ (Total area of sediments)
[4] Annual cyst prod. (previous yr.) × prop. oxygenated sediments.
[5] (First generation adult abundance ÷ cysts in oxygenated sediments) × 100.

cent confidence intervals were based on these 1000 pseudo-replicates.

Results

Annual cyst production

Annual cyst production calculated from *Artemia* population data ranged from 2–5 million cysts m^{-2} yr^{-1} for the years 1983–1987 (Table 1). Annual estimates from sediment traps in 1986 and 1987 were 3 and 7 million cysts m^{-2} compared to 2 and 5 million cysts m^{-2} calculated from the population data. Assuming that sediment traps accurately measure cyst production, derived estimates underestimated cyst production by about 30% in both years. Average daily rates of cyst production were determined from sediment traps and reached a maximum of 46 000 cysts m^{-2} day^{-1} in 1986, and 59 000 cysts m^{-2} day^{-1} in 1987 (Fig. 1). Cyst produc-

tion rates for individual females ranged between a mean of 2.6 and 3.6 cysts female^{-1} d^{-1} (Table 1).

Annual cyst production was highly correlated with both the number of female-days ($r^2 = 0.86$) and the ratio of summer to first generation peak abundance of adults ($r^2 = 0.92$). However, cyst production was not significantly correlated with the actual peak abundance of either summer or first generation adults.

An inspection of the 95% confidence intervals generated by the Monte Carlo simulations indicate that annual cyst production was not significantly different for the years 1983–1986 (Fig. 2). Cyst production in 1987 was significantly higher than in 1986 (but not other years), since the means of those two years generated from sediment trap data do not lie within each other's confidence intervals. The Monte Carlo simulations do not entirely corroborate sediment trap data for 1986 and 1987. Although mean cyst production in 1987 lies outside the 95% confidence intervals of 1986,

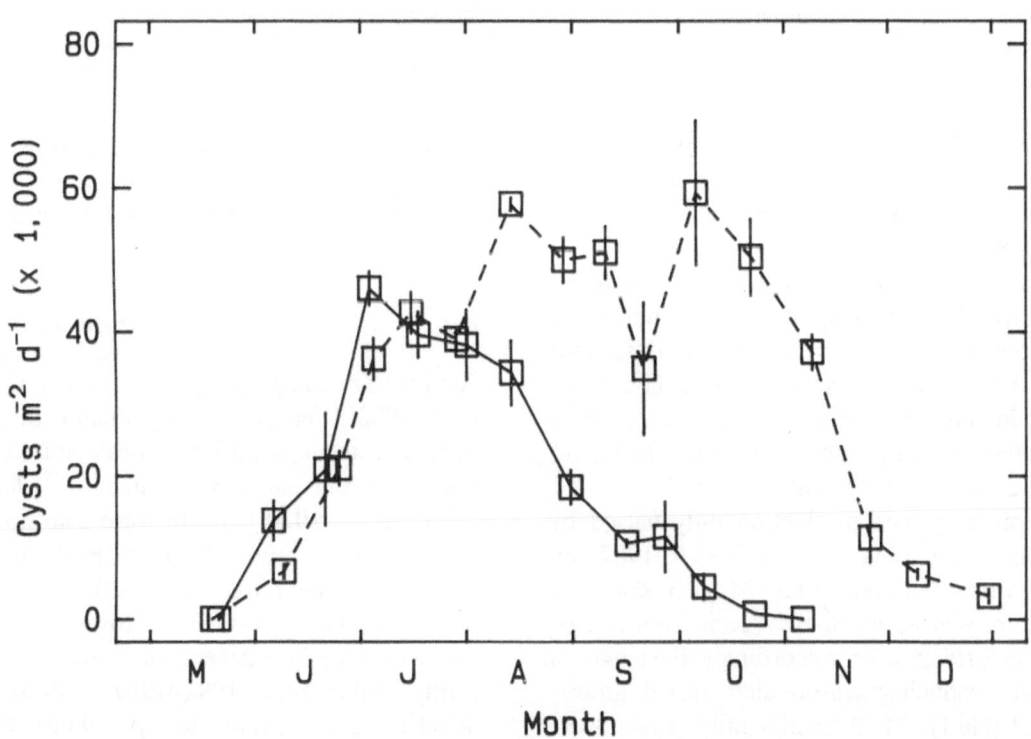

Fig. 1. Artemia cyst (oviparous) production in 1986 (solid line) and 1987 (dashed line), calculated from sediment trap data. Vertical lines are the standard errors.

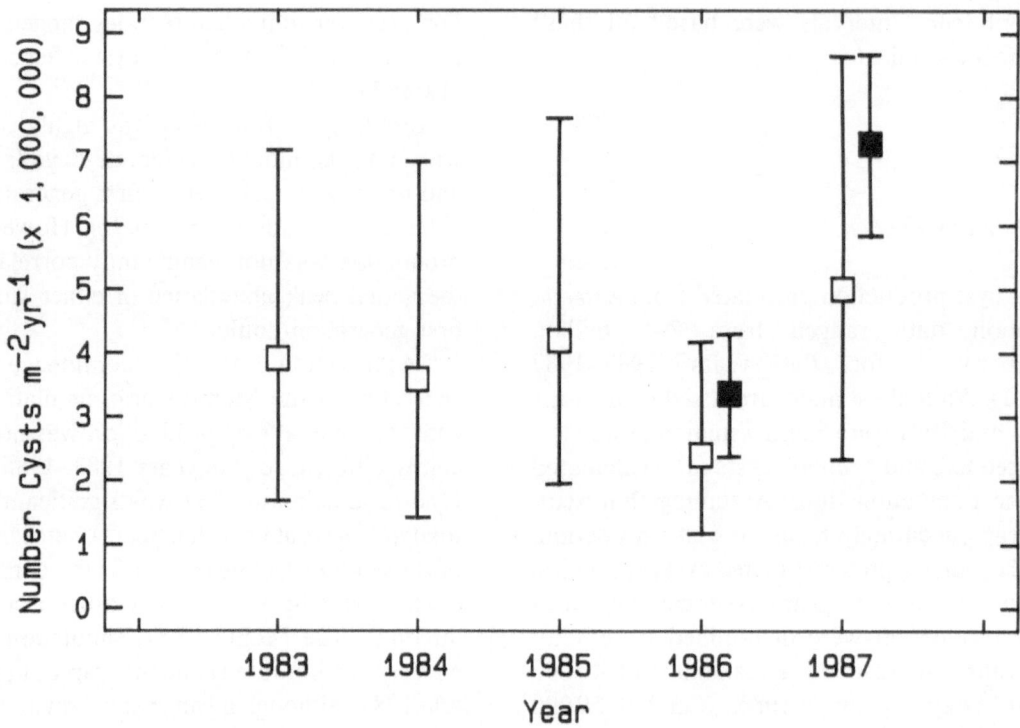

Fig. 2. Annual *Artemia* cyst production 1983–1987. Open squares are the means of 1000 pseudo-replicates generated from Monte Carlo simulations (see methods for explanation of technique). Solid squares (1986 and 1987 only) are the lakewide means from sediment trap data. Vertical lines are the 95% confidence intervals.

the converse is not true; mean cyst production in 1986 lies just inside the confidence intervals of 1987. Annual production in 1986 was only 47% that of 1987 (see Table 1). The confidence intervals associated with sediment trap estimates derived from replicate traps are smaller than those calculated from *Artemia* population data and Monte Carlo simulation techniques. The latter method includes the uncertainties associated with the number of cyst producing females, the brood size, and the brood interval.

The onset of cyst production only varied by approximately two weeks from 1983 to 1987 and occurred between mid- to late May. In contrast, the time of year at which cyst production ceased varied by three months. Accordingly, the length of the cyst producing season also varied among years (Table 1). Most significantly, cysts were produced for approximately 272 days in 1987 compared to 191–212 days from 1983–1986. This

contrast is clearly seen when comparing seasonal patterns of daily cyst production rates in 1986 and 1987, the years with the shortest and longest reproductive seasons (Fig. 1).

Artemia population and reproductive characteristics

Summer *Artemia* population size was stable from 1983–1988 (Fig. 3). No significant differences were detected among mean peak abundance of adults for the summer population, 1983–1988 (ANOVA, $p < 0.48$). There were also no significant differences among the mean peak abundance of adults in the 1st generation from 1984–1987 (Fig. 3; ANOVA, $p < 0.33$). However, the abundance of 1988 first generation adults was significantly higher than 1984–1987 (ANOVA, and Scheffe's multiple range test, $p < 0.05$). The ratio of summer to first generation adult peak abundance is an indicator of the relative dominance of

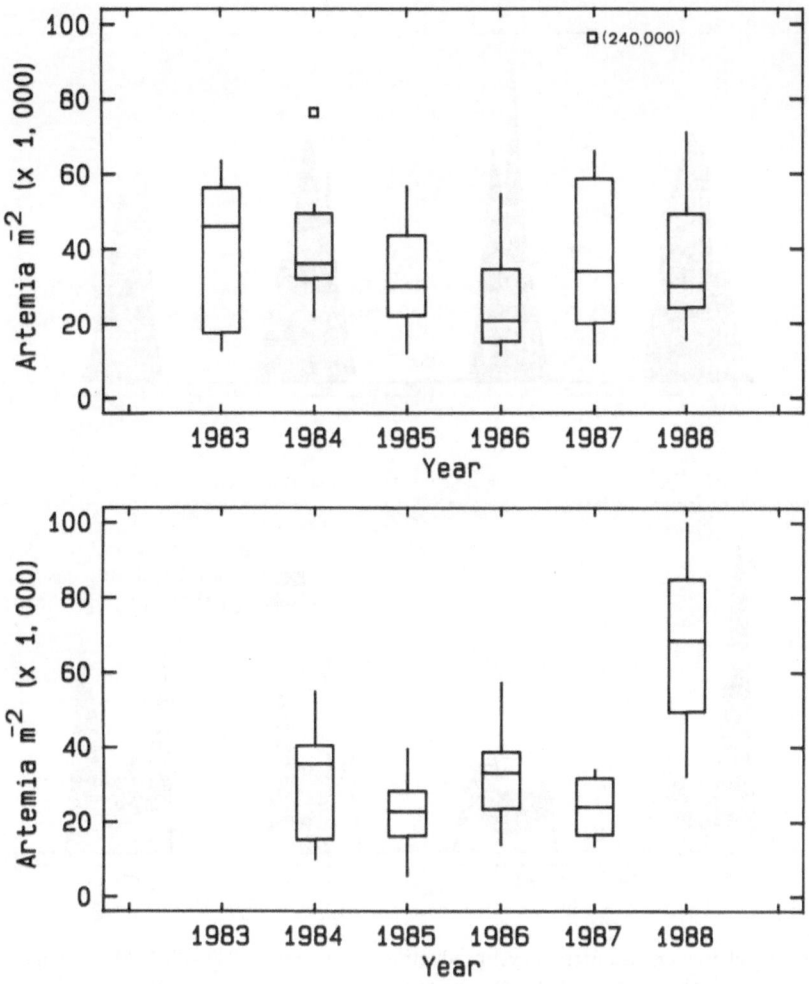

Fig. 3. Box and Whisker plots of peak adult abundances of the summer (combined first and second generation) 1983–1988 (top plot), and spring (first generation) *Artemia* population 1984–1988 (bottom plot). Each plot is composed of mean abundances from ten stations. Central box covers the middle 50% of the data values (interquartile range); central line is the median. Whiskers extend out to the extremes (minimum and maximum data values). Data points which are over 1.5 the interquartile range are plotted as separate points. Note datum in top graph, 1987 is off-scale.

these two populations. This ratio ranged between 0.5 and 2.3 from 1983–1988 (Table 1).

Peak summer abundance of oviparous females varied between 9000 and 15000 ind m^{-2} from 1983 to 1987 (Fig. 4). However, an ANOVA on these data showed that there were no significant differences among these mean abundances ($p < 0.07$). Female-days, which incorporates both the abundance of oviparous females and the length of the reproductive season, varied considerably over the five year period, from 770000

to 1400000 (Table 1). Female-days in 1987 was almost twice that observed in 1986.

Summer brood size of *Artemia* was consistently 30–40 eggs brood^{-1} from 1983 to 1987 (Fig. 4). In contrast, spring and autumn brood size varied widely, between 30 and 140 eggs brood^{-1} during these years. Brood size was noticeably depressed in the spring of 1984 and 1985, and in the autumn of 1983 and 1984. Brood size is positively correlated with chlorophyll *a* and reflects the seasonal changes in algal abundance ($r^2 = 0.61$).

240

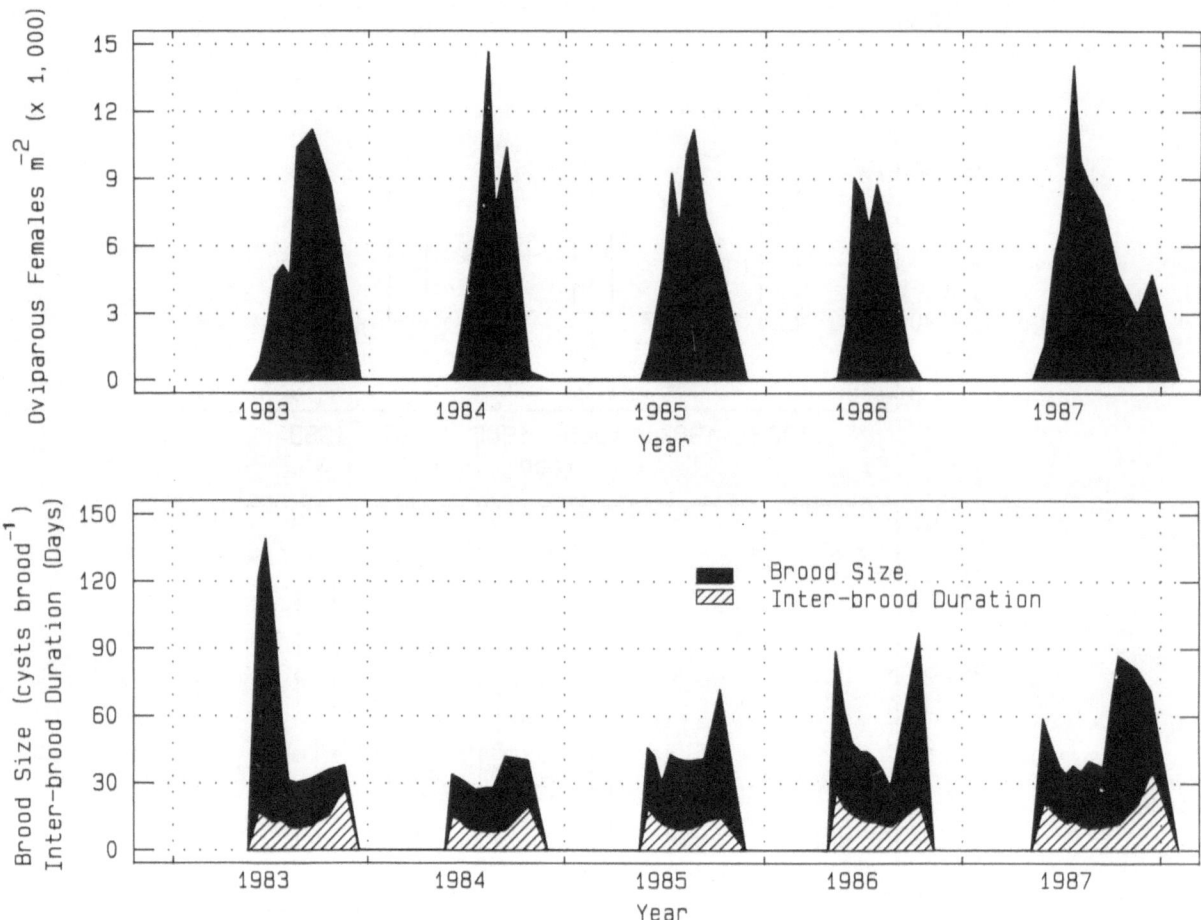

Fig. 4. Artemia life-history characteristics used in cyst production calculations, 1983–1987: Mean oviparous female abundance (top plot); oviparous brood size and inter-brood duration (bottom plot).

Mixolimnetic temperatures yielded inter-brood durations between 8 and 36 days (Fig. 4). Inter-brood durations were shorter during the warmer summer months and longer during the cooler spring and autumn seasons since inter-brood duration is highly dependent on temperature (see Methods). Higher than average mixolimnetic temperatures resulted in shorter inter-brood durations in the summer of 1984.

Discussion

Stability in year-to-year annual cyst production results from constancy of its individual components or compensatory effects when com-

ponents differ. These components include the abundance of oviparous females and the length of time these individuals reproduce, brood size of oviparous females, and inter-brood duration. Differences in individual components of cyst production were observed during the period of study: lower spring brood size in 1984 and 1985, shorter inter-brood duration in summer 1984 due to higher mixolimnetic temperatures, and varying number of female-days. Despite year-to-year variation in individual components, we were unable to detect significant differences in annual cyst production between 1983 and 1986. However, cyst production in 1987 was significantly higher than 1986.

The feature which distinguishes 1987 from all

the other years was the larger reproductive population which extended late into the year. The maintenance of a high abundance of ovigerous females later into the year is determined by a suite of factors which affect mortality, including natural senescence and starvation. A relatively large third generation observed in autumn of 1987 may have also contributed reproducing females to the population late in the year.

Abundance may decline earlier when the population is dominated by first generation shrimp which hatch early in the year and experience mortality in late summer due to natural senescence. If this is true, then we are likely to see an earlier die-off of shrimp in years when the first generation dominates, and the ratio of peak summer to first generation adult abundances is low. This hypothesis is supported by the strong correlation observed between female-days and the population ratios over all five years ($r^2 = 0.86$). The ratios of the summer population to the spring generation in 1986 and 1987 were 0.8 and 2.3, respectively. In keeping with our hypothesis, the reproductive population declined early in the year in 1986, and late in 1987.

Evidence for food limitation causing starvation and an early population decline is lacking. Chlorophyll a concentrations in August and September of most years have been low ($1-3 \mu g \, l^{-1}$). In 1986, the year in which the shrimp decline was earlier relative to other years, chlorophyll a levels during the time of die-off were higher ($7-16 \mu g \, l^{-1}$).

Brood size is an important component of cyst production. During much of the cyst producing season (summer), brood size does not vary substantially from year-to-year. However, lower spring and autumn brood sizes were observed following the onset of meromixis, from autumn 1983 through spring 1985. These lower brood sizes can be attributed to the effects of meromixis. Reduced mixing of nutrients into the mixolimnion resulted in lower algal levels during this period (Jellison & Melack, 1988), and brood size is significantly correlated with chlorophyll a concentrations in the mixolimnion ($r^2 = 0.61$). Summer brood size was not noticeably influenced by meromixis, since summer chlorophyll a concentrations have been consistently low both before and during meromixis. As the chemocline gradually deepened each year, and monimolimnetic ammonium concentrations increased, chlorophyll a concentrations also increased in spring and autumn. At the same time, brood size has slowly returned to higher levels. The variation in spring brood size observed over the five year period may be more important for ovoviviparous reproduction which occurs for a short period during the spring.

Estimates of cyst production calculated from *Artemia* population data were 30% lower than those obtained from sediment traps. Since it is unlikely that the sediment traps oversample cyst deposition, the calculated values are probably underestimates. Two of the factors in the cyst production calculations, female abundance and brood size, are based on replicated field samples. The third factor, inter-brood duration, is based on a regression equation derived from the literature for a closely related species, *A. franciscana*. The inter-brood durations calculated from this equation are in agreement with independent estimates for *A. monica* (see Methods). The remaining possibility is that the average mixolimnetic temperature used in the equation does not accurately reflect the temperature environment of the females in the lake. If females spend more time in the surface waters, where temperatures are warmer, rather than spending equal time throughout the mixed layer, then inter-brood duration would be shorter than we estimated. This would serve to increase the rate of cyst production and hence, annual cyst production.

Recruitment into the adult population can be calculated using annual cyst production estimates from the preceding year, the percent of oxygenated sediments, and the peak abundance of first generation adults. The number of adults observed in the first generation was small relative to the annual cyst production during the previous year. Recruitment varied between 1.4 and 3.2% from 1984 to 1988 (Table 1). Given the large confidence intervals associated with cyst production estimates, differences between these recruitment values cannot be shown to be significant. These values most likely overestimate recruitment by 30% since they were derived from cyst production estimates based on *Artemia* population data (see above).

Low recruitment may result from low hatching success, premature hatching, and/or naupliar mortality. Low hatching success can result from inviability or extended dormancy. For example, *Artemia* spp. is known to produce infertile eggs (L. Drinkwater, pers. commun.). Experimentally determined hatching success of *A. monica* cysts ranges between 60 and 90% (Dana, 1981; Dana & Lenz, 1986) and it is possible that hatching success in the field is potentially this high. Dana *et al.* (1988) measured *in situ* hatching rates in 1985 using emergence traps. Based on these numbers and the number of cysts present in oxygenated sediments in 1984, it is estimated that hatching success in 1985 was at least 25%. These data suggest that mortality plays a significant role in the observed low recruitment of cysts into the next generation.

Extended dormancy may occur due to the burial of cysts or deposition onto sediments below the chemocline. The presence of viable cysts in the sediments (below the sediment-water interface) indicates that there are cysts which do not hatch the year following their production (Dana & Lenz, unpublished data). Cysts also get 'trapped' below the chemocline which has been present during the entire study period. The lake had not completely mixed since 1982 and approximately half of the sediments remained anoxic. Cysts require oxygen to undergo hatching. Since the depth of the chemocline and winter mixing varied during the period of study, the area of oxygenated sediments and available cysts also varied. A cyst bank in the sediments could result in an uncoupling of first generation abundance from the previous years' cyst production.

Recruitment into the adult population in the following year may be reduced if cysts hatch in the late autumn since these individuals die during the winter. Evidence of premature hatching comes from the presence of a small third generation often seen in the autumn at Mono Lake (Lenz, 1982;

and Dana, unpublished data). Since there are virtually no ovoviviparous females present at this time, these shrimp must hatch from cysts. This third generation dies off by February, rarely matures to produce cysts and thus does not contribute to annual cyst production (1987 was a possible exception).

Mortality of first generation nauplii contributes substantially to the observed low recruitment. Recent laboratory experiments indicate that naupliar mortality is extremely high (>95%) when water temperatures are 5 °C or less (Dana & Jellison, unpublished data). Newly hatched nauplii are often observed in the early spring when lake temperatures are less than 5 °C and it is unlikely that these individuals attain maturity. Another indication that mortality of first generation nauplii may be high comes from the *in situ* hatching study of Dana *et al.* (1988), which indicates that the number of nauplii emerging from cysts in spring 1985 was much higher than the observed adult population. Less than 8% of the observed hatch in 1985 reached adulthood, based on field population data.

Annual cyst production was highly correlated with both the number of female-days and the ratio of the summer to first generation size of adult *Artemia*. These two factors co-vary since a high ratio of summer to first generation population size results in the maintenance of a large reproductive population late in the year. When the ratio of the summer population to the first generation is lower, *Artemia* decline earlier in the year and cyst production is reduced. A higher ratio results in an extended period of reproduction and higher cyst production. This compensatory effect tends to dampen year-to-year fluctuations in the summer population, but other environmental factors may override this effect. In 1981 and 1982 the summer population was two to three times as large as those observed during this study. The ratio of the summer population to the first generation was 42 and 36 for 1981 and 1982, respectively (Lenz, 1984; Dana & Lenz, unpublished data). Despite large summer generation sizes, the first generation size was reduced following both of these years. This occurred during a period when lake levels were at a historic low and salinities were consequently increased. Environmental effects may therefore have overridden compensatory effects on the population in 1981 and 1982.

The episode of meromixis, which occurred during this study, terminated at the end of 1988 when the lake completely mixed (Dana & Jellison, unpublished data). This mixing event resulted in a large influx of ammonium into the mixed layer and resulted in elevated chlorophyll *a* concentrations. The effect of this environmental change on the *Artemia* population dynamics may override the compensatory effects observed during the study period.

Acknowledgements

We thank Lee Dyer, Margaret Palchak, and Rebecca Todd for their assistance in the laboratory and field. The following people provided critical review of the manuscript: Dr. Andrew Jahn, Orlando Sarnelle, Randall Orton, and an anonymous reviewer. Laboratory work was performed at the Sierra Nevada Aquatic Research Laboratory. This work was supported by grants from the Los Angeles Department of Water and Power, Mono County, California, and the Lucile and David Packard Foundation.

References

Bloesch, J. & N. M. Burns, 1980. A critical review of sedimentation trap technique. Schweiz. Z. Hydrol. 42: 15–55.

Checkley, D. M. Jr., 1980. Food limitation of egg production by a marine, planktonic copepod in the sea off southern California. Limnol. Oceanogr. 25: 991–998.

Conte, F. P., R. S. Jellison & G. L. Starrett, 1988. Nearshore and pelagic abundances of *Artemia monica* in Mono Lake, California. Hydrobiologia 158: 173–181.

Dana, G. L., 1981. Comparative population ecology of the brine shrimp *Artemia*. M.A. thesis, San Francisco State University. 125 pp.

Dana, G. L. & P. H. Lenz, 1986. Effects of increasing salinity on an *Artemia* population from Mono Lake, California. Oecologia 68: 428–436.

Dana, G. L., C. J. Foley, G. L. Starrett, W. M. Perry & J. M.

Melack, 1988. *In situ* hatching of *Artemia monica* cysts in hypersaline Mono Lake, California. Hydrobiologia 158: 183–190.

Durbin, E. G., A. G. Durbin, T. J. Smayda & P. G. Verity, 1983. Food limitation of production by adult *Arcatia tonsa* in Narragansett Bay, Rhode Island. Limnol. Oceanogr. 28: 1199–1213.

Dutrieu, J. & D. Chrestia-Blanchine, 1966. Resistance des œufs durables hydrates *Artemia salina* A Lanoxie. Cr. Acad. Sci. Paris DD 263: 998.

Edmondson, W. T., 1960. Reproductive rates of rotifers in natural populations. Mem. Ist. ital. Idrobiol. 12: 21–77.

Edmondson, W. T., 1964. The rate of egg production by rotifers and copepods in natural populations as controlled by food and temperature. Verh. int. Ver. Limnol. 15: 673–675.

Edmondson, W. T., G. W. Comita & G. C. Anderson, 1962. Reproductive rate of copepods in nature and its relation to phytoplankton populations. Ecology 43: 625–634.

Gardner, W. D., 1980. Field assessment of sediment traps. J. mar. Res. 38: 41–52.

Jellison, R. & J. M. Melack, 1988. Photosynthetic activity of phytoplankton and its relation to environmental factors in hypersaline Mono Lake, California. Hydrobiologia 158: 69–88.

Lenz, P. H., 1982. Population studies on *Artemia* in Mono Lake California. Ph. D. Dissertation, University of California, Santa Barbara, 230 pp.

Lenz, P. H., 1984. Life-history analysis of an *Artemia* population in a changing environment. J. Plankton Res. 6: 967–983.

Lovejoy, C. & G. Dana, 1977. Primary producer level. In D. W. Winkler, (ed.), An ecological study of Mono Lake, California. Institute of Ecology Publication No. 12. Davis, California: Institute of Ecology: 42–57.

McLaren, I. A., 1965. Some relationships between temperature and egg size, body size development rate, and fecundity, of the copepod *Pseudocalanus*. Limnol. Oceanogr. 10: 528–538.

Mason, D. T., 1967. Limnology of Mono Lake, California. Univ. Calif. Publ. Zool. 83. 110 pp.

Melack, J. M., 1983. Large, deep salt lakes: a comparative limnological analysis. Hydrobiologia 105: 223–230.

Uye, S., 1981. Fecundity studies of neritic calanoid copepods *Arcatia clausi* Giesbrecht and *A. steueri* Smirnov: A simple empirical model of daily egg production. J. exp. mar. Biol. Ecol. 50: 255–271.

Weglenska, T., 1971. The influence of various concentrations of natural food on the development, fecundity, and production of planktonic crustacean filtrators. Ekol. pol. 19: 427–473.

Wear, R. G., S. J. Haslett & N. L. Alexander, 1986. Effects of temperature and salinity on the biology of *Artemia franciscana* Kellogg from Lake Grassmere, New Zealand. 2. Maturation, fecundity and generation times. J. exp. mar. Biol. Ecol. 98: 167–183.

Hydrobiologia **197**: 245–256, 1990.
F. A. Comin and T. G. Northcote (eds), Saline Lakes.
© 1990 Kluwer Academic Publishers.

Comparative limnology of Sambhar and Didwana lakes (Rajasthan, NW India)

G.R. Jakher, S.C. Bhargava & R.K. Sinha
Department of Zoology, University of Jodhpur, Jodhpur, India

Key words: saline lakes, plankton, primary productivity, salt tolerance

Abstract

Two alkaline saline inland lakes of Indian arid region were studied during 1984 and 1985, to assess functioning and interaction of various environmental and biological factors. Changes in physical and chemical variables, planktonic composition, chlorophyll content and phytoplankton primary productivity were examined.

Salinity in both lakes fluctuated from almost fresh water (1.80‰), to hypersaline (300‰) and acted as the main controlling factor for almost all the biotic parameters. Maximum total alkalinities were 2162 mg l^{-1} and 2090 mg l^{-1}, respectively in Sambhar and Didwana lakes. Dissolved oxygen ranged from completely anoxic conditions to maxima of 11.68 and 7.29 mg l^{-1}, respectively in Sambhar and Didwana lakes. Nutrient enrichment in the lakes was low.

The phytoplankton species composition of Sambhar lake was reduced from an earlier reported 20 genera to only 11 (*Nostoc, Microcystis, Spirulina, Aphanocapsa, Oscillatoria, Merismopedia, Nitzschia, Navicula, Synedra, Cosmarium* and *Closterium*). Phytoplankton of Didwana was composed of only 9 genera including *Anabaena* and *Nodularia*. Sambhar lake, which once contained *Artemia*, is now totally devoid of them. On the other hand, *Artemia* was the most dominant zooplankter in Didwana lake at a salinity range of 15–288‰. Other zooplankters such as *Moina, Cyclops* and *Brachionus* flourished at lower salinity levels in Didwana lake. The seasonal quantitative and qualitative phyto- and zooplankton changes in relation to salinity are documented.

Introduction

The main inland alkaline and saline water bodies of the Indian arid region are Sambhar and Didwana lakes, of which the former is the largest. Sambhar Lake has been subjected to extreme conditions varying from immense floods to long dry periods. This lake, which once supported a rich population of brine shrimp, *Artemia*, became totally devoid of it during 1977–78 (Alam, 1980). This has been a consequence of a drastic decrease

in salinity due to repeated floods in the region during the last decade, which also resulted in the introduction of a variety of freshwater biota including predatory fishes from tributary rivers. According to Williams (1964), this lake was hypersaline (50‰) during 1956–67 (Baid, 1962, 1969) but it became hyposaline (16‰) in 1977–79 (Alam, 1980) and no has again regained high salinity levels.

Didwana Lake is of great interest as it is the only existing natural inland biotope of the brine shrimp *Artemia* in India (Bhargava *et al.*, 1987a) and the importance of *Artemia* as a high quality live food in aquaculture is well known.

The present study of Sambhar and Didwana lakes was undertaken to follow the important changes in the biocoenosis of these lakes in relation to changes in salinity and other environmental factors and to explore the possibility of reintroducing *Artemia* to Sambhar Lake.

Description of sites studied

Sambhar lake

Sambhar Lake (27° 58′ N, 75° 55′ E) is in a shallow depression, in the Nagaur and Jaipur districts of State of Rajasthan, NW India (Fig. 1). It is about 22.47 km long and 3.21 to 11.23 km in width. At maximum depth its total surface area is about 190 km². The maximum depth of the lake was 3 m, recorded in the early part of this study. The surrounding area is sandy and sterile, with the Indian desert on the western side. The lake is rainfed, and receives water from a large catchment area of about 471 km².

Besides a rich planktonic flora and fauna, eight freshwater fish species (*Puntius sophore*, *Labeo bata*, *Labeo gonius*, *Oxygaster bacaila*, *Channa punctatus*, *Mystus seenghala*, *Saccobranchus fossilis* and *Chanda nama*) were recorded from the lake just after the monsoon of 1983, when the salinity declined to 1.84‰. These fishes found their way into the lake through incoming streams carrying water of overflowing freshwater ponds and reservoirs in the catchment area. However, the fishes could not survive after the salinity increased above 9‰.

Didwana Lake

Didwana Lake is 70 km west of Sambhar Lake (Fig. 1). At maximum depth it is 6.5 km long and 2.5 km wide, covering an area of 16.5 km². Maximum depth of the lake was 5 m during this study. The lake was devoid of macrophytes and nektonic

fauna. Flocks of flamingos (*Phoenicopterus roseus*) and solitary Indian black winged stilts (*Himantopus himantopus*) inhabit the lake.

Material and methods

The investigations were carried out at Sambhar Lake from February 1984 to January 1985 and at Didwana Lake from December 1983 to January 1985, though it was dry from May to August in 1984. Water samples were collected monthly from each lake (Fig. 1). Chemical determinations were done according to APHA (1975). Phytoplankton were studied after sedimentation and zooplankton after filtering water through a standard bolting silk plankton net (0.3 mm mesh size). Primary production of the phytoplankton was estimated *in situ* by the ^{14}C method (Vollenweider, 1969) and the chlorophyll content determined according to Strickland & Parsons (1972).

Results

Sambhar Lake

Table 1 provides mean monthly values of various physical and chemical variables of Sambhar Lake. No data could be recorded during June 1985, as the lake was dry.

The pH remained highly alkaline, ranging from 8.8 to 9.7. Carbonate alkalinity increased with the rise in salinity and varied from 64 mg l^{-1} (February 1984) to 872 mg l^{-1} (May 1985). Bicarbonate alkalinity dominated throughout the study period, ranging from 144 mg l^{-1} (February 1984) to 1290 mg l^{-1} (May 1985).

Between 1974 and 1983 repeated floods in the region resulted in the dilution of the lake brine and the salinity along with the potential for salt production went down remarkably. During the present study the salinity fluctuated widely, from a initial value of 5.14‰ (February 1984) to a maximum of 267.3‰ (May 1985). The term salinity herein refers to that derived from chlorinity values. Dissolved oxygen varied in relation to

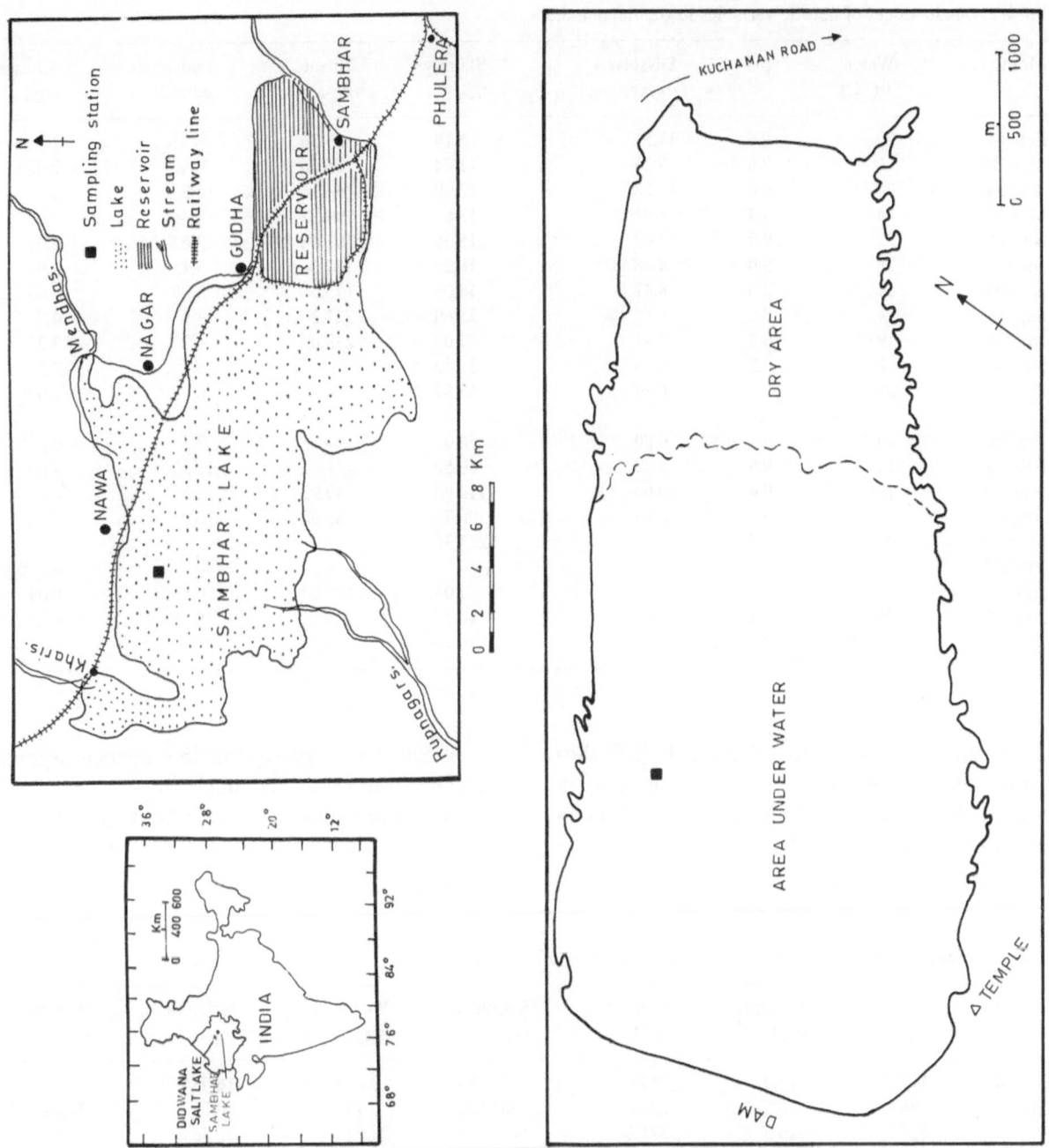

Fig. 1. Sambhar and Didwana lakes. Geographical setting and sampling location.

salinity, with a maximum of 11.52 mg l⁻¹ when salinity was low and was completely depleted when salinity increased above 100‰. The ionic fractions of cations, Na, K, Ca and Mg and the

anion SO_4 and Cl^- estimated from October 1984 to September 1985, are summarized in Table 2.

Nutrient enrichment of the lake was rather low, except that of nitrate which exhibited a wide range

Table 1. Mean values of abiotic variables in Sambhar Lake.

Month	Water T (°C)	pH	Dissolved oxygen mg l^{-1}	Salinity ‰	Nitrate μg-at l^{-1}	Phosphate μg-at l^{-1}	Silicate mg l^{-1}
Feb '84	20	9.6	11.52	5.14	29.36	–	1.74
Mar '84	26	9.6	9.26	11.74	36.66	0.45	2.42
Apr '84	26	9.0	11.58	12.90	64.16	1.35	3.6
May '84	24	8.8	1.91	13.8	90.83	2.40	4.23
Jun '84	32	9.0	4.63	15.35	101.66	2.62	4.61
Jul '84	28	9.0	4.48	16.25	137.2	3.15	4.90
Aug '84	24	8.8	8.47	18.06	175.0	3.00	4.45
Sep '84	24	8.8	6.87	21.60	221.2	3.6	4.7
Oct '84	19	9.2	4.31	27.09	238.0	4.2	3.25
Nov '84	18	9.2	6.86	33.40	–	5.7	2.3
Dec '84	21	9.4	6.24	47.87	–	6.96	2.62
Jan '85	20	9.4	4.80	56.0	–	3.3	5.02
Feb '85	14	9.5	2.40	60.50	–	13.50	7.05
Mar '85	19	9.4	0.64	112.90	49.99	4.5	6.6
Apr '85	23	9.7	0.85	198.72	69.90	24.0	4.15
May '85	22	9.7	–	267.36	–	–	–
Jun '85	–	–	–	–	–	–	–
Jul '85	29	9.0	–	2.06	106.6	0.8	0.09
Aug '85	25	9.1	–	48.77	–	8.15	0.26
Sep '85	31	9.1	–	176.13	–	7.96	–

of fluctuation from nil to 238 μg-at l^{-1}. Ortho-phosphate was initially absent but appeared in March 1984 and gradually increased to attain its maximum concentration by April 1985 (24 μg-at l^{-1}). Silica was present in low concentrations, ranging from nil to 8.75 mg l^{-1}.

The maximum phytoplankton population (5256 × 10^3 cells l^{-1}) was observed in July 1984

Table 2. Mean monthly values of cations and anions in Sambhar Lake.

Month	Alk. mg l^{-1}	Chloride mg l^{-1}	Sulphate mg l^{-1}	Sodium mg l^{-1}	Potassium mg l^{-1}	Calcium mg l^{-1}	Magnesium mg l^{-1}
Oct '84	1172	15.0	249.0	3793.85	0.273	16.03	10.94
Nov '84	962	18.5	321.80	6092.35	0.93	80.16	34.04
Dec '84	1247	26.5	326.6	31609.46	23.06	32.06	58.36
Jan '85	1301	31.0	273.77	35748.08	32.06	40.08	51.07
Feb '85	1620	33.5	297.78	41325.74	16.42	60.12	116.73
Mar '85	1750	62.5	220.93	53789.18	19.15	36.07	63.23
Apr '85	2045	110.0	312.990	78723.75	29.12	36.07	104.57
May '85	2162	148.0	307.39	14012.65	31.28	36.07	107.00
Jun '85	–	–	–	–	–	–	–
Jul '85	234	1.0	–	899.8	1.95	20.04	–
Aug '85	640	27.0	–	10046.65	14.46	32.06	65.55
Sep '85	965	67.0	–	57431.50	17.59	29.30	–

at the salinity of 16.25‰ (Fig. 2). Blue-greens formed the dominant group and contributed the most genera. Among the blue-green algae *Nostoc* tolerated a salinity range from 5.1 to 47.8‰ and formed the major part (45.8%) of the total phytoplankton population during April to October 1984. *Spirulina* showed the maximum salinity tolerance (5.1–198.7‰) and was found throughout the study period except near the end of the investigation. *Spirulina* formed the highest percent composition (33.4–99.6%) in the salinity range of 33 to 198.7‰. *Aphanocapsa* and *Oscillatoria* first appeared when salinity reached 27‰ and disappeared when salinity exceeded 112.9‰ (Fig. 3). Other two blue-green algae, *Microcystis* and *Merismopedia* were observed only during early part of study when salinity ranged

5.1–13.8‰. The diatoms *Nitzschia*, *Navicula* and *Synedra* occurred in a salinity range of 15.3 to 47.8‰ and represented only a small portion (0.34–7.34%) of the total population. Chlorophyceans were represented only by two genera, *Cosmarium* and *Closterium*, appeared only once during the study period and *Cosmarium* was only found in few 1984 samples, forming 0.1–4.6% of the total density.

Primary production was maximum (gross production 11.52 g C m^{-3} day^{-1}; net production 7.88 g C m^{-3} day^{-1}) in April 1984 at 12.9‰ salinity (Fig. 2). Primary production was related more to chlorophyll concentration than to number of cells of phytoplankton (Fig. 2).

Total chlorophyll concentration ranged from nil (last part of the study) to 30.30 mg m^{-3} in

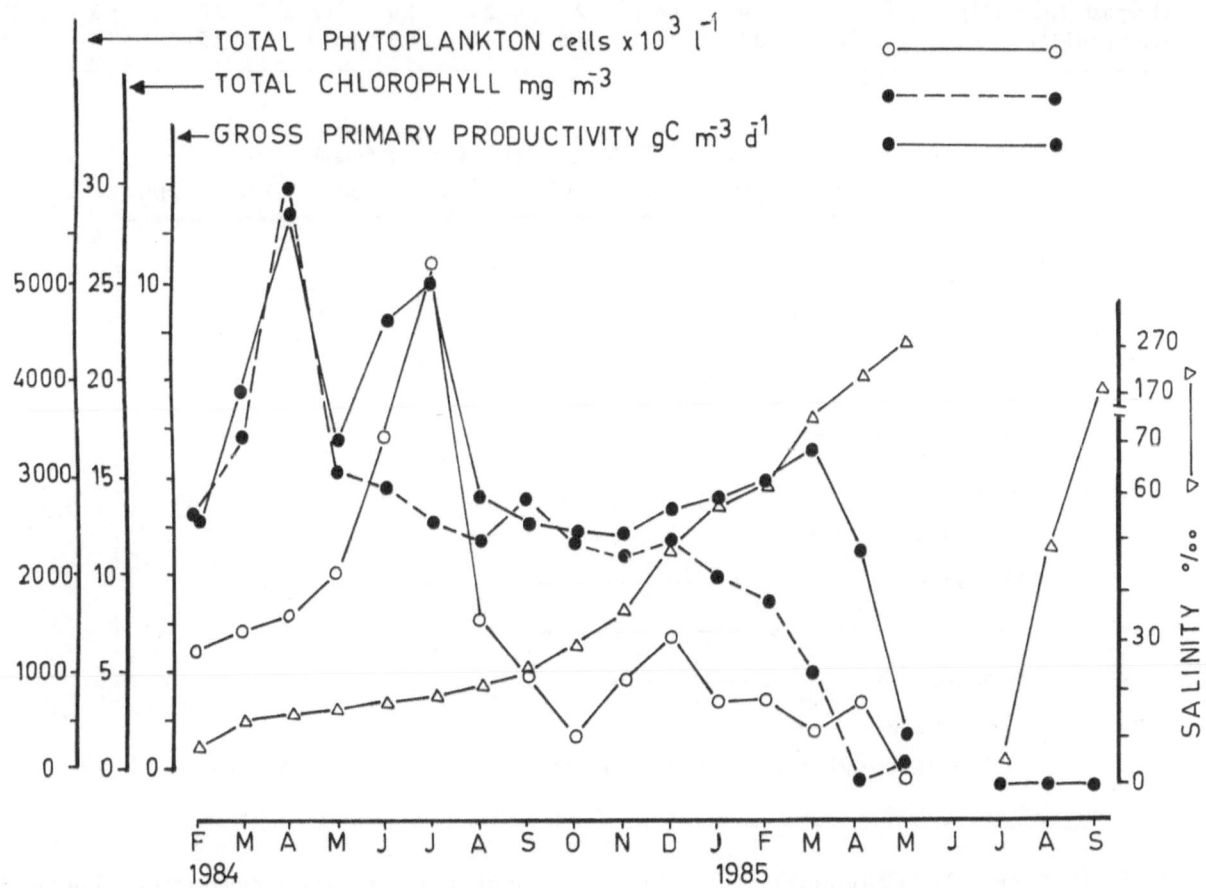

Fig. 2. Monthly variations of total phytoplankton density, total chlorophyll concentration and primary productivity in Sambhar lake.

Table 3. Mean monthly values of various variates at three stations in Didwana Lake from december 1983 to January 1985. Lake dry from May to August.

Parameters	Months									
	D	J	F	M	A	S	O	N	D	J
Water Temperatura (°C)	16.0	13.0	15.3	20.0	25.0	24.0	24.0	18.5	17.7	19.0
pH	8.4	8.7	9.0	9.5	10.0	8.7	8.99	8.6	8.4	7.8
CO_3 (mg l^{-1})	146.0	177.0	209.5	290.5	498.5	48.5	145.5	150.5	226.7	
HCO_3 (mg l^{-1})	372.5	450.0	626.0	1064.5	1574.0	158.5	180.0	275.0	374.2	391.5
Alkalinity (mg l^{-1})	518.5	627.0	835.5	1355.0	2072.5	207.0	325.5	425.5	600.9	391.5
Dissolved oxygen (mg l^{-1})	7.1	5.0	0.5	0.6		5.6	4.7	4.2	1.8	
Sodium (mg l^{-1})								13983.7	55946.3	98923.7
Potassium (mg l^{-1})								38.0	79.1	101.7
Calcium (mg l^{-1})								51.2	86.2	115.5
Magnesium								56.1	109.7	139.4
Chloride ‰ (ppt)	21.6	35.8	45.7	67.6	121.5	8.1	19.2	39.0	112.3	183.0
Sulphate (mg l^{-1})								228.1	309.7	298.1
Salinity (‰)	46.3	64.8	82.6	120.4	237.5	15.7	34.7	70.4	203.0	336.0
Nitrate (μg-at l^{-1})		28.9	33.6	56.8	62.4	51.9	40.7	36.4	73.9	119.9
Nitrite (μg-at l^{-1})		0.1	0.0	0.2	0.9	0.0	0.1	0.0		
Phosphate (μg-at l^{-1})		1.0	1.6	2.1	2.9	2.0	0.9	3.8	6.8	10.3
Silica (mg l^{-1})		0.4	0.4	1.6	4.0	0.3	1.1	2.5	2.5	5.0

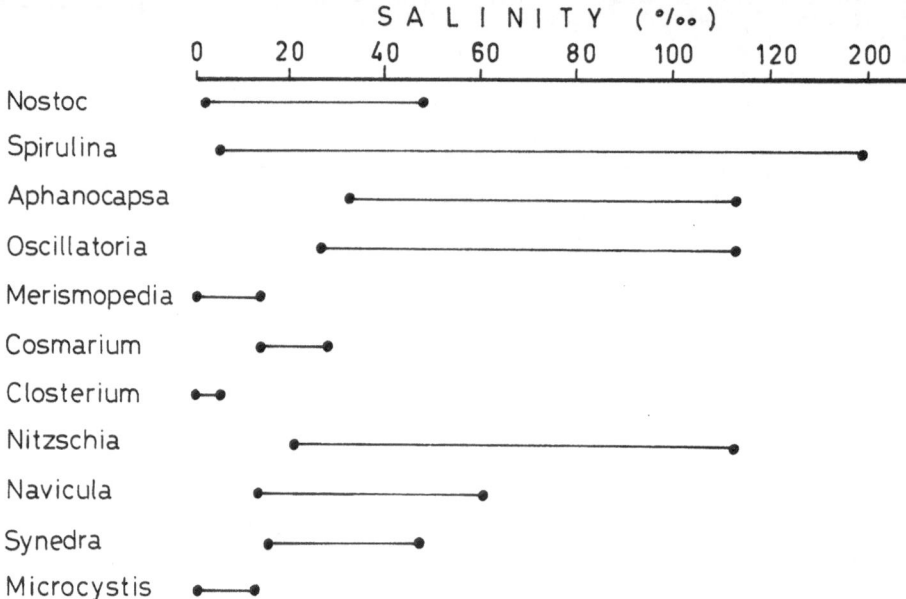

Fig. 3. Occurrence of various phytoplankton taxa in relation to salinity in Sambhar lake.

April 1984 (Fig. 2). Chlorophyll *a* was the dominant pigment whose concentration ranged from nil to 15.87 mg m^{-3}. Chlorophyll *c* was found in the lowest concentration among the chlorophylls. Carotenoids also occurred regularly except during the last four months of the study.

The maximum value ($37.45 \, \text{mg m}^{-3}$) was recorded in April 1984, coinciding with the highest peak of total chlorophyll.

The lake supported a poor population of zooplankton, quantitatively as well as qualitatively. The zooplankton density was low in April 1984, when the salinity was less than 13‰ (Fig. 4), but increased to the highest level in July (315 individuals 1^{-1}) when fishes were no longer present. *Cyclops* and *Brachionus* were the dominant genera and occurred the whole salinity range observed. Other forms like *Moina* and nauplii showed an erratic distribution and *Diaptomus* was found only once (July 1984). However, when salinity exceeded 56‰, all zooplanktonic forms disappeared.

Didwana Lake

Didwana Lake was highly alkaline (Table 3), with pH ranging from 7.6 to 10.2. The highest pH was recorded in April 1984, probably an effect of high carbonate and bicarbonate content. Carbonate and bicarbonate concentrations varied from 18 to 510 $\text{mg} \, 1^{-1}$ and 70 to 1580 $\text{mg} \, 1^{-1}$ respectively. Total alkalinity was high (437 to 2090 $\text{mg} \, 1^{-1}$). Dissolved oxygen concentration varied from complete anoxia to 7.29 $\text{mg} \, 1^{-1}$, declining with increased salinity.

Salinity ranged from hyposaline (7.5‰, September 1984) to hypersaline (336‰, January 1985). Highest salinities were recorded when the lake had nearly dried up and approached the saturation point. The major ions present were sodium and chloride with the sequence of ionic dominance being $Na > Mg > Ca > K$ and $Cl > HCO_3 > CO_3 > SO_4$. Ionic fractions in milliequivalent percentages were as follows: Cations, Na 98.61, K 0.21, Ca 0.40, Mg 0.76; anions, Cl 98.60, SO_4 0.59, HCO_3 0.39 and CO_3 0.30.

Nutrients were found in low concentrations. Nitrate ranged from nil to 120 $\mu\text{g-at} \, 1^{-1}$. The maximum values were recorded on two occasions (April 1984 and January 1985) just before the lake dried up. This was perhaps due to algae decay following high salinities and subsequent nitrification. Orthophosphate concentrations showed wide fluctuations, particularly after the dry period (May–August 1984). It ranged from nil to 11.15 $\mu\text{g-at} \, 1^{-1}$, and like that of the nitrate, the maximum concentration was recorded when the lake was on the verge of drying up. Reactive silicate was the dominant nutrient (0.50–5.02 $\text{mg} \, 1^{-1}$).

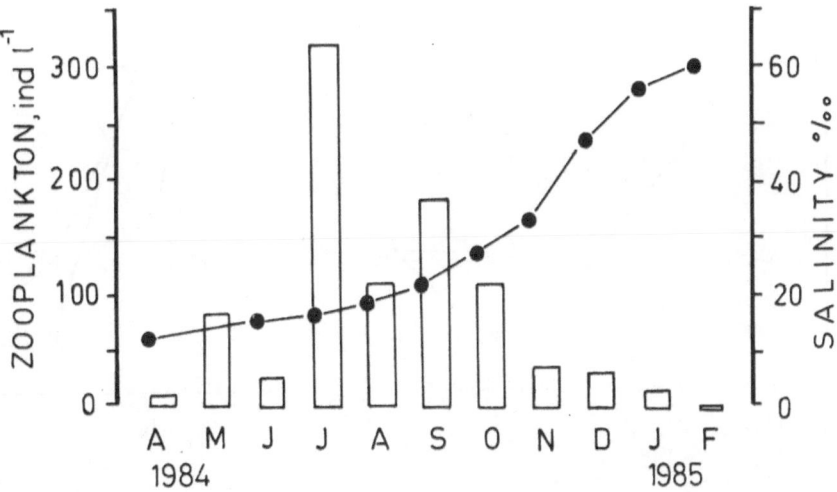

Fig. 4. Monthly variations of total zooplankton density in relation to salinity changes in Sambhar lake.

252

In Didwana Lake there were relatively few phytoplankton taxa and their total density ranged up to 3991×10^3 cells l^{-1} (Fig. 5). Only 9 genera were recorded from this lake. Six were blue-greens (*Aphanocapsa*, *Microcystis*, *Nostoc*, *Anabaena*, *Nodularia* and *Spirulina*), two diatoms (*Nitzschia* and *Synedra*) and one chlorophycean (*Closterium*). The maximum density of phytoplankton was observed at 34.7‰ salinity in October 1984 (Fig. 5). A *Spirulina* bloom occurred just preceding the dry period in April 1984.

Aphanocapsa was the most abundant alga in this lake and it occurred over a wide range of salinity. Maximum counts (1539×10^3 cells l^{-1}) were recorded in November 1984, when salinity was 62‰ and it occurred even up to a salinity of 146‰. *Microcystis* was the next dominant alga numerically. *Spirulina* showed the greatest salinity tolerance, from 7.54 to 178.84‰ (Fig. 6), and *Nitzschia* also represented up to a salinity level of 146.6‰. *Synedra* had a low salinity range (Fig. 6).

The order of dominance of chlorophyll pigments was $a > c > b$. Highest total chlorophyll content (29.33 mg m^{-3}) was concomittant with the maximum total phytoplankton population (Fig. 5).

Gross phytoplanktonic productivity was low, the maximum being 5.1 g C m^{-3} d^{-1} (Fig. 5). The average daily gross production was estimated 2.12 g C m^{-3} d^{-1} in 1984. Net primary productivity ranged from nil to 2.8 g C m^{-3} d^{-1}. On an annual basis (1984) gross and net primary

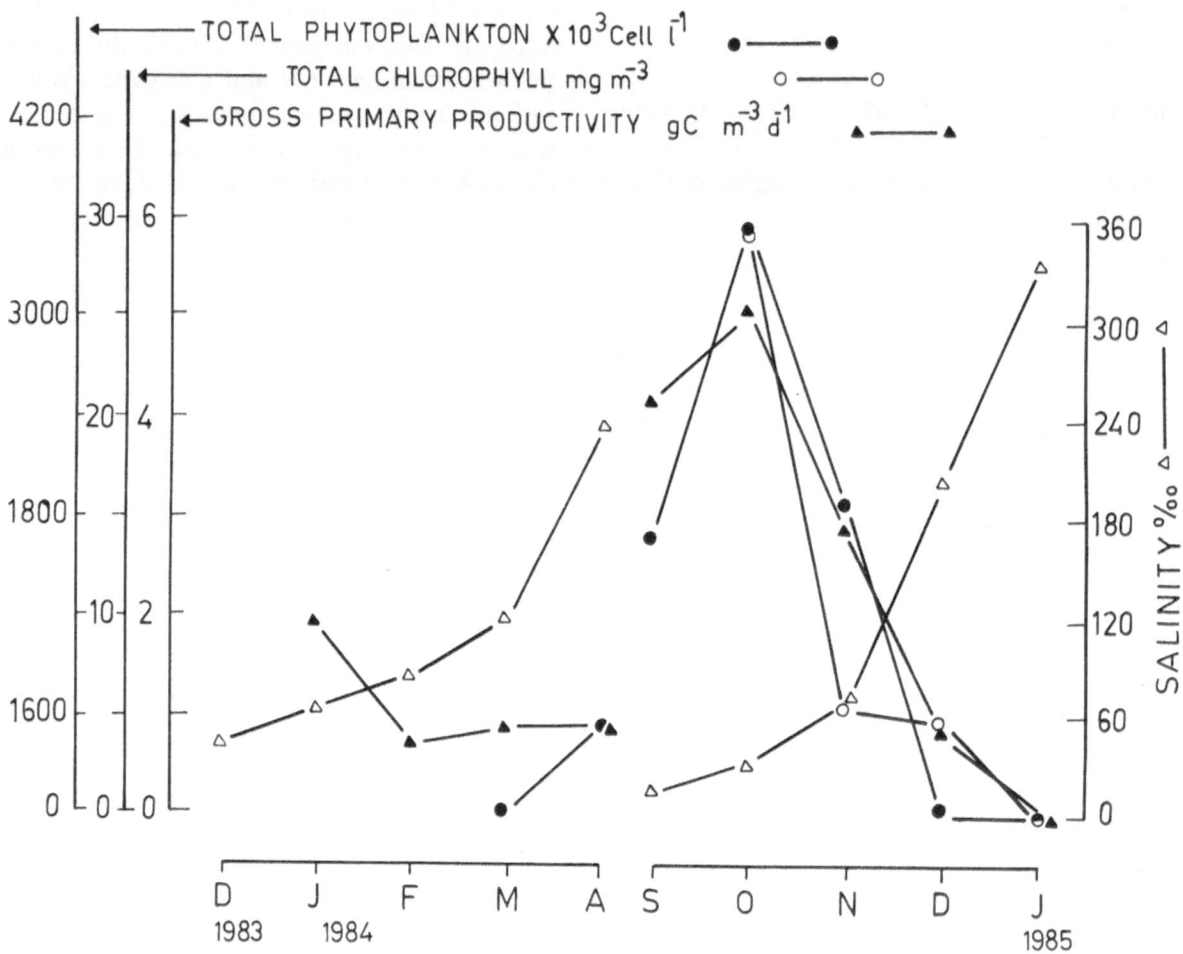

Fig. 5. Monthly variations of total phytoplankton density, total chlorophyll concentration and primary productivity in Didwana lake.

productions were $781.10 \, \mathrm{g \, C \, m^{-3} \, y^{-1}}$ and $416.10 \, \mathrm{g \, C \, m^{-3} \, y^{-1}}$ respectively.

Salinity and temperature appeared to have important effects on the occurrence, abundance and growth of zooplankton, particularly of the brine shrimp, *Artemia*. In the beginning of the study (December 1983), when the salinity of the lake was 39.7‰, two crustaceans *Moina* sp. and *Cyclops* sp. and a rotifer *Brachionus* sp. were present but in small numbers. As the salinity increased to 87.9‰ (late February 1984), these forms disappeared and were succeeded by *Artemia* in its various stages of development (Fig. 7). Its numbers increased rapidly attaining a high population of 1308 individuals l^{-1} by March 1984 (salinity 94.2‰). However, brine shrimp population declined when the salinity reached 200‰ and at 270‰ no adults or juveniles were found (Fig. 7). At that time a dense accumulation of *Artemia* cysts was seen along the shore of the lake.

After a dry period of 4 months, young stages of brine shrimp appeared soon after rains began in September 1984, when the salinity had dropped to 15.7‰. At this low salinity other zooplanktonic forms such as *Moina* sp., *Cyclops* sp., *Brachionus* sp. and insect larvae reappeared and replaced *Artemia* which apparently could not survive at the low salinity (Fig. 7). When the salinity increased to 77.68‰ (November 1984), all the other zooplankters disappeared and were succeeded again by *Artemia* which persisted this time until the salinity reached 288‰.

Discussion

High salinities in both Sambhar (267.3‰) and Didwana (336‰) lakes in 1985 can be attributed to extremely low rainfall and high rate of evaporation. During the previous thirty years, the salinity of Sambhar Lake fluctuated turning the lake from hypersaline to subsaline and back to hypersaline levels (Baid, 1958, 1969; Alam, 1980). Baid (1962) reported maximum salinity in Sambhar lake to be 164‰ which contrasts with 15.22‰ during 1977–78 (Alam, 1980).

Various salts dissolved in water lower its ability to absorb and hold oxygen (Cole, 1979). Hence highly saline waters have low dissolved oxygen -D.O.- concentration. During the present study also D.O. was found inversely correlated with salinity. Dissolved oxygen concentrations recorded ranged from nil to as high as 11.68 mg l^{-1} in Sambhar Lake. In the earlier studies on Sambhar

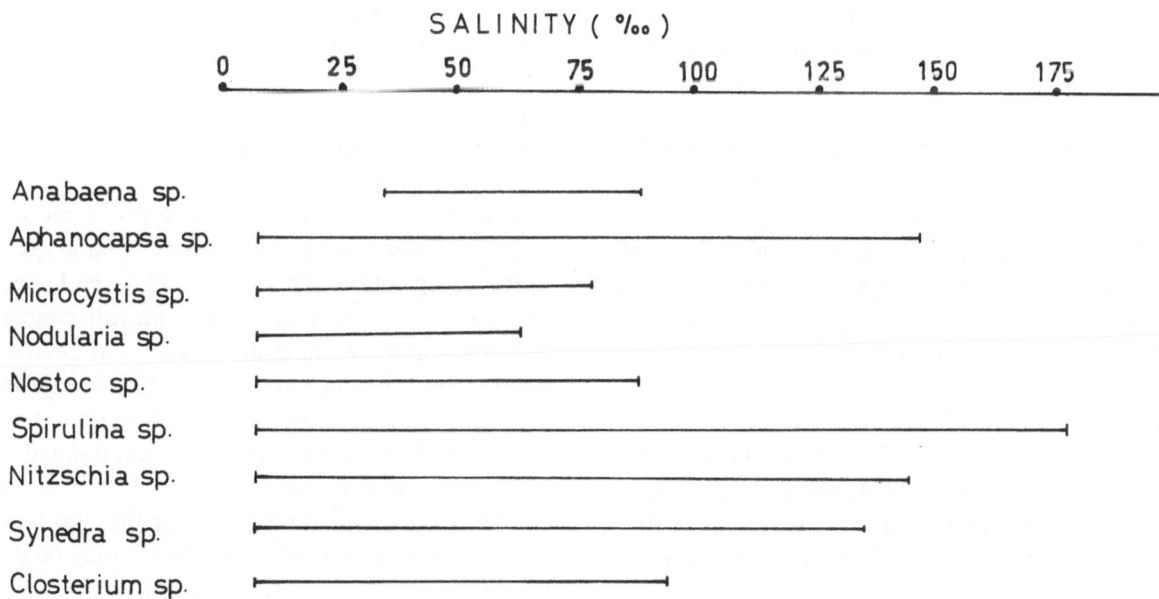

Fig. 6. Occurrence of various phytoplankton taxa in relation to salinity in Didwana lake.

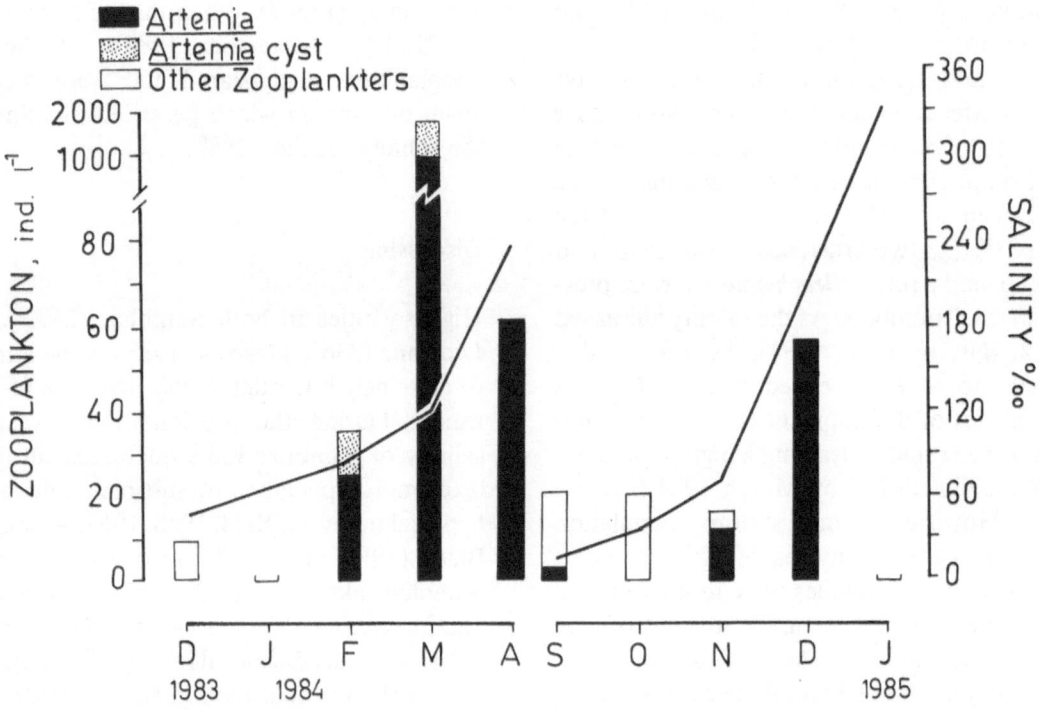

Fig. 7. Monthly variations of total zooplankton density in relation to salinity changes in Didwana lake.

Lake, Baid (1962) reported a range of 9.68–31.46 mg l^{-1} (although Baid himself was doubtful about such high abnormal values) and Alam (1980) 1.57 to 7.28 mg l^{-1}. But complete depletion of D.O. was found for the first time during the present study. The very high salinity may prevent the healthy growth of phytoplankton and thereby lowering the photosynthetic activity and release of oxygen, increasing oxygen consumption for organic matter decomposition.

The ionic composition of the total salinity of Sambhar and Didwana lakes are very similar. Other saline lakes all around the world show a high dominance of sodium and chloride (Bayly, 1969; Topping & Scudder, 1977; Williams, 1978; Comín *et al.*, 1983) at different salinities. Goldman & Horne (1983) gave support to the idea of biological diversity limited by high sodium concentrations due to osmotic stress, which also holds good in the case of Sambhar and Didwana lakes.

Nutrient concentrations in Sambhar Lake

showed higher range during the present study in comparison to earlier one (Alam, 1980). In both Sambhar and Didwana lakes phosphate showed an increasing trend with the increase of salinity, while at the same time it was paralleled by decrease of phytoplankton populations, due to high salinity. Silicate concentrations were lower than in other saline lakes elsewhere (e.g. Rawson & Moore, 1944).

Sambhar Lake had higher phytoplankton density than Didwana Lake and also varied in qualitative composition. This agrees with the heterogeneity observed in salinity and chloride concentrations that were generally higher and changed quicker in Didwana Lake. Alam (1980) recorded 20 phytoplanktonic genera from Sambhar Lake during 1977–78. Total phytoplankton population over that period ranged from 150×10^3 to 9230×10^3 cell l^{-1}. The lower density and number of species observed in our study may be due to greater change in salinity range.

Comparing data from Baid (1962, 1969) and

Alam (1980) with our data it is evident a change in phytoplankton composition of Sambhar Lake. Genera *Chlamydomonas*, *Dunaliella* and *Aphanotece* which inhabited the Sambhar Lake during 1962–1969 are not found thereafter. During 1977–78, when salinity ranged only 6.86 to 15.22‰, maximum diatoms genera (11) were reported, out of which only three genera were found in Sambhar Lake and two in Didwana Lake during the present study. One species of the blue green *Oscillatoria*, which prevailed during 1962–69 (Baid, 1969), has been observed during present study from Sambhar Lake.

Species number decreased in Sambhar and Didwana lakes as salinity increased. A high carbonate concentration has also been mentioned as a limiting factor for many algal species, because it can be the reason for absence of free CO_2 (Hammer *et al.*, 1983). Low species number has also been reported from many other saline lakes around the world (Gessner, 1957; Walker, 1973; Post, 1975; Vareschi, 1982).

Primary productivity of Sambhar Lake showed two peaks below 20‰ salinity but thereafter declined. The maximum of both gross and net primary production synchronized with the monsoon peak of phytoplankton in Didwana Lake (Fig. 5). Similar relationship was indicated by Haynes & Hammer (1978) for saline lakes of Canada. Total chlorophyll also showed inverse relationship with the salinity. In present study of Sambhar Lake different chlorophyll pigments a, b and c showed higher concentrations than those reported earlier by Alam (1980).

The disappearance of *Artemia* and *Branchinella* from Sambhar Lake (Alam, 1980) can be attributed to dilution of brine or to the arrival of predators and change in food resources. Lakes Didwana and Sambhar have similar crustacean and rotifer species. *Artemia* is also present in Lake Didwana but not in Sambhar Lake. However, Tiwari (1958) and Baid (1958) recorded, respectively, *Branchinella* and *Artemia* in Sambhar Lake. Since the salinity of Sambhar Lake has regained a moderately high level and no predators occur, an attempt can be made to reintroduce the brine shrimp in this lake environment. To begin with,

rearing trials with *Artemia* from Didwana Lake have been carried out in a salt pan near Sambhar Lake (Bhargava *et al.*, 1987b).

Acknowledgements

This study was carried out under a Man and Biosphere Project sponsored by the Department of Environment, Government of India, New Delhi. The authors are thankful to the National MAB Committee -India- for the financial support, and to the Head of the Department of Zoology, University of Jodhpur, Jodhpur, for providing laboratory facilities.

References

Alam, M., 1980. Limnological studies of Sambhar salt lake and its reservoir. Ph.D. Thesis, University of Jodhpur, Jodhpur, India. 158 pp.

American Public Health Association, American Water Work Association & Water Pollution Control Federation, 1975. Standard methods for the examination of water and wastewater. 14th ed. APHA, AWWA, WPCF, Wash. D.C., 769 pp.

Baid, I. C., 1958. On the occurrence of *Artemia salina* in Sambhar Lake (Rajasthan). Curr. Sci. 27: 58–59.

Baid, I. C., 1962. Ecological studies on crustacean and insect fauna of the Sambhar Lake, Rajasthan with special reference to *Artemia salina* L. Ph.D. Thesis, University of Rajasthan.

Baid, I. C., 1969. The arthropod fauna of Sambhar salt lake, Rajasthan, India. Oikos 19(2): 292–303.

Bayly, I. A. E., 1969. Symposium on salt and brackish inland waters. Introductory comments. Verh. int. Ver. Limnol. 17: 419–420.

Bhargava, S. C., G. R. Jakher, M. M. Saxena & R. K. Sinha, 1987a. Ecology of *Artemia* in Didwana salt lake (India) p. 127–133. In: *Artemia* Research and its Applications. Vol. 3. Ecology, Culturing, Use in aquaculture. P. Sorgeloos, D. A. Bengtson, W. Decleir & E. Jaspers (eds.). Universa Press, Wetteren, Belgium, 556 pp.

Bhargava, S. C., G. R. Jakher, M. M. Saxena & R. K. Sinha, 1987b. Rearing *Artemia* in a salt pan near Sambhar Lake (India). In: *Artemia* research and applications. vol. 3: 271–274. Ecology, culturing, use in aquaculture. P. Sorgeloos, D. A. Bengtson, W. Decleir & E. Jaspers (eds.). Universa Press. Wetteren, Belgium, 556 pp.

Cole, G. A., 1979. Text book of Limnology. 2nd ed. C.V. Mosby Company, Missouri, USA. 426 pp.

Comín, F. A., M. Alonso, P. López & M. Comelles, 1983.

Limnology of Gallocanta Lake, Aragón, NE Spain. Hydrobiologia 105: 207–223.

Gessner, F., 1957. Van Golu zur limnologic des grossen soda sees in Ostanatolien (Turkei) Archiv. Hydrobiol. 53: 1–22.

Goldman, C. R. & A. J. Horne, 1983. Limnology. Tosho Printing Co. Ltd., Tokyo, Japan. 464 pp.

Hammer, U. T., J. S. Shamess & R. C. Haynes, 1983. The distribution and abundance of algae in saline lakes of Saskatchewan, Canada. In: Saline lakes. Proc. 2nd Int. Symp. Athalassic (Inland) saline lakes. U. T. Hammer (ed.). Dev. Hydrobiologia 16. Junk. The Hague: 1–26.

Haynes, R. C. & U. T. Hammer, 1978. The saline lakes of Saskatchewan. IV. Primary production by phytoplankton in selected saline ecosystems. Int. Revue ges. Hydrobiol. 63: 337–351.

Post, F. J., 1975. Life in the Great Salt Lake. Utah Sci. 36: 43–47.

Rawson, D. S. & J. E. Moore, 1944. The saline lakes of Saskatchewan. Can. J. Res. 22: 141–201.

Strickland, J. D. H. & T. R. Parsons, 1972. A practical handbook of seawater analysis. Bull. Fish. Res. Bd Can. 167, 310 pp.

Tiwari, K. K., 1958. Diagnosis of a new species of the genus *Branchinella* Saycee (Crustaceae, Brachiopoda, Anostraca) from Sambhar Lake, Rajasthan. J. Bombay Natl. Hist. Soc. 55: 585–588.

Topping, M. S. & G. G. E. Scudder, 1977. Some physical and chemical features of saline lakes in Central British Columbia. Syesis 10: 145–166.

Vareschi, E., 1982. The ecology of lake Nakuru (Kenya). III. Abiotic factors and primary production. Oecologia (Berlin) 55: 81–101.

Vollenweider, R. A. (ed.)., 1969. Primary production in aquatic environments. I.B.P. Handbook, 12: 213 pp.

Walker, K. F., 1973. Studies on saline lakes ecosystem. Aust. J. mar. Freshwat. Res. 24: 21–27.

Williams, W. D., 1964. A contribution to lake typology in Victoria, Australia. Verh. int. Ver. Limnol. 15: 158–163.

Williams, W. D., 1978. Limnology of Victorian salt lakes, Australia. Verh. int. Ver. Limnol. 20: 1165–1174.

Hydrobiologia **197**: 257–266, 1990.
F. A. Comin and T. G. Northcote (eds), Saline Lakes.
© 1990 Kluwer Academic Publishers.

Salinity as a determinant of salt lake fauna: a question of scale

W. D. Williams, A. J. Boulton & R. G. Taaffe
Department of Zoology, University of Adelaide, North Terrace, Adelaide, S.A. 5000, Australia

Key words: ecological scale, salt lakes, species richness, species composition, salinity, macroinvertebrates

Abstract

High and often variable salinity is an obvious feature of salt lakes. Correspondingly, salinity is usually assumed to be an important ecological determinant in such lakes. An investigation of the macro-invertebrate fauna of 79 lakes (salinities from 0.3 to 343 g l^{-1}) in the Western District of Victoria, Australia, examined this assumption. Over the total range of salinity, species richness and composition are highly correlated with salinity. However, these relationships become nonsignificant over intermediate ranges of salinity. Furthermore, many taxa have very broad tolerances to salinity at these intermediate ranges, implying that factors other than salinity may determine their distribution. An appreciation of scale (that is, the range of salinity over which observations are considered) resolves the paradox that, despite these broad tolerances by most taxa, species richness and composition strongly reflect salinity over the entire salinity range.

Introduction

The negative correlation observed between species richness and salinity over a broad range of salinities (cf. Hammer, 1986) underlies the general assumption that salinity is an important determinant of the fauna of saline localities (e.g. Remane & Schlieper, 1971). However, regional differences in the composition of the macro-invertebrate fauna of Australian salt lakes (Williams, 1984; Williams & Kokkinn, 1988) led Williams (1984) to suggest that salinity *per se* is relatively unimportant in determining what species occurs in a particular lake. This paradox exemplifies the importance of scale in ecological research, an issue not yet addressed in studies of salt lakes.

In the Western District of Victoria, Australia, many salt lakes of widely differing salinity but relatively homogeneous ionic composition provide an ideal natural laboratory for an analysis of the relationship between salinity and faunal composition. All but the least saline lakes are dominated by Na and Cl ions (Williams & Buckney, 1976; Williams, 1981; De Deckker & Williams, 1988), and it may confidently be assumed that differences among species in the lakes do not reflect major ionic differences.

Here, using data from an investigation of these lakes, we demonstrate the significance of scale for interpretation of the relationship between salinity and certain ecological attributes of salt lake fauna.

Material and methods

Seventy-nine lakes (Fig. 1) spanning the widest possible salinity range (0.3 to 343 g l^{-1}) were sampled in January 1980, a month when field salinities usually are maximal. From near the edge of each lake, benthic and planktonic macro-invertebrates were sampled for a standard time (5

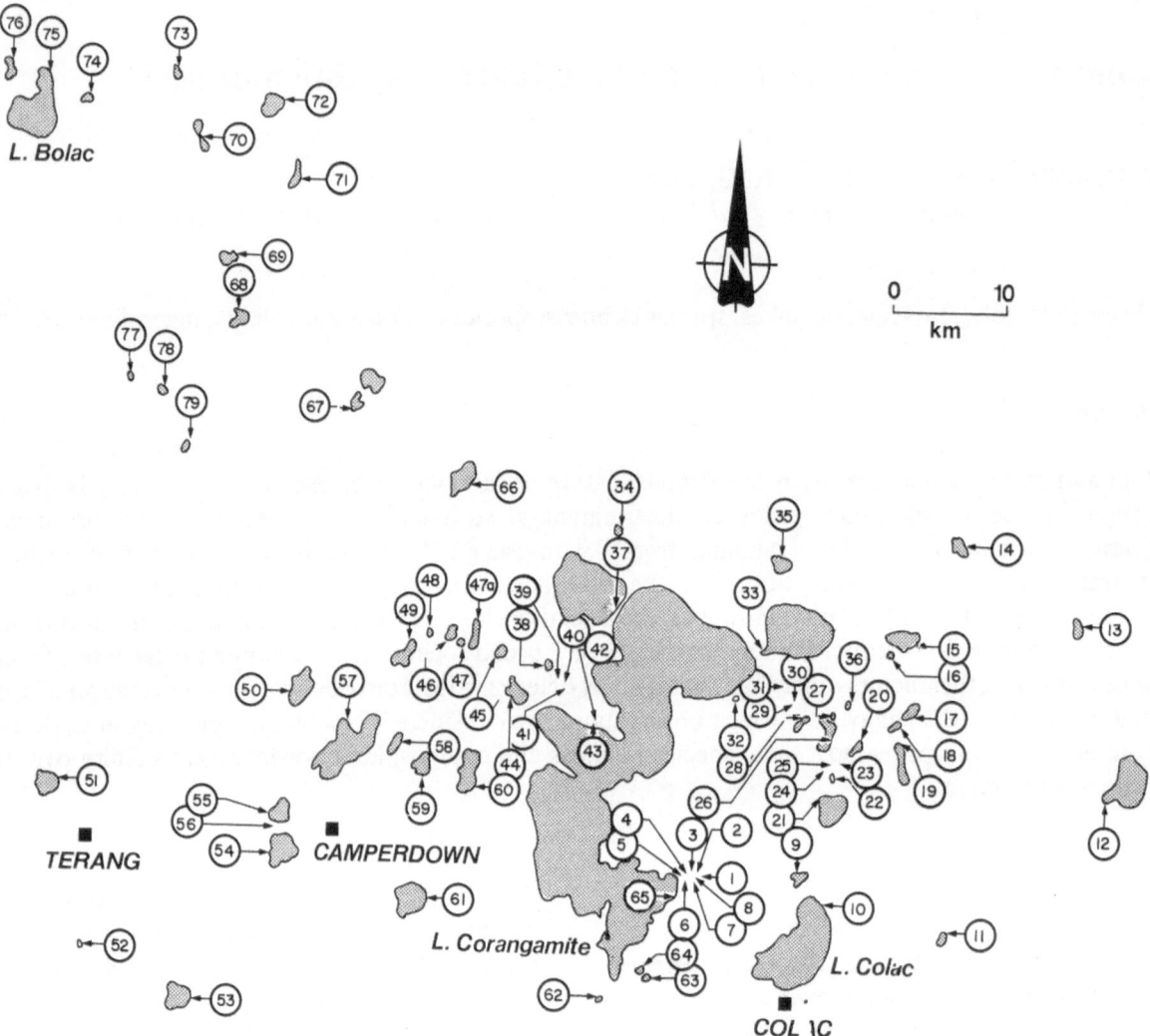

Fig. 1. Position of 79 localities studied. The region lies some 150 km WSW of Melbourne, Victoria.

minutes) with hand-held nets of aperture sizes ≈ 400 mm and ≈ 160 mm, and a littoral water sample was collected in a polyethylene bottle. In the laboratory, conductivities were determined using a Radiometer conductivity meter, and salinities derived using the formula of Williams (1986). Most taxa were identified to species (Table 1).

Spearman rank correlations and multivariate analyses were used to investigate the relationship between salinity and species richness, and between salinity and faunal composition. The data matrix (79 lakes by 147 taxa) was classified by two-way indicator species analysis (TWIN-

SPAN, Hill *et al.*, 1975; Hill, 1979), a polythetic divisive method recommended for hierarchical classification because of its effectiveness and robustness (Gauch & Whittaker, 1981; Gauch, 1982). The resulting dendrogram was truncated at level 5 or when the groups contained fewer than six elements and further subdivision provided little additional information.

The data also were ordinated using hybrid multidimensional scaling (HMDS), computed using the program KYST in the ecological database package ECOPAK (Minchin, 1986). HMDS is a robust technique that has been dem-

Table 1. Range of field salinities and number of localities at which taxa were collected.

Taxon	Salinity range (g l⁻¹)	Number of locs.
CNIDARIA:		
Cordylophora cf. lacustris	0.4–35.1	6
MOLLUSCA:		
Coxiella sp.	0.3–53.3	19
Glyptophysa aliciae (Reeve)	0.3, 0.6	2
Segnitila victoriae (Smith)	0.4, 1.0	2
Physa sp.	1.9, 3.2	2
Corbiculina australis (Deshayes)	1.6, 1.9	2
ANNELIDA:		
Hirudinea (1 sp.)	1.5–5.9	4
Oligochaeta sp. 1	0.4–57.4	13
Oligochaeta sp. 2	1.9–14.9	5
HYDRACARINA (spp.)	0.3–6.2	9
CRUSTACEA: COPEPODA		
Boeckella triarticulata (Thomson)	0.3–30.1	27
B. symmetrica Sars	1.0–17.5	4
Calamoecia clitellata Bayly	1.3–157	16
C. salina (Nicholls)	16.5–131	6
C. australica Sars	0.3, 1.3	2
C. ampulla (Searle)	1.5, 2.0	2
Halicyclops ambiguus Kieffer	9.3–30.1	4
Australocyclops australis (Sars)	0.3–2.0	3
Mesocyclops sp.	8.1, 13.4	2
M. noticus Kiefer	0.6, 2.0	2
Eucyclops serrulatus (Fisher)	0.5–2.0	4
Macrocyclops sp.	0.5	1
Microcyclops sp. 1	1.3–186	5
M. sp. 2	1.3	1
M. sp. 3	4.2–104	5
M. sp. 4	0.3, 3.6	2
M. sp. 5	1.3	1
M. arnaudi Sars	20.9–102	4
Harpacticoida (spp.)	1.5–131	9
CRUSTACEA: CLADOCERA		
Daphnia carinata King	1.9–5.9	5
D. cephalata King	1.3	1
D. longicephala Hebert	0.3	1
Daphniopsis cf. pusilla Serventy	0.5–186	18
D. sp. 1	10.2–159	6
D. sp. 2	8.6–186	9
Bosmina meridionalis Sars	17.5	1
Echinisca capensis capensis Sars	2.0, 4.2	2
Macrothrix breviseta Smirnov	0.6–17.5	3

Table 1. (continued).

Taxon	Salinity range (g l⁻¹)	Number of locs.
Chydorus sphaericus O. F. Müller	0.4–1.5	3
Pleuroxus jugosus (Henry)	1.5, 4.9	2
Alona rectangula novae-zealandiae Sars	0.5–2.0	3
Alonella sp.	2.0	2
Moina sp.	4.2	1
M. micrura Kurz	4.8	1
Simocephalus acutirostris (King)	0.6	1
S. vetulus elisabethae (King)	0.4–2.0	5
Ceriodaphnia quadrangula O. F. Müller	1.5	1
CRUSTACEA: OSTRACODA		
Platycypris baueri Herbst	4.8–288	22
Mytilocypris henricae (Chapman)	1.5–20.0	6
M. praenuncia (Chapman)	4.8–41.3	7
M. splendida (Chapman)	0.3–21.6	18
M. mytiloides (Brady)	16.5, 17.5	2
Australocypris robusta De Deckker	20.9–288	15
Diacypris compacta (Herbst)	0.3–186	22
D. dietzi (Herbst)	0.3–131	6
D. spinosa De Deckker	4.8–157	8
Reticypris clava De Deckker	4.8–41.3	7
R. herbsti McKenzie	102–177	4
Ilyocypris austaliensis Sars	1.6, 4.8	2
I. bradyi Sars	2.0, 4.9	2
Cyprinotus edwardi McKenzie	0.6	1
Candonocypris novaezelandiae (Baird)	0.4–1.0	3
Newnhamia fenestrata King	0.6–2.0	4
Cypricercus salinus De Deckker	0.3–12.7	8
Limnocythere milta De Deckker	15.9	1
L. dorsicula De Deckker	2.0	2
L. sp.	0.4	1
Sarscypridopsis aculeata (Costa)	1.3	1
Cypretta sp.	0.4	1
Cytheridella sp.	0.4	2
Ilyodromus viridulus Sars	0.4	1
CRUSTACEA: MALACOSTRACA		
Parartemia zietziana Sayce	102–343	15
Haloniscus searlei Chilton	7.7–157	23
Austrochiltonia subtenuis (Sayce)	0.3–46.8	35
A. australis (Sayce)	0.4–59.8	19
Paratya australiensis Kemp	0.4–3.2	3
Amarinus lacustris (Chilton)	1.5, 3.2	2
INSECTA: EMPHEMEROPTERA		
Tasmanocoenis tillyardi (Lestage)	1.5, 3.2	2
Atalophlebia australis (Walker)	0.3–3.2	3
Cloeon sp.	0.6	1

Table 1. (continued).

Taxon	Salinity range (g l^{-1})	Number of locs.
INSECTA: ODONATA		
Hemianax papuensis (Burmeister)	0.6	1
Hemicordulia australiae (Rambur)	0.3	1
Ischnura heterosticta (Burmeister)	1.5–4.2	3
Austroaeschna parvistigma (Selys)	0.6	1
Austrolestes annulosus (Selys)	1.5–20.0	8
A. psyche (Hagen)	1.0–20.0	5
Austroagrion watsoni Lieftinck	0.6	1
Oristicta sp.	0.6	1
Xanthagrion erythroneurum Selys	2.0, 4.2	2
INSECTA: HEMIPTERA		
Sigara sublaevifrons (Hale)	0.4, 1.5	2
S. australis (Fieber)	0.4–21.6	18
S. truncatipala (Hale)	0.5, 1.5	2
Agraptocorixa hirtifrons (Hale)	0.3–20.9	6
A. eurynome (Kirkaldy)	0.6–39.3	4
Micronecta robusta Hale	0.3–9.9	10
M. gracilis Hale	0.3	1
M. annae Kirkaldy	0.4, 3.2	2
Anisops gratus Hale	0.6–17.5	7
A. thienemanni Lundblad	0.6–30.4	7
Enithares woodwardi Lansbury	0.5, 1.0	2
Veliidae (spp.)	1.9–39.7	4
Plea brunni Kirkaldy	0.4–2.0	4
P. halei Lundblad	0.5	1
Naucoris congrex Stal.	0.6–1.5	3
Nerthra sp.	57.4, 59.8	2
Diplonychus eques (Dufour)	1.5	1
INSECTA: DIPTERA		
Chironomus occidentalis Skuse	4.2–57.4	3
Paratanytarsus sp.	1.5–57.4	3
Corynoneura sp.	0.5	1
Pentaneura sp.	0.5–2.0	3
Tanytarsus sp. 1	14.5–102	4
T. fuscithorax Skuse	16.5	1
T. barbitarsis Freeman	21.6	1
Polypedilum sp.	4.9–104	3
Procladius paludicola Skuse	1.5–59.8	9
Ceratopogonidae sp. 1	21.6–86	4
Ceratopogonidae sp. 2	1.6–84	5
Ceratopogonidae sp. 3	0.4, 57.4	2
Ceratopogonidae sp. 4	57.4	1
Ephydridae sp.	1.0–39.7	9
INSECTA: LEPIDOPTERA		
Pyralidae sp.	16.5, 20.0	2

Table 1. (continued).

Taxon	Salinity range (g l^{-1})	Number of locs.
INSECTA: TRICHOPTERA		
Ecnomus sp.	0.5–3.2	3
E. cygnitus Neboiss	1.5	1
E. pansus Neboiss	3.2	1
Notalina sp.	0.6–16.5	8
Condoceras sp.	8.1	1
INSECTA: COLEOPTERA		
Megaporus sp.	0.3–4.2	5
Berosus australiae Mulsent	0.4–39.7	24
Necterosoma penicillatum (Clark)	1.5–30.1	10
Limnoxenus sp.	0.3–20.9	5
Hydrochus sp.	0.3–20.9	3
Octhebius sp.	6.2	1
Limnodessus sp.	14.2	1
Paracymus pygmaeus Macleay	4.2, 6.2	2
Enochrus sp.	16.5, 20.0	2
Helochares sp.	1.3	1
Antiporus femoralis (Boheman)	0.4	1
Helminthidae sp.	4.2	1
Liodessus sp.	1.3–14.9	5
Hyphydrus elegans (Montrouzier)	6.2	1
Rhantus suturalis (Macleay)	0.4	1
Sternopriscus multimaculatus (Clark)	1.5	1

onstrated to recover ecological structure more effectively than popular ordination methods such as detrended correspondence analysis (Faith *et al.*, 1987; Minchin, 1987).

Compositional dissimilarities could not be calculated for three lakes that contained no taxa and these were excluded from the analyses. The relationship between salinity and ordination scores in two dimensions of HMDS was examined by superimposing arbitrarily defined contours of salinity upon the ordination diagram.

Since only absence or presence data were recorded, it is not possible to consider the relationship of species diversity to salinity.

Results

Salinity and species richness

The negative correlation between species richness and salinity over a broad scale of salinities (Fig. 2) is very highly significant ($r^2_{77} = -0.442$, $p < 0.001$). However, at a finer scale (i.e. within particular salinity ranges), the correlation is non-significant ($p > 0.05$) at intermediate salinities (10–30, 30–50, 50–100 and 100–200 g l^{-1}) often used to categorise salt lake biota (Hammer, 1986). At the lowest extreme of the range (0.3–10 g l^{-1}), the relationship remains very highly significant ($p < 0.001$), indicating that salinity is closely associated with the species richness of freshwater species.

Our data do not conform well with Hammer's (1986) model (Fig. 2: broken line). The overall decline in species numbers with salinity in the lakes we investigated is much less abrupt; indeed, in all but the most saline lakes (salinity > 100 g l^{-1}), species richness in lakes whose salinities lie between 3 and 100 g l^{-1} is not markedly lower than it is in freshwater lakes.

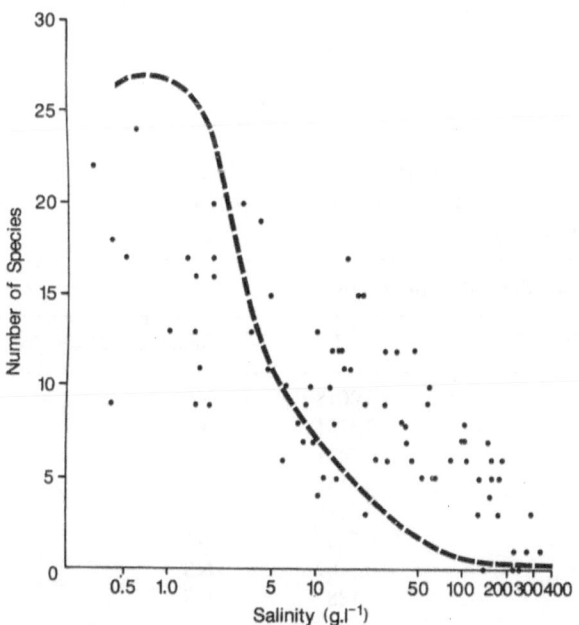

Fig. 2. Number of species : salinity in the 79 localities studied.

Salinity and community composition

Multivariate analyses indicated a strong relationship between community composition (on a taxonomic basis) and the broad range of salinities, reiterating the trend observed in species richness. TWINSPAN classification initially split 48 fresh or moderately saline lakes (0.3–41.3 g l^{-1}, 9 groups) from 28 more saline lakes (30.1–343 g l^{-1}, 9 groups). Subsequent divisions within these groups also reflected differences in salinity. Other factors such as geographical location may be relevant, especially at the lower salinities, and may contribute to the relatively higher coefficients of variation obtained for the first set of groups.

The lake groups yielded by the TWINSPAN appear to lie along a gradient related to their salinity, a gradient even more apparent when salinity contours are super-imposed upon the ordination diagram produced by HMDS (Fig. 3). Although some lakes fall outside their expected contours, the trend along the first axis is very highly significant (Table 2). No marked discontinuity along the gradient is evident on the ordination diagram. The community composition in the most saline lakes is quite distinctive; these lakes contain only the brine shrimp *Parartemia zietziana* Sayce.

When the relationship between community composition and salinity is examined on a finer scale (within particular ranges of salinity), non-significant or just significant ($0.01 < p < 0.05$) correlations are found at low salinities. However, in hypersaline waters (*sensu* Hammer, 1986), a very highly significant correlation exists between community composition and salinity (Table 2). This significant correlation disappears when finer ranges are examined (Table 2); either the relationship becomes more complex at a finer scale, where the importance of salinity appears diminished, or the sample sizes become too small to detect significant trends. These results indicate that attention should be focussed at a finer scale upon the relationship between community composition and salinity.

It is important to caution that our analyses are

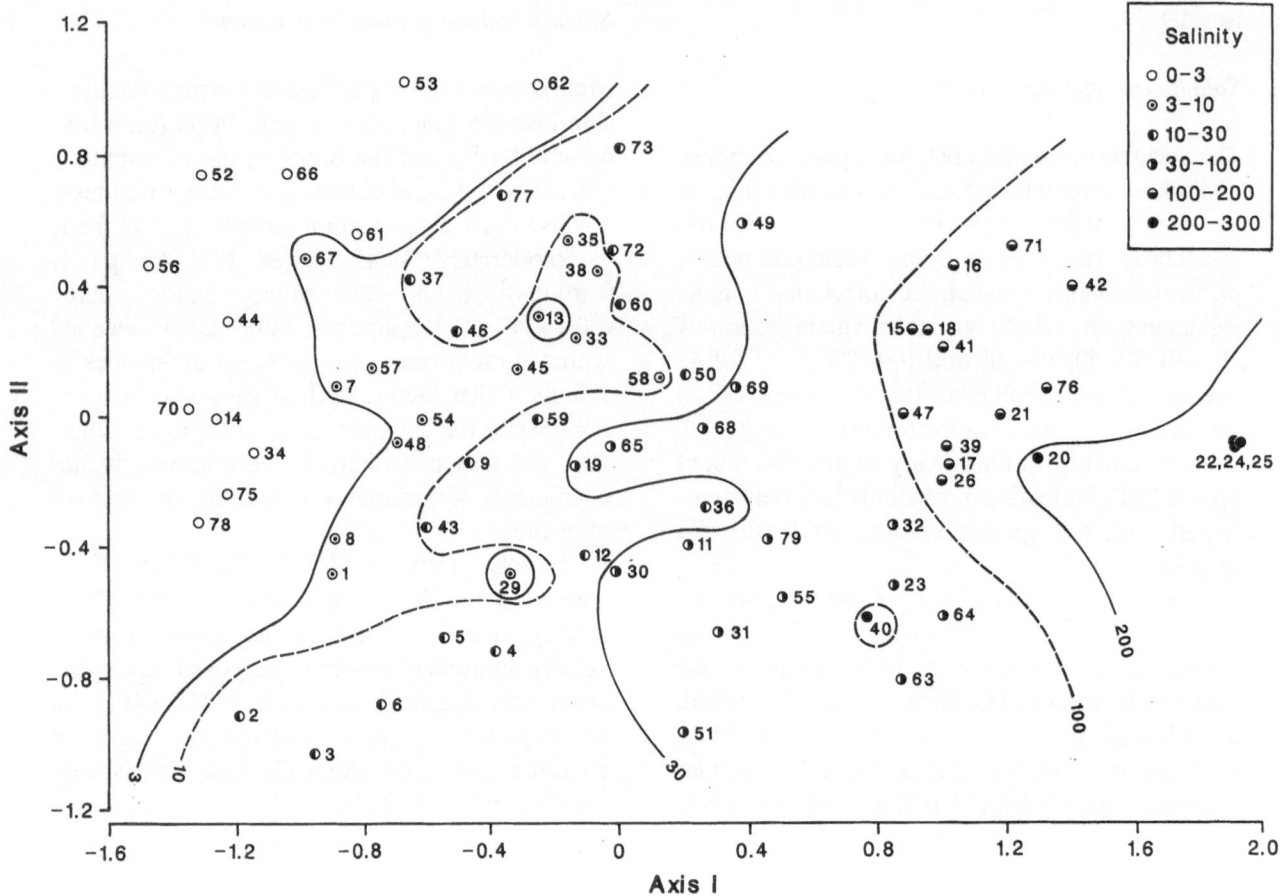

Fig. 3. General pattern of species number : salinity in western Victoria waters.

Table 2. Spearman rank correlation coefficients between ordination scores on the HMDS axes and salinity. Too few lakes fell within the salinity ranges of 30–50, 50–100 and >200 g l^{-1} for separate consideration. (NS = not significant; * = significant; ** = highly significant; *** = very highly significant; n = number of lakes).

Salinity range (g l^{-1})	n	Correlation between salinity and ordination score			
		Axis I	Significance	Axis II	Significance
0–3 [†]	15	− 0.0630	NS	− 0.0018	NS
0–10	27	0.4273	*	0.3151	NS
3–20 [†]	25	0.0285	NS	− 0.5431	**
10–30	17	− 0.4975	*	− 0.1152	NS
20–50 [†]	13	− 0.2751	NS	0.5117	*
> 50 [†]	23	− 0.8350	***	0.2053	NS
100–200	13	− 0.2063	NS	0.3109	NS
Total	76	0.8750	***	− 0.1882	NS

[k] These ranges are those referred to by Hammer (1986) as, respectively, subsaline, hyposaline, mesosaline and hypersaline.

based upon single collections and salinity measurements from each lake. Also, the data are only presence-absence data which give rare and abundant species equal importance and further limits the extent to which we may interpret analyses. And, finally, the conclusions that we draw from the multivariate analyses are based entirely upon correlative evidence; no causality can be inferred.

Salinity and species occurrence

In total, at least 142 taxa were collected, 93 from lakes with salinities exceeding $3 \, g \, l^{-1}$ (Table 1). Many salinity ranges in Table 1 exceed previously published tolerances. An important point documented in the Table is that many species have broad ranges of salinity tolerance. The narrowest range of tolerance is shown by those species collected from lakes of low salinity ($< 10 \, g \, l^{-1}$). At intermediate salinities, species tolerances encompass a broader range (Table 2). At high salinities, very wide ranges of species tolerance to salinity are displayed.

Discussion

The inverse correlation between species numbers and salinity is well established in the literature, though few authors have provided any quantitative data on the relationship, as noted by Lancaster and Scudder (1987). Exceptions include De Deckker & Geddes (1980), Timms (1983) and Lancaster & Scudder (1987). Most, like Hammer (1986), have been content with presenting general conceptual models. In these, species numbers are seen as declining rapidly in the salinity range $0-10 \, g \, l^{-1}$, and less rapidly in waters more saline than $10 \, g \, l^{-1}$. Our own data clearly support the general model, though, as noted, correlations at wide intermediate ranges (e.g. $10-30$, $100-200 \, g \, l^{-1}$) are weak or absent.

A similar picture emerges from consideration of the correlations between salinity and community composition: correlation is unequivocal over the broad range of salinity, but weak or non-significant over intermediate ranges. We add, however, that given the constraints of our data, as indicated previously, and the limitations of multivariate analyses (see Gauch, 1982; Austin, 1985; Minchin, 1987; Faith et al., 1987), the relationship between salinity and community composition noted here requires further examination.

The weak or insignificant correlations between salinity and species richness and community composition at intermediate salinity levels, taken with the broad salinity tolerance ranges of most of the taxa collected, suggest that salinity is not the only, or perhaps not the most important factor which determines the occurrence of a particular taxon in a salt lake. That is, whilst salinity is undoubtedly an important correlative over a broad range of salinity, over smaller ranges, its importance is not certain. Thus, an appreciation of scale is fundamental to any interpretation of the relationship between salinity and various faunal attributes.

At the very least, therefore, our data suggest that assumptions of causal relationships between salinity and various ecological attributes of salt lakes need more rigorous analysis, and need to be tested at the appropriate scale. Relationships at the community level (species richness, composition) will not necessarily be matched by responses from individual taxa. Indeed, lumping the taxa from a salt lake (or any other macrohabitat) and expecting the integral to respond in the same way to an environmental variable such as salinity is over-simplistic.

Additionally, our results highlight the importance of and need for experimental manipulations involving examination of community composition and dynamics at a fine scale. The coarse relationship between salinity and faunal composition is unquestioned, but the mechanism behind this relationship is unknown and can only be resolved by experiments designed and analyzed at a range of scales. Experiments of this sort are already in progress in our laboratory.

Other authors (e.g. Beadle, 1959) have made similar observations to ours but have not made the connection between their apparently para-

264

doxical observations and the importance of scale. Thus Geddes (1976) stated that the occurrence of halobiont species in 16 western Victorian saline lakes appeared to be controlled by salinity. Later, he claimed that 'the communities in the less saline localities, although containing *some* [our stress] different species, appear no more diverse or complex than those in the more saline localities and, although no exact figures are available, plankton standing crops in all localities appeared to be of a similar high level'. At first these statements appear hard to reconcile, but they actually represent examination of the response to salinity at two quite different ecological scales. Likewise, Timms (1983), who studied the benthos of 24 saline lakes in the same area studied by Geddes (and coincidentally at almost the same time as our study), noted that in his study (and others in Australia) attempts to quantify relationships between species diversity and salinity had not produced statistically significant correlations. He added (p. 175) for benthic communities, however, that 'perhaps factors other than salinity are important'.

Brief consideration of work by others, involving investigations of relationships between salinity and species occurrence of disparate groups of biota, is of interest in this connection. We confine our attention to papers given at the previous international symposium on salt lakes held at Nairobi (Melack, 1988a) except for reference to the recent paper of Lancaster & Scudder (1987).

Lancaster & Scudder (1987) studied communities of Coleoptera and Hemiptera in eight lakes in British Columbia, Canada, with mean conductivities of 56–13,115 mS cm^{-1}. For both groups of insects, species richness decreased with increasing salinity. The authors concluded that water chemistry had a significant effect but exactly how was not manifest. They suggested that apart from direct physiological processes, indirect phenomena such as macrophyte community structure, food resources or biological interactions could be important.

Melack (1988b) studied changes in the biota of Lake Elmenteita, Kenya, during a period of salinity increase. He offered eight hypotheses to explain changes, most of which involved biologi-

cal interactions rather than the direct impact of salinity. His rather cautious 'summary suggestion' was that the rate of salinity change affected a key macrofaunal element (*Paradiaptomus africanus* Daday) and an important primary producer (*Anabaenopsis arnoldii* Aptekais). Other biota were affected in cascade fashion.

Wood & Talling (1988) investigated chemical and algal relationships in a 'salinity series' of Ethiopian salt lakes. They drew attention to the need for caution in relating the algal flora to salinity. Considering previous work (both experimental and field) on algal salinity relationships, they noted (p. 63) that: 'at present these types of evidence suggest that salinity itself is not the most important factor, an interpretation supported by the contrast between the present algal-salinity distributions [in Ethiopia] and the much wider ones in the low alkalinity saline lakes of Saskatchewan'.

Hammer & Heseltine (1988), who studied macrophytes in a series of saline lakes in Canada, found that (p. 114): 'the only major factor apparently affecting aquatic vascular vegetation and its distribution in Saskatchewan saline lakes appears to be specific conductance of the substrate ... [but] ... substrate conductivity cannot be taken as equivalent to water conductivity'. They further found that there were considerable differences in plant cover between and within lakes but that these differences were apparently unrelated to salinity differences (in fact, the greatest differences in plant cover were within lakes where salinity was essentially constant throughout). Macrophyte biomass in the littoral also was not directly related to salinity.

Galat *et al.* (1988) clearly have similar reservations to our own concerning the degree to which salinity has been uncritically taken to be a significant ecological factor in salt lakes. They stress that causality is more usually inferred than rigorously demonstrated in studies of salinity tolerances. Their own studies of this sort, which involved three abundant invertebrate species in Pyramid Lake, Nevada, yielded equivocal results. Nevertheless, they concluded that if the salinity of Pyramid Lake doubled, populations of the species investigated would decrease in density (but not be extirpated) and this would be reflected in shifts of food habits by fish.

The work of Herbst (1988) on the brine fly, *Ephydra hians* Say, in North American lakes is also relevant. Larvae of this species are found over a wide range of salinity: from ≈ 20 to $200 \, \text{g} \, \text{l}^{-1}$. On the basis of careful studies of the population dynamics of the species, Herbst hypothesised that, at the lower end of the salinity spectrum, population densities are limited by biotic factors (e.g. by predation, competition), and by physiological constraints at the upper end. Thus, over most of the salinity range occupied by this species, it seems that salinity *per se* is not an important determinant of species abundance. Herbst refers to his hypothesis as the 'intermediate salinity hypothesis'.

Finally, Colburn (1988), in a study of the fauna of saline waters in Death Valley, U.S.A., concluded that whilst 'abiotic factors such as salinity' are important in limiting species diversity, other factors, both abiotic and biotic (e.g. predation), may explain the particular distributions of individual species.

It appears, then, that the very factor which most distinguishes salt lakes from other bodies of inland water, namely salinity, may not be an important determinant of species occurrence over most of the salinity spectrum occupied by the biota of salt lakes. This conclusion leads naturally to the question of what factors are important. A number of possible ones spring to mind, not all of which can be supported at present by evidence.

Chance (stochasticity) is likely to be important in some if not all cases. Prior occupancy of a habitat (competition) by a particular species may be another factor, exemplified by the dearth of localities containing more than one species of *Parartemia* in Australia. Predation is certainly important, as Kokkinn & Williams (1988) found with regard to *Tanytarsus barbitarsis* Freeman. Fish predation may limit brine shrimp to highly saline waters (i.e. waters lacking fish) worldwide, as argued long ago by Whittaker (1940) and more recently (and generally) by Kerfoot & Lynch (1987). Certainly the lowest field salinity at which

Parartemia zietziana was found in the 79 lakes we investigated, viz. $102 \, \text{g} \, \text{l}^{-1}$, is well above the lowest salinity it can tolerate ($\approx 8 \, \text{g} \, \text{l}^{-1}$; Geddes, 1975), and well above the maximum salinity at which fish have been recorded in western Victoria ($31 \, \text{g} \, \text{l}^{-1}$; Chessman & Williams, 1974). Parasitism may also sometimes be important (Scudder, 1983).

Thus, in summary, once a salt lake species 'solves' the initial physiological problem of osmoregulatory stress, it is able to occupy a wide salinity range; within this range, predation, food availability, competition and other forms of biological interaction, and perhaps interactions between particular physical and chemical factors (including salinity), rather than salinity *per se*, seem more likely to determine its extinction or persistence.

Vareschi (1987) has recently drawn attention to the substantial role that the study of saline ecosystems can play in the debate concerning the ecological importance of abiotic and/or biotic factors. We agree, but emphasise that the conclusions drawn from such studies will only be productive if the significance of ecological scale is recognised at the outset.

Acknowledgements

Several colleagues have helped us. We sincerely thank them. We particularly thank Dr. P. De Deckker (Australian National University) who accompanied one of us (W.D.W.) on the field trip to western Victoria and who identified ostracod material. We also thank for identifications: Dr. V. Sergeev (formerly, University of Adelaide) for Cladocera and Copepoda; Mr. I. Lansbury (University of Oxford) for Hemiptera; Dr. P. Suter (Engineering and Water Supply Department, Adelaide) for confirmation of ephemeropteran species; and Ms. Ros St Clair (James Cook University) for confirmation of trichopteran species. Dr. K. F. Walker (University of Adelaide) is thanked for commenting on a draft manuscript. This paper was completed when the senior author (W.D.W.) was in receipt of a

Scientific Merit Grant from AWRAC; the support of that Council is acknowledged. Finally, we thank Miss Sandra Lawson (University of Adelaide) who typed the original manuscript at short notice.

References

Austin, M. P., 1985. Continuum concept, ordination methods, and niche theory. Annu. Rev. Ecol. Syst. 16: 39–61.

Beadle, L. C., 1959. Osmotic and ionic regulation in relation to the classification of brackish and inland saline waters. Arch. Oceanogr. Limnol. (Suppl.): 143–151.

Chessman, B. C. & W. D. Williams, 1974. The distribution of fish in inland saline waters in Victoria, Australia. Aust. J. mar. Freshwat. Res. 25: 167–172.

Colburn, E. A., 1988. Factors influencing species diversity in saline waters of Death Valley, USA. Hydrobiologia 158: 215–226.

De Deckker, P. & M. C. Geddes, 1980. Seasonal fauna of ephemeral saline lakes near the Coorong Lagoon, South Australia. Aust. J. mar. Freshwat. Res. 31: 677–699.

De Deckker, P. & W. D. Williams, 1988. Physico- chemical limnology of eleven, mostly saline permanent lakes in western Victoria, Australia. Hydrobiologia 162: 275–286.

Faith, D. P., P. R. Minchin & L. Belbin, 1987. Compositional dissimilarity as a robust measure of ecological distance. Vegetatio 69: 57–68.

Galat, D. L., M. Coleman & R. Robinson, 1988. Experimental effects of elevated salinity on three benthic invertebrates in Pyramid Lake, Nevada. Hydrobiologia 158: 133–144.

Gauch, H. G., 1982. Multivariate Analysis in Community Ecology. Cambridge University Press, Cambridge.

Gauch, H. G. & R. H. Whittaker, 1981. Hierarchical classification of community data. J. Ecol. 69: 135–152.

Geddes, M. C., 1975. Studies on an Australian brine shrimp *Parartemia zietziana* Sayce (Crustacea: Anostraca). I. Salinity tolerance. Comp. Biochem. Physiol. 51A: 553–560.

Geddes, M. C., 1976. Seasonal fauna of some ephemeral saline waters in western Victoria with particular reference to *Parartemia zietziana* Sayce (Crustacea: Anostraca). Aust. J. mar. Freshwat. Res. 27: 1–22.

Hammer, U. T., 1986. Saline Lake Ecosystems of the World. Junk, Dordrecht.

Hammer, U. T. & J. M. Heseltine, 1988. Aquatic macrophytes in saline lakes of the Canadian prairies. Hydrobiologia 158: 101–116.

Herbst, D. B., 1988. Comparative population ecology of *Ephydra hians* Say (Diptera: Ephydridae) at Mono Lake (California) and Albert Lake (Oregon). Hydrobiologia 158: 145–166.

Hill, M. O., 1979. TWINSPAN – A FORTRAN program for arranging multivariate data in an ordered two-way table by classification of the individuals and attributes. Cornell University, Ithaca, New York.

Hill, M. O., R. G. H. Bunce & M. W. Shaw, 1975. Indicator species analysis, a divisive polythetic method of classification and its application to a survey of native pinewoods in Scotland. J. Ecol. 63: 597–613.

Kerfoot, W. C. & M. Lynch, 1987. Branchiopod communities: Associations with planktivorous fish in space and time. In: W. C. Kerfoot & A. Sih (Eds) Predation: Direct and Indirect Impacts on Aquatic Communities. University Press of New England, Hanover & London.

Kokkinn, M. J. & W. D. Williams, 1988. Adaptations to life in a hypersaline water-body: adaptations at the egg and early embryonic stage of *Tanytarsus barbitarsis* Freeman (Diptera: Chironomidae). Aquatic Insects 10(4): 205–214.

Lancaster, J. & Scudder, G. G. E., 1987. Aquatic Coleoptera and Hemiptera in some Canadian saline lakes: patterns in community structure. Can. J. Zool. 65: 1383–1390.

Melack, J. M., 1988a (ed.). Saline lakes. Developments in Hydrobiology, 44. Dr. W. Junk Publishers, Dordrecht. 316 pp.

Melack, J. M., 1988b. Primary producer dynamics associated with evaporative concentration in a shallow, equatorial soda lake (Lake Elmenteita, Kenya). Hydrobiologia 158: 1–14.

Minchin, P. R., 1986. How to use ECOPAK: an ecological database system. Tech. Mem. 86/6. CSIRO, Australia.

Minchin, P. R., 1987. An evaluation of the relative robustness of techniques for ecological ordination. Vegetatio 69: 89–107.

Remane, A. & C. Schlieper, 1971. Biology of Brackish Water. Schweitzerbart'sche, Stuttgart.

Scudder, G. G. E., 1983. A review of factors governing the distribution of two closely related corixids in the saline lakes of British Columbia. Hydrobiologia 105: 143–145.

Timms, B. V., 1983. A study of benthic communities in some shallow saline lakes of western Victoria, Australia. Hydrobiologia 105: 165–177.

Vareschi, E., 1987. Saline lake ecosystems. In: E.-D. Schultze & H. Zwölfer (eds.). Potentials and Limitations of Ecosystem Analysis. Springer Verlag, Berlin.

Whittaker, D. M., 1940. The brine shrimp *Artemia* and its environment. Science Monthly 51: 192–193.

Williams, W. D., 1981. The limnology of saline lakes in western Victoria. Hydrobiologia 82: 233–259.

Williams, W. D., 1984. Chemical and biological features of salt lakes on the Eyre Peninsula, South Australia, and an explanation of regional differences in the fauna of Australian salt lakes. Verh. int. Ver. Limnol. 22: 1208–1215.

Williams, W. D., 1986. Conductivity and salinity of Australian salt lakes. Aust. J. mar. Freshwat. Res. 37: 117–182.

Williams, W. D. & R. T. Buckney, 1976. Stability of ionic proportions in five salt lakes in Victoria, Australia. Aust. J. mar. Freshwat. Res. 27: 367–377.

Williams, W. D. & M. J. Kokkinn, 1988. The biogeographical affinities of the fauna in episodically filled salt lakes: a study of Lake Eyre South, Australia. Hydrobiologia 158: 227–236.

Wood, R. B. & J. F. Talling, 1988. Chemical and algal relationships in a salinity series of Ethiopian inland waters. Hydrobiologia 158: 29–67.

Hydrobiologia **197**: 267–290, 1990.
F. A. Comin and T. G. Northcote (eds), Saline Lakes.
© 1990 *Kluwer Academic Publishers.*

Multivariate analysis of diatoms and water chemistry in Bolivian saline lakes

S. Servant-Vildary[1] & M. Roux[2]

[1] *Antenne ORSTOM, Laboratoire de Géologie, Museum National d'Histoire Naturelle, 43 Rue Buffon, 75005, Paris;* [2] *CEPE. CNRS Route de Mende, BP 5051, 34033 Montpellier*

Key words: Bolivia, salt lakes, chemistry, diatoms, multivariate analysis

Abstract

Diatom assemblages are described from surface sediments in thirteen salt lakes located in the southern Bolivian Altiplano. Factor analysis of correspondences and cluster analysis are used to classify the diatom assemblages. New methods are proposed to establish the qualitative and quantitative relationships between diatom floras and ecological parameters. Diatom assemblages are linked more to the ionic elements than to the salinity, pH, depth, temperature or elevation. Environmental variables are divided into three modalities which allow considerations of many different variables not under the same units.

Introduction

Diatoms in hypersaline lakes from South America have been little studied (Hustedt, 1927; Frenguelli, 1936; Patrick, 1961; Lopez, 1980; Servant-Vildary, 1978, 1983, 1984). However, stratigraphy (Servant & Fontes, 1978; Fernandez, 1980), geochemistry, (Carmouze *et al.*, 1978; Miranda, 1978; Risacher, 1978a; Risacher & Eugster, 1979; Ballivian & Risacher, 1981), clay neoformation (Badaut *et al.*, 1979), hydrobiology (Iltis *et al.*, 1984), ornithology (Hurlbert & Keith, 1979), geocryology (Hurlbert & Chang, 1984, 1988) have been subjects of deep interest. Dissolution and transformation of diatoms in the sites studied are intense and begin in the three first centimeters of the sediments (Badaut *et al.*, 1979).

The objective of the research presented in this paper is to provide information concerning the response of diatoms to salt concentrations in order to reconstruct past environments. It involves, (1) an inventory of the diatom species, (2) a classification of the lakes based on the diatom flora using correspondence factor and cluster analysis, (3) an ordering of the species inside a cluster, using an 'interpreting help' method to determine the degree of relation of the subdominant or scarce species to a given milieu, (4) the determination of the ionic elements which mostly influence a diatom assemblage; using the 'variables/classes' and 'classes/variables' contributions program, (5) the quantification of the chemical variables, using a 'interactions species/ionic variables' program which puts into the same graph, samples and/or species and ionic elements whose concentration is expressed in categories (Roux, 1985).

Description of the sites studied

The volcanic area, called Lipez, is broken up into a large number of small salt lakes located at between 4000 and 5000 m of altitude. The climate is characterized by strong winds, high insolation and evaporation (1400 mm year^{-1}), and low pluviosity (100–200 mm year^{-1}). Average daily air and water temperature amplitude is in the order of 15 °C. Rain water additions were negligible in the two years of our study. Similar general climatic conditions affect to the 13 lakes (Fig. 1), which have waters in a wide range of salinities.

The ionic composition of the lakes mainly depends on the nature of the surrounding rocks, the interactions between water and sediments and the origin of influent waters. Ballivian & Risacher (1981) based a chemical classification on these three parameters and separated the lakes into four groups.

1. *Na-SO$_4$-Cl lake*: Hedionda is essentially fed by spring waters. Chiar Kota is fed by running water from the north, whose composition is already of high salinity (TDS: 1.88 g l^{-1}). There is deposition of gypsum near the margins. Honda is fed by water springs whose chemical composition is affected by the volcanic and sedimentary rocks. There are deposits of halite on the margins. Lakes Pujio and Puripica Chico are very small; see Hurlbert & Keith (1979) and Hurlbert & Chang (1983) for details. Ballivian is isolated from Ramaditas basin by a high sill, but they could have been connected during highest water-level periods. Ramaditas is fed by running water springs from the south, and diffuse springs in the north-western and eastern margins. Laguna Verde is characterized by lack of salt deposits in the margins.

2. *Na-SO$_4$ lakes*: Canapa and Chulluncani are the northernmost studied lakes in this series. Canapa is a very small lake, evaporites are thick and the underground water is independent of the superficial lake. The lake is fed by spring water in the northern part and by the Tapaquillcha River in the eastern part. Chulluncani, fed by a small river, is only 15 cm deep, and can be completely dry. There are thenardite deposits on the margins.

3. *NA-Ca-Cl lakes*: The lake surface of Pastos Grandes is small compared to the evaporitic crust. It is composed of a 'central lake' and marginal lakes which can reunite after the wet season to form a continuous ring around the salar. The lake is fed by small rivers and spring waters. Laguna Colorada typically belongs to the sodium chloride group. Although the chemical characteristics are variable within the lake according to the chemistry of inflow waters, only one sample was studied. However, as the water supply is bicarbonated, a sample for diatom study was taken outside the lake, near the natron deposits. Unfortunately, no chemical analyses at the same sample site was available, so we used the nearest water sample CLD 4 which gives the best idea of the outside water chemistry.

4. *Na-CO$_3$ lake*: In Cachi Laguna, 90% of its water supply comes from the western part of the lake. The sample for diatom study was taken from the part of the lake where surficial water has a subterranean origin.

Detailed discussions about seasonal and inter-annual variations of chemistry and diatom floras are given in Iltis *et al.* (1984).

Materials and methods

Sampling and chemical analysis

Water samples from the water column near the shoreline were collected with plastic bottles and preserved with chloroform. Surface sediment samples were collected at the same sites (Fig. 1) and time (May and November 1978) as water samples, and preserved with formalin (4%), both at room temperature in Paris. The problem of poor representativeness by dissolution and transformation of diatoms was avoided using these diatoms samples collected just in the water-sediment interfaces.

Thirty diatom samples and the corresponding water data were used for multivariate analysis. Samples of water and sediment were submitted to acid digestion in order to obtain clean and free of organic matter diatom frustules. The number of

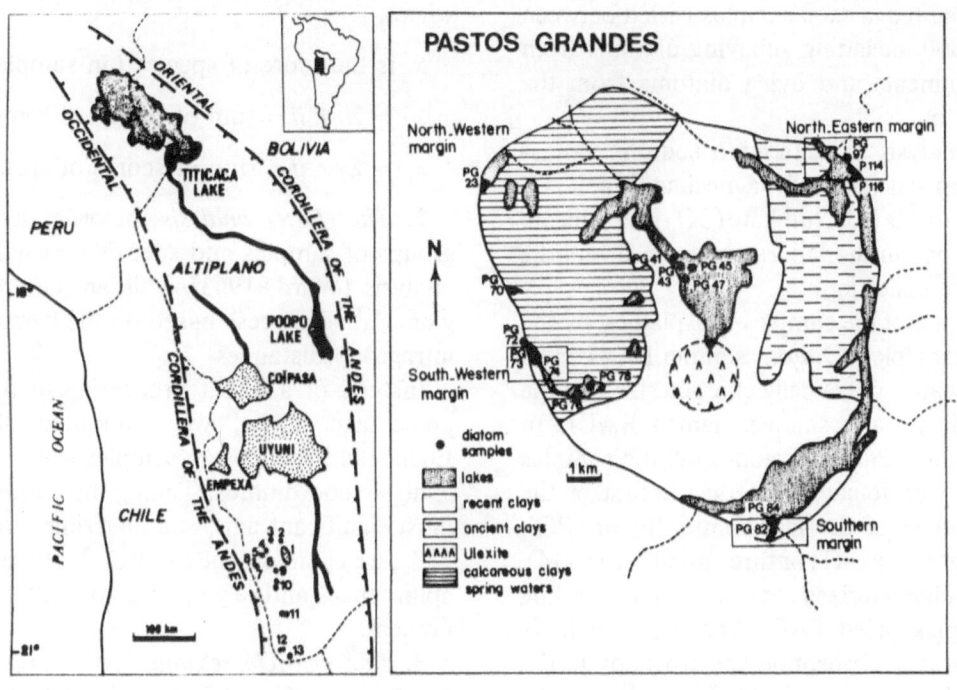

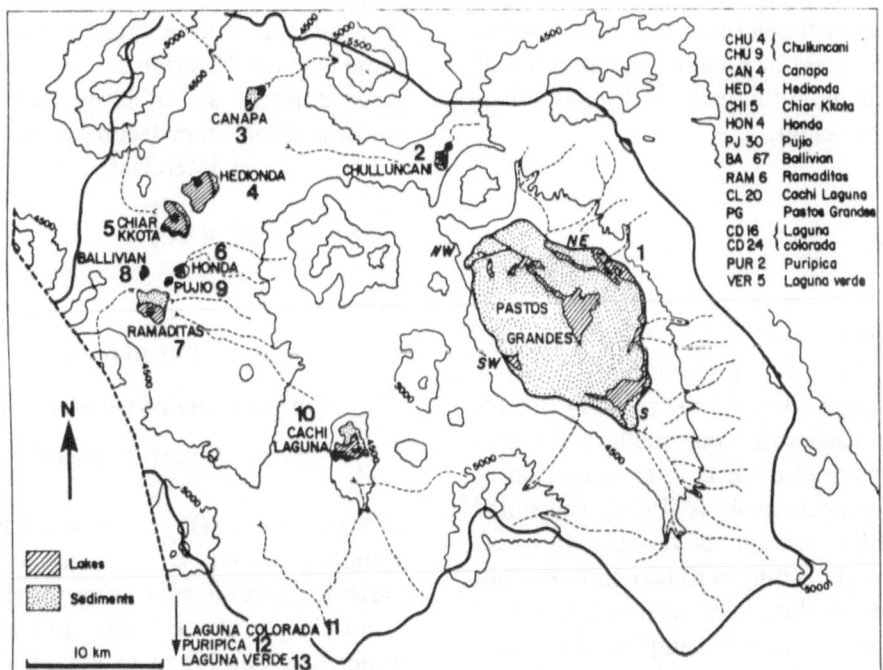

Fig. 1. Geographical location of the sites studied. Samples for diatom analysis with the list of the codes are located for the small lakes in the lower part, and for Pastos Grandes in the right upper part.

frustules counted in each samples varied between 251 and 4500, including subliving diatoms from surface sediments and living diatoms from the water column.

Water analysis to determine sodium, potassium, lithium, calcium and magnesium (Na, K, Li, Ca, Mg), chloride (Cl), sulphate (SO_4) and silicate (Si) were done in the laboratory several months after sample collection.

Codification of the samples is explained by the following examples. Diatom sample BA67 from Lake Ballivian is chemically characterized by the corresponding water sample, named BAL1. In the largest lake, Pastos Grandes, all the samples are named with four characters: the first or the two first letters of the lake name, 'P' or 'PG', followed by two (PG41) or three numbers (P114). They are characterized by the corresponding water samples called PAG. Analytical methods used are: atomic absorption spectroscopy to determine sodium, potassium, lithium, calcium and magnesium; mercurium thiacyanate colorimetry for chloride; indirect colorimetry with methylthymol blue upon excess of Ba after precipitation of $BaSO_4$ for sulfate; colorimetry with ammonium molybdate for silicium.

Statistical analysis

1. *The factor analysis of correspondences* (FAC) (Greenacre, 1984; Lebart *et al.*, 1984) or reciprocal averaging (Hill, 1973), is currently used for species and sample data (Gauch, 1982). In this case we have used this same method for processing the chemical data. It is an extension of principal components analysis (PCA) well suited to deal with either categorical variables or count variables. Its main feature is to take into account the margins of the data table, that is the sum of the scores for species and for samples. The underlying distance function is said to be double weighted. For the distance between samples i and i', the formula is:

$$D_{ii'}^2 = \Sigma_j \frac{1}{x_j} \left(\frac{x_{ij}}{x_i} - \frac{x_{i'j}}{x_{i'}} \right)^2$$

where

x_{ij} is the score of species j in sample i

$x_i = \Sigma_j \times ij$ = sum of scores of sample i

$x_j = \Sigma_i \times ij$ = sum of scores of species j

2. *The cluster analysis*: in order to make up groups of samples and species we used a cluster analysis (Ward, 1963). It is an hierarchical agglomerative process based on the variance of the intragroup distances.

Instead of a direct processing of the species percentage table, we computed the usual Euclidean distances on samples from their factor analysis coordinates. Taking into account the 6 most significant axis, summarizing a variance of 58%, we eliminate the random fluctuations of the abundances and rely upon an overall more stable process.

3. *Variables/classes and classes/variables contributions program*. After having condensed the samples into groups or classes of homogeneous flora, we felt it necessary to determine whether some chemical variables were responsible for these different assemblages. The computations performed consist in decomposing the generalized sum of squares interclasses deviations (SSID):

$$\text{SSID} = \Sigma J \, \Sigma k \, nk \, (\bar{x}_{kj} - \bar{x}_j)^2$$

where:

nk = number of samples in group k

x_{kj} = mean value of variable j within group k

x_j = mean value of variable j over all samples.

The quantity, $nk \, (\bar{x}_{kj} - \bar{x}_j)$, which weights the deviation of group k from the overall mean of variable j, may be regarded as the share of variable j and class k in the SSID. But such a figure is easier to group when expressed as a ratio. We computed two types of ratios, filling up two different tables as follows.

In the first table we put up the ratios (as percentages):

$$100 \, (\bar{x}_{kj} - \bar{x}_j)^2 / \overline{\Sigma} J \, (\bar{x}_{kj} - \bar{x}_j)^2$$

For each particular group k, these figures allow for determining the most discriminant variables. The second table is filled with the following values:

$$100 \, nk \, (\overline{x}_{kj} - \overline{x}_j)^2 / \overline{\Sigma} k \, \Sigma nk \, (\overline{x}_{kj} - \overline{x}j)^2$$

which indicates the most typical class with regard to a particular variable J.

4. *Variables/classes and classes/variables interactions program*. This program is used to quantify the ionic variables which mostly influence each group of samples and each species. We have separated the values of each chemical variable into three classes or modalities corresponding to low, medium, high concentrations. Each of them becomes a new variable, a *category*, and a new table is built up, the rows of which are the species and the columns are the categories. Each cell (i, j) of the table contains the mean percentage of species i, among all the samples falling in the category j. In other words the cells of the table are made of mean floristical abundances, but the columns of the table represents modalities (low, medium, and high) of chemical variables. In this way, we enforce the factor analysis of correspondences (FAC) to work out the relationship between species and chemical variables. This method gives us the possibility to see in the same graph both the classes and the chemical variables, ordinated from lower (1) to higher values (3).

Combining the methods (CA, CVC, CVI and FAC), we can determine the chemical variables which characterize each cluster and we can quantify the relative role of a chemical variable on the cluster but also on each sample and each species. This method also computes and submits to factor analysis the environmental data which are not in the same units, such as pH, depth, temperature, elevation.

Results

The diatom flora

Ninety four Pennatophycideae species (Table 2) were identified in the 30 samples. Good preser-vation of the frustules can be observed (Fig. 9–56) which indicates that dissolution of the frustules in the sediment water interface can be neglected and that our samples can be considered as good representatives of the sampling sites. Basic environmental data are given in Table 1.

According to many authors (Frenguelli, 1945; Krammer, 1980; Krammer & Lange-Bertalot, 1985, 1986; Osada & Kobayasi, 1985), the species identified can be distributed in four general groups: Endemic, athalassic and/or sea-related, athalassic saltwater and fresh-oligohaline species.

Most of the species are linked to low water level habitats. Most of them are benthic, epiphytic and aerophilous. There are no euplanktonic species but some of them can tolerate depth variations (eurytopic) and can be planktonic or tychoplank-tonic: *Ceratoneis arcus, Fragilaria brevistriata, Fragilaria pinnata, Synedra rumpens, Nitzschia palea*. The pH is always higher than 7, thus most of species have to be alkaliphilous or alkalibiontic, but some are pH-indifferent.

(i) Comparisons between published and local affinities to the salinity.
Achnanthes linearis, Amphora libyca, Cymbella affinis, Cymbella norvegica, Neidium apiculatum, Neidium bisulcatum, Rhopalodia gibba considered as halophobous or oligohalines were only found in the low salinity samples (PG72, PG74, PG82, PG97, PG23). The local ecological affinity agrees with the literature data.

Few species were found in samples with salinities between 10 and 30 g l^{-1} (RAM6, VER5, HON4, CAN4, CHU9, PG70, PG73, PG76, PG41, PG43). *Achnanthes chilensis* mesohalobous according to Hustedt (1927), is abundant in VER5, 13.5 g l^{-1}, but two oligohalines species live here in saline waters: *Caloneis silicula* in CAN4 13.8 g l^{-1} and *Amphora ovalis* in CHU9 11.2 g l^{-1}.

Among the species restricted to the hyperhaline samples (BA67, HED4, PJ30, PUR2, CHI5, CHU4, CL20, CD16, CD24, PG78, P114, P116, PG45, PG47), *Mastogloia smithii amphicephala* was considered as mesohalobous by De Wolf

Table 1. Physical and chemical characteristics of the sites studied. Lakes BA-Ballivian, RAM-Ramaditas, VER-Laguna verde, PJ-Pujio, HON-Honda, CHI-Chiar Kkota, CAN-Canapa, CHU-Chulluncani, CD-Laguna Colorada, CL-Cachi laguna, PG and P-Pastos Grandes. Numbers after the lake code mean the samples chemically well characterized. Elev.: Elevation (m), W. Long.: West longitud, SW Lat.: Southwest latitude, Water L.: water level (cm), Temp.: temperature (°C), Dens.: density g cm^{-3}, pH, Alk: alkalinity (meq l^{-1}), ionic contents (mM l^{-1}), TDS: total dissolved salts (g l^{-1}).

| Diatom s. | BA67 | RAM6 | VER5 | HED4 | PJ30 | PUR2 | HON4 | CHI5 |
Water s.	BAL1	RAM5	VER2	HED3	PUJ5	PUR4	HON3	CHI4
Elev.	4117	4120	4350	4120	4110	4393	4110	4112
W. Long.	68°05	68°05	67°48	68°03	68°04	67°30	68°04	68°04
SW. Lat.	21°38	21°38	22°48	21°34	21°37	22°31	21°37	21°35
Water L.	–	30	<100	20	100	100	20	20
Temp.	5	1	2	8	1	4	6	8
Dens.	1.03	1.02	1.01	1.05	1.02	1.02	1.01	1.05
pH	8.18	8.15	8.72	8.5	8.85	8.52	8.28	8.05
Alk	4.88	2.93	7.25	10	7.22	7.8	4.4	8.05
Cl	620	392	182	693	400	430	290	1090
SO$_4$	59.4	32	24	186	45	48.5	27.1	42.5
B	13.9	7.12	11.6	21.7	13.4	22	5.27	23.1
Si	0.9	1.48	1.02	0.983	0.933	0.735	1.13	1.23
Na	591	330	196	885	435	415	293	900
K	43.5	26.3	7.88	53.7	26.1	44	25.3	63.9
Li	3.67	1.7	5.26	17.6	5.33	15.7	6.77	25.4
Ca	29.9	34.2	5.44	13	9.98	11.6	4.99	33.4
Mg	24.9	13.4	10.8	26.7	8.64	11.3	5.76	46.9
TDS	45.2	29	13.5	72.3	35.9	32.7	25.6	69.4

| Diatom s. | CAN4 | CHU4 | CHU9 | CD16 | CD24 | CL20 |
Water s.	CAN3	CHU3	CHU2	CLD4	CLD33	CAL19
Elev.	4140	4430		4278		4495
W. Long.	68°01	67°53		67°47		67°57
SW. Lat.	21°·	21°32		22°11		21°43
Water L.	15	15		20		
Temp.	6	5	8	6	10	21
Dens.	1.009	1.087	1.008	1.08	1.04	1.02
pH	9.18	8.8	10.2	8.4	8.5	10.38
Alk	2.15	35	11.4	31.5	12.9	355
Cl	63.4	1240	55.5	1830	831	128
SO$_4$	52.8	277	45.4	61.1	62.3	38.6
B	23.1	1.2	88.7	56.6	2403	13.2
Si	1.12	0.783	0.667	1.63	1.16	5.75
Na	156	1310	101	1770	865	460
K	5.42	327	46	109	54	73
Li	2.81	3.24	0.396	28.2	1205	7.8
Ca	1.62	18.2	4.99	6.48	2.57	0.06
Mg	1.4	78.2	2.18	37.2	15.7	6.03
TDS	13.8	144.4	11.2	120.35	59.16	36.27

Table 1. (continued).

Pastos Grandes: Central lake

Diatom s.	PG23	PG41	PG43	PG45	PG47
Water s.	PAG22	PAG40	PAG30	PAG44	PAG48
Elev.	4400				
W. Long.	67°47				
SW. Lat.	21°39				
Temp.	1	4	6	5	5
Dens.	1.001	1.02	1.01	1.073	1.211
pH	9.35	8.52	8.05	7.4	7.2
Alk	1.51	4.25	3.21	9.08	22.9
Cl	19.7	470	227	1730	5470
SO_4	0.75	4.84	2.6	13.2	25.6
B	0.323	5.55	2.63	26.8	87.3
Si	1.37	1.13	0.617	0.733	1.12
Na	19.6	403	196	1480	4480
K	1.1	26.1	12.8	101	363
Li	0.692	16.9	7.57	72.1	236
Ca	0.612	8.98	4.99	27.4	77.3
Mg	0.453	10.9	5.43	4.53	143
TDS	1.4	26.7	14.2	103	371.2

Pastos Grandes: W–N lake   W–S lake

Diatom s.	PG70	PG72	PG74	PG73	PG76	PG78
Water s.	PAG69	PAG72	PAG74	PAG73	PAG75	PAG77
Temp.	5	10	1	1	10	10
Dens.	1.009	1	1.009	1.001	1.01	1.098
pH	8.42	6.95	8.15	7.85	8.35	7.91
Alk	5.09	3.3	5.13	4.2	3.85	9.42
Cl	204	2.54	190	16.9	234	2420
SO_4	2.76	1.09	3.15	1.35	3.18	30.4
B	2.77	0.092	2.36	0.27	2.96	29.6
Si	1.3	1.33	2.1	1.08	1.15	1.22
Na	174	3.48	170	15.2	196	2000
K	13.6	0.422	10.2	1.23	14.8	128
Li	6.12	0.073	6.12	0.56	8.29	86.5
Ca	4.99	1.27	4.74	0.815	4.99	37.4
Mg	3.7	0.831	5.68	2.28	5.43	49.4
TDS	13.1	0.6	12.1	1.54	14.3	144.1

Pastos Grandes: Southern lake   Eastern lake

Diatom s.	PG82	PG84	PG97	P114	116
Water s.	PAG81	PAG83	PAG96	PAG124	PAG115
Temp.	10	15	10	10	7
Dens.	1	1.14	1	1.16	1.17
pH	9.62	7.46	8.92	6.95	7.5
Alk	0.523	9.7	1.36	7.68	13.1
Cl	1.78	3770	4.23	4330	4450
SO_4	0.146	35.1	0.052	33.8	33.1
B	0.065	37.4	0.1	50.4	48.1
Si	0.567	1.18	0.8	0.6	1.12
Na	1.82	3350	3.7	3520	4000
K	0.113	165	0.322	251	3520
Li	0.05	97.3	0.17	167	124
Ca	0.152	41.1	0.312	59.4	62.3
Mg	0.132	51.4	0.305	105	85.6
TDS	0.19	225.3	0.40	255.23	267.36

274

Table 2. The diatom flora. Column 1, Ecology: E = endemic species, A = athalassic saltwater species, M = marine or sea-related species, F = freshwater species. O = oligohaline species. Column 2, Alphabetical list of taxa. Column 3, Codes of taxa.

Ecology	Taxa	Codes	Ecology	Taxa	Codes
FO	*Achnanthes arenaria* Amossé	AL	FO	*N. gastrum* Ehrenberg	NG
AM	*A. breviceps* Agardh	ABR	F	*N. mutica binodis* Hustedt	NMB
EA	*A. chilensis* Hustedt	ACI	F	*N. mutica nivalis* (Ehr.) Hustedt	NMN
A	*A. delicatula* (Kütz) Grunow	AD		*N. nov. sp.*	NNS
FO	*A. linearis* W. Smith	ALI	F	*N. pseudolitoricola* Häkansson	NLI
FO	*A. speciosa* Hustedt	AS	FO	*N. pseudolanceolata* Lange-Bertalot	NL
EA	*Amphora atacamae* Frenguelli	AAM	FO	*N. pupula* Kützing	NPP
EA	*A. boliviana* Patrick	ABE	AM	*N. pygmaea* Kützing	NPY
EA	*A. boliviana f. elongata*	ABE	FO	*N. rhynchocephala* Kützing	NR
EA	*A. carvajaliana* Patrick	AC		*N. sp.*	NS
EA	*A. chilensis* Hustedt	ACI	F	*Neidium apiculatum* Reimer	NEA
FO	*A. libyca* Rhrenberg	AY	F	*N. bisulcatum* (Lagerstedt) Cleve	NEB
AM	*A. lineolata* Ehrenberg	AML	EA	*Nitzschia accedens chilensis* Patrick	NAC
FO	*A. ovalis* Kützing	AO	FO	*N. alpina* Hustedt	NA
E	*A. platensis* Frenguelli	AP	AM	*N. denticula* Grunow	ND
AM	*Anomoeoneis sphaerophora* var. *angusta* Frenguelli	ASA	A	*N. epithemioides* Grunow	NE
AME	*A. sph. navicularis* (O. Muller) Frenguelli	ASN	AM	*N. frustulum* (Kütz.) Grunow	NF
AM	*A. sph. platensis* Frenguelli	ASP	AM	*N. grunowii* (Cleve) Hasle	NIG
AM	*A. sph. polygramma* (Ehr.) O. Muller	ANS	FO	*N. hantzschiana* Rabenhorst	NIH
AM	*Brachysira aponina* Kutzing	BA	AM	*N. hungarica* Grunow	NHU
FO	*Caloneis silicula* (Ehr.) Cleve	CS	OA	*N. inconspicua* Grunow	NI
	C. sp.	CSP		*N. minutula* Grunow	NM
AM	*C. westii* (W. Smith) Hendey	CW		*N. nov. sp.*	NINS
FO	*Ceratoneis arcus* Kützing	CA	AM	*N. palea* (Kütz.) W. Smith	NPA
FA	*Cocconeis placentula* Ehrenberg	CP	AM	*N. punctata* (W. Smith) Grunow	NP
FO	*Cymbella affinis* Kützing	CYA	AM	*N. pusilla* (Kutz.) Grunow	NPU
FO	*C. lunata* W. Smith	CYL	AM	*N. quadrangula* Lange-Bertalot	NQ
FO	*C. norvegica* Grunow	CYN	A	*N. valdecostata* Lange-Bertalot	NV
	C. sp.	CYS	FO	*Opephora martyi* Heribaud	OM
AM	*Denticula elegans* Kützing	DE	F	*Pinnularia bogotensis* Grunow	PB
AM	*D. elegans f. valida* Pedicino	DEV	FO	*Rhopalodia gibba* (Ehr.) O. Muller	RG
AM	*D. thermalis* Kützing	DT	EA	*R. wetzeli* Hustedt	RW
A	*Entomoneis alata* Kützing	AAL	A	*Scoliopleura peisonis* Grunow	SP
F	*Fragilaria brevistriata* Grunow	FB	F	*Stauroneis anceps* Ehrenberg	SA
F	*F. contruens venter* (Ehr.) Grunow	FCV	A	*S. bathurstensis* Giffen	SB
F	*F. elliptica* Schumann	FE	A	*S. gregorii* Ralfs	SG
FO	*F. pinnata* Ehrenberg	FP	A	*S. legleri* Hustedt	STL
F	*F. zeilleri* Heribaud	FZ		*S. spd.*	SSPD
F	*Gomphonema parvulum* Kützing	GP		*S. sp.*	SSP
F	*Hantzschia amphioxys maior* Grunow	HN	EA	*Surirella chilensis* Janish	SC
	H. nov. sp.		FO	*S. oregonica* Ehrenberg	SO
EA	*Mastogloia atacamae* Frenguelli	MA	A	*S. ovata utahensis* Grunow	SOU
	M. smithii amphicephala Grunow	MSA	A	*S. peisonis* Hustedt	SUP
A	*Navicula cari cincta* (Ehr.) Lange-Bertalot	NCC	EA	*S. sella* Hustedt	SUS
AM	*N. cincta* (Ehr.) Kützing	NCI	EA	*S. wetzeli* Hustedt	SW
FO	*N. cryptocephala* Kützing	NC	A	*Synedra pulchella* Kützing	SYP
			FO	*S. rumpens* Kützing	SYR

Figs. 9–26. 9, *Amphora carvajaliana* Patrick; 10, *Amphora boliviana* Patrick; 11, *Amphora atacamae* Frenguelli, a: external valve view, b: LM; 12, *Amphora atacamae* ⟶ Frenguelli, a: internal valve view, b: LM, c: this small form is called var. *minor* in the countings (Table 4); 13, *Amphora boliviana* Patrick, var. *elongata* in the countings; 14, *Cymbella gracilis* (Ehr.) Kützing, *Cymbella lunata* in Table 4; 15–17, *Rhopalodia wetzeli* Hustedt, 15, internal valve view, 16, LM, 17, detail of the costae; 18, *Stauroneis bathurstensis* Giffen, 19, *Surirella sella*, Hustedt; 20, *S. wetzeli* Hustedt; 21, *S. wetzeli*, twisted specimen; 22, *Navicula* sp., *Navicula* nov. sp. in Table 4; 23, *Navicula* sp., internal valve view; 24, *Nitzschia* sp., 25, *Brachysira aponina* Kützing; 26, *Surirella ovata* var. *utahensis* Grunow.

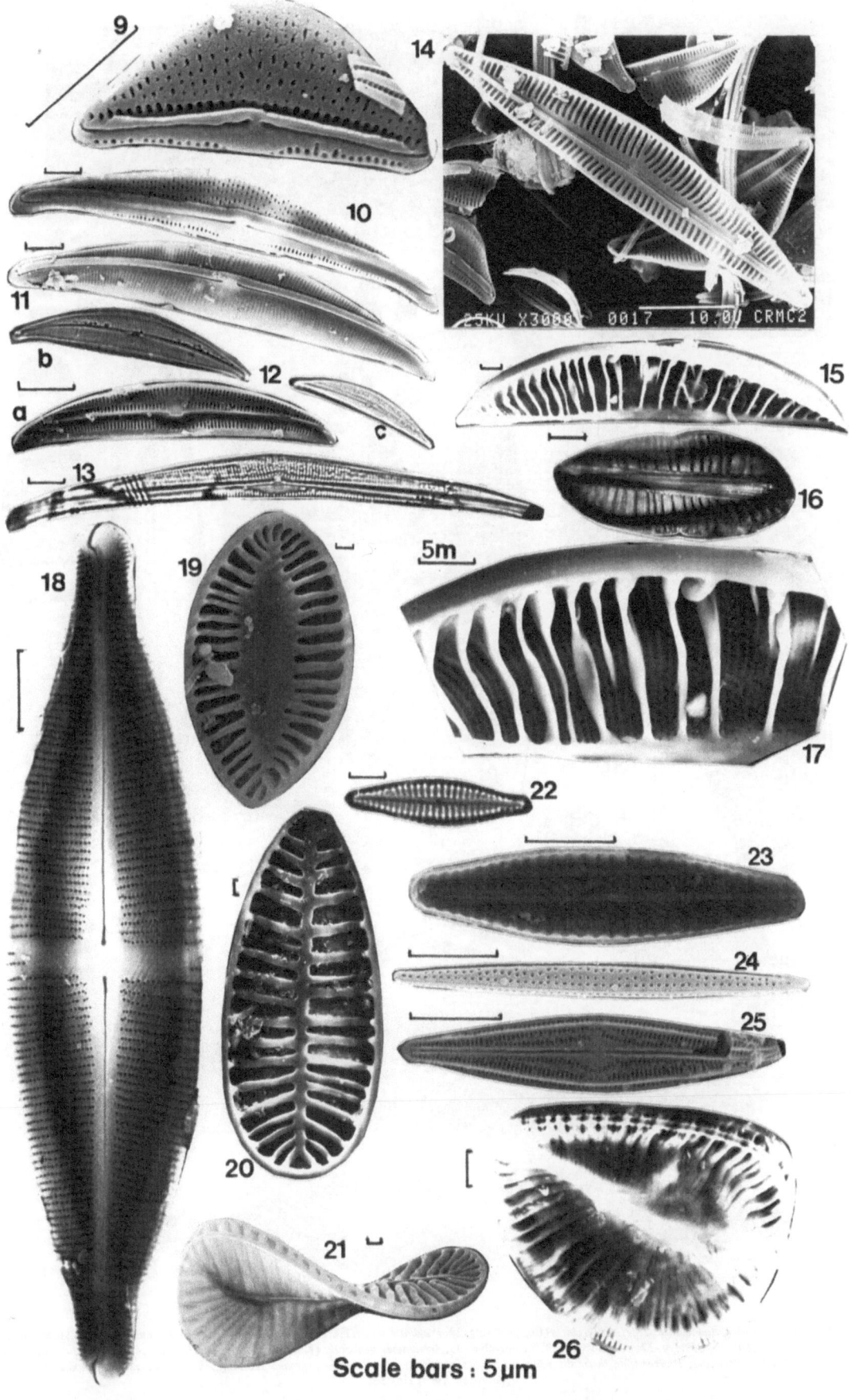

Scale bars : 5 µm

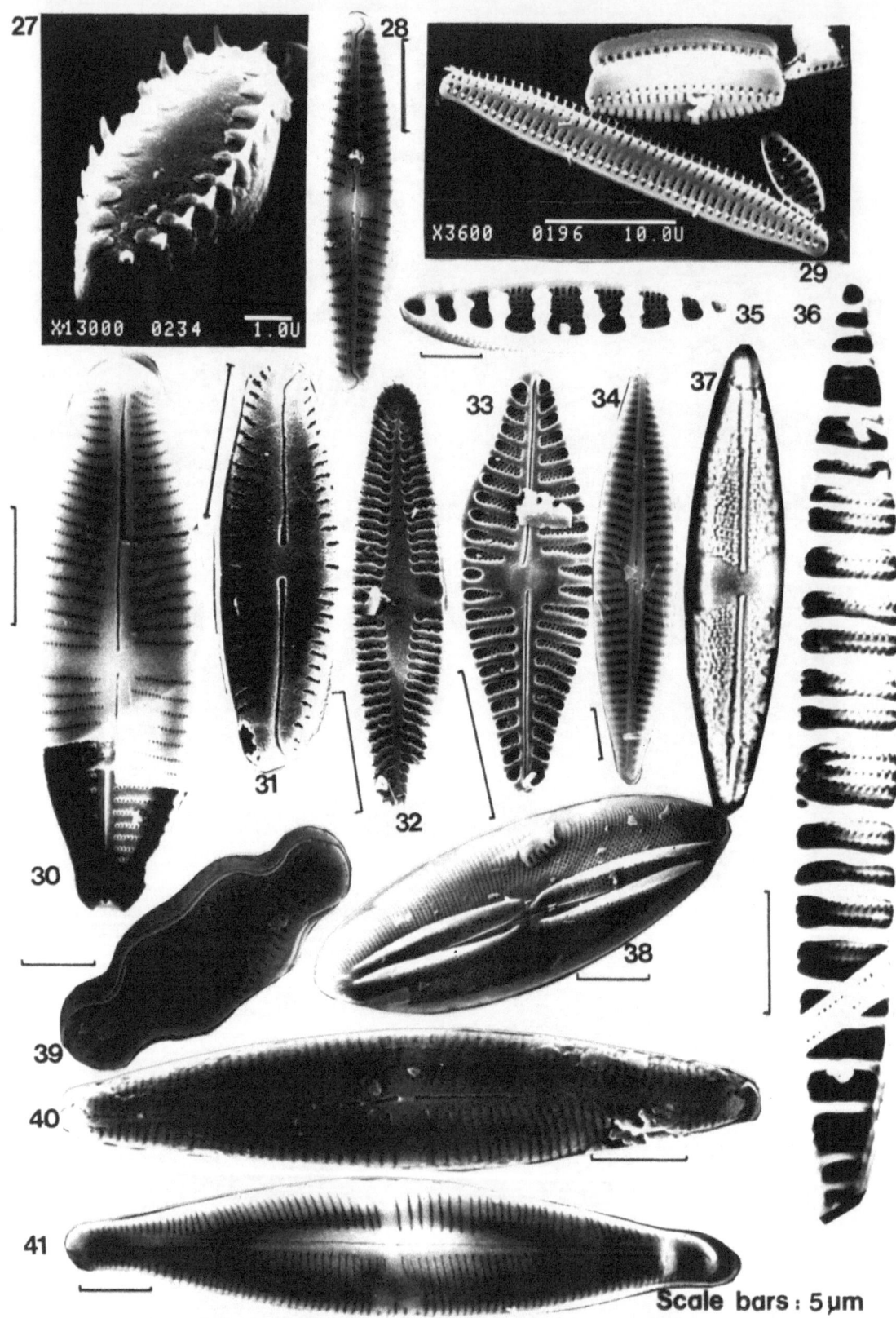

Figs. 27–41. 27, *Fragilaria brevistriata* Grunow; 28, *Navicula cincta* (Ehr.) Kützing; 29, *Fragilaria zeilleri* Heribaud; 30, *Stauroneis anceps* Ehrenberg; 31–32, *Achnanthes chilensis* Hustedt, 31, external view of the hypovalve, 32, internal view of the epivalve; 33, *Achnanthes delicatula* (Kütz.) Grunow; 34, *Navicula pseudolanceolata* Lange-Bertalot; 35, *Denticula elegans* f. *valida* Pedicino; 36, *Denticula thermalis* Kützing; 37, *Stauroneis* sp.; 38, *Navicula pygmaea* Kützing; 39, *Navicula mutica binodis* Hustedt; 40–41, *Stauroneis legleri* Hustedt.

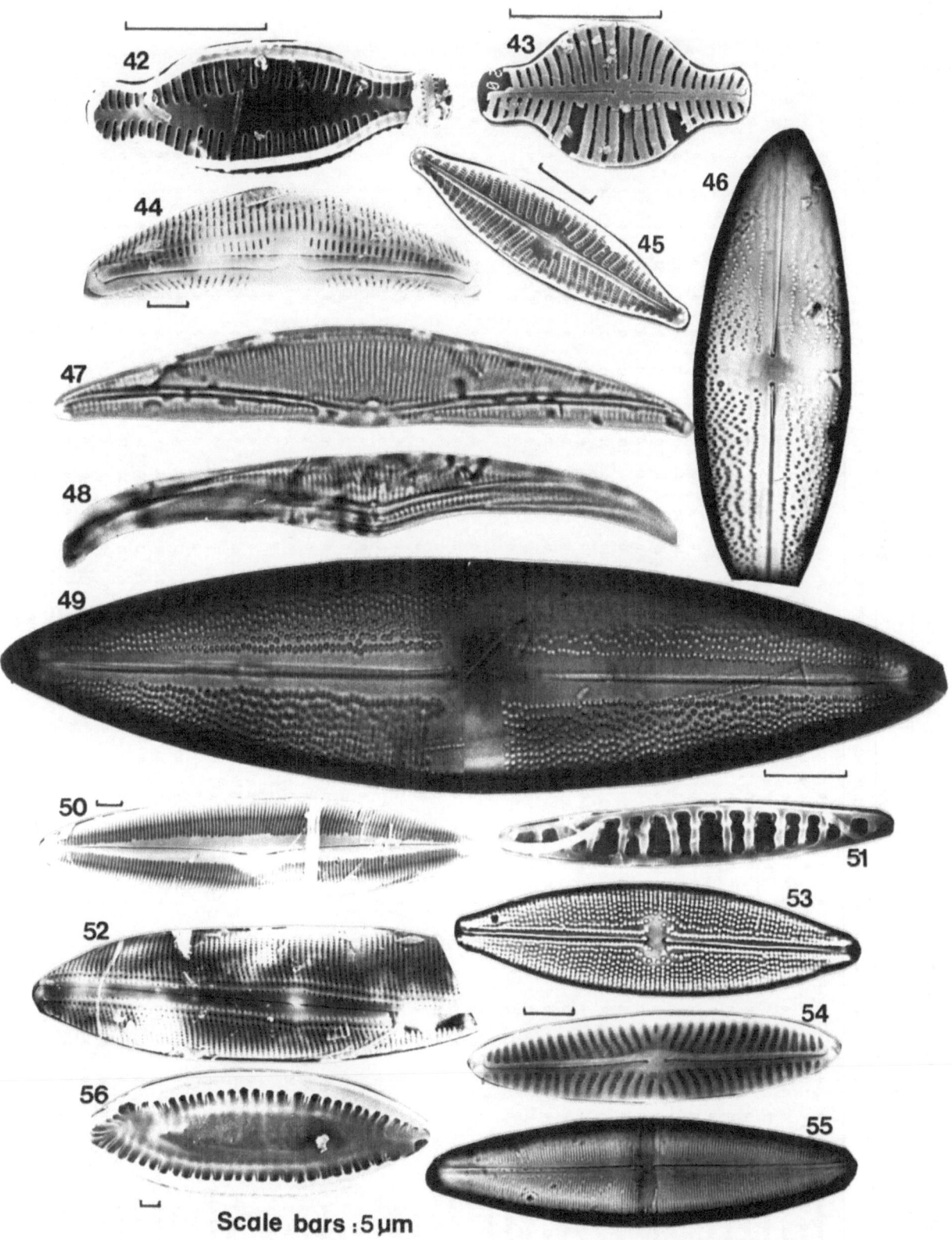

Scale bars : 5 μm

Figs. 42–56. 42–43, *Achnanthes arenaria* Amossé, 42: internal view of the epivalve, 43: external view of the hypovalve; 44, *Amphora libyca* Ehrenberg; 45, *Navicula crypto-cephala* Kützing; 46, *Anomoeoneis sphaerophora* var. *polygramma* (Ehr.) O. Müller; 47–48, *Amphora lineolata* Ehrenberg, 49, *Anomoeoneis sphaerophora* var. *platensis* Frenguelli; 50, Caloneis westii (W. Smith) Hendey; 51, *Denticula elegans* Kützing; 52, *Scoliopleura peisonis* Grunow; 53, *Mastogloia smithii* var. *amphicephala* Grunow (cf. *M. patens* Frenguelli); 54, *Navicula cari* var. *cincta* (Ehr.) Lange-Bertalot (cf. *N. cari* (Ehr.) Ralfs); 55, *Stauroneis gregorii* Ralfs (cf. *S. amphioxys* Gregory); 56, *Surirella chilensis* Janish.

Table 3. Relative abundances of diatom taxa, the basic units for the analysis (in per mil).

Taxa	Samples																													
	BA67	RAM6	VER5	HED4	P130	PUR2	HON4	CHI5	CAN4	CHU4	CHU9	CL20	CD24	CD16	PG70	PG23	PG41	PG43	PG45	PG47	PG72	PG73	PG76	PG74	PG78	PG82	PG84	PG97	P114	P116
Achnanthes breviceps	0	0	0	0	0	0	3	0	0	0	0	0	0	0	0	0	0	0	0	0	0	0	0	0	0	2	0	18	0	0
Achnanthes chilensis	0	20	0	0	0	0	0	0	0	0	0	0	0	0	0	0	0	0	0	0	0	0	0	0	0	0	0	0	0	0
Achnanthes delicatula	0	0	12	0	0	0	0	0	0	0	0	0	0	0	0	0	0	0	0	0	0	0	0	0	0	0	0	4	4	3
Achnanthes arenaria	23	5	0	0	0	126	0	0	0	0	0	0	0	1	0	0	0	0	0	0	0	0	0	0	0	0	0	0	0	0
Achnanthes linearis	40	1	0	0	0	0	0	0	0	0	0	0	0	0	0	0	0	0	0	0	0	0	0	0	0	0	0	50	0	0
Achnanthes speciosa	0	0	0	0	0	0	0	0	0	0	0	0	0	7	7	0	0	0	0	0	0	15	0	0	0	0	0	0	0	0
Amphiprora alata	0	0	0	2	0	0	0	0	0	0	0	0	0	0	0	0	0	0	0	0	0	0	0	0	0	0	0	0	0	0
Amphora libyca	0	0	0	0	0	0	0	0	0	0	0	0	0	0	0	2	0	0	0	0	0	0	0	0	0	0	0	0	0	0
Amphora atacamae	0	0	8	0	10	0	0	0	17	0	31	0	31	7	0	0	44	67	49	0	0	31	0	115	0	7	87	0	4	18
Amphora atacamae minor	0	0	0	34	0	19	0	0	0	0	13	0	0	0	0	0	45	45	0	22	0	0	0	0	53	0	0	0	0	38
Amphora boliviana	10	4	0	0	0	2	0	0	0	0	1	0	27	27	0	0	0	0	0	0	0	0	0	0	0	0	0	0	0	0
Amphora boliviana elongata	0	0	0	0	0	0	0	0	0	0	6	0	0	0	0	8	0	0	6	13	0	32	0	39	0	0	2	0	0	0
Amphora carvajaliana	23	0	0	874	541	632	715	895	96	0	0	0	64	126	0	0	60	61	24	31	0	15	52	18	13	11	14	0	0	0
Amphora chilensis	0	0	0	0	0	0	0	0	0	0	1	0	0	0	0	0	0	0	0	0	0	0	0	0	0	0	0	0	0	0
Amphora lineolata	16	0	0	0	0	0	0	0	0	0	0	0	0	0	0	0	0	0	0	0	0	16	0	0	0	0	0	0	0	0
Amphora ovalis	0	0	0	0	0	0	0	0	0	14	0	0	0	0	0	0	0	0	0	0	0	0	0	0	0	0	0	0	0	0
Amphora platensis	1	0	0	0	0	0	0	0	0	4	0	0	0	0	0	0	0	0	0	0	0	0	0	0	0	0	0	0	0	0
Anomoeoneis sphaerophora angusta	0	0	2	2	0	0	0	0	7	1	0	2	1	7	7	0	0	0	0	0	0	0	0	0	0	2	0	6	0	0
Anomoeoneis sphaereophora polygramma	0	0	0	0	0	0	0	0	0	0	0	0	0	0	0	0	0	0	0	0	0	0	0	0	0	0	0	0	0	0
Anomoeoneis sphaerophora platensis	0	0	0	0	0	0	0	0	0	0	0	0	0	0	0	0	0	0	0	0	0	3	44	0	0	0	0	0	0	0
Anomoeoneis sphaerophora navicularis	0	0	0	0	0	0	0	0	0	0	0	0	0	0	0	0	0	0	0	0	0	0	0	0	0	0	10	0	0	0
Brachysira aponina	3	39	0	0	0	0	0	0	0	0	0	0	1	0	0	2	0	0	0	0	0	0	0	0	0	0	0	0	0	0
Caloneis silicula	0	28	0	0	0	0	0	0	0	0	0	0	0	0	0	0	0	0	0	0	5	0	0	0	0	0	0	0	0	0
Caloneis sp.	0	0	0	0	0	0	0	0	0	0	0	0	0	3	3	0	0	0	0	0	0	0	0	0	0	0	0	0	0	0
Caloneis westii	1	1	0	0	0	0	0	0	0	0	0	0	0	0	0	0	0	0	0	0	0	0	0	0	0	0	0	0	0	0
Ceratoneis arcus	1	0	0	0	0	0	0	0	0	0	0	0	0	0	0	0	0	0	0	0	0	0	0	0	0	0	0	0	0	0
Cocconeis placentula	0	10	0	4	0	0	0	0	0	0	1	0	0	0	0	1	0	0	0	0	0	0	0	0	0	2	0	2	0	0
Cymbella affinis	44	5	0	0	0	0	0	0	0	0	0	0	0	1	0	0	0	0	0	0	0	0	2	0	4	0	0	0	0	0
Cymbella lunata	0	0	0	118	0	25	13	0	0	0	0	0	0	0	0	0	0	0	6	0	0	94	34	80	26	4	0	0	0	0
Cymbella norvegica	0	0	0	0	0	0	0	0	0	0	0	0	0	0	0	0	0	0	0	0	0	0	0	0	0	0	0	30	0	0
Cymbella sp.	0	0	0	0	0	2	0	0	0	0	0	0	0	3	0	2	2	0	0	0	0	0	0	0	0	2	0	0	0	0
Denticula elegans	47	7	0	0	0	0	0	0	0	0	0	0	0	0	23	28	0	0	0	0	0	31	0	33	0	2	2	0	0	0
Denticula elegans valida	0	148	0	0	0	0	0	0	0	0	0	0	0	15	0	0	0	0	0	0	0	0	0	0	0	0	0	0	0	0
Denticula thermalis	6	5	0	0	0	2	0	0	0	0	1	0	0	0	0	0	0	0	0	0	0	0	0	0	0	0	0	0	0	0
Fragilaria brevistriata	0	0	0	0	0	0	40	0	15	0	0	0	0	0	346	735	47	0	0	0	66	4	56	13	13	0	0	104	0	52
Fragilaria construens venter	0	0	0	0	0	0	0	0	0	0	0	0	0	0	0	0	83	73	120	58	11	0	14	17	0	0	0	18	18	0
Fragilaria elliptica	0	0	0	0	0	0	0	0	0	0	0	0	0	0	0	0	15	0	0	24	0	0	0	0	0	0	0	0	0	0
Fragilaria pinnata	0	0	0	0	0	0	0	0	17	0	0	0	0	7	7	0	15	6	0	13	0	0	0	0	0	123	0	0	0	90
Fragilaria zeilleri	0	0	0	0	0	0	0	0	0	0	0	0	0	0	0	100	0	0	0	6	0	0	0	18	0	426	0	0	0	0
Gomphonema parvulum	0	0	0	0	0	0	0	0	23	0	0	0	0	0	0	2	0	0	0	0	0	0	0	0	0	0	18	18	0	0
Hantzschia amphioxys major	0	0	0	0	0	0	0	0	0	0	0	0	0	0	0	1	0	0	0	0	0	0	0	0	0	2	6	6	4	0
Hantzschia nov. sp.	0	0	0	0	0	0	0	0	0	0	0	1	0	0	0	0	0	0	0	0	120	0	0	6	2	2	130	130	4	0
Mastogloia atacamae	0	0	0	0	0	0	0	0	0	0	0	0	0	19	0	5	0	0	12	0	0	156	78	177	6	2	8	12	0	6
Mastogloia smithii amphicephala	20	0	0	0	0	0	0	0	0	0	0	0	0	0	0	0	0	0	0	0	0	0	0	0	0	0	0	0	0	0

Species	S1	S2	S3	S4	S5	S6	S7	S8	S9	S10	S11	S12	S13	S14	S15	S16	S17	S18	S19	S20	S21	S22	S23	S24	S25	S26	S27	S28	S29	S30
Navicula cari cincta	0	0	0	0	0	0	0	0	0	0	0	0	0	0	0	0	0	0	0	0	0	0	0	0	0	65	0	0	0	0
Navicula cincta	0	0	0	0	0	0	0	0	0	0	0	0	0	0	23	0	0	0	6	0	9	500	62	189	12	12	8	40	34	22
Navicula cryptocephala	16	0	2	4	12	89	12	1	63	1	51	0	89	73	15	30	46	54	78	123	85	0	0	12	27	90	320	90	92
Navicula gastrum	0	0	0	0	0	3	3	0	0	0	0	0	0	0	0	0	0	0	0	125	0	0	0	2	6	0	0	0	0
Navicula mutica binodis	0	0	0	0	2	0	0	0	0	2	0	0	0	0	0	0	0	0	0	0	0	10	0	0	2	6	54	4	6
Navicula mutica nivalis	0	0	0	0	2	0	0	0	0	0	0	0	0	0	0	0	0	0	0	0	0	0	0	0	0	0	110	0	3
Navitula nov. sp.	3	5	0	59	16	15	0	0	0	0	228	20	121	214	192	255	31	0	19	126	434	324	236
Navicula pseudolitoricola	3	1	0	0	0	0	0	0	0	0	0	0	0	0	9	0	4	0	2	0	0	2	0
Navicula pseudolanceolata	0	0	0	0	0	1	0	13	260	514	13	0	161	0	0	0	0	0	0	0	0	0	0
Navicula pupula	0	0	0	0	0	0	0	0	0	0	0	0	0	6	0	0	0	0	4	4	0	0	0
Navicula pygmaea	1	1	0	0	0	1	1	0	0	0	0	3	0	0	0	0	0	12	0	0	0	0	0
Navicula rhynchocephala	6	3	4	22	38	79	38	0	0	0	55	70	103	129	15	30	256	56	331	0	60	60	110
Navicula sp.	0	0	0	0	0	0	0	0	0	0	0	0	3	16	0	0	0	6	0	0	0	0	0
Neidium apiculatum	0	0	0	0	0	0	0	0	0	0	7	0	0	0	185	0	0	0	0	0	0	0	0
Neidium bisulcatum	0	0	14	0	0	0	2	5	0	0	0	0	57	85	0	0	4	0	0	0	0	18	82
Nitzschia accedens chilensis	0	0	0	0	0	2	1	0	0	0	62	0	0	0	0	73	4	22	9	0	71	0	0
Nitzschia alpina	0	0	0	0	0	0	0	0	37	0	48	0	0	22	0	0	0	0	0	0	0	0	0
Nitzschia denticula	0	0	0	0	0	3	0	0	23	0	1	0	0	0	0	0	9	0	0	0	0	0	0
Nitzschia epithemioides	0	0	0	0	21	0	0	0	17	0	17	0	28	101	0	54	0	0	17	0	90	41	73
Nitzschia frustulum	0	0	0	16	0	0	0	0	0	4	35	0	15	0	30	18	25	60	7	13	14	13	0
Nitzschia grunowii	6	2	0	24	0	0	0	8	0	0	0	3	12	0	0	0	0	0	16	0	0	0	0
Nitzschia hantzschiana	13	14	4	19	66	7	15	0	0	0	0	11	0	0	185	0	51	24	14	0	6	4	6
Nitzschia hungarica	0	0	0	0	29	0	47	26	150	0	13	0	0	12	4	23	46	12	34	13	50	4	0
Nitzschia inconspicua	0	0	1	0	0	0	0	0	1	0	0	0	0	0	0	0	0	0	0	0	60	0	0
Nitzschia minutula	0	1	0	0	0	0	0	0	0	0	0	0	0	0	0	0	0	0	0	0	0	0	0
Nitzschia nov. sp.	654	839	711	21	0	0	0	0	0	0	0	3	0	0	0	20	0	70	0	0	0	0	0
Nitzschia palea	0	0	0	0	0	2	0	0	0	0	0	0	0	0	0	60	0	10	7	0	0	0	6
Nitzschia punctata	0	0	2	16	0	0	0	0	0	0	10	0	0	25	0	4	2	0	16	4	2	0	0
Nitzschia pusilla	0	0	2	0	0	0	5	8	0	3	0	23	0	0	0	0	0	87	0	0	106	106	139
Nitzschia quadrangula	0	0	0	0	0	0	0	0	80	0	0	0	11	0	145	116	14	0	0	0	12	8	0
Nitzschia valdecostata	0	0	0	8	19	0	0	0	0	1	20	0	0	4	0	14	12	12	0	2	6	3	3
Opephora martyi	0	0	0	0	0	0	0	7	0	0	0	3	0	19	0	0	18	0	4	0	8	3	0
Pinnularia bogotensis	0	1	0	0	0	0	0	0	0	0	0	0	0	0	0	0	11	0	9	0	2	6	3
Rhopalodia gibba	3	10	0	0	20	6	26	6	0	0	15	0	6	22	4	36	34	6	2	4	11	6	3
Rhopalodia wetzeli	3	1	19	0	0	13	0	0	0	0	10	0	11	11	0	12	31	0	9	17	0	0	0
Scoliopleura peisonis	0	0	0	16	12	1	3	0	34	0	20	24	22	12	18	52	31	4	2	58	4	0	0
Stauroneis anceps	0	0	0	0	0	3	0	0	0	0	0	0	0	0	0	0	0	9	0	0	4	0	0
Stauroneis bathurstensis	0	0	16	0	0	0	0	0	0	0	0	0	0	0	0	0	0	2	0	0	8	8	15
Stauroneis gregorii	0	0	0	14	0	0	0	0	123	0	22	778	22	0	60	0	0	0	0	0	42	42	0
Stauroneis legleri	0	0	0	0	0	5	0	5	7	0	0	5	0	0	440	0	0	0	0	0	0	0	0
Stauroneis spd.	0	0	0	0	0	0	0	0	0	0	15	196	0	0	0	0	0	12	0	0	0	0	0
Stauroneis sp.	0	0	0	0	0	0	0	0	0	0	0	0	2	0	0	0	0	0	0	0	3	0	0
Surirella chilensis	0	0	0	0	0	0	0	0	0	0	0	0	2	0	0	0	0	0	0	0	7	0	0
Surirella oregonica	0	0	0	0	0	6	0	0	0	0	0	0	0	0	4	0	0	0	0	0	21	13	82
Surirella ovata utahensis	3	0	0	0	0	13	0	15	0	0	0	0	0	0	0	0	0	0	0	0	0	0	0
Surirella peisonis	3	0	0	48	19	0	1	0	48	1	90	0	6	8	1	42	14	26	4	4	0	6	0
Surirella sella	60	30	19	0	0	10	1	4	0	3	35	0	0	36	0	39	15	82	1	1	2	0	0
Surirella wetzeli	0	0	0	20	0	1	268	0	0	0	0	0	0	0	2	0	0	35	0	0	0	4	0
Synedra pulchella	0	0	8	0	0	0	0	0	0	0	0	0	0	0	0	0	0	0	0	0	6	0	0
Synedra rumpens	0	0	0	152	0	0	0	0	0	0	0	0	0	0	0	0	0	0	0	0	2	18	0
Total frustules	590	1102	700	982	469	604	856	1076	347	251	478	514	888	577	1404	495	627	354	333	444	400	517	402	423	452	821	938	330	431	552

(1982) and halophile by Hustedt (1927), *Amphiprora alata* as mesohalobous by De Wolf (1982) and Frenguelli (1936), *Anomoeoneis sphaerophora polygramma* as oligohalobe by De Wolf (1982), *Surirella ovata utahensis* as euryhaline by Wornardt (1964), *Anomoeoneis sphaerophora navicularis* as brackwasser by Cholnoky (1968), *Navicula pupula* as oligohaline by Hustedt (1959).

The last group concerns the halophobous or oligohaline species which are here living in saline waters and, consequently, must be considered as euryhalines: *Cymbella lunata, Fragilaria brevistriata, Gomphonema parvulum*. Six freshwater diatoms occur in small quantity in only one high salinity sample. They are considered as allochtonous: *Hantzschia amphioxys maior, Ceratoneis arcus, Navicula mutica nivalis, Navicula mutica*

binodis, Pinnularia bogotensis, Synedra rumpens.

These general comments show that we cannot use only published ecological data for past reconstructions because even for some well-known species, local adaptation to salinity is variable. It is recommended to use living diatoms from the area where paleoecological studies take place. But, even in the best conditions, the applicability of our own data is limited 'to lakes that have remained within the represented ranges of environmental modern parameters' as suggested by Anderson *et al.* (1986).

(ii). Relationships between salinity, number of species and diversity in the Lipez area.
It is generally admitted that the number of species and the diversity are lower in salt than in fresh-

Table 4. Pastos Grandes lake: relation between salinity changes and number of species, salinity changes and dominant species. 2a) No relation between salinity and number of species along a gradient of salinity in the SW margin. 2b, c, d,e) The low salinity samples are characterized by different dominant oligohaline species according to the area. Decrease in number or disappearance of these oligohaline species are observed; they are replaced by a meso-polyhalobous species (*Navicula* nov. sp.) when the salinity increases.

a. South-western area					
Samples	PG72	PG73	PG74	PG76	PG78
Salinity g l^{-1}	0.6	1.6	12.1	14.4	144
Number of species	9	25	26	21	23
b. South-western area					
Samples	PG72	PG73	PG74	PG76	PG78
Salinity g l^{-1}	0.6	1.6	12.1	14.3	144
% *Navicula* nov. sp.	0	0	0	3.1	12.6
% *Nitzschia hantzschiana*	18.5	1.4	5.1	0	0
c. Southern area					
Samples	PG82	PG84			
Salinity g l^{-1}	0.19	225			
% *Navicula* nov. sp.	1.9	43.4			
% *Fragilaria zeilleri*	42.6	0			
d. North-eastern area					
Samples	PG97	P114	P116		
Salinity g l^{-1}	0.4	255	267		
% *Navicula* nov. sp.	0	32.4	23.6		
% *Navicula cryptocephala*	32	9	9.2		
e. North-western area					
Samples	PG23	PG43	PG41	PG45	PG47
Salinity g l^{-1}	1.4	14.2	26.7	103	371
% *Navicula* nov. sp.	0	21	12.1	19.2	25.5
% *Fragilaria brevistriata*	73.5	0	4.7	0	9.4

water lakes. In the Bolivian freshwater montane lakes, several hundred species were identified (Servant-Vildary, 1986), against 94 in these saline lakes.

In low salinity samples the Shannon diversity index is generally higher (3.7) than it is in high salinity samples (0.8). However, no defined relationship correlation was observed between decreasing Shannon diversity index and increasing salinity. In Pastos Grandes, on the southern margin, the index value is the same (3.39) in a very low salinity sample PG82 ($0.4 \, \mathrm{g \, l^{-1}}$), and in a very high salinity sample PG84 ($225 \, \mathrm{g \, l^{-1}}$). Correlation between decreasing number of species and increasing salinity, generally emphasized by many authors, has not been observed in Pastos Grandes. Along a salinity gradient there from the NE margin to the centre, no correlation appeared (Table 4a).

(iii). Relationships between salinity and dominant species.

Different salinities are not often linked to different dominant species. Most of these lakes are characterized by two or three species which represent more than 70% of the total flora. Same dominant species are found in lakes with different salinities. Thus, *Amphora carvajaliana* is the dominant species in five lakes, CHI5, HED4, PJ30, PUR2, HON4, the salinity of which is respectively 77, 72, 35, 32, 25 $\mathrm{g \, l^{-1}}$. Furthermore, the changes of the salinity observed between 1980 and 1983 are not related either to diatom assemblage composition or to dominant species: In Hedionda, the measured salinity was $72 \, \mathrm{g \, l^{-1}}$ in 1980 and $57 \, \mathrm{g \, l^{-1}}$ in 1983 and *Amphora carvajaliana* was still the dominant species, the same phenomenon has been observed for Honda (salinity from 25 to $21 \, \mathrm{g \, l^{-1}}$), Puripica (salinity from 32 to $28 \, \mathrm{g \, l^{-1}}$) and Chiar Kkota (salinity from 77 to $119 \, \mathrm{g \, l^{-1}}$). It appears that other parameters in addition to salinity control the growth of a few species.

Different dominant species are found in low salinity samples, as for example in Pastos Grandes below $1.4 \, \mathrm{g \, l^{-1}}$. This can be explained by the origin and chemistry of the freshwater inputs in the lake. *Nitzschia hantzschiana* is dominant in the western part of the lake, *Fragilaria zeilleri*, in the southern part, *Navicula cryptocephala*, in the north-eastern part and *Fragilaria brevistriata* in the north-western part (Table 4b, c, d, e).

As freshwater-oligohaline conditions change to meso-polyhaline conditions, the relative abundance of freshwater diatoms slowly decreases and saltwater diatoms slowly increase. In Pastos Grandes, the percentage of *Navicula* nov.sp. increases when the salinity increases, at the same time the freshwater-oligohaline species disappear or decrease.

Our observations show that strong modifications of the diatom flora can be explained by the salinity only when it crosses the limit between fresh-oligohaline to meso-hyperhaline conditions (around $2 \, \mathrm{g \, l^{-1}}$). Within these limits, the diatom flora changes are better explained by other parameters.

Classification of the lakes, based on the diatom flora

(i). Definition of groups (or classes) of samples.

The FAC applied to the relative abundances of the 94 taxa (Table 3) in the 30 samples ($j = 94$ and $i = 30$) produces 10 factors which account for 78,8% of the distributional variance of the species data base. The factor matrix gives the composition of each sample in terms of the ten factors. There is an important difference between the 4th

Table 5. Factor analysis of correspondences of taxa, inertia values.

Axis	Eigenvalues	Inertia	Cumulate inertia
1	0.86	13.31	13.31
2	0.70	10.77	24.08
3	0.64	9.94	34.02
4	0.60	9.24	43.26
5	0.48	7.44	50.70
6	0.46	7.06	57.76
7	0.38	5.96	63.72
8	0.37	5.72	69.44
9	0.33	5.12	74.56
10	0.27	4.23	78.79

282

and the 5th axis, but the first five axes have 50.70% of the variance (Table 5).

Factors 1 and 2 represent 24.08% of the total variance. According to these two factors two groups are separated from all others. Along axis 1, group 1 (VER5, BA67, RAM6); group 1 explains 88% of the factor 1. Along axis 2, group 2 (CL20, CD16) explains 88% of the factor 2 (Fig. 2).

In the space defined by axis 3 and 4 (Fig. 3) the remaining 25 samples, located close to the origin

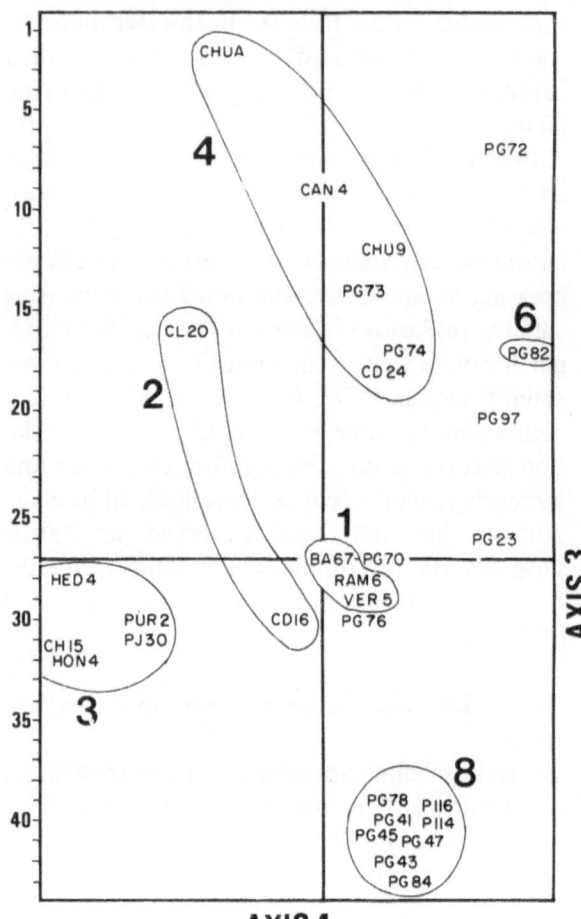

Fig. 3. Projection of the samples points on factorial plane 3–4.

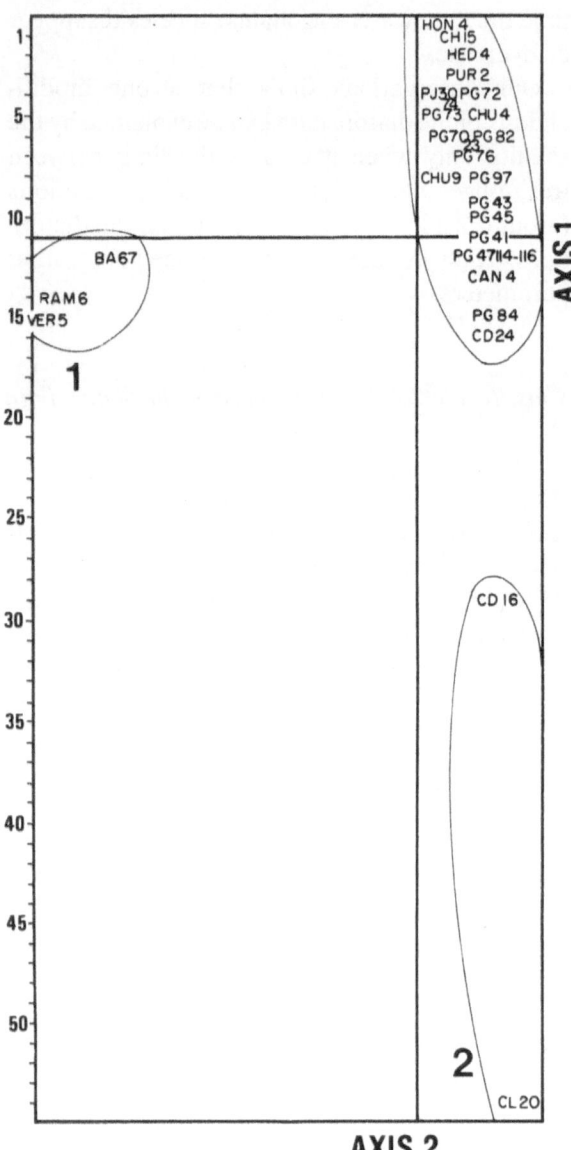

Fig. 2. Projection of the samples points on factorial plane 1–2.

of axis 1 and 2, are distributed in 4 groups. Axis 3 is explained by group 3 composed of HED4, CHI5, HON4, PUR2, PJ30 which accounts for 56% of the inertia and by group 6 composed of one sample PG82 (8% of the inertia) opposed to the former one. Factor 3 explains 64% of the total inertia. Axis 4 is explained by group 4 composed of CHU4, CAN4, CHU9, PG73, PG74, CD24 (39% of the inertia) opposed to the group 8 composed of PG78, PG41, PG45, PG47, PG43, P116, P114, PG84 (38% of the inertia). Factor 4 is explained by 77% of the total inertia.

Group 5 (PG23, PG70) which explains 29% of the total inertia, is opposed to group 7 (PG72, PG97), which explains 50% of the inertia, along axis representing factor 5 (Fig. 4), which accumu-

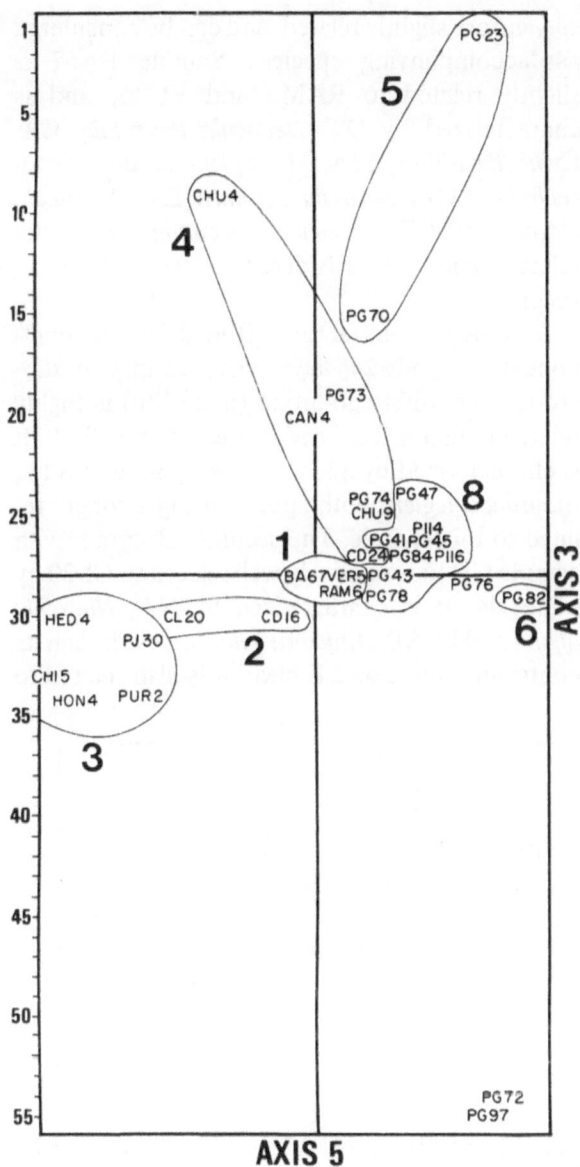

Fig. 4. Projection of the samples points on factorial plane 3–5.

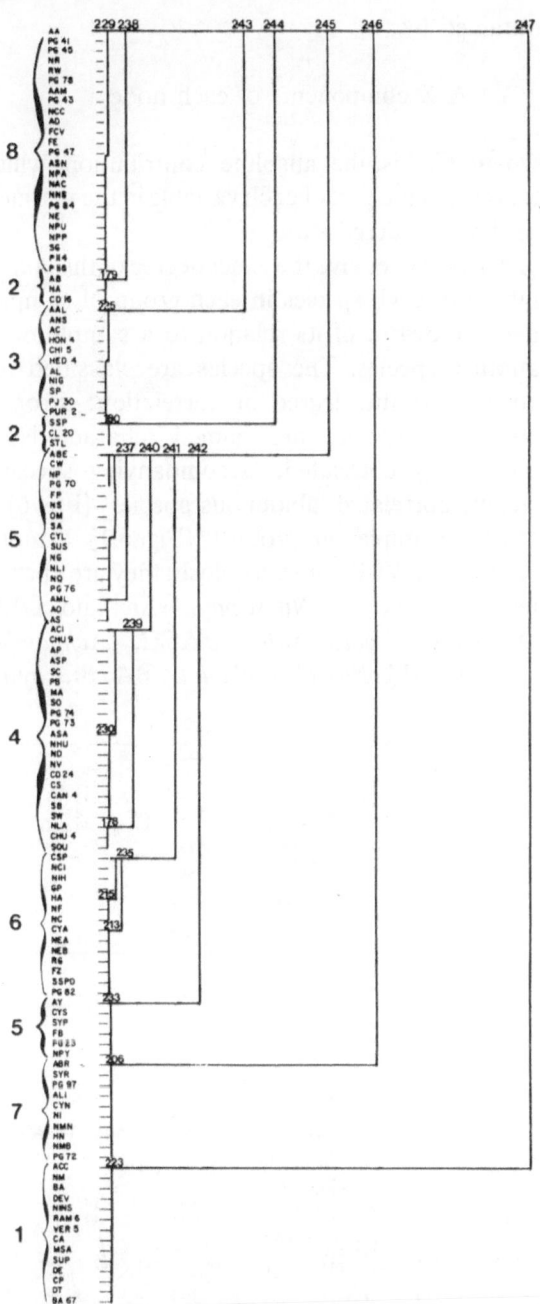

Fig. 5. Cluster Analysis (CA) of samples and taxa. Codes for taxa are given in Table 2.

lates 79% of the inertia. Considering the five first factors, only sample PG76 cannot easily be related to any factor.

(ii). Diatom species ordination.

Cluster analysis of taxa yielded 8 major groups (Fig. 5) composed by species and samples linked by a node: group 1 (node 223), group 2 (244), group 3 (node 225), group 4 (node 239), group 5 (node 237), group 6 (node 235), group 7 (node 206), group 8 (node 229). In order to show the relationships between species and samples in a group, we are able to calculate the values of all the nodes which compose a group. The values are obtained by

$$\Sigma\, CA/\Sigma\ \text{components of each node}$$

where CA is the absolute contribution which represents the part of each variable in the variance of the considered axis.

These values give the exact degree of the contribution of each species in each group of samples and the degree of its relation to a sample or to another species. The species are classified according to the degree of correlation: strongly correlated species are named 'characteristic', moderately correlated, 'accompanying species', weakly correlated 'ubiquitous species' (Fig. 6).

For example, in group 1 (Fig. 6,1) samples RAM6 and VER5 are very close; they are characterized by NINS, *Nitzschia nov.sp.* and DEV, *Denticula elegans valida*. ACC, *Achnanthes chilensis*, NM, *Nitzschia minutula*, BA, *Brachysira*

aponina are slightly related, and can be considered as 'accompanying species'. Sample BA67 is slightly related to RAM6 and VER5, and is characterized by DT, *Denticula thermalis*. CA, *Cymbella affinis*, MA, *Mastogloia smithii amphicephala*, SUP, *Surirella peisonis*, DE, *Denticula elegans* and CP, *Cocconeis placentula* are slightly related, and are considered as 'accompanying species'.

In group 2 on factor 2 (Fig. 6,2), the components of node 244 are separated in two subgroups. The first sub-group (node 180) is highly related to factor 2 defined by the sample CL20. It is characterized by SSP, *Stauroneis sp.* and STL, *Stauroneis legleri*, both species being strongly related to this sample. The second sub-group with sample CD16 is relatively well related to CL20 by node 244. It is characterized by NA, *Nitzschia alpina* and by AB, *Amphora boliviana*. Although its contribution to axis 2 is clear, it is also related to

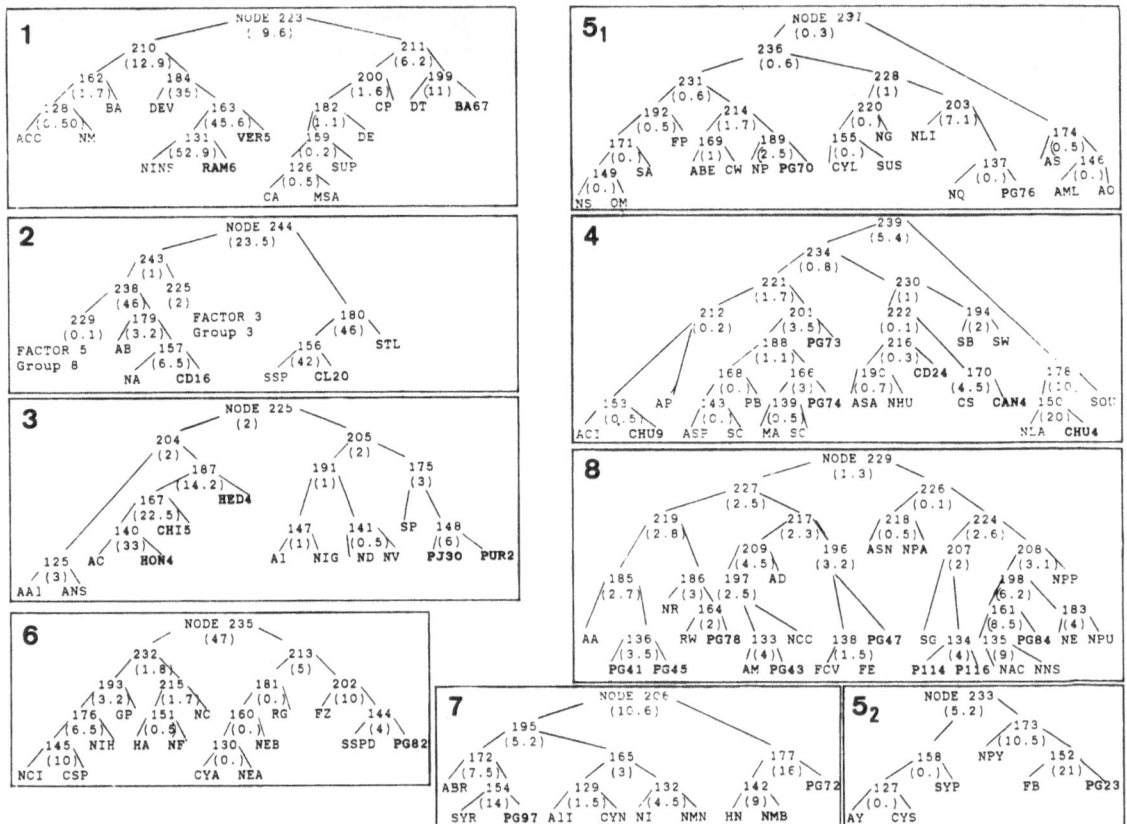

Fig. 6. Hierarchical classification of samples and taxa in Groups 1 to 8. Main values of the nodes are in brackets.

factor 5 by node 229 and to factor 3 by node 225. Sub-group 2 is slightly related to group 3 by co-occuring species AB, *Amphora boliviana* by the node 243 and slightly related to group 8 by the same species. We can conclude that the accompanying species are co-occuring species. This fact explains the close position of sub-group 2 to the group 8 in the hierarchical classification (Fig. 5). Similar comments can be done about group 5 (Figs. 5 and 6).

Table 6. Chemical variables/classes, classes/chemical variables contributions program. A – The positive correlation indicates the ion which mostly affected the considered group of samples, the negative correlation indicates the ion which, by its low content or its lack, mostly separates the considered group from the others. B – The positive correlation indicates the most typical group with regard to a variable.

A – Chemical variables/classes contributions.

	Positive correlation		Negative correlation	
	Strong	Weak	Strong	Weak
Group 1	SO_4, Cl	Na, Ca	Alk	Si, K
Group 2	Alk	Na, K	Cl	Ca, Si, SO_4
Group 3	SO_4, Na	–	Alk	Cl, Si
Group 4	SO_4	K	Cl, Na	–
Group 5	Cl	Na, Alk, Li	SO_4	–
Group 6	Si, Alk	–	Cl	Na, SO_4
Group 7	Alk, Si	Ca	Cl, Na	SO_4
Group 8	Cl	Na	SO_4	Alk, Si

B – Classes/Chemical variables contributions

	Positive contribution		Negative contribution	
	Strong	Weak	Strong	Weak
Alk	2, 7, 6	5	8	3, 1
Ca	7, 1	6	2	8, 3, 4
Cl	8	5	7, 4	2, 3
K	4, 2	–	7	1, 6, 8, 5
Mg	7, 1	4, 6	2, 8	5, 3
Na	8	3, 2, 1, 5	7, 4	6
Si	6, 7	–	–	8, 3, 1, 2, 4
SO_4	4	3, 1	8	5, 7, 6
Li	5	8	–	1, 4, 3, 2, 7, 6

Diatom assemblages as affected by chemical variables

(i). The variables/classes, classes/variables contributions method

For each of the 8 clusters of samples, we determined its correlation (positive or negative) with each ionic variable (Table 6).

Group 1 is essentially influenced by sulfate and chloride and to a lesser degree by sodium and calcium and is characterized by low alkalinity and to a lesser extent by the absence of silicium and potassium, in contrast with groups 2, 6 and 7 which are positively influenced by alkalinity. Group 1 is close to groups 3 and 8 by low alkalinity and silicium content, but differs from

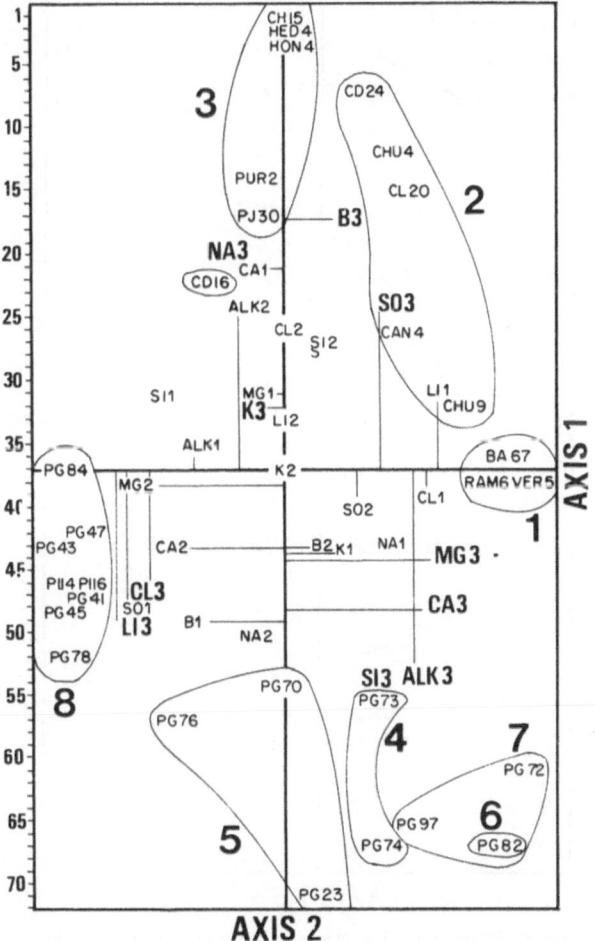

Fig. 7. Projection of samples and chemical variables points on factorial plane 1–2.

class 8 by sulfate. Highly related to sulfate, it is however less determined by this anion than group 4 is, and less related to chloride as group 8 is. Finally, the diatom assemblage of group 1 is linked to the presence of chloride and sulfate, but under a lower concentration of sulfate than is needed by the diatom assemblage of group 4; as well as by a concentration of chloride than is needed by group 8 (Table 6a).

(ii). The interactions species/ionic variables method
We use this method to quantify the ionic variables related to each cluster, and then present a FAC graph of the established relations (Fig. 7). We can

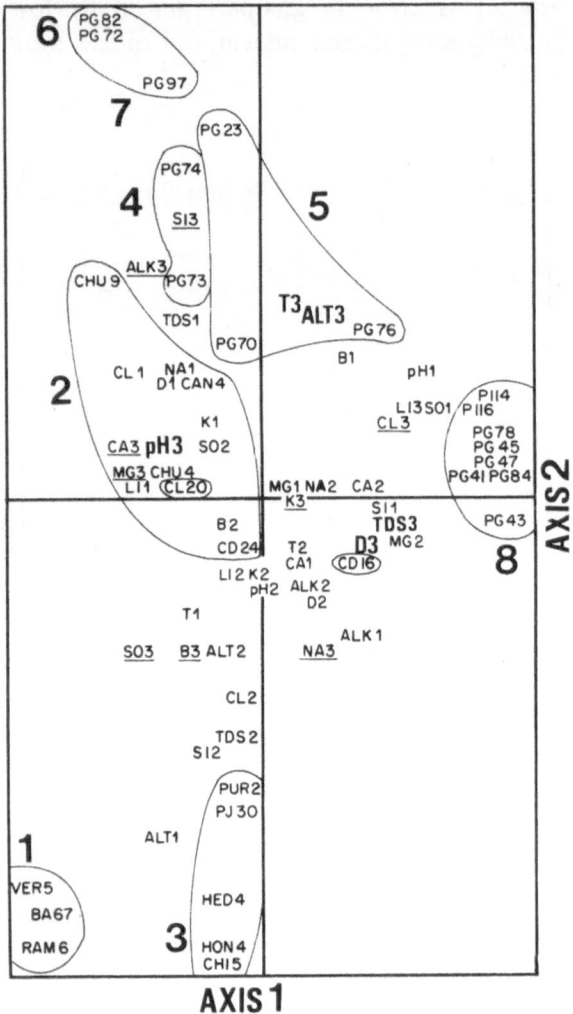

Fig. 8. Projection of samples, chemical variables, salinity TDS, physical parameters depth (D), pH (PH), temperature (T) and elevation (ALT) points on factorial plane 1–2.

observe that the results are similar to those obtained by the former method, but the advantage is that the results can be shown in a graph were the ionic components are quantified using for low content values < 34%, category 1, medium content, values between 34 and 68% (category 2), and high content, values > 68% (category 3).

The last step of these different approaches is to verify if the ionic contents are really the main factors influencing the composition of diatom assemblages. For such a purpose we use another FAC where other environmental variables such as total dissolved salts (TDS), and physical parameters such as temperature, (T), altitude, (ALT), pH (pH) and depth (D) are included and split into three modalities as for ionic components (Fig. 8).

Comparing the results obtained with only the ionic variables (Fig. 7) to those obtained with all the environmental factors (Fig. 8), it is clear that the last parameters are only secondary factors, because they do not strongly modify the clusters neither their location on the main factorial axis. The composition of each cluster is similar and it is related to the same ionic elements in both analyses. The ordination along axis 1 and axis 2 in both graphs has changed a little. Opposition along axis 1 between clusters 1 and 8 and along axis 2, between cluster 3 and clusters 5, 6 and 7 are unchanged, cluster 4 remains intermediate, but there is a shift of cluster 1 along axis 2 due to parameter altitude.

Discussion

We may suppose that eurytherm species are the best adapted to the hard climatic conditions as wide daily temperature variations occur in the Lipez area.

Qualitative and quantitative relationships between diatom species, groups of samples and environmental variables obtained by different statistical approaches are summarized in Table 7 together with measured data and ecological informations.

Group 1 is characterized by *Nitzschia* nov. sp. whose development is due to high content of

Table 7. Groups of environmental parameters and species set up using measured published and inferred data for each group of samples. Classes of salinity, pH and temperature are according to Lowe's system (1973). Line 1, 8 groups of samples (classes) obtained from FAC of taxa. Line 2, samples which composed the groups 1 to 8. Line 3, characteristic species. Line 4a, accompanying species. Line 5, depth. Line 6, salinity. Line 7, pH: Acb = % of acidobiontic species, Ac = % of acidophilous species, Ind = % of pH-indifferent (circumneutral) species, Al = % of alkaliphilous species, Alb = % of alkalibiontic species, U = % of species whose affinity to pH is unknown. Line 8, groups of species according to temperature: TE = true eurytherms (tolerate temperature fluctuations of 20 °C or over), ME = mesoeurytherms (tolerate fluctuations of about 15 °C), MEW = warm (living in waters from 20 to °C), MST = temperature (15–25 °C), MSC = cold (10–20 °C), ES = Eustenotherms (tolerate temperature fluctuations of about 5 °C), ESW = warm (living at or over 25 °C), EST = temperate (15–25 °C), ESC = cold (at or below 15 °C), U-unknown affinity to temperature. Line 9, elevation.

1 GROUPS		1	2	3	4	Units
2 SAMPLES		RAM 6, VER5, BA 67.	CL 20, CD 16.	HED 4, HON 4, PUR 2, PJ 30, CHI 5.	(1)CHU 4, CD 24, CAN 4, CHU9. (2) PG 73, PG 74.	
3 CHARACTERISTIC SPECIES		Nitzschia nov. sp. / Denticula elegans valida / Denticula thermalis	Stauroneis legleri / Stauroneis sp.	Amphora carvajaliana	Navicula pseudolanceolata / Surirella ovata utahensis / Anomoeoneis sphaerophora angusta / Nitzschia hungarica / Caloneis silicula	
4 FELLOW SPECIES		Brachysira aponina / Nitzschia minutula / Achnanthes chilensis / Mastoglia smithii amphicephala / Ceratoneis arcus / Surirella peisonis / Denticula elegans / Cocconeis placentula	Amphora boliviana / Nitzschia alpina	Amphiprora alata / Anomoeoneis sphaerophora polygramma / Nitzschia denticula / Nitzschia valdecostata / Nitzschia grunowii / Scoliopleura peisonis / Achnanthes lemmermanii	(1) Stauroneis bathurstensis / Surirella wetzeli / Amphora platensis / Amphora chilensis / (2) Mastoglia atacamae / Surirella oregonica / Surirella chilensis / Anomoeoneis sphaerophora / Pinnularia bogotensis / (1) FIRST SUB-GROUP / (2) SECOND SUB-GROUP	Units

5 DEPTH

		1	2	3	4	Units
A	range	30-100	20-50	20–100	15-20	cm
	mean	65	35	40	17	
B	a	36.4	—	96.3	52	%
	e	0.8	—	0.92	0.22	
	b	41.2	—	0.01	23	
	l	—	—	—	—	
	p	—	—	—	5.4	
	E	21.2	—	—	17.2	
	U	0.4	100	2.7		
C	D1	+	+	+		modalities 1→<20 cm, 2→20-30, 3→>30
	D2				+	
	D3					

6 SALINITY

		1	2	3	4	Units
A	range	13 to 45	36 to 120	25 to 77	1.6 to 144	g/l
	mean	29	78	48.2	40.3	
B	<0,2	—	—	0.05	5.43	%
	0,2-10	0.4	—	0.2-10	—	
	10-20	14.1	—	—	90.1	
	20-30	31	70.3	96.4	0.25	
	>30	31	—	—	2.4	
	Eurys	20.8	29.7	2.7		
	U	12		0.71		

Salinity B — ion affinities by group:

Group 1: $SO_4^-(2)$, $Cl^-(1)$, $Ca^{++}(3)$, $Mg^+(3)$, $Na^+(1)$ / NINS: $Ca^{++}(3)$, $Mg^+(3)$ / DEV: $Alk(3)$, $Ca^{++}(3)$, $Mg^+(1)$

Group 2: $B^+(3)$, $K^+(3)$, $Na^+(3)$, $SO_4^-(3)$, $Si(2)$, $Alk(2)$, $Ca^{++}(1)$, $Mg^+(1)$, $Li(1)$ / STL: $B^+(3)$, $Mg^+(1)$, $SO_4^-(2)$ / SSP: $Alk(3)$, $Si(3)$, $SO_4^-(2)$

Group 3: $SO^{--}(3)$, $Na^+(3)$, $B(3)$, $Alk(2)$, $Si(2)$, $Cl^-(2)$, $Ca^{++}(1)$ / AC: $Cl^-(2)$, $Li(2)$, $Na^+(3)$, $Si(2)$, $SO_4^-(3)$

Group 4: for the first sub-group group: $SO_4^-(3)$, $B(3)$, $Si(2)$, $Li(1)$, $K(3)$ / for the second sub-group: $Si(3)$ / NLA: $Li(1)$, $Alk(2)$, $SO_4^-(3)$, $Cl^-(1)$

		1	2	3	4	Units
C	TDS1	+	+	+	+	modalities 1→0.2-20 g/l, 2→20-50, 3→>50
	TDS2					
	TDS3					

7 pH

		1	2	3	4	Units
A	range	8.18 to 8.78	10.3	8.05-8.85	7.8 to 10.2	
	mean	8.35		8.4	8.7	
B	Acb	—	—	—	—	%
	Ac	—	—	—	—	
	Ind	—	3	—	—	
	Al	30	15.5	0.7	—	
	Alb	1.33	—	0.32	0.89	
	U	68.6	81.5	99	99.11	
C	pH1	+	+	+		modalities 1→6.9-7.5, 2→7.5-8.5, 3→8.5-10.3
	pH2					
	pH3				+	

8 TEMPERATURE

		1	2	3	4	Units
A	range	1. to 5	6–21	1–8	1–10	°C
	mean	2.6	13.5	6	5.1	
B	TE	26.1		0.2		%
	ME — MEW	—				
	ME — MET	—				
	ME — MEC	—				
	MS — MSW	—				
	MS — MST	1.33		1.4		
	MS — MSC	—			0.25	
	ES — ESW	—				
	ES — EST	—				
	ES — ESC	31.4	⌐	98.4	99.75	
	U	41	100			
C	T1	+		+	+	modalities 1→1-4°C, 2→4-8, 3→8-20
	T2		+			
	T3					

9 ELEVATION

		1	2	3	4	Units
	range	4117-4350	4278-4495	4110-4393	4140-4400	m
	mean	4195	4386	4169	4312	
	Alt 1	+	+	+		modalities 1→4100-4200m, 2→4200-4300, 3→>4300
	Alt 2		+			
	Alt 3					

Table 7. (continued).

1 GROUPS		5	6	7	8	
2 SAMPLES		1rst sub.group : PG 70, PG 76 2nd sub-group : PG 23	PG 82	PG 72, PG 97.	PG 84, PG 41, PG 45, PG 43, PG 47, P114, P116.	
3 CHARACTERISTICS SPECIES		(1) Nitzschia punctata Nitzschia quadrangula Navicula pseudolittoricola (2) Fragilaria brevistriata Navicula pygmaea	Fragilaria zeilleri	Navicula mutica binodis Hantzschia nov. sp. Achnanthes breviceps	Navicula nov. sp. Nitzschia epithemioides Nitzschia accedens chilensis Nitzschia pusilla Amphora atacamae Amphora atacamae minor Navicula cari cincta Fragilaria construens venter. Fragilaria elliptica Stauroneis gregorii	
4a ACCOMPANYING SPECIES		(1) Navicula gastrum Surirella sella Cymbella lunata (2) Synedra pulchella Cymbella sp. Amphora lybica	Neidium bisulcatum Neidium apiculatum Cymbella affinis Rhopalodia gibba Stauroneis spd	Achnanthes linearis Cymbella norvegica Synedra rumpens Navicula mutica nivalis Nitzschia inconspicua	Navicula pupula Achnanthes delicatula Anomoeoneis sphaerophora navicularis Nitzschia palea	
4b UBIQUIST SPECIES		Navicula sp. Amphora ovalis Achnanthes speciosa Amphora boliviana elongata Amphora lineolata Caloneis westii Opephora martyi Stauroneis anceps Fragilaria pinnata	Gomphonema parvulum Nitschia hantzschiana Navicula cincta Nitzschia frustulum Hantzschia amphioxys			
						Units
5 DEPTH	A range mean	20-20 20	 20	 20	 100	cm
	B a e b l p E U	— 57 11.6 5.1 — 2.9 11.6	0.29 29.2 — 6.7 63.7	25 3.7 1.2 — 20.4 49.5	1.3 0.14 17.4 2.3 18.1 60.3	%
	C D1 D2 D3	+	+	+ 	 +	modalities 1 → <20cm 2 → 20-30 3 → >30
6 SALINITY	A range mean	1.4-14.3 9.6	0.19	0.4 to 0.6 0.5	13-324 169	g/l
	B <0.2 0.2-10 10-20 20-30 >30 Eurys U	— 10.9 28.5 — — 58.7 1.85	— 3.4 — — 29 63.7	— 96.2 — 3.75 0	11.7 — 51.6 18.1 8.5	%
	C TDS 1 TDS 2 TDS 3	+	+	+	 +	modalities 1 → 0.2-20g/l 2 → 20-50 3 → >50
7 pH	A range mean	8.4 to 9.35 8.7	9.62	6.95-8.92 7.93	6.95-9.7 7.9	
	B Acb Ac Ind Al Alb U	— 4.7 5.1 53.05 1.96 30.14	— — — 29.5 1.9 68.4	— — 62 — 237 14.1	— 2.46 19.4 — 77.9	%
	C pH 1 pH 2 pH 3	+	 +	 +	+	modalities 1 → 6.5-7.5 2 → 7.5-8.5 3 → 8.5-10.3
8 TEMPERATURE	A range mean	1.10 5.3	10	10 10	5-15 8.8	°C
	B TE ME MEW/MET/MEC MS MSW/MST/MSC ES ESW/EST/ESC U	63 1 — — — — 2 96 — 5.5 28	33.8 — — — — 2 — 2.9 63.7	 3 3 16.6 75.8	27.5 — — — 3.7 — — 68.5	%
	C T 1 T 2 T 3	+	+	+	+	modalities 1 → <4°C 2 → 4-8 3 → 8-20
9 ELEVATION	range mean Alt 1 Alt 2 Alt 3	4400 +	4400 +	4400 +	+	m modalities 1 → 4100-4200m 2 → 4200-4300 3 → >4300

Group 5 salinity B notes: for the first sub-group: CL⁻(3), Li(3), Na⁺(2), B⁺(1) for the second sub-group: Si(3), Alk(3) NP: Ca⁺⁺(2), B⁺(1), Cl⁻(3) NQ: Ca⁺⁺(2), Cl⁻(3) K⁺(3) FB : Si(3), Alk(3), B⁺(2), Mg(2)

Group 6 salinity B notes: Si(3), Alk(3), Ca⁺⁺(3), Mg⁺(3). FZ : Alk(3), Ca⁺⁺(3), K⁺(1), Si(3) SO₄⁻(2).

Group 7 salinity B notes: Alk(3), Si(3), Ca⁺(3), Mg⁻(3). NMB : Alk(3), Cl⁻(1) Ca⁺⁺(3), Si(3), Mg⁺(3)

Group 8 salinity B notes: Cl⁻(3), Na⁺(3), Li(3) NNS: SO₄⁻(1), Si(1) Alk(1), B⁺(1), Na⁺(3), Li(3), Ca⁺⁺(3), Cl⁻(3)

calcium (CA3) and magnesium (MG3). As indicated by measured values, this group is related to higher concentration of sulfate (SO2) than chloride (Cl1), to medium salinity (TDS2), (the measured salinity is $29 \, g \, l^{-1}$) to medium pH (pH2), (the measured pH 8.35) to low temperature (T1), (the measured temperature is 2.6 °C) and low elevation. Except for depth and salinity, where 41% of benthic and 31% of polyhalobous species are present and give good ecological information which correspond to those obtained from measured data, the other ecological parameters should not have been determined with only ecological informations from literature because 68.1% of the species have unknown pH affinity and 41% of the species have unknown temperature affinity. This lack of ecological data is due to the abundance in this group of *Nitzschia* nov. sp. (Fig. 24) whose ecology is unknown.

Group 2: ecological data are lacking for depth, pH and temperature. Data upon salinity affinity of characteristic species of *Stauroneis* were given by Patrick (1961).

Group 3: there is a complete lack of data for temperature and pH, but some useful ones have been found concerning depth and salinity affinities.

Group 5, 6, 7: Based only on published ecological data, depth and salinity should have been determined (essentially separation between groups 5, 6, 7 and group 8), but without any possibility of quantification. The low salinity groups 5, 6 and 7 contain different 'characteristic' oligohaline species *Fragilaria brevistriata*, *Fragilaria zeilleri* and *Navicula mutica binodis*, each one is related to the chemistry of the freshwater. *Fragilaria brevistriata* develops in high concentrations of silica, high alkalinity and medium concentration of boron; *Fragilaria zeilleri* develops better in relation to silica, high alkalinity and calcium; *Navicula mutica binodis* develops better in high alkalinity, silica, calcium and magnesium. Thus, differences in the development of the three oligohaline species should be mainly related to differences in the concentrations of boron, calcium, and magnesium.

Group 8: Based only on ecological data, only depth and salinity should have been determined. As indicated by measured data, this group is characterized by high salinity, with high concentration of chloride, sodium and lithium. The development of characteristic species *Navicula* nov. sp. is due to high content of calcium.

Acknowledgements

The study, proposed and encouraged by Dr. M. Servant, has been funded by O.R.S.T.O.M., GEOCIT (UR 103).

References

Anderson, D. S., R. B. Davis & F. Berge, 1986. Relationships between diatom assemblages in lakes surface-sediments and limnological characteristics in southern Norway. Diatoms and Lake Acidity. Smol, J. P., Battarbee, R. W., Davis, R. B. & J. Merilaïnen (eds.): 97–113.

Badaut, D., F. Risacher, H. Paquet, J. P. Eberhart & F. Weber, 1979. Néoformation de minéraux argileux à partir de frustules de diatomées: le cas de lacs de l'Altiplano bolivien. C. r. Acad. Sci., Paris, 289D: 1191–1193.

Ballivian, O. & F. Risacher, 1981. Los salares del Altiplano, boliviano. Metodos de estudio y estimacion economica. Orstom, Paris: 246 p.

Carmouze, J. P., C. Arze & Y. Miranda, 1978. Estudio de la regulacion hidroquimica del sistema fluvio-lacustre del Altiplano. Revista boliviana de Quimica, 2, 1. La Paz.

Cholnoky, B. J., 1968. Die Ökologie der Diatomeen in Binnengewässer. J. Cramer, Lehre. 699 p.

De Wolf, H., 1982. Method of coding of ecological data from diatoms for computer utilization. Rijks Geologische Dienst, 13 p.

Fernandez, G., 1980. Evolucion cuaternaria de las Cuencas lacustres del Sud Oeste boliviano en la region de Mina Corina (Sud Lipez). Thesis, UMSA, La Paz: 103 p.

Frenguelli, J., 1936. Diatomeas de la caliza de la Cuenca de Calama. Revista del Museo de La Plata, seccion Paleontologia, I: 3–120.

Frenguelli, J., 1945. Las diatomeas del Platense. Revista Museo de La Plata, seccion Paleontologia, III: 77–221.

Gauch, H. G., 1982. Multivariate analysis in community ecology. Cambridge University Press, Cambridge, London: 248 p.

Greenacre, M. J., 1984. Theory and applications of correspondence analysis. Acad. Press, New York, London: 364 p.

Hill, M. O., 1973. Reciprocal averaging: an eigenvector method of ordination. J. Ecol. 61: 340–354.

290

Hurlbert, S. H. & O. Keith, 1979. Distribution and spatial patterning of flamingoes in the Andean Altiplano. Auk 96: 328–342.

Hurlbert, S. H. & C. C. Chang, 1983. Ornithology: Effects of grazing by the Andean flamingo (Phoenicoparrus andinus). Proc. nat. Acad. Sci. USA, Ecology 80: 4766–4769.

Hurlbert, S. H. & C. C. Chang, 1984. Ancient Ice Islands in Salt Lakes of the Central Andes. Science 244: 299–302.

Hurlbert, S. H. & C. C. Chang, 1988. The distribution, structure and composition of freshwater ice deposits in Bolivian salt lakes. Hydrobiologia 158: 271–299.

Hustedt, F., 1927. Fossile Bacillariaceen aus dem Loa – Becken in der Atacama – Waste, Chile. Archiv. f. Hydrobiologie XVIII: 224–251.

Hustedt, F., 1959. Die Kieselalgen. Dr. L. Rabenhorsts Kryptogamen-flora von Deutschland, Österreich und der Schweiz. Akad. Verlaggesellschaft. Geest & Portig K. G. Leipzig C 1.: 737–843.

Iltis, A., F. Risacher & S. Servant-Vildary, 1984. Contribution à l'étude hydrobiologique des lacs salés du Sud de L'Altiplano bolivien. Revue Hydrobiol. trop. 19, 3: 259–273.

Krammer, K., 1980. Morphologic and taxonomic investigations of some freshwater species of the diatom Genus Amphora Ehr., Bacillaria, J. Cramer, Braunschweig, 3, 197–225.

Krammer, K. & H. Lange-Bertalot, 1985. Naviculaceae. Bibliotheca diatomol. 9: 230 p.

Krammer, K. & H. Lange-Bertalot, 1986. Bacillariophyceae. 1. Teil: Naviculaceae. Susswasserflora von Mitteleuropa. Herausgegeben von H. Ettl., J. Gerloff., H. Heynig & D. Mollenhauer: 876 p.

Lebart, L., A. Morineau & K. Warwick, 1984. Multivariate descriptive statistical analysis, J. Wiley. New York.

Lopez, M. M., 1980. Un nuevo subgenero de Surirella en sedimentos del Salar Carcote, Chile. Museo Nacional Historia Natural, noticario mensual 3–7: 281–282.

Lowe, R. L., 1972. Diatom populations dynamics in a Central Iowa ditch. Iowa State J. of Research 47, 1: 7–59.

Miranda, Y., 1978. Evolucion de aguas dulces a salmueras en presencia de boro y litio para la boratera de Rio Grande. Revista Boliviana de Quimica, 2, 1: La Paz.

Osada, K. & H. Kobayasi, 1985. Fine structure of the brackish water pennate diatom Entomoneis alata (Ehr.) Ehr. ver. japonica (Cl.) comb. nov. Sorui. The Jap. J. Phycol. XXXIII, 3: 215–224.

Patrick, R., 1961. Diatoms (Bacillariophceae) from the alimentary tract of Phoenicoparrus Janesi (Sclater). Postilla, Yale Peabody Museum of Natural History, 49, 1: 43–55.

Risacher, F., 1978a. Le cadre géochimique des bassins a évaporites des Andes boliviennes. Cah. ORSTOM, ser. Geol. X, 1: 37–46.

Risacher, F., 1978a. Genèse d'une crôute de gypse dans un bassin de l'Altiplano bolivien. Cah. ORSTOM, ser. Géol. X, 1: 91–100.

Risacher, F. & H. P. Eugster, 1979. Holocene pisolithes and encrustation associated in the springs. Pastos Grandes, Bolivia. Sedimentology, 26: 253–270.

Roux, M., 1985. Algorithmes de classification. Méthodes + Programmes. Masson, Paris: 151 p.

Servant, M. & J. C. Fontes, 1978. Les lacs quaternaires des hauts plateaux des Andes boliviennes. Premières interprétations paléoclimatiques. Cah. ORSTOM, ser. Geol. X, 1: 9–23.

Servant-Vildary, S., 1978a. Etude des diatomées et paleolimnolo du Bassin tchadien au Cenozoique supérieur. Travaux et Documents de l'ORSTOM, 2 tomes: 346 p.

Servant-Vildary, S., 1978b. Les diatomées des sédiments superficiels d'un lac salé, chloruré, sulfaté sodique de l'Altiplano bolivien, le lac Poopo. Cah. ORSTOM, ser. Géol. X, 1: 79–97.

Servant-Vildary, S., 1982. Altitudinal zonation of mountainous diatom flora in Bolivia: application to the study of the Quaternary. Acta Geol. Acad. Scient. Hungaricae, 25 (1–2): 179–210.

Servant-Vildary, S., 1983. Les diatomées des sédiments superficiels de quelques lacs salés de Bolivie. Sciences Géologiques Bull. 36, 4: 249–253.

Servant-Vildary, S., 1984. Les diatomées des lacs sursalés boliviens. Sous classe des Pennatophycidees. I – Famille des Nitzschiacées. Cah. ORSTOM, ser. Geol. XIV, 1: 35–53.

Servant-Vildary, S., 1986. Les diatomées actuelles des Andes de Bolivie (Taxonomie, écologie). Cahiers de Micropaléontologie, CNRS, 1, 3–4: 99–124.

Servant-Vildary, S. & M. Blanco, 1984. Les diatomées fluviolacustres plio-pleistocènes de la Formation Charana (Cordillère occidentale des Andes, Bolivie). Cah. ORSTOM, ser. Géol. XIV, 1: 55–102.

Ward, J. H., 1963. Hierarchical grouping to optimize an objective function. J. am. Stat. Assoc. 58: 236–244.

Wornardt, W. W., 1964. Pleistocene diatoms from Mono and Panamint Lake Basins California. Oc. Papers California. Ac. of Sciences 46: 27 p.

Hydrobiologia **197**: 291–303, 1990.
F. A. Comín and T. G. Northcote (eds), Saline Lakes.
© 1990 *Kluwer Academic Publishers.*

Hydrochemistry from Sr and Mg contents of ostracodes in Pleistocene lacustrine deposits, Baza Basin (SE Spain)

P. Anadón & R. Julià
Institut de Geologia 'J. Almera' (C.S.I.C.), c/ Martí i Franqués s/n, 08028 Barcelona, Spain

Key words: paleohydrochemistry, ostracode shells, magnesium, strontium, salinity

Abstract

A reconstruction of the early Pleistocene paleohydrochemistry based on the Mg, Sr and Ca content of the *Cyprideis* valves is presented for shallow lacustrine sequences of the Baza basin. A large number of environmental changes in this marginal area has been recorded by the recurrent alternation of two fossil assemblages which differ in their salinity requirements. Measurements of the Sr/Ca and Mg/Ca ratios of individual calcite shells of *Cyprideis* show that the water in the higher saline stages (with thalassic organisms indicating marine-like conditions) was of non-marine origin. The Sr/Ca values of *Cyprideis* valves from sands deposited during a saline water phase show lower values than those from an overlying carbonate sequence which was formed under lower salinity conditions. These unexpected values are assumed to be the result of major changes in the chemical composition of the water in shallow, littoral ponded areas of a hydrologically complex lake. In the sequences that originated in these areas, Sr/Ca values may be used only as salinity indicators within each portion of the sequence formed in a single, continuous evolution. In more open areas, the wide fluctuations of Sr/Ca and Mg/Ca recorded in ostracodes from individual layers of rippled ostracode-shell sands probably reflect the mixing of valves from changing short-term environmental conditions.

Introduction

Many limnic ostracode species have precise ecological requirements and the relationships between ostracodes and their chemical environment have been emphasized in recent years (see Carbonel *et al.*, 1988 for a review). One of the best known species is *Cyprideis torosa* Jones, an abundant ostracode in the Pleistocene lacustrine deposits of the Baza basin. For this species the relationship between geochemical properties of its aquatic habitat and the morphological response of its caparace have been discussed by Vesper (1972), Peypouquet (1977) and Carbonel *et al.* (1988).

Furthermore, the incorporation of trace elements in biogenic carbonates through the formation of solid solution (coprecipitation) has been the subject of several studies – summarized by Renard (1985) – over the last decades. Some authors have attempted to determine the relationships existing between host water and ostracode valve composition (Chave, 1954; Cadot *et al.*, 1972; Bodergat & Andreani, 1980; Bodergat, 1983, 1985; Chivas *et al.*, 1983, 1985, 1986a, b). Recent works by Chivas *et al.* (1983, 1985, 1986a, b) both in lakes and controlled aquaria have permitted the calculation of the partition coefficients of Sr and Mg in ostracode shells where Ca is the carrier element. These authors have shown that the Sr distribution coefficient in non-marine ostracodes (including *Cyprideis*) is independent of temperature and that, for a particular species, the individual ostracode valves contain Sr in proportion to the Sr content of their aquatic habitat. Thus, Sr content of ostracodes has been proposed to be an indicator of water salinity for hydrologically simple lakes (Chivas *et al.*, 1985). Mg incorporation in ostracodes is dependent both on temperature and Mg/Ca ratio of the host water (Chivas *et al.*, 1986a, b). Thus a combination of $\delta^{18}O$ and Mg measurements on non-marine ostracodes would enable a joint resolution of paleotemperature and paleosalinity variations (Chivas *et al.*, 1986a, b).

Previous works in the Pleistocene lacustrine sequence of the NE Baza basin based on the paleoecology of ostracodes (Anadon *et al.*, 1986) and on the paleoecology of the fossil assemblages (Anadon *et al.*, 1987) have documented salinity variations as well as alternations of saline Na-Cl dominant water episodes and slightly saline to fresh, bicarbonate-rich water episodes.

The aim of this paper is to reconstruct some of the past water chemistry conditions in the aquatic environment of the Baza basin during the Early Pleistocene by means of the application of the Sr and Mg contents in ostracode shells as hydrologic indicators. Analyses of the Ca, Mg and Sr contents in the *Cyprideis* individual valves, coupled with the experimentally determined distribution coefficient values for this genus calculated by Chivas *et al.* (1986b), enable us to determine the Mg/Ca and Sr/Ca of the waters in which the ostracodes lived.

Description of site studied

The Guadix-Baza intramontane basin (Fig. 1) is located in the Alpine Betic Chain (Southern Spain). This basin yields a sedimentary sequence ranging from Lower Miocene to Pleistocene. During the Pliocene and Pleistocene an extensive lacustrine sedimentation took place in the eastern part of the basin (Vera, 1970; Peña, 1985). These lacustrine deposits can exceed 100 m thickness in

the central part of the Baza basin and belong to the Baza Formation as defined by Vera (1970).

In the NE part of the basin (Orce-Venta Micena region) a lower Pleistocene (Upper Villanyian-Lower Biharian) lacustrine sequence up to 25 m thick, crops out extensively (Vera *et al.*, 1984; Agusti *et al.*, 1984; Soria *et al.*, 1987). These lacustrine deposits overlie calcareous and dolomitic mudstones with interbedded sandstones which have been interpreted as mud flat facies of the Baza lacustrine system deposited during a major water level recession (Anadon *et al.*, 1986). The overlying lacustrine sequence is formed by a variety of lithologies showing a complex lateral and vertical arrangement: calcareous and dolomitic mudstones, sands, sandstones, gravels, limestones and dolostones. This littoral lacustrine sequence has been interpreted to represent a main phase of lacustrine expansion of one hundred thousand years during the early Pleistocene. Nevertheless, minor contractions and expansions of the lake system have also been recorded in this littoral area by the recurrent alternation of two fossil assemblages which differ in their salinity requirements (Anadon *et al.*, 1986, 1987). Three sections (Fig. 1), including these two fossil assemblages, were selected for collecting ostracode shells in order to analyze their Ca, Sr and Mg contents and determine the Mg/Ca and Sr/Ca of their host waters.

Materials and methods

The Mg, Sr and Ca content of 78 individual valves of *Cyprideis torosa* were calculated. The individual valves were collected from *Cyprideis*-bearing beds in three sections of the Orce-Venta Micena area. These sections were selected because, apart from *Cyprideis*, other fossils of paleoecological interest were found (Anadon *et al.*, 1986, 1987). Although up to 10 individual valves were selected from each level, only results from 3 to 6 valves were obtained for each sample (one sample corresponding to one level).

The analytical procedure followed a similar pattern to that described by Chivas *et al.* (1986a).

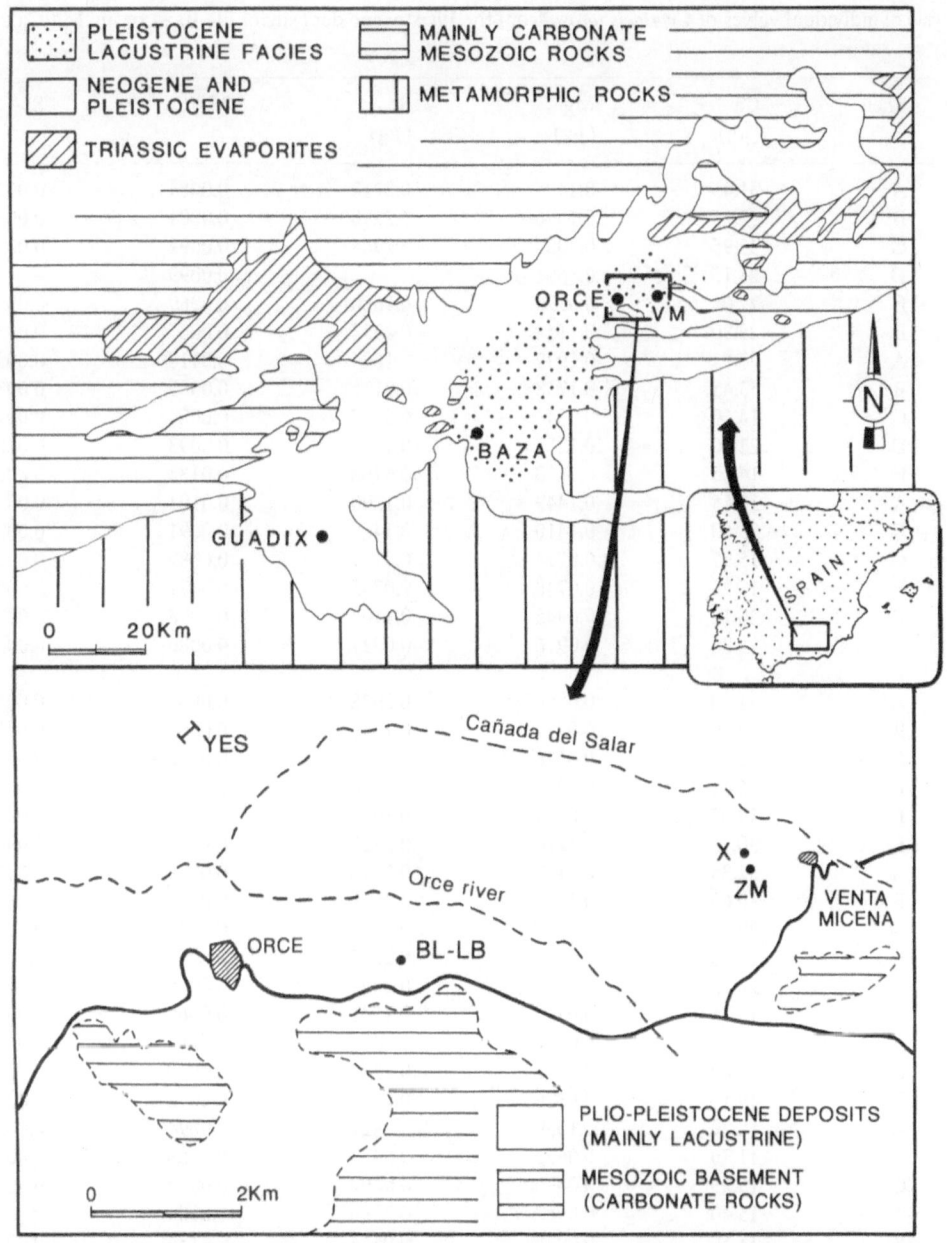

Fig. 1. Areal extent of the Pleistocene lacustrine deposits in the Guadix-Baza Basin. VM: Venta Micena.

Ca, Mg and Sr analyses were performed on individual ostracode valves, which had been previously cleaned and dried. The individual valves were immersed in 5 ml 2% 'Suprapur' Merck HCl for at least 24 h. De-ionized water from Millipore Q (18 m Ω quality) was used. The solutions were analysed by means of an inductively coupled argon-plasma atomic emission spectrometer (ICPAES) with limits of detection of Ca at 0.13 ppb, Mg at 0.1 ppb and Sr at 0.28 ppb. The Ca, Mg and Sr content of the dissolutions were determined and expressed as μg of the element present in each valve. Standard solutions of $CaCO_3$ JMC-90, $SrCO_3$ JMC-91 and MgO

Table 1. Analysis of individual valves of *Cyprideis torosa* from the Pleistocene deposits of the Baza basin. Mg/Ca and Sr/Ca are expressed as atom ratios.

Sample	Shell	Ca (μg)	Mg (μg)	Sr (μg)	Mg/Ca	Sr/Ca
X-9	A	31.45	0.1595	0.2212	0.0084	0.0032
	B	32.35	0.1596	0.2270	0.0081	0.0032
	C	35.95	0.2002	0.2944	0.0092	0.0037
	D	25.15	0.1503	–	0.0099	–
	E	22.05	0.0920	0.1680	0.0069	0.0035
	F	29.30	0.1154	0.2333	0.0065	0.0036
X-7	A	27.65	0.1972	0.1604	0.0118	0.0027
	B	33.45	0.1758	0.2175	0.0087	0.0030
	C	14.60	0.0746	0.0817	0.0084	0.0026
	D	22.30	0.1253	0.1298	0.0093	0.0027
	E	14.45	0.1170	0.0768	0.0133	0.0024
X-6	A	23.65	0.1449	0.1290	0.0101	0.0025
	B	25.50	0.1410	0.1466	0.0091	0.0026
	C	15.95	0.0794	0.0737	0.0082	0.0021
	D	17.25	0.0748	0.0795	0.0071	0.0021
	E	20.10	0.0945	0.1103	0.0078	0.0025
	F	18.80	0.0976	0.0937	0.0086	0.0023
BL-13	A	21.50	0.0800	0.2038	0.0061	0.0043
	B	19.20	0.0737	0.1541	0.0063	0.0037
	C	15.50	0.0526	0.1328	0.0056	0.0039
	D	24.45	0.1005	0.1910	0.0068	0.0036
	E	19.05	0.0754	0.1919	0.0065	0.0046
	F	30.75	0.1216	0.2820	0.0065	0.0042
LB-3	A	32.35	0.1129	0.2770	0.0058	0.0039
	B	30.60	0.1031	0.2619	0.0056	0.0039
	C	29.36	0.1000	0.2369	0.0056	0.0037
	D	26.40	0.0975	0.2207	0.0061	0.0038
	E	28.15	0.1127	0.2427	0.0066	0.0039
LB-2	A	19.55	0.0715	0.1541	0.0060	0.0036
	B	28.25	0.0810	0.2428	0.0047	0.0039
	C	15.80	0.0571	0.1363	0.0060	0.0039
	D	19.15	0.0836	0.1678	0.0072	0.0040
BL-5	A	18.29	0.1039	0.0827	0.0094	0.0021
	B	11.70	0.0623	0.0435	0.0088	0.0017
	C	15.25	0.0896	0.0602	0.0097	0.0018
BL-4	A	15.40	0.0967	0.0580	0.0104	0.0017
	B	11.50	0.0938	0.0444	0.0134	0.0018
	C	6.30	0.0332	0.0203	0.0087	0.0015
BL-3i	A	17.65	0.1096	0.0778	0.0102	0.0020
	B	10.60	0.0675	0.0546	0.0105	0.0024
	C	26.85	0.1077	0.1251	0.0066	0.0021
	D	12.00	0.0612	0.0482	0.0084	0.0018
	E	15.20	0.0767	0.0746	0.0083	0.0022
YES-24	A	1.93	0.0218	0.0051	0.0186	0.0012
	B	8.85	0.1312	0.0219	0.0243	0.0011
	C	4.27	0.0450	0.0155	0.0174	0.0017
	D	7.01	0.0982	0.0186	0.0231	0.0012

Table 1. (continued).

Sample	Shell	Ca (µg)	Mg (µg)	Sr (µg)	Mg/Ca	Sr/Ca
YES-23	A	1.82	0.0260	0.0051	0.0236	0.0013
	B	3.82	0.0950	0.0226	0.0410	0.0027
	C	1.93	0.0586	0.0162	0.0501	0.0038
	D	1.91	0.0342	0.0057	0.0295	0.0014
	E	2.62	0.0662	0.0140	0.0417	0.0024
	F	4.74	0.0703	0.0496	0.0245	0.0048
YES-22	A	10.05	0.1091	0.0313	0.0179	0.0014
	B	11.55	0.0995	0.1470	0.0142	0.0060
	C	4.78	0.0300	0.0309	0.0103	0.0030
	D	10.15	0.0757	0.0434	0.0123	0.0020
YES-21	A	7.85	0.0581	0.0250	0.0122	0.0015
	B	8.20	0.0669	0.0248	0.0134	0.0014
	C	7.95	0.0486	0.0277	0.0101	0.0016
	D	11.45	0.0759	0.0331	0.0109	0.0013
YES-C	A	8.37	0.2408	0.0149	0.0474	0.0008
	B	10.35	0.1419	0.0164	0.0226	0.0007
	C	7.59	0.1061	0.0131	0.0230	0.0008
YES-20	A	10.87	0.1651	0.0412	0.0250	0.0017
	B	9.69	0.1390	0.0351	0.0236	0.0017
	C	13.93	0.2350	0.0716	0.0278	0.0024
	D	7.02	0.1210	0.0238	0.0284	0.0016
YES-19	A	3.56	0.0653	0.0130	0.0302	0.0017
	B	5.32	0.1013	0.0246	0.0314	0.0021
	C	13.55	0.1912	0.0692	0.0233	0.0023
YES-18	A	3.30	0.0532	0.0125	0.0266	0.0017
	B	2.01	0.0309	0.0040	0.0253	0.0009
	C	4.44	0.0622	0.0191	0.0231	0.0020
YES-17	A	2.19	0.0861	0.0161	0.0648	0.0034
	B	2.32	0.0745	0.0096	0.0529	0.0019
	C	5.08	0.1052	0.0263	0.0341	0.0024
	D	1.06	0.0296	0.0038	0.0460	0.0016

JMC-130 were used. Atomic ratios (Mg/Ca and Sr/Ca) for ostracode valves were also calculated. The results of the chemical analyses of individual ostracode valves are shown in Table 1.

Results

Venta Micena X section (Fig. 2)

Description and fossil content
The X section is located in a marginal zone of the Orce-Venta Micena ancient embayment, near the Venta Micena fossil-mammal site (lower Pleistocene).

The lower part of the section (a) is formed by bioturbated dolomitic mudstones and marls, without fossils, interpreted as a mudflat deposit.

The overlying b1 bed, 0.3 m thick, is formed by greenish grey marls and dark silty clays. The marls contain an abundant fresh-water fauna (molluscs and ostracodes) in association with some species which tolerate slightly saline waters (*Candona neglecta* Sars, Hydrobiidae). This bed is overlain by 0.85 m of white or grey laminated sands with interbedded green mudstones (b2). A

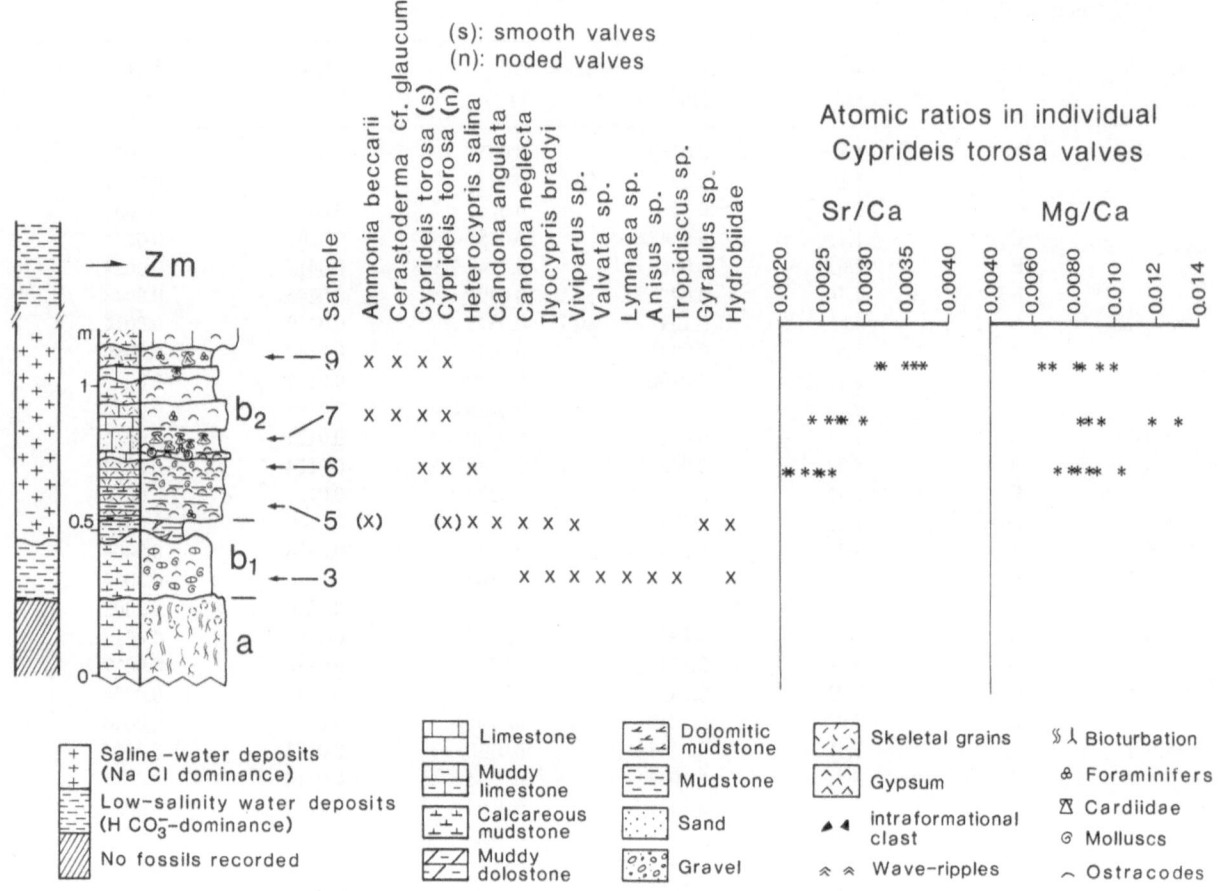

Fig. 2. The X section, near the Venta Micena mammal site. Fossil fauna distribution and plots of the Sr/Ca atom ratios of individual *Cyprideis torosa* shells. Arrow and Zm on top of the lithologic log indicate that the fresh-water condition for the X section top level has been reported from the close Zm section.

sample from the base contains mainly fresh-water organisms, some of them preferring bicarbonate-rich waters, together with some fossils of euryhaline (eurytopic) organisms: *Ammonia beccarii* (Linné), *Cyprideis torosa*. The overlying sediments contain abundant *C. torosa* and *A. beccarii*. The layer of X-7 sample is formed by a lumachelle of the thalassic bivalve *Cerastoderma*. The b2 level originated in a lacustrine environment with permanent saline water. The fossil content of the sample from the lower part of this level may be interpreted to represent a lower salinity episode or, alternatively, the mixing of chemically different waters.

The X section top level *c*, not represented in Fig. 2, is formed by white, chalky limestones which contain an abundant fresh-water ostracode and mollusc fauna in neighbouroughing outcrops.

The described sequence is assumed to have originated during a minor expanding-receding cycle in the marginal zone of the Baza basin (Anadon *et al.*, 1987) The *c* bed has been interpreted as the deposit of fresh-water ponds surrounding the main body of the lake after a receding phase.

Geochemistry of Cyprideis torosa valves.
The Sr/Ca and Mg/Ca plots (Fig. 2) indicate that the Sr/Ca atomic ratio increases from the lower sample (X.6) to the uppermost one (X.9) suggesting a relative increase in the Sr content of the host waters. As they are located in a continuous

sequence, that originated during a minor expanding-receding cycle, these values have been interpreted in terms of an increase in the salinity of the water. This interpretation is supported by the presence of the thalassic organisms *Ammonia beccarii* and *Cerastoderma* in the X.7 and X.9 beds and the absence of these in the lowermost X.6 bed which, in turn, contains *Heterocypris salina* (Brady). A Na-Cl dominant water is deduced from the biota composed by *C. torosa*, and the thalassic organisms *A. beccarii* and *Cerastoderma*.

The Mg/Ca variations are more difficult to interpret because of strong thermodependence of the distribution coefficient of Mg in calcitic shells (Chivas *et al.*, 1986a, b). A Mg/Ca trend that is not parallel to the Sr/Ca trend may be attributed to the variation in temperature in accordance with Chivas *et al.* (1986b).

Barranco Leon (BL-LB) sections (Fig. 3)

Description and fossil content
The lower part of the section (a) is formed by bioturbated red sandy mudstones interpreted as mud flat deposits. They can be correlated with the 'a' bed from the X section.

The overlying (b) beds are formed by sands and conglomerates in a coarsening upwards sequence (Fig. 3). The sands display ripple cross lamination and contain abundant ostracodes and a smaller amount of foraminifers, indicating a saline water of Na-Cl type. An episode of lower salinity is reflected by the BL-3 sample characterised by the presence of fresh-water species such as *Ilyocypris bradyi* Sars and *I. gibba* (Ramd) and the absence of *A. beccarii*. The (c) beds are formed by brecciated and nodulized azoic limestones where the uppermost level corresponds to a paleosoil. The upper sequence of this section (d) is formed by a lithologically complex succession: sands, mudstones, marls and limestones, the beds displaying lenticular geometry. Although a low-salinity episode is recorded at the base of this level (d) by the presence of *A. beccarii*, the bulk of this upper succession is considered to be deposited in a fresh-water environment as is suggested by the presence of a polyspecific fauna comprising fresh-water molluscs and ostracodes.

Geochemistry of Cyprideis valves
The plots of Sr/Ca and Mg/Ca atom ratios allow us to differentiate two groups of data (Fig. 3). One of the groups is formed by the lower samples

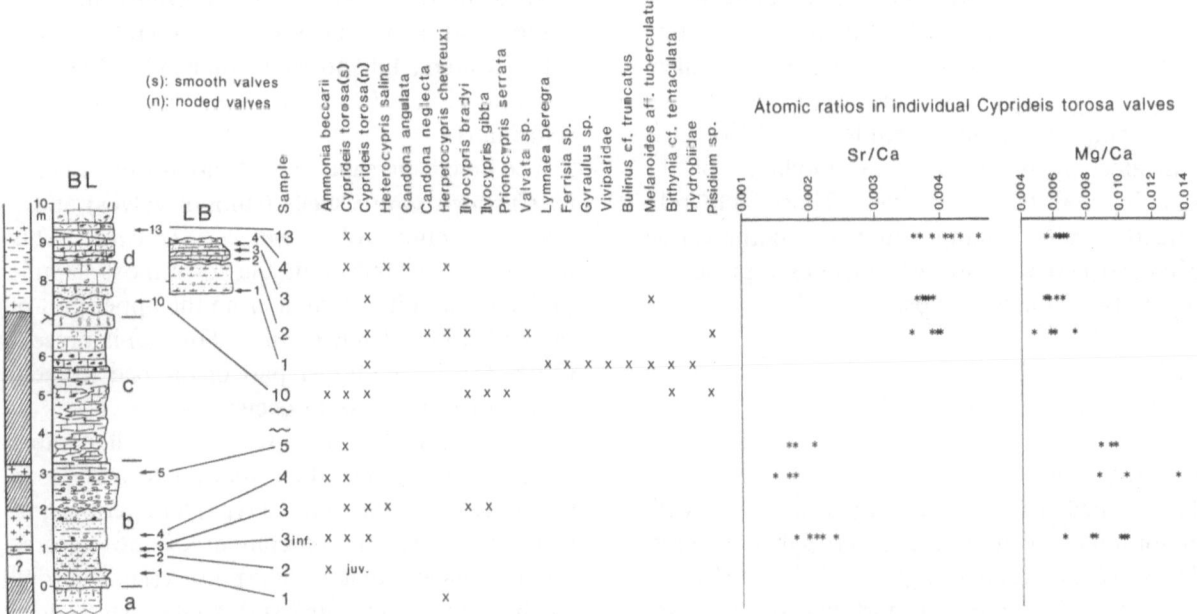

Fig. 3. Barranco León (BL and LB) sections. For legend see Fig. 2.

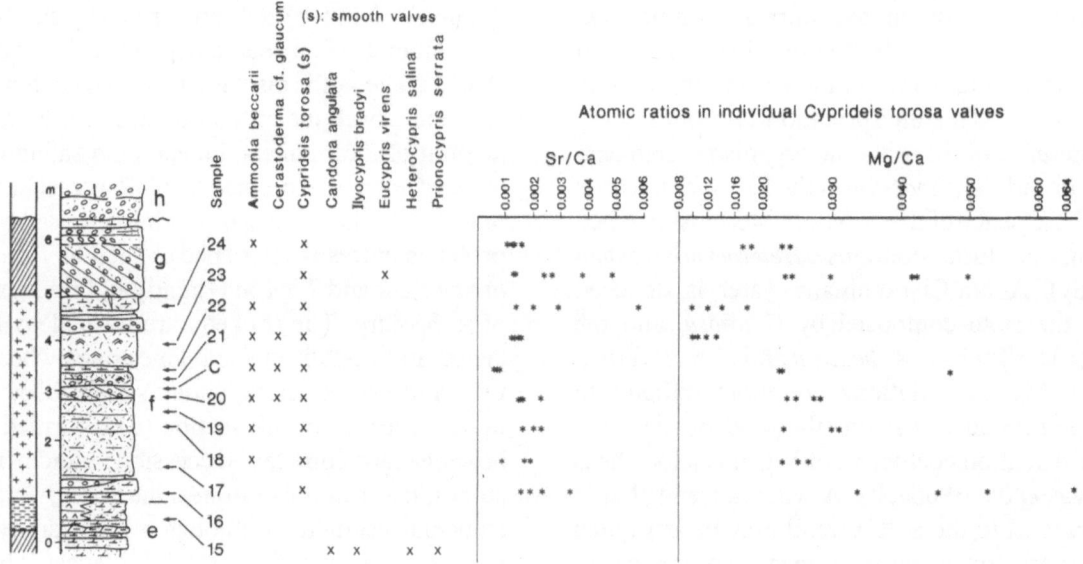

Fig. 4. Yeseras (YES) section. For legend see Fig. 2.

(BL-3 inf, BL-4, BL-5) which show lower Sr/Ca values and higher Mg/Ca values than the other group which is formed by the upper samples (LB-2, LB-3, BL-13). Thus, two kinds of waters may be deduced during the early Pleistocene in the BL zone.

The low Sr/Ca and high Mg/Ca values in the ostracodes correspond to the highest salinity water, as is suggested by the ostracode paleoecology. The high Sr/Ca and low Mg/Ca values correspond to the water with lower salinity, as is indicated by ostracodes and molluscs of the samples from the upper levels. These facts indicate that major changes in water composition occurred between the two episodes represented by the two sample groups.

Yeseras (YES) section (Fig. 4)

Description and fossil content
The studied succession is formed by sands with minor interbedded marls and rare gypsum layers. The sands are formed by quartz grains and ostracode shells in varying proportions. In the lower beds, the sands contain up to 50% of quartz

grains but they decrease noticeably towards the upper beds. The ostracode sands present current and wave-ripples, and locally, large-scale cross bedding. The sandy sequence (e and f beds) is overlain by a tabular cross bedded gravel bed (g).

The biota from the lower bed (e) consists of fresh-water ostracode species (*I. bradyi*) in association with some species which tolerate slightly saline waters, but no more than 5%: *Candona angulata* Müller, *H. salina* and *Prionocypris serrata* (Norman).

The bioclastic sand beds (f) are formed mainly by *Cyprideis torosa* shells (smooth valves) and a minor fraction of *Ammonia beccarii*. This foraminifer is absent in the lowermost sample (YES-17) and in a sample from the upper part of the bed, where *Eucypris virens* (Jurine) is present (YES-23). In the upper part of the bed isolated valves of *Cerastoderma* occur.

The sedimentary features and fossil content suggest that the f bed deposits were formed along the shore of a saline (Na-Cl type) and permanent-water lake. The fossil content of the Yeseras section suggests that in some phases (i.e. recorded by the samples 15 and 23) this lake experienced low salinity (<5%).

Geochemistry of Cyprideis valves

The absolute values for Ca (Table 1) are lower when compared with the values from other sections. This is due to the smaller size and thinness of the *Cyprideis* valves from Yeseras. This thinness indicates that the system was probably rather depleted in dissolved carbonate ion or in Ca^{+2} or in both (Gasse *et al.*, 1987; Carbonel *et al.*, 1988).

Although the fossil content from this section shows certain affinities with that of the X section, these similarities are not reflected in the Mg and Sr content of the ostracode valves. The Sr/Ca and Mg/Ca plots indicate that although some samples display a relatively high grouping of individual values, other samples show a more scattered pattern. Some of the levels reveal wide fluctuations in Sr/Ca reflecting similar salinity fluctuations. The wide fluctuations in the Mg/Ca ratios of *Cyprideis* from individual layers suggest a shallow water body where recurrent changes in temperature and in the Mg/Ca of the waters occurred. Moreover, in some levels of this section, formed by rippled ostracode sands, a single sample probably integrates different stages of lacustrine evolution. When for one specific bed, Sr/Ca values are grouped and Mg/Ca are scattered, it may be suggested that this Mg/Ca scattering is related to temperature fluctuations during phases of unchanged salinity. The wide Mg/Ca variations in the waters will be discussed below.

Discussion

The Mg, Sr and Ca contents of Cyprideis

Most samples show a grouping in the Sr versus Ca diagram (Fig. 5) with a linear distribution indicating that the Sr content is, in general, a function of the Ca content which in turn reflects the ostracode weight. The values from the different sections may be separated into minor linear distributions with an almost parallel slope.

The pattern for Mg versus Ca content shows a wider scattering than the Sr versus Ca plot

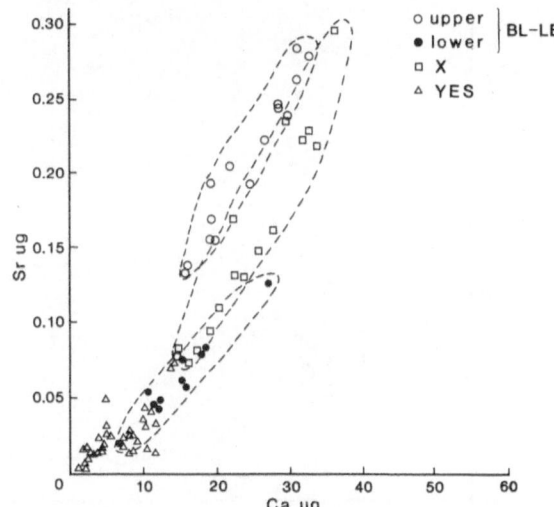

Fig. 5. Sr versus Ca contents diagram for all the studied individual valves referred to in Table 1. Observe the bulk linear distribution and the minor linear distributions displayed by the samples from different sections. Further explanations in the text.

(Fig. 6). Nevertheless, this pattern indicates that the Mg content is a function of the ostracode weight within the group of samples from a particular section. Thus, the Mg versus Ca diagram shows that the data from each section display a

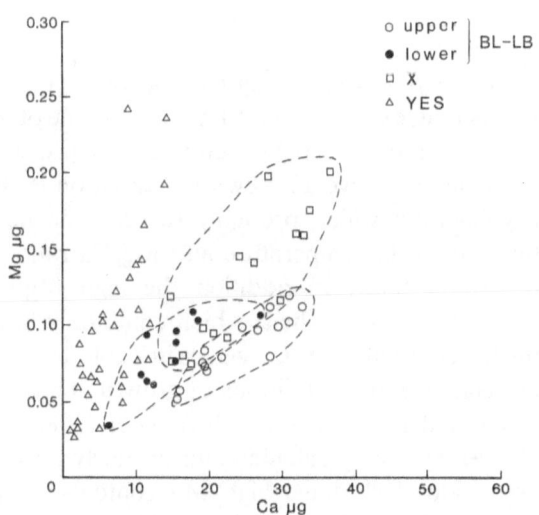

Fig. 6. Mg versus Ca contents diagram for all the studied individual valves referred to in Table 1. See text for explanation.

300

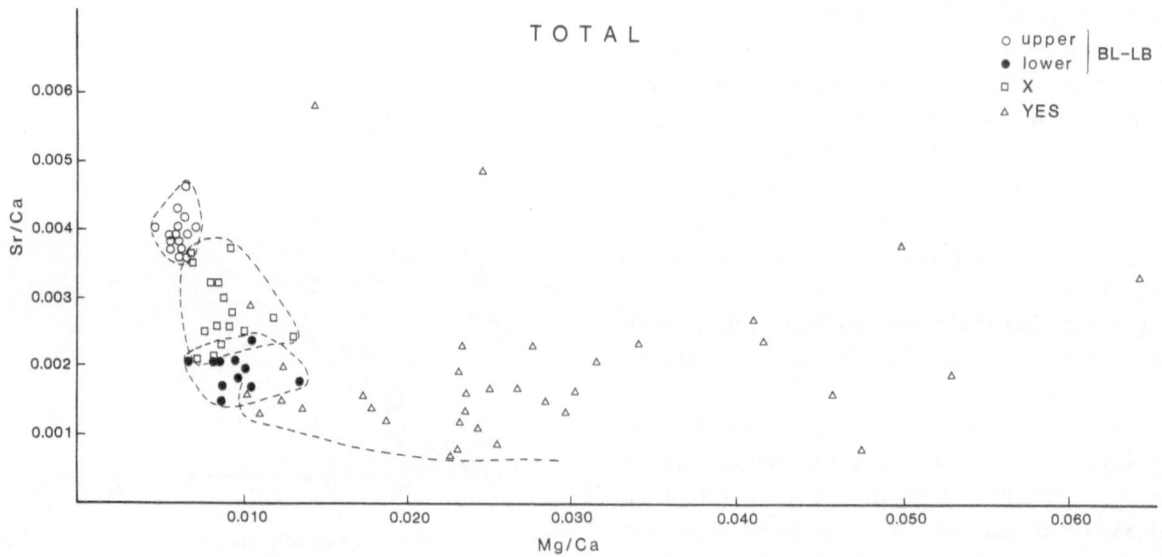

Fig. 7. Sr/Ca versus Mg/Ca atom ratios plot for all the studied individual valves referred to in Table 1. Note the noticeable variation in Mg/Ca ratio displayed by the YES valves. See text for explanation.

linear distribution with different increasing slopes from low Mg/Ca values (BL-LB section, upper levels) to high Mg/Ca values (YES section). The distribution for Mg versus Ca content in each group of samples (Fig. 6) has a wider scattering than the Sr versus Ca plots (Fig. 5). This pattern probably results from the fact that there are two variables controlling Mg uptake: temperature and water composition whereas Sr uptake is only controlled by water composition, as has been reported by Chivas *et al.* (1986a, b).

The Sr/Ca versus Mg/Ca diagram (Fig. 7) shows that, except for the YES samples, the other samples studied display section groupings. The YES samples have a noteworthy variation in the Mg/Ca ratio which probably reflect wide fluctuations in the temperature and Mg/Ca ratio of the water body. In addition the high Mg/Ca values in the YES shells which commonly have the lower weights (see Ca values in Table 1), must be related to the conditions of growth. Thus, as has been demonstrated by Chivas *et al.* (1983 and 1986a) the poorly calcified and/or newly formed ostracode shells have high Mg contents. Only after reaching adult size is the Mg content an indication of the geochemical conditions of the water. Furthermore, thin, poorly calcified and

poorly ornamented shells in *Cyprideis torosa* are attributed to CO_2 rich host waters that are undersaturated with respect to calcite (Gasse *et al.*, 1987). The low weight may also be attributed to a high salinity in the host waters.

Paleohydrochemistry

The application of the Sr and Mg distribution coefficient relationships to the Sr/Ca and Mg/Ca ratios obtained in the ostracode valves allow us to calculate the Sr/Ca and Mg/Ca ratios in their host waters. These distribution coefficient values are $Kd(Sr) = 0.474$ and $Kd(Mg) = 0.0046$ (Chivas *et al.*, 1986b). As the Mg content of ostracodes depends on the water temperature, this calculation has been made assuming that $T = 25\ °C$, (this is the temperature from laboratory culture where $Kd\ (Mg)$ for *Cyprideis* was deduced by Chivas *et al.*, 1986b). The Sr/Ca versus Mg/Ca diagram (Fig. 8) for the deduced geochemistry of the lake waters shows that the molar ratio distributions are removed from the typical values for marine water, either evaporated, or diluted, and differ also markedly from the mean values for continental, dilute, non-marine waters.

A mainly bicarbonate freshwater has been de-

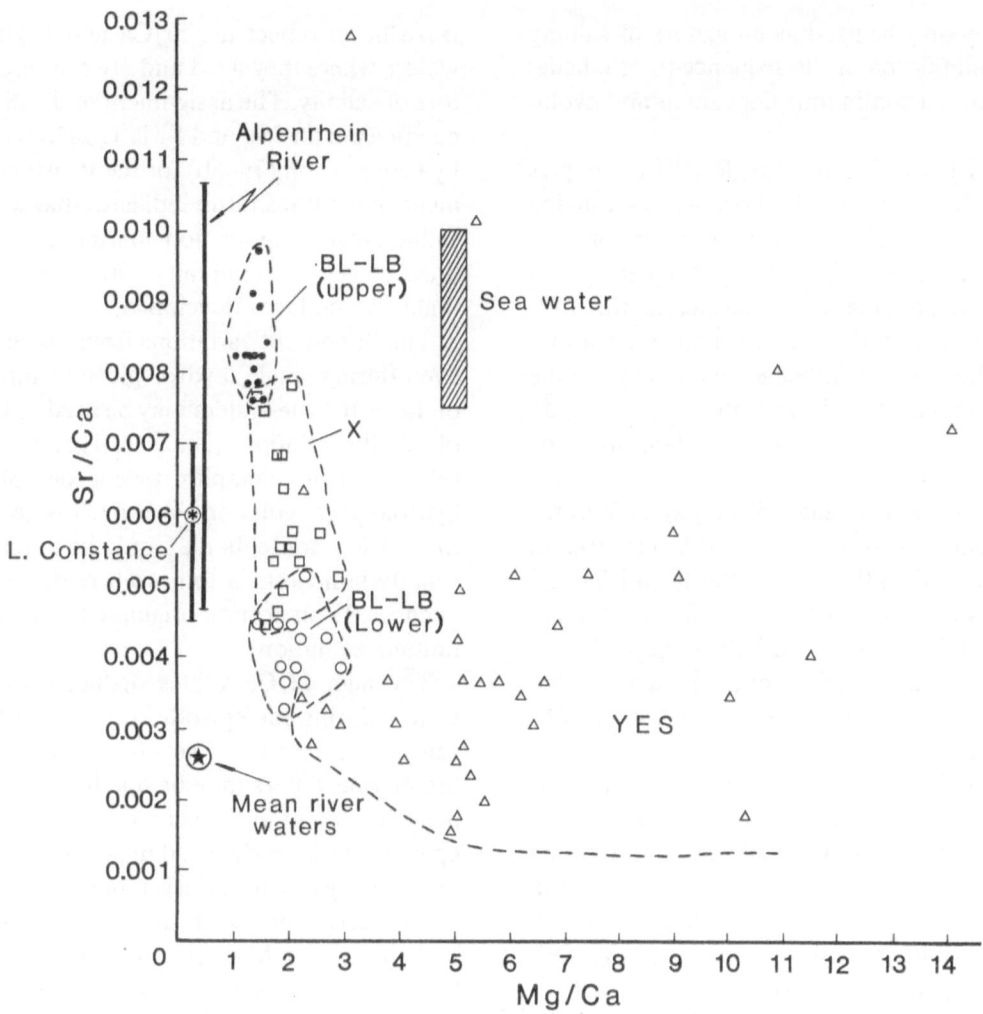

Fig. 8. Deduced Sr/Ca and Mg/Ca ratios for host waters from ostracodes applying the Sr and Mg distribution coefficients. Mean river values and sea water range from Renard (1985). Lake Constance and Alpenrhein River values from Müller (1968). See the text for explanation.

duced for the marginal zone of the lake during the deposition of the BL-LB upper levels, where a mainly carbonate deposition took place (Anadon *et al.*, 1986, 1987). The low Mg/Ca values deduced for lake water are in agreement with this assumption. The Sr/Ca values deduced for the lake during this episode are high, and similar to those for sea water. These high values may be explained by an abnormally high Sr content in the source area. Gypsum, which has a relatively high Sr content is a frequent rock type in the Keuper (Triassic) of the surrounding area of the Baza basin. In addition high Sr content anomalies have

been reported from this source area (Sebastian, 1979). A similar phenomenon has been described by Müller (1968, 1969) in lake Constance where high Sr/Ca values, unusual in a fresh-water environment, fall almost into the range of marine waters. Thus, these high Sr/Ca values in BL-LB upper level shells most probably are the result of a major change in the water composition experienced in the shallow, marginal ponded area of a hydrologically complex lake system. This major change is reflected both by the changes in the fauna and by the emersion episode recorded in the sequence (Fig. 3). Thus, in this section the Sr/Ca values may only be used as indicators of salinity within each portion of the sequence (b or c beds) which shows a continuous deposition and evolution.

All the BL-LB, X and some few YES samples indicate Mg/Ca values for host waters ranging between 1 and 3. These values, common for non-marine waters, are differentiated from the mean sea-water values (Fig. 8). Nevertheless, the YES samples pattern in the Sr/Ca versus Mg/Ca diagram for host waters indicate a wide range in the Mg/Ca deduced for these waters. Some additional remarks may be made to explain this wide Mg/Ca variation.

The mean Mg/Ca values displayed by Yeseras samples gathered from individual layers present higher values than those from the X and BL-LB sections; the lower mean values from YES are similar to those from X and BL (Fig. 7). These higher mean Mg/Ca values could indicate higher temperature in the saline (Na-Cl) lake waters from more open and shallower areas, given that the Sr/Ca display similar values in all these sections. Another reason for higher mean Mg/Ca values is that the Yeseras shells are thinner than those from the X and BL-LB sections which indicates poorly calcified shells. A probable preferential uptake of Mg may have occurred in the YES shells during the early stages of calcification.

The high Sr/Ca and Mg/Ca variations displayed by the YES samples, gathered from individual layers, indicate wide Sr/Ca and Mg/Ca fluctuations in the lake waters from a more open area. These fluctuations probably reflect the mixing of valves from different short-term environmental conditions (depth, salinity, temperature, etc). The Mg/Ca fluctuation in the lake waters may also be attributed to simultaneous gypsum precipitation. This precipitation in $SO_4^{2-} > Ca^{2+}$ water would produce a sharp rise in the Mg/Ca ratio.

Conclusions

The Sr and Mg content of the *Cyprideis* valves from the Pleistocene marginal sequences of the Baza basin reflect the Sr/Ca and Mg/Ca in the waters where they lived and are not direct indicators of salinity. The assignment of the distribution coefficients for Mg and Sr in *Cyprideis* calculated by Chivas *et al.* (1986b) for the Pleistocene specimens of the Baza basin indicates that water in the saline stages was of non marine origin. During these stages, a community with foraminifers and thalassic molluscs developed.

The Sr content variations from ostracodes that lived during single, hydrologically simple phases of the lacustrine system may be used as indicators of salinity variations (i.e. X section). In sequences resulting from complex paleogeographical and hydrological evolutions (i.e. BL-LB section), the Sr and Mg contents may only be used as broad salinity indicators within each portion of the sequences. Each portion originated in a single, continuous evolution.

The high Sr/Ca values deduced for the lake waters during the episode recorded in the upper samples of the BL-LB section within the range of the marine waters may be due to the abnormally high Sr content in the source area. A fresh water episode has been deduced in accordance with the paleoecology of fossil assemblages.

Cyprideis valves with the highest Mg contents and the widest Mg content variations in valves from the same level in the YES section display a smaller mean size and thinness than all the other valves studied. These morphotype characters probably recorded depletion in dissolved carbonated ion or in Ca^{2+} as well as a relatively high salinity. As the Mg/Ca measurements are only indicative of the environment if restricted to fully calcified specimens (as demonstrated by Chivas *et al.*, 1986b) the Mg content of most of the *Cyprideis* valves from the YES section probably reflects higher values of the distribution coefficient and is not indicative of the Mg/Ca of their host waters.

The combined paleoecological studies and geochemical analyses of fossil-shell carbonates provide useful data on total salinity, anionic dominance and cation molar ratios in the waters where they lived.

Acknowledgements

We gratefully acknowledge M. Baucells, M. Roura and G. Lacort (Servei d'Espectroscopia, Universitat de Barcelona) for analyses done on the ostracode shells and P. Carbonel and P. De Deckker for constructive criticism and suggestions on the manuscript.

References

Agusti, J., P. Anadón, J. Gilbert, R. Julià, E. Martín-Suárez, E. Menéndez, S. Moyà-Solà, J. Pons-Moyà, P. Rivas & I. Toro, 1984. Estratigrafía y paleontología del Pleistoceno inferior de Venta Micena (Orce, depresión Guadix-Baza, Granada). Resultados preliminares. Paleont. i Evol. 18: 19–38.

Anadón, P., P. De Deckker & R. Julià, 1986. The Pleistocene lake deposits of the NE Baza Basin (Spain): salinity variations and ostracod succession. Hydrobiologia 143: 199–208.

Anadón, P., R. Julià, P. De Deckker, J. -C. Rosso & I. Soulié-Märsche, 1987. Contribución a la Paleontología del Pleistoceno inferior de la cuenca de Baza (sector Orce-Venta Micena). Paleont. i Evol. Mem. esp. 1: 35–72.

Bodergat, A. -M., 1983. Les ostracodes témoins de leur environment: approche chimique et écologique en milieu lagunaire et océanique. Docum. Lab. Géol. Lyon 88: 1–246.

Bodergat, A. -M., 1985. Composition chimique des caparaces d'ostracodes. Paramètres du milieu de vie. Bull. Centres Rech. Explor. Prod. Elf-Aquitaine, Mém. 9: 379–386.

Bodergat, A. -M. & A. -M. Andreani, 1980. Mise en évidence de la réponse adaptive d'une espèce euryhaline, *Cyprideis torosa* (Jones, 1850) à des conditions écologiques difficiles par l'analyse multi-élémentaire en spectrometrie de masse à étincelle. In J. Martinell (ed.); Concept and method in Paleontology. Departamento de Paleontología, Univ. Barcelona: 135–139.

Cadot, H. M., W. R. van Schmus & R. L. Kaesler, 1972. Magnesium in calcite of marine Ostracoda. Geol. Soc. Am. Bull. 83: 3519–3522.

Carbonel, P., J. -P. Colin, D. L. Danielopol, H. Löffler & I. Neustrueva, 1988. Paleoecology of limnic ostracodes: a review of some major topics. Palaeogeogr., Palaeoclimatol., Palaeoecol. 62: 413–161.

Chave, K. E., 1954. Aspects of the biogeochemistry of magnesium. I. Calcareous marine organisms. J. Geol. 62: 266–283.

Chivas, A. R., P. De Deckker & J. M. G. Shelley, 1983. Magnesium, strontium and barium partitioning in nonmarine ostracode shells and their use in paleoenvironmental reconstructions – a preliminary study. In R. F. Maddocks (ed.), Applications of Ostracoda. Univ. Houston Geosci.; Houston: 238–249.

Chivas, A. R., P. De Deckker & J. M. G. Shelley, 1985. Strontium content of ostracods indicates lacustrine palaeosalinity. Nature 316: 251–253.

Chivas, A. R., P. De Deckker & J. M. G. Shelley, 1986a. Magnesium content of non-marine ostracod shells: a new palaeosalinometer and palaeothermometer. Palaeogeogr., Palaeoclimatol., Palaeoecol. 54: 43–61.

Chivas, A. R., P. De Deckker & J. M. G. Shelley, 1986b. Magnesium and strontium in non-marine ostracod shells as indicators of palaeosalinity and palaeotemperature. Hydrobiologia 143: 135–142.

Gasse, F., J. C. Fontes, J. C. Plaziat, P. Carbonel, I. Kaczmarska, P. De Deckker, I. Soulié-Märsche, Y. Callot & P. A. Dupeuble, 1987. Biological remains, geochemistry and stable isotopes for the reconstruction of environmental and hydrological changes in the Holocene lakes from North Sahara. Palaeogeogr., Palaeoclimatol., Palaeoecol., 60: 1–46.

Müller, G., 1968. Exceptionally high strontium concentrations in fresh water onkolites and mollusk shells of lake Constance. In G. Müller & G. M. Friedman (eds.). Recent Developments in Carbonate Sedimentology in Central Europe. Springer. Berlin-Heidelberg-New York: 116–127.

Müller, G., 1969. High Strontium Contents and Sr/Ca – ratios in Lake Constance Waters and Carbonate and their Sources in the Drainage Area of the Rhine River (Alpenrhein). Mineral Deposita. 4: 75–84.

Peña, J. A., 1985. La depresión de Guadix-Baza. Estudios Geol. 41: 33–46.

Peypouquet, J. P., 1977. Les ostracodes et la connaissance des paléomilieux profonds. Application au Cénozoique de l'Atlantique nord-oriental. Thesis. Univ. Bordeaux I: 1–443.

Renard, M., 1985. Géochimie des carbonates pélagiques. Docum. Bur. Rech. Geol. Min. (France) 85: 1–650.

Sebastián, E., 1979. Mineralogia de los materiales Pliocено-Pleistocenos de la Depresión de Guadix-Baza. Thesis Univ. Granada I: 1–311.

Soria, F. J., A. C. López Garrido & J. A. Vera, 1987. Análisis estratigráfico y sedimentológico de los depósitos neógeno-cuaternarios en el sector de Orce (depresión de Guadix-Baza). Paleont. i Evol. Mem. esp. 1: 11–34

Vera, J. A., 1970. Estudio estratigráfico de la Depresión Guadix-Baza. Bol. Geol. Minero 81: 429–462.

Vera, J. A., J. Fernández, A. C. López-Garrido & J. Rodríguez-Fernández, 1984. Geología y estratigrafía de los materiales plio-pleistocenos del sector de Orce-Venta Micena (Prov. Granada). Paleont. i Evol. 18: 3–11.

Vesper, B., 1972. Zur Morphologie und Ökologie von Cyprideis torosa (Jones, 1850) (Crustacea, Ostracoda, Cytheridae) unter besonderer Beruck sichtigung seiner Biometrie. Mitt. hamb. zool. Mus. Inst. 68: 21–77.

Hydrobiologia **197**: 305–308, 1990.
F. A. Comin and T. G. Northcote (eds), Saline Lakes.
© 1990 *Kluwer Academic Publishers.*

Absolute dating of sedimentation on Lake Torrens with spring deposits, South Australia

R. M. Schmid
University of Zambia, School of Mines, PO Box 32379, Lusaka, Zambia

Key words: [14]C dating, travertine, playa, deflation

Abstract

The travertine structure containing Mountford Spring is in the only surface accumulation of $CaCO_3$ on Lake Torrens, a playa in South Australia. Here the first ^{14}C data on Lake Torrens is presented. [14]C ages for travertine samples surrounding Mountford Spring, range from 13 770 ± 130 years BP to 22 700 ± 290 years BP, giving a time frame for lacustrine sediments wedged in between. In the absence of recent tectonic movements, the old travertine (22 700 ± 130 years) on the clifftop, surrounded by younger spring deposits (13 770 ± 130 years) at a lower level, prove a deflation event at this playa.

Introduction

Lake Torrens, surface area 5932 km², is a physiographic depression, located in South Australia, the driest state in the driest continent of the world (Fig. 1). It has been filled with lacustrine sediments since Mid-Tertiary. The mineral assemblage at the surface of the playa is restricted to quartz, gypsum and clay (Schmid, 1985). Travertine only occurs at the travertine structure containing Mountford Spring (Fig. 1), offering the possibility to apply ^{14}C dating of sedimentation on Lake Torrens. This may then be compared with the sedimentation in other basins such as Lake Eyre, Lake Frome and Lake Gairdner. The travertine structure was first recorded by Madigan (1930).

Description of the study site

The circular travertine structure of Lake Torrens with its hard, partly vegetated surface, stands out amongst the flat, soft, moist and vegetation-free surface of the surrounding lake area composed mainly of quartz, gypsum and clay. The travertine structure, diameter 1300 m containing Mountford Spring, is located in the southern part of Lake Torrens (Fig. 1). The travertine occurs at two levels, an elevation of 2.5 meters above the surrounding playa surface and at the present playa level. Present spring activity is confined to a small outlet near the centre of the travertine structure. Mountford Spring has a diameter of 10 meters.

Material and methods

X-ray diffraction (XRD) on unoriented powder samples identified the minerals across the playas: quartz, gypsum, clay (Schmid, 1985). Samples from the travertine structure on Lake Torrens were collected in August 1984 and analysed through thin section interpretation (Schmid, 1985). Algal remains were found surrounded by

306

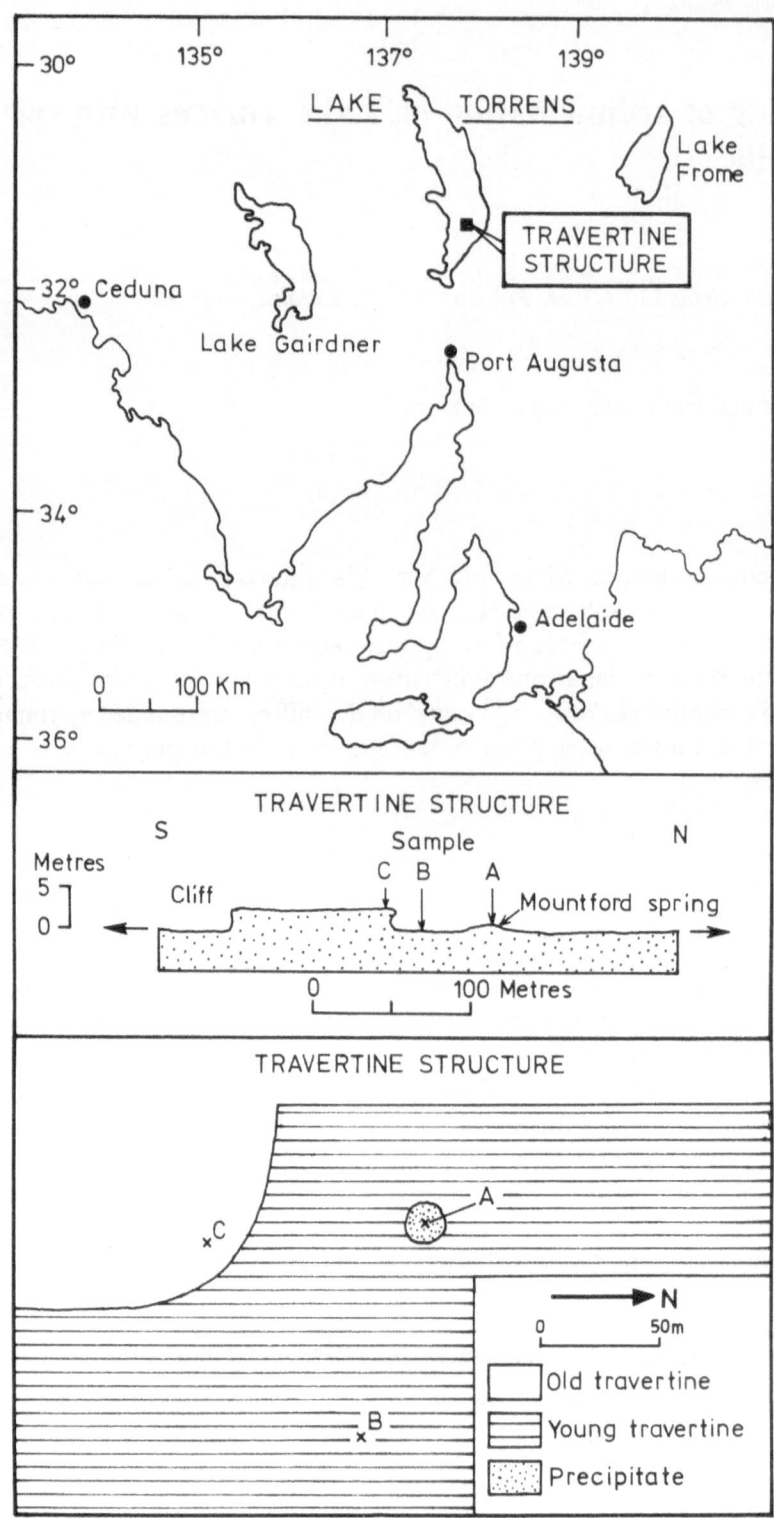

Fig. 1. Upper: General location of the Lake Torrens study area in South Australia. Centre and lower: Cross section and plan view, respectively, of the travertine structure at Lake Torrens.

calcite in all samples (A, B, C, Fig. 1). The ^{14}C analysis on three travertine samples (A, B, C, Fig. 1) were performed as percent modern carbon (Hendy, 1970) in December 1984. The travertine samples B and C (Fig. 1) were collected well away from the active spring (Mountford Spring) to minimise contamination.

Results

Sedimentary and biological facies of the travertine

An unconsolidate, gypsiferous, quartz slit rim defines the outer margin of the circular travertine structure. Inside this rim an essentially flat surface of travertine has precipitated at the present playa level, where sample B (Fig. 1) was taken. In thin section no recrystallisation of the pure calcite was evident. The enclosed carbon traces suggested precipitation of calcite in a spring setting, in conjunction with algal growth. The weathered looking travertine of the clifftop represents another older level of active spring discharge (sample C, Fig. 1). The present day precipitate (sample A, Fig. 1) is associated with green algal growth. All three samples show the same sedimentary structures, thin laminae of algal related $CaCO_3$ precipitation.

Dating

The travertine (sample C, Fig. 1) of the clifftop was determined as $22\,700 \pm 290$ years BP, activity $5.9 \pm 0.2\%$. The travertine (sample B, Fig. 1), at the present playa level was $13\,770 \pm 130$ years BP, activity $18.0 \pm 0.3\%$. The present day precipitate of Mountford Spring (sample A, Fig. 1) was $15\,840 \pm 270$ years BP, activity $13.9 \pm 0.5\%$.

Discussion

In the absence of any $CaCO_3$ at the surface of Lake Torrens, the travertine of the clifftop represents an active period of spring discharge with

$CaCO_3$ precipitation. The Ca could be supplied by groundwater from the surroundings of Lake Torrens. This clifftop layer, 0.3 meters thick, protects underlying gypsiferous, quartz and silt sediments, 2.2 metres thick, of playa origin. Then, 80 metres of Quaternary sediments in Lake Torrens (Johns, 1967) do not represent a complete sedimentary record. Two metres and twenty centimeters, preserved at the travertine structure, towards the southern part of Lake Torrens, have been deflated on the remaining playa.

The three radiocarbon dates give the opportunity to date this deflation event at Lake Torrens. A time-lag induced by the time difference between groundwater recharge and discharge, supplying Mountford spring, must be considered. The order of the time-lag, taking into account (1) sediment porosity, (2) distance to recharge area, (3) hydraulic conductivity of the sediments, (4) hydraulic head difference, could be as high as 20 000 years. In addition to the time lag, the ratio $^{14}C/^{12}C$, precipitated as travertine, will depend on the extent of the isotopic exchange between carbon dioxide in the atmosphere and the carbon in the solution. The present day precipitate at Mountford Spring in conjunction with the travertine at the playa level may represent a compromise of time-lag and isotopic exchange. The ages of $13\,770 \pm 130$ years BP and $15\,840 \pm 270$ years BP of present day precipitates become calibration values. Substracting the calibrating figures from the age of the clifftop travertine gives the oldest (ca. 8000 years) age limit of the deflation event on Lake Torrens. This is the time when spring discharge stopped and deflation commenced. This compares well with a period of alluvation and dune building in the surroundings of Lake Torrens, occurring between 5000 – 12 000 years BP (Williams & Polach, 1971).

Acknowledgement

The CSIRO, Division of Water Resources Research, Adelaide Laboratory, determined [14]C.

308

References

Hendy, C. H., 1970. The use of ^{14}C in the study of cave processes. In: Olsen, I. U. (ed.). Radiocarbon variation and absolute chronology. Proceedings of the 12th Nobel Symposium held at the Institute of Physics at Uppsala University: Wiley Interscience Division, New York: 419–443.

Madigan, C. T., 1930. An aerial reconnaissance into the south-eastern portion of Central Australia. Proc. Geogr. Soc. Aust. (S. Aust. Branch) 30: 83–108.

Johns, R. K., 1967. Investigation of Lake Torrens and Gairdner. Geol. Surv. S. Aust. Report of Investigation No. 31, 90 pp.

Schmid, R. M., 1985. Lake Torrens, sedimentation and hydrology. Ph.D. Thesis, Flinders University, unpubl. 250 pp.

Williams, G. E. & H. A. Polach, 1971. Radiccarbon dating of arid zone calcareous paleosols. Geol. Soc. of Am. Bull. 82: 3069–3086.